SEXTANTS AT GREENWICH

SEXTANTS AT GREENWICH

A Catalogue of the Mariner's Quadrants, Mariner's Astrolabes, Cross-staffs, Backstaffs, Octants, Sextants, Quintants, Reflecting Circles, and Artificial Horizons in the National Maritime Museum, Greenwich

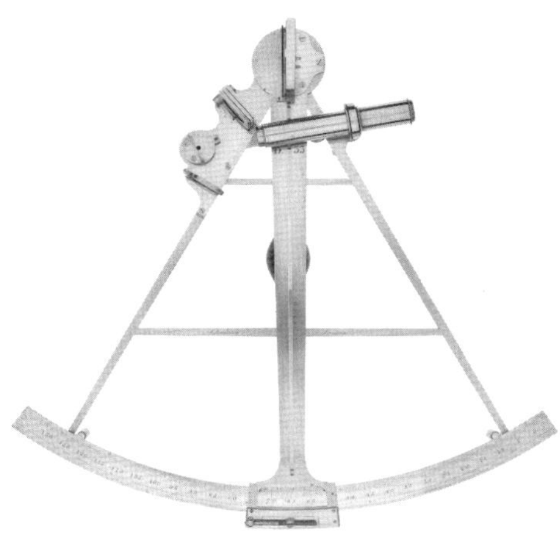

W. F. J. MÖRZER BRUYNS

With a contribution by

RICHARD DUNN

OXFORD
UNIVERSITY PRESS

and the

NATIONAL MARITIME MUSEUM

OXFORD
UNIVERSITY PRESS

Great Clarendon Street, Oxford OX2 6DP

Oxford University Press is a department of the University of Oxford.
It furthers the University's objective of excellence in research, scholarship,
and education by publishing worldwide in

Oxford New York

Auckland Cape Town Dar es Salaam Hong Kong Karachi
Kuala Lumpur Madrid Melbourne Mexico City Nairobi
New Delhi Shanghai Taipei Toronto

With offices in

Argentina Austria Brazil Chile Czech Republic France Greece
Guatemala Hungary Italy Japan Poland Portugal Singapore
South Korea Switzerland Thailand Turkey Ukraine Vietnam

Oxford is a registered trade mark of Oxford University Press
in the UK and in certain other countries

Published in the United States
by Oxford University Press Inc., New York

British Library Cataloguing in Publication Data
Data available

Library of Congress Cataloging in Publication Data
Data available

Printed in China through Asia Pacific Offset

ISBN 978-0-19-953254-4 (hbk)

1 3 5 7 9 10 8 6 4 2

FOREWORD

Museums have been called the temples of a secular age, and one of their purposes is certainly for people to step briefly out of the daily rush and gain perspectives on the present and future through contemplating aspects of the past. It is also one of the strengths of great museums, not least when their buildings match the quality of great churches, as ours do, that their purposes are long-term: both serve changing needs, generations and ages, while asserting constant beliefs. For museums these are in the continuing significance of the collections they hold in trust, and a steady determination to preserve and interpret them for general public benefit.

At its highest levels – the scholarly cataloguing and recataloguing of collections – such interpretation is never quickly done and is rarely the last word: often it too spans decades and generations. In this light, *Sextants at Greenwich* is the fourth volume in a uniform series published with the Oxford University Press, of which the first, on the Museum's globe collection, appeared in 1999 under the Directorship of Richard Ormond. A catalogue of sundials and horary quadrants, and one on astrolabes, followed in 2002 and 2005 under my predecessor Roy Clare. At least one more is planned to cover our marine chronometers.

The present work, however, differs from all these. They cover essentially similar instruments. In contrast, this deals with several very different types that none the less served the same purpose, from the sixteenth century to the onset of modern electronic positioning systems: that of measuring the positions of the Sun, Moon and stars by visual means, for the purpose of safe navigation at sea. It is thereby also the first which covers the vital tools of practical seamen, rather than instruments of more general scientific use.

No-one is better qualified to have undertaken this project than Dr Willem Mörzer Bruyns, formerly Senior Curator of Navigation of the Nederlands Scheepvaartmuseum in Amsterdam, and an international authority on the history of that subject. He is also an old friend to many at Greenwich, and other British maritime history colleagues. Willem has been a valued colleague and personal friend of mine ever since he and I served together on the Executive of the International Congress of Maritime Museums in the early 1990s. It is a pleasure to thank him for his work on this project. Like all three previous compilers – two of whom were also from the historic Low Countries (the Netherlands and Belgium) – he has been able to do so as the Museum's Sackler Fellow in the History of Astronomy and Navigational Sciences (2005–7), and we again thank Dr Mortimer and Mrs Therese Sackler for their generous and continuing support of this programme. We are also grateful for Dutch assistance with this book, in a grant from the Commissarissen van het Vaderlandsch Fonds ter Aanmoediging van 's Lands Zeedienst.

The navigational history to which the instruments covered here relate is now largely told at the Royal Observatory, Greenwich, part of this Museum since the 1950s. Thanks are also due to Dr Richard Dunn, based there as our Curator of the History of Navigation, for his contribution; also to other members of Museum staff who have helped in practical ways. These include Ken Hickey and his colleagues in the Photographic Studio, the staff of the Library and Museum archive, the Publications department and others acknowledged by the author below.

While those who own this book are most likely to be specialists, collectors and libraries, its very readable introductory chapters can be recommended to anyone interested in the history of the instruments it describes, or how the Museum acquired this area of its pre-eminent collections. As with its forerunners, how- ever, the individual instruments also appear for wider general reference, based on Dr Mörzer Bruyns's work, on the 'Collections Online' pages of the Museum's website (www.nmm.ac.uk). Here too, 'the times change and we change with them', but always aiming for consistently high quality in the message, whatever the medium.

Kevin Fewster
Director
National Maritime Museum, Greenwich

PREFACE

The history of navigation and navigational instruments is part of a wider history of science that includes, for example, astronomy, biology, chemistry, medicine, and physics. Objects pertaining to these subjects are often preserved in science museums that cover the entire field, but substantial collections of historic navigational instruments are usually found in specialist maritime museums. Even here, size is relative. Maritime museums deal with a wide variety of subjects such as warfare at sea, the merchant marine, shipbuilding, yachting, and cartography. The number of ship models and paintings of sea battles is often considerable; they appeal to the general public and are superficially also easier to understand than navigational instruments. A separate department of navigation is often not justified by the size of the collection, and the specialist knowledge that is required to collect, catalogue, and interpret historic navigational objects is not easily available.

The National Maritime Museum at Greenwich has been an exception. From early on it had a Department of Navigation, to which 'and Astronomy' was added in 1960. Since that year the Museum has played a leading role in the field, providing expert knowledge through exhibitions, symposiums, catalogues, and other publications. The series of volumes that have appeared since 1999 as a result of Sackler Research Fellowships is a good example of this policy. After the catalogues of globes, sundials, and astrolabes, *Sextants at Greenwich* is the first volume to deal with a purely navigational part of the collection.

The 347 catalogue entries here demonstrate that the collection of octants, sextants, and related instruments in the National Maritime Museum is by far the largest public holding of its kind. In the past this collection has led to *An Inventory of the Navigation and Astronomy Collections in the National Maritime Museum Greenwich*, a three-volume compilation that appeared in 1970–73 and was updated in 1977 and in 1983, but which was not meant to provide the scholarly depth that the collection justifies. Since the *Inventory* appeared, new research has been published on historic navigational instruments and their makers, and in some cases this has led to different conclusions in this catalogue, for example in the dating of objects. Nevertheless, the descriptions and documentation of the objects by staff of the Department of Navigation and Astronomy forms the basis for *Sextants at Greenwich*.

A general overview of the history of navigation, and of the various instruments that were used for celestial navigation, is given in the introductory chapters. Where possible, this is linked to the Greenwich collection. There is an emphasis on the octants and sextants, as these classes of instrument are by far the largest and also the most complete. Even so, not every variation is represented. In order to present a balanced history, a number of inventions and patents that are not found at Greenwich have therefore been included in the survey. This still does not mean, however, that *Sextants at Greenwich* is a complete history of instruments for altitude measurement at sea. The account of artificial horizons, for example, deals with some highlights only, because this section of the collection is far from comprehensive. Another less well-balanced group is the reflecting circles: British examples are well represented but not circles by French manufacturers.

The first chapter introduces the history of celestial navigation since the beginning of European expansion in the fifteenth century. This was when the earliest instruments for altitude measurement were developed, and I begin with those preceding the invention of the octant in 1731: the mariner's quadrant, the mariner's astrolabe, the cross-staff, and the backstaff or Davis quadrant. Hadley's quadrant, now better known as the octant, was the first significant improvement in navigational instruments after the Longitude Act of 1714, and the principle of Hadley's catoptrics remained in use at sea for the following 250 years. Similar contemporary inventions are also dealt with in this chapter, as is the octant's diffusion in the British Isles and beyond, from the eighteenth century onwards. In 1760, John Dollond called the octant a 'noble instrument': for Squire Lecky, in 1891, it was the 'hardier and more humble relative' of the sextant. That instrument is the subject of the next chapter, which also includes the reflecting circle, the quintant, and various artificial horizons. Then follow developments in the nineteenth and twentieth centuries, up to the advent of satellite navigation in the 1970s. This was a period of technological improvements, for instance through hydrography and aviation, but it also saw the end of the octant and the reflecting circle.

Richard Dunn has contributed the last introductory chapter and he describes how the collection at Greenwich was formed, expanded, exhibited, and first researched. He discusses the role of E.T. Adams, to whom the navigational collection of the National Maritime Museum owes so much. Richard also compiled the first appendix, which lists the instruments acquired in the Adams, Gabb, and Landau collections, including those presented by Adams that are not catalogued in this volume.

The catalogue itself consists of ten sections, in each of which the objects have been arranged chronologically. The more detailed organization is described in the 'Explanation' on page 85. The basis of the descriptions was the typed records for each instrument, which were compiled by the Department's curators in the late 1960s and 1970s.

ACKNOWLEDGEMENTS

I am grateful to Dr Mortimer and Mrs Theresa Sackler for establishing their Research Fellowship in the History of Astronomy and Navigational Sciences at the National Maritime Museum, and want to thank its Trustees and Director for granting it to me for the period 2005–7. I thank Dr Nigel Rigby, Head of the Centre for Imperial and Maritime Studies at the Museum, and responsible for the research fellowships, for his support.

During my work at the Museum, I received support and encouragement from many other staff. Foremost I want to thank Dr Richard Dunn, Curator of the History of Navigation, and my daily guide to the collection. He generously agreed to contribute to this volume and also selected the majority of the illustrations. I am grateful to Dr Gloria Clifton, Head of the Royal Observatory, Greenwich, for her guidance during the entire project. Gloria and Richard read the manuscript at an early stage and I thank them for their comments.

Many days were spent in the photographic studio with Tina Warner and Ben Gilbert, and I am grateful for their commitment to give the photography for this volume a high priority, and with a splendid result. I also want to thank Gillian Hutchinson, Curator of the History of Cartography, and Jonathan Betts, Curator of Horology, for advising me in areas of their specialist knowledge. I thank Bernard Bryant at the Museum's Kidbrooke store for his assistance, and the Museum's handling staff, who took objects from display for me to catalogue. Staff in the Caird Library were always helpful in retrieving books and manuscripts for my research, and I am grateful for the administrative support provided during my fellowship by Janet Norton and Janet Small.

The help and advice of friends and colleagues in various parts of the United Kingdom has been indispensable. I am especially grateful to Dr Anita McConnell, an independent researcher at the University of Cambridge. Anita introduced me to the Board of Longitude papers that are preserved in the Cambridge University Library; she kindly supplied me with a number of 'snippets' that I would otherwise have missed, and let me use her biography of Jesse Ramsden before its publication. She also commented on the introductory chapters, as did Dr Jim Bennett, Director of the Museum of the History of Science in Oxford, and Dr David Bryden, Pershore, Worcestershire. Jim and I discussed various aspects of the reflecting circle and David generously provided information on British patents for scientific instruments up to about 1830. Throughout the fellowship my friend Alan Stimson, formerly Curator of Navigation at the National Maritime Museum, has been helpful and supportive. Alan too read the introductory chapters and I am particularly grateful for his wise comments on the history of the Museum's Department of Navigation.

I also thank staff at the British Museum, the Kent Maritime Trust at Ramsgate, the National Maritime Museum Cornwall, and the Captain Cook Birthplace Museum in Middlesbrough for taking instruments – on loan from Greenwich – off display for me to catalogue, and later again for photography. Staff in the manuscript room of Cambridge University Library were helpful in my work on the Board of Longitude papers. I also thank Dr Alison Morrison-Low, Principal Curator at the National Museums Scotland, in Edinburgh, and her assistant Chris Craig, for showing me the relevant instruments in their collection.

I have exchanged much useful e-mail with George Huxtable of Abingdon, Oxfordshire, on the construction and use of quintants, reflecting circles, micrometers, and dip sectors. Dr Deborah Jean Warner, Curator at the National Museum of American History, Washington, DC, advised me on American octant makers and their trade labels, and Dr Randall Brooks, Curator at the National Museum of Science and Technology in Ottawa, advised me on dividing engines. Dr Wolfgang Köberer of Frankfurt am Main, Germany was helpful with my research into German SOLD sextants, and Dr Paolo Brenni of the Museo di Storia della Scienza in Florence answered a number of queries concerning instrument makers.

As a Sackler Fellow, I also visited collections in Switzerland and the United States. I want to thank Stéphane Fischer, Assistant Curator at the Musée d'Histoire des Sciences in Geneva, for allowing me to study relevant instruments in his institution. The second largest public collection of octants and sextants is preserved in the Mystic Seaport Museum, Mystic, Connecticut. I spent four days in its storage area, generously assisted by Georgina York, Registrar, and Don Treworgy, Head of the Planetarium at Mystic. I am grateful to Dr Stuart M. Frank, Senior Curator at the New Bedford Whaling Museum in New Bedford, Massachusetts, for letting me study the reserve collection in his institution. I thank Dr Daniel Finamore, Russell W. Knight Curator, and George Schwartz, Assistant Curator at the Peabody Essex Museum in Salem, Massachusetts, for providing the same facilities in their museum. Diederick Wildeman, my successor at the Netherlands Maritime Museum in Amsterdam, was supportive throughout the fellowship, and supplied me with a flow of in formation and photocopies from the collection for which he is now responsible.

Three institutions generously provided illustrations for this volume free of charge; the Altonaer Museum in Hamburg, The Mariners' Museum in Newport News, Virginia, and the Netherlands Maritime Museum in Amsterdam.

Finally, I want to thank Dr Pieter van der Merwe, General Editor at the National Maritime Museum, for helping to bring my written English up to standard.

I sincerely hope that *Sextants at Greenwich* will contribute to our knowledge of the history of navigation and navigational instruments, and that it will be of use to those collecting in this field both professionally and privately. I dedicate it to the men who by day and night found their way over the sea with the help of cross-staffs, backstaffs, octants, and sextants, under often far more difficult circumstances than I had to endure, studying the instruments on which their lives depended, in 'normal working hours' only and in the quiet surroundings of a museum store.

Bussum and Greenwich
December 2007 *W.F.J.M.B*

CONTENTS

ABBREVIATIONS

BOL	Board of Longitude
CHSI	Collection of Historical Scientific Instruments, Harvard University, Cambridge, MA, USA
ERO	Essex Record Office
FMB	Focke-Museum, Bremer Landesmuseum, Bremen, Germany
HO	Hydrographic Office
HNC	Hart Nautical Collection (MIT Museum), Cambridge, MA, USA
JUM	Jagiellonian University Museum, Cracow, Poland
MBL	Museum Boerhaave, Leiden, Netherlands
MHG	Museum für Hamburgische Geschichte, Hamburg, Germany
MMB	Museu Maritim, Barcelona, Spain
MMH	Maritime Museum, Haifa, Israel
MMNN	The Mariners' Museum, Newport News, VA, USA
MNDLM	Musée national de la Marine, Paris, France
MSM	Mystic Seaport Museum, Mystic, CT, USA
MSN	Museo Storico Navale, Venice, Italy
MSS	Museo di Storia della Scienza, Florence, Italy
MWC&S	Museum of Wellington City and the Sea, Wellington, New Zealand
NBWM	New Bedford Whaling Museum, New Bedford, MA, USA
NMAH	National Museum of American History, Washington, DC, USA
NMM	National Maritime Museum, Greenwich, London, UK
NMS	National Museums Scotland, Edinburgh, UK
NSA	Nederlands Scheepvaartmuseum, Amsterdam, Netherlands
NZNMM	New Zealand National Maritime Museum, Auckland, New Zealand
PEM	Peabody Essex Museum, Salem, MA, USA
QMM	Queensland Maritime Museum, Brisbane, Qld, Australia
RN	Royal Navy
ROE	Royal Observatory, Edinburgh, UK
ROG	Royal Observatory, Greenwich, London, UK
R.U.S.I.	Royal United Service Institution
ScML	Science Museum, London, UK
SNR	Society for Nautical Research
VOC	Verenigde Oost-Indische Compagnie (Dutch East India Company)
YPM	Yale Peabody Museum, New Haven, CT, USA

PART ONE

HISTORICAL BACKGROUND

CHAPTER ONE

AN INTRODUCTION TO THE HISTORY OF CELESTIAL NAVIGATION

Navigation is the knowledge that is required to sail a ship safely from one port to another, along the shortest practical route and in the least possible time. The Elizabethan mathematician John Dee defined it in 1570 as follows: The art of navigation demonstrateth how by the shortest good way, by the aptest direction, and in the shortest time, a sufficient ship . . . be conducted.[1] The basis for traditional navigation is experience that is handed down from generation to generation; but a seaman also requires a certain amount of theoretical knowledge, which he must be able to interpret and apply. This includes the coordinates of his position at departure, and those of his destination. These are defined by latitude, north and south of the equator, and by longitude, east and west of a prime meridian. A seaman must also know which course or courses to steer by the ship's magnetic compass in order to reach his destination. By the sixteenth century it was realized that the magnetic north and south poles did not coincide with the geographic or true poles, and that their locations shifted over time. Seamen therefore had to be able to determine the variation, or angle, between the magnetic pole, indicated by the ship's compass, and the geographic pole. They also needed – and still do depending on circumstances – to be able to interpret the data in a sea chart and a pilot book or sailing directions, to avoid both hidden and visible dangers, and to take advantage of prevailing winds, sea, and tidal currents.

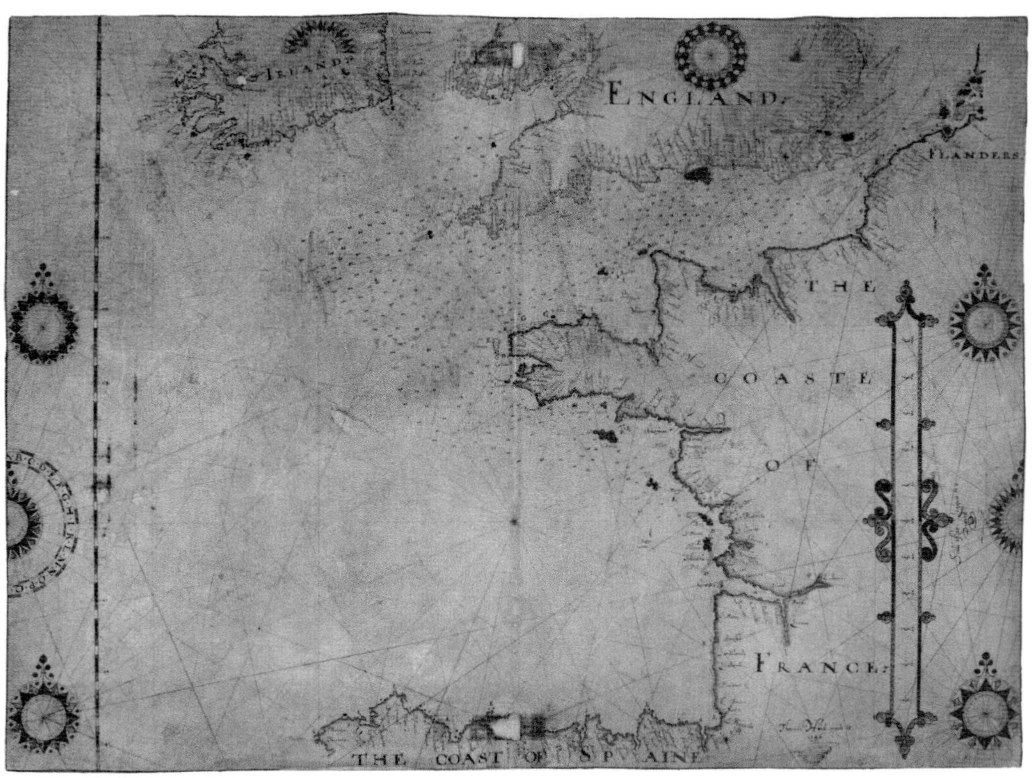

FIGURE 1.1 Chart of the Bay of Biscay and the English Channel by Thomas Hood, London, 1596. Manuscript on vellum. National Maritime Museum; G224:1/2

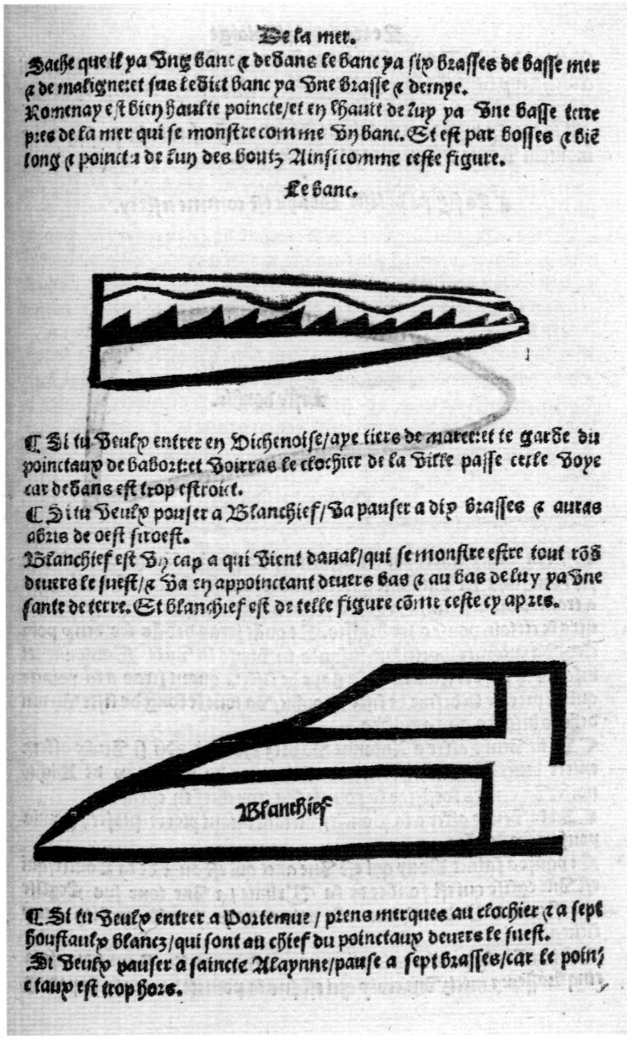

De la mer.

Sache que il ya vng banc & de banc le banc ya six brasses de basse mer & de malignes: et sas le dict banc ya vne brasse & demye.
Romenay c'st bien haulte poincte/et en hault de luy ya vne basse terre pres de la mer qui se monstre comme vng banc. Et est par bosses & bië long & poinctu de luy des bouts Ainsi comme ceste figure.

Le banc.

Si tu veulx entrer en Dichenoise/aye l'ictre de mareè: et te garde du poinctaux de babort: et voirras le clochier de la ville passe ceste voye car dedans est trop estroict.
Si tu veulx pouser a Blanchief/va pauser a dix brasses & autres obzis de oest suroest.
Blanchief est vng cap a qui vient dauaal/qui se monstre estre tout rõs deuers le suest/& va en appoinctant deuers bas & au bas de luy ya vne fante de terre. Et blanchief est de telle figure cõme ceste cy apres.

Blanchief

Si tu veulx entrer a Dorlemue / prens merques au clochier & a sept houstaulx blancs/qui sont au chief du poinctaulx deuers le suest.
Si veulx pauser a saincte Alaynne/pause a sept brasses/car le poinctaux est trop hors.

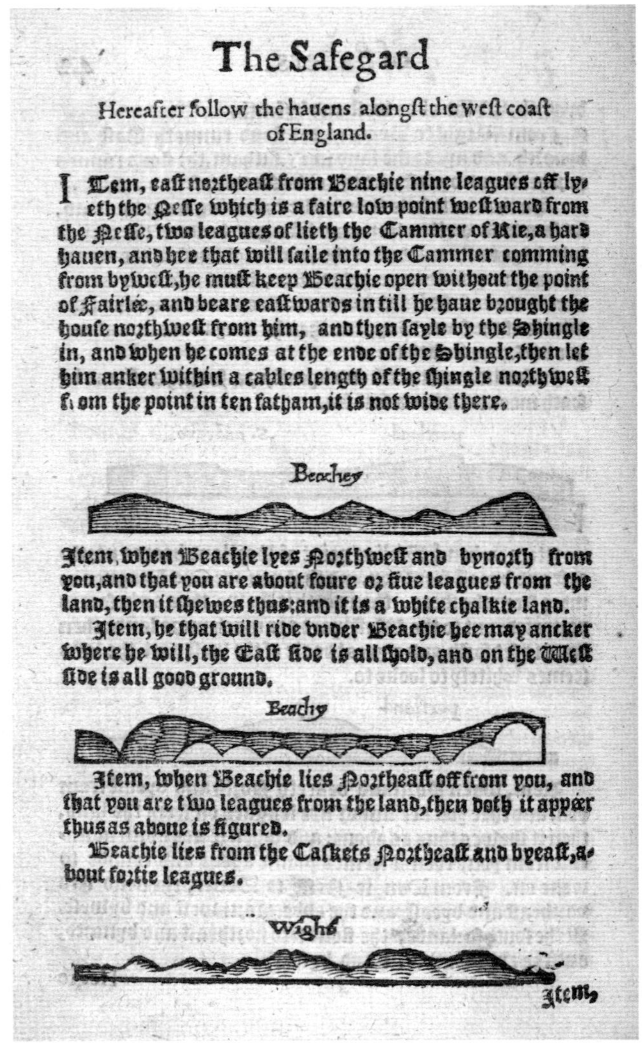

The Safegard

Hereafter follow the hauens alongst the west coast of England.

Item, east northeast from Beachie nine leagues off lyeth the Nesse which is a faire low point westward from the Nesse, two leagues of lieth the Cammer of Rie, a hard hauen, and hee that will saile into the Cammer comming from bywest, he must keep Beachie open without the point of Fairlee, and beare eastwards in till he haue brought the house northwest from him, and then sayle by the Shingle in, and when he comes at the ende of the Shingle, then let him anker within a cables length of the shingle northwest from the point in ten fatham, it is not wide there.

Beachey

Item, when Beachie lyes Northwest and bynorth from you, and that you are about foure or fiue leagues from the land, then it shewes thus: and it is a white chalkie land.
Item, he that will ride vnder Beachie hee may ancker where he will, the East side is all shold, and on the West side is all good ground.

Beachy

Item, when Beachie lies Northeast off from you, and that you are two leagues from the land, then doth it appear thus as aboue is figured.
Beachie lies from the Caskets Northeast and byeast, about fortie leagues.

Wight

Item,

FIGURE 1.2 (*above*) Coastal view of Beachy Head near Eastbourne, East Sussex. From *Le grant routtier ert pilotage et encrage de mer*, a sailing guide by Pierre Garcie, published in Rouen, 1557. National Maritime Museum; PDB6066.

FIGURE 1.3 (*above right*) Coastal views of Beachy Head near Eastbourne, East Sussex, and the Isle of Wight. From *The safegarde of saylers: or great rutter*, a sailing guide by Cornelis Anthonisz. of Amsterdam, translated by Robert Norman and published in London, 1590. National Maritime Museum; PDB6057.

There are two main categories of navigation: 'coastal', within sight of land, and 'astronomical', which is practised out at sea. As long as a seaman can see land, he can find his position by observing conspicuous landmarks that are identified on the chart and in the sailing directions. Bearings are taken of these landmarks by means of the compass; the intersection of the bearing lines, corrected for variation, is pricked on the chart and provides the fix. The depth of water beneath the keel is also an indication of the position. This is measured with a sounding rod or with a lead and line, and cross-checked with the data found in the chart. Often the lead is 'armed' with some tallow in a hollow in its base. This brings up a sample from the sea bottom, the nature of which is indicated on the chart.

Out of sight of land, the position is found by dead reckoning. The ship's speed is regularly estimated or measured with a log, and over a certain period of time the courses steered, corrected for variation, are recorded. Leeway must also be taken into account; this is the difference between the steered and the actual course due to the wind pushing the ship sideways. However faithfully data are recorded, uncertain factors such as sea currents and prolonged periods of bad weather can divert a ship miles from its intended course. The dead-reckoned position must therefore be verified as often as possible by observing celestial bodies. That subject is the central theme of this chapter.

The first seaman whom we are told steered by the stars was Odysseus.[2] At one stage, as he crossed the sea during his imaginary voyage, Odysseus had to keep the stars of the Great Bear on his left hand night after night. This was the method of steering by horizon stars. A star was observed shortly after rising (or before setting) and the direction in which it was seen was used as a compass bearing. After a while the next rising or setting star, with more or less the same amplitude as the first, was used to steer by

FIGURE 1.4 Sounding with lead and line. Detail from a view of the city of Olinda in Pernambuco, Brazil. Engraving by Claes Jansz. Visscher, Amsterdam, 1620. Courtesy of the Nederlands Scheepvaartmuseum, Amsterdam; A.0145(130).

in the same way, and thus the ship was held on the same course without the use of a compass.[3]

Celestial navigation was developed independently by Arab sailors, the Chinese, and Pacific islanders.[4] Western navigation originated before the Christian era, when Phoenician seamen sailed the Indian Ocean and the enclosed and therefore comparatively safe Mediterranean, where land is never far off. Greek seamen traversed the Mediterranean using handwritten sailing guides and orientated themselves on the four main cardinal directions; east, south, and west, where the Sun rises, reaches its highest, and sets; and north, the apparent direction in which the Pole Star is seen at night. The directions of a number of seasonal winds completed the 'compass rose'.[5] In the Middle Ages, Irish monks sailed to Iceland and Norsemen went even further, to North America. They had an elementary knowledge of the tides, and the passage of the Moon. Natural phenomena, such as migrating birds, were used for orientation, and cloud formations on the horizon indicated the presence of land. A primitive kind of compass, consisting of a lodestone or a magnetized needle, was used by Arab seamen in the twelfth century and probably brought to the Mediterranean by Crusaders.[6] It was almost immediately introduced into middle and northern Europe. In the thirteenth century manuscript charts came into use in the Mediterranean now generally known as 'portulan charts' from the written sailing directions (*portolani* in Italian) on which they were based. When printing became known in Europe in the fifteenth century, charts came into more general use.[7]

IBERIAN EXPANSION

Around the middle of the fifteenth century, Portuguese and Spanish seamen embarked on a series of southbound voyages along the west coast of Africa, sailing into unknown waters in search of a passage to the Indian Ocean. They hoped to make contact with the legendary Christian priest-king, sometimes known as Prester John, who supposedly reigned on the east coast of Africa, and to encourage the adoption of Christianity on that continent. At the same time they wanted to find a sea route leading to the commercial wealth of Asia, because the expanding Ottoman Empire blocked the overland route. Iberian seafarers soon found that the navigational instruments and techniques used in Mediterranean and European waters were inadequate for this purpose.[8]

These seamen recorded their progress southward down the coast of Africa using their knowledge of latitude. This was found at night by observing the altitude of the Pole Star, which is situated almost directly over the North Pole, making its altitude nearly equal to latitude of the observer. A correction of a few degrees must be applied to the observation, called the 'regiment for the North Star', which was found by observing the position of the Guards of the constellation Little Bear relative to the Pole Star.[9] The earliest instrument used to measure altitudes at sea was the mariner's quadrant, a simple wooden or brass quarter-circle, with a division along the arc and a plumb bob suspended from the apex.[10] (see for example cat. no. 001). South of the equator, which was reached by 1474, the Pole Star is no longer visible. Iberian navigators had first seen the constellation of the Southern Cross in 1454 but had found that there was no southern pole star. The Southern Cross stands at an angular distance of approximately 27 degrees of arc from the celestial South Pole, and by 1508 a Portuguese navigator had worked out a rule similar to the 'regiment for the North Star' to find latitude from the altitude of certain stars of the Southern Cross.[11]

Before that, in the 1480s, a method of finding latitude from the Sun's altitude at noon was developed. Unlike the Pole Star, the Sun is not in a fixed position above the Earth's surface, but apparently moves between the tropics of Capricorn (south) and Cancer

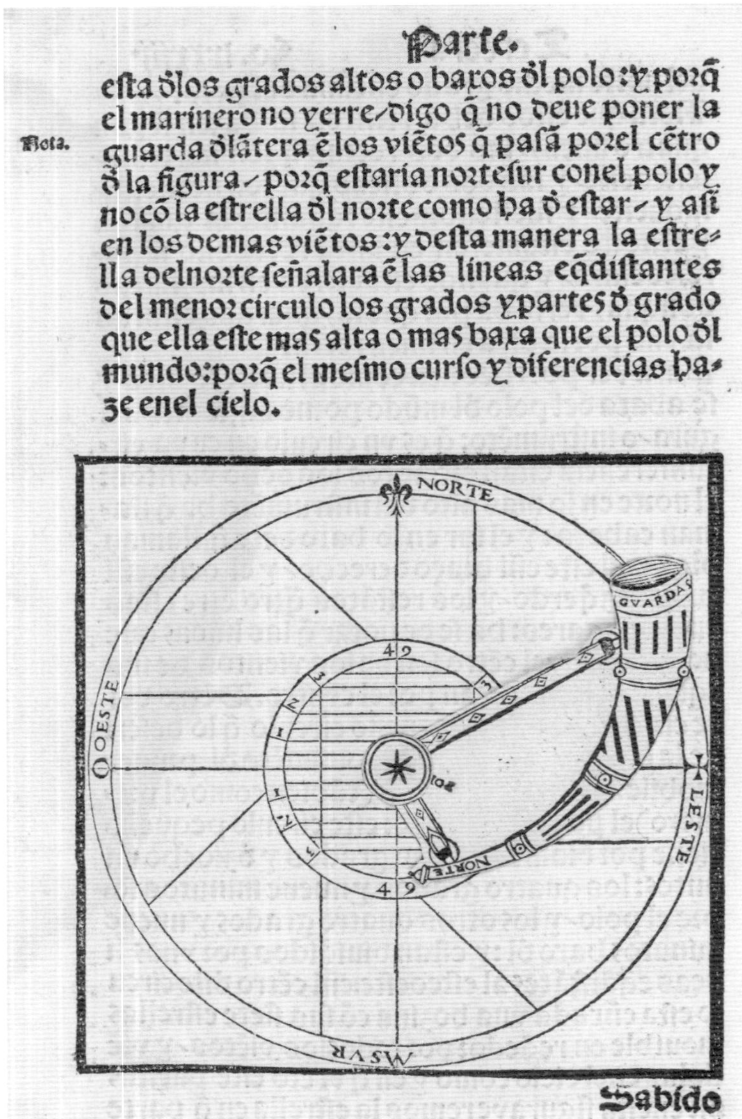

Parte.

esta dlos grados altos o baxos dl polo: y porq̃ el marinero no yerre, digo q̃ no deue poner la
Nota. guarda dlãtera ẽ los viẽtos q̃ pasã porel cẽtro d̃ la figura, porq̃ estaria nortesur conel polo y no cõ la estrella dl norte como ha d̃ estar, y asi en los demas viẽtos: y desta manera la estrella delnorte señalara ẽ las lineas eq̃distantes del menor circulo los grados y partes d̃ grado que ella este mas alta o mas baxa que el polo dl mundo: porq̃ el mesmo curso y diferencias haze enel cielo.

FIGURE 1.5 'The Regiment for the North Star', with which the correction could be found that had to be applied to the altitude observation of the Pole Star. From *Breve compendio de la sphera de la arte de navegar* by Marten Cortes, Seville, 1551. National Maritime Museum; PBE5140.

(north). The number of degrees that the Sun is to the north or south of the equator is called the declination and its value changes by about a degree in a few days. Declination can be calculated in advance, and the first day-to-day declination tables of the Sun for use at sea, including the rule on how to find latitude, was titled *Regimento do Astrolabio e do Quadrante.* When Christopher Columbus sailed towards America in 1492 he was provided with such tables in manuscript. The only known extant printed copy dates from about 1509, and the only known surviving copy of the similar *Regimento da Declinaçam do Sol* appeared in about 1516.[12] Besides the Sun's declination, this rule required the observer to calculate the zenith distance, the complement of the altitude. The Sun's semi-diameter also had to be taken into account, in order to reduce the height to its centre.[13]

As the mariner's quadrant was not practical for observations of the Sun, sophisticated astronomers' instruments were simplified for use on the decks of small, unsteady vessels. The first was the mariner's astrolabe, which evolved in the 1480s as a simplified version of an astronomical instrument.[14] Some years later, the Arab *kamal* inspired the adaptation of the astronomer's cross-staff for use at sea in the same way. Unlike the quadrant and mariner's astrolabe, the cross-staff used the horizon for altitude measurement. If fog or land obscures this, however, an artificial horizon is required and it was Pedro de Medina who probably described the first such instrument for use at sea, in his manual *Arte de Navegar* of 1545.

Encouraged by the commercial success of their trade with Asia, Portugal and Spain created institutions where the knowledge of astronomy, cartography, and navigation was brought together and improved, through cooperation between practitioners and thinkers in this field. The *Casa de Guinea e India* was established in Lisbon, and the *Casa de Contratación* in Seville.[15] These institutions also had to ensure that sea routes and the related cartographic and navigational knowledge were kept secret from foreign competitors. Despite these measures, the French, English, and Dutch also found the way towards Asia in the second half of the fifteenth century. Their knowledge came from disloyal Iberian pilots and cartographers, and through Spanish and Portuguese manuals, such as those by Martin Cortes, Medina, and Rodrigo Zamorano, which were translated into French, English and Dutch. Soon scholars in these countries, including the Englishmen Thomas Blundeville and William Bourne, and the Fleming Michiel Coignet, published their own manuals. These included descriptions of the mariner's astrolabe and the cross-staff, tables of the Sun's declination, and the 'regiment for the North Star'.[16] Northern-European navigators adopted these instruments and techniques, and the transferred knowledge soon brought on innovations. In the 1590s a notable improvement for measuring the Sun's altitude at sea was the invention, by the English pilot John Davis, of the backstaff or Davis quadrant.

LONGITUDE AT SEA

When sailing from Europe down the west coast of Africa, the observer's latitude changes rapidly, while his longitude remains more or less the same. In the 1450s seamen had learned to 'run down the latitude', a method by which a ship heading for the Azores, for example, would proceed along the coast of Africa until she had reached the latitude of the islands. She would then sail due west to make landfall on them, checking the latitude regularly at noon from an altitude of the Sun or at night by the Pole Star. The latitude could be corrected by altering the course to west-by-north or west-by-south.[17] After the Atlantic Ocean was crossed in 1492, and the Indian Ocean in 1497, east–west courses were sailed more frequently and over greater distances, and the

FIGURE 1.6 Observing the Sun's altitude with a mariner's astrolabe. From *Arte de navigacion* by Pedro de Medina, Valladolid, 1545. National Maritime Museum; PBE5221.

need to measure longitude at sea increased. Various astronomical methods were put forward, all based on measuring the difference in time between two meridians. Columbus attempted to find his longitude in the Caribbean by observing lunar eclipses, in 1494 and 1504. He compared the time of the eclipse with the predicted time that he assumed it would occur in Spain. The effort failed by many tens of degrees, mainly because of the inaccurate prediction arising from imperfect knowledge of the Moon's motion.[18]

In 1514 Johann Werner, a Nuremberg astronomer, was probably the first to suggest that longitude could be determined by measuring lunar distances, that is, the angular distance between the Moon and a fixed star, by using a cross-staff.[19] The Moon moves across the sky relatively fast, and at any given time the angle between it and any star, including the Sun, is the same for every observer who can see both celestial bodies, except that some allowance has to be made for parallax.[20] In theory, 'lunar distances' can be calculated in advance for certain stars, dates, and times and for a particular meridian, and can be recorded in tables. An observer at sea would find the time corresponding to this angle at the standard meridian, and compare this with his local time. The difference between the two times provided the difference in longitude. Ferdinand de Magellan's astrologer, Andrés de San Martín, tried to find the longitude in South America by the lunar-distance method in 1519. Failure was blamed on errors in the predicted astronomical ephemeris. Although Werner's method is indeed feasible, at that time the two conditions for its success could not yet be fulfilled. It was not possible to calculate sufficiently accurate lunar tables, and the cross-staff was not accurate enough for the measurement needed to produce usable results.

In 1530 the originally Frisian scholar Gemma Frisius (*Jemme Reinderszoon*), a professor at Louvain University, suggested that longitude could be found by using a clock to measure the difference in time between two meridians. That clock would be taken to sea where it would have to keep the time of the meridian of departure during the voyage, without being influenced by the ship's movement and changes in temperature. An observer would find his longitude from the difference between his celestially observed local time and that kept by the clock. While the theory was impeccable, it was at that period impossible to construct a sufficiently accurate clock.[21]

As the attempts to find longitude at sea failed, King Philip II of Spain offered a prize for a practical solution in 1598. This example was soon followed by the States-General of the Dutch Republic, the States of Holland, the governments of France and Venice, and finally, in 1714, by the British Parliament.[22] These prizes encouraged inventors, astronomers, mariners, mathematicians, and instrument makers to seek a solution. Among them was the astronomer Galileo Galilei, who proposed that the eclipses and occultations of Jupiter's satellites could provide an answer, as these occurrences are visible at the same instant for every observer. Sufficiently powerful telescopes were available and the strict regularity of the eclipses allowed accurate prediction. The method was indeed practicable on land but not on the deck of a moving vessel, and also because the two largest satellites are not visible on the daily basis needed by navigators.[23]

IMPROVEMENTS IN NAVIGATION IN THE SEVENTEENTH CENTURY

Although the longitude problem was not solved during the seventeenth century, the rise of Dutch and English shipping brought on new and improved navigating techniques and instruments. Dutch and English manuals gradually replaced the older Iberian texts. The first books to include a comprehensive chapter on the art of navigation and navigational instruments were sea-atlases and sailing directions, which started to appear at the end of the sixteenth century. The *Spiegel der Zeevaerdt* by Lucas Janszoon Waghenaer of 1584 was translated into English, as *The Mariner's Mirror*, in 1588. It also appeared in French and German, and was soon followed by other sea-atlases.[24] Mathematics was made easier when in 1614 John Napier, in Scotland, invented calculation by logarithms. Henry Briggs in London, improved Napier's tables, and in 1628 Adriaen Vlacq, in the Netherlands, refined Briggs's work, by which logarithms were available for navigational calculations.[25]

Northern-European navigating manuals included the *Vergulden Licht der Zee-vaert* by Claes Hendrikszoon Gietermaker, which appeared in Amsterdam in 1661 and saw many editions including a French translation. In London, John Seller's *Practical Navigation, or, An Introduction to the Whole Art* appeared in 1669 and was republished many times. In France, the *Traité Complet de la Navigation* by Jean Bouguer appeared in Paris in 1698. These books contain a general description of the art of navigation, arithmetic, basic mathematics, trigonometry, and astronomy; logarithm and declination tables, and a section dealing with navigational instruments. They show that the mariner's astrolabe gradually disappeared in

FIGURE 1.7 Navigational instruments used at the end of the sixteenth century, including a mariner's quadrant, a mariner's astrolabe, a cross-staff, and a sounding lead. Title page of *The Mariners Mirror*, a sea atlas by Lucas Jansz. Waghenaer of Enkhuizen, translated and revised by Anthony Ashsley, and published London, 1588. National Maritime Museum; PBD8264.

the 1670s owing to improvements that made the cross-staff and the backstaff superior instruments for altitude measurement. These developments will be dealt with in the following chapter.

Another notable improvement was the introduction of a method to convert an observed altitude into the true altitude. The Sun's semi-diameter has already been mentioned, and this correction must also be applied for the Moon. Another correction is that for astronomical refraction, the diversion of light as it passes through the Earth's atmosphere, as a result of which a star is seen slightly higher than its actual position. Refraction is not constant but decreases as a star's altitude increases. The Englishman Edward Wright was probably the first to bring this phenomenon to the attention of seafarers, in *Certaine Errors of Navigation* of 1599. Willem Janszoon Blaeu, in *Het Licht der Zee-Vaart* of 1608, published tables for refraction that were based on work by his former teacher, the astronomer Tycho Brahe.[26] The third major correction to be applied to a measured altitude is for dip, the angle between the horizontal plane through the eye of the observer and the direction in which he sees the horizon. It is caused by – and increases with – the height of the observer's eye above sea level. Dip is influenced by the effect of terrestrial refraction, the apparent lowering or elevation of the sea horizon caused by the prevailing difference in temperature between the air and the seawater.[27]

LONGITUDE FOUND

A practical solution for determining longitude at sea was one of the many subjects tackled by the Royal Society of London, founded in 1660, and the Académie Royale des Sciences in Paris, founded in 1666. Following these, the Observatoire de Paris was founded in 1667 and the Royal Observatory at Greenwich in 1675, and both institutions were commissioned to rectify the existing tables of stars in order to establish methods of determining longitude for navigation. The Observatoire published new tables of astronomical data from 1678 onwards, the *Connoissance* [later *Connaissance*] *des temps ou des Mouvements Celestes*.

The 'Longitude Prize' offered by the British Parliament in 1714 amounted to £20,000, and the Commissioners of the Board of Longitude were appointed to judge applications. The reward stimulated a number of improvements in navigational instruments and techniques. The first practical achievement was the invention of the octant, a double-reflecting instrument designed by John Hadley, vice-president of the Royal Society, and announced there by him in 1731.[28] It was more accurate and far easier in use than the cross-staff and backstaff. It was soon improved and within a few decades was sufficiently accurate to be used for the lunar-distance method, which meant that the first of the two conditions for bringing this method into practice was fulfilled. The invention

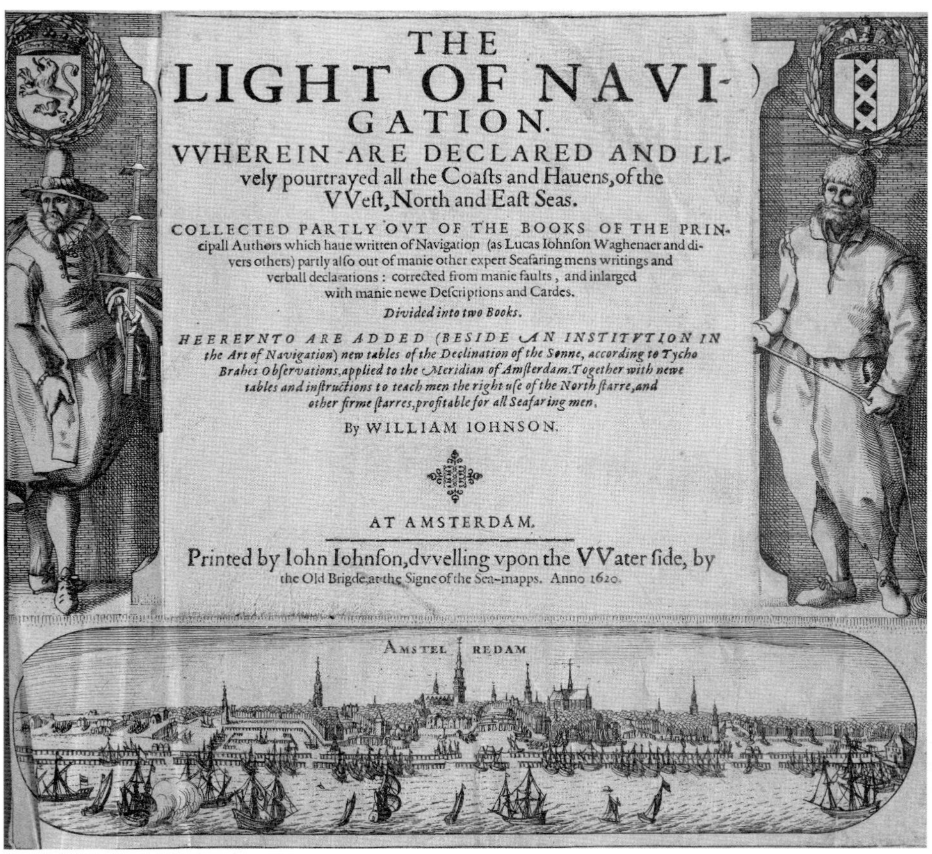
THE LIGHT OF NAVIGATION.

VVHEREIN ARE DECLARED AND LIVELY pourtrayed all the Coasts and Hauens, of the VVest, North and East Seas.

COLLECTED PARTLY OVT OF THE BOOKS OF THE PRINCIPALL Authors which haue written of Navigation (as Lucas Iohnson Waghenaer and divers others) partly also out of manie other expert Seafaring mens writings and verball declarations: corrected from manie faults, and inlarged with manie newe Descriptions and Cardes.

Divided into two Books.

HEEREVNTO ARE ADDED (BESIDE AN INSTITVTION IN the Art of Navigation) new tables of the Declination of the Sonne, according to Tycho Brahes Observations, applied to the Meridian of Amsterdam. Together with newe tables and instructions to teach men the right use of the North starre, and other firme starres, profitable for all Seafaring men.

By WILLIAM IOHNSON.

AT AMSTERDAM.

Printed by Iohn Iohnson, dvvelling vpon the VVater side, by the Old Brigde at the Signe of the Sea-mapps. Anno 1620.

FIGURE 1.8 The *Light of Navigation*, a sailing guide by Willem Jansz. Blaeu (1620), the English translation of *Het Licht der Zee-Vaart* (Amsterdam, 1608). The guide contains a chapter on navigation and tables for the refraction of the Sun. National Maritime Museum; PBE8259.

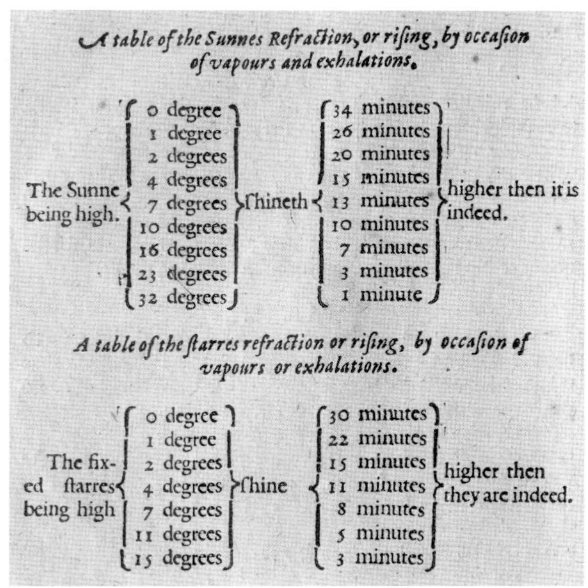
A table of the Sunnes Refraction, or rising, by occasion of vapours and exhalations.

The Sunne being high.	shineth	higher then it is indeed.
0 degree		34 minutes
1 degree		26 minutes
2 degrees		20 minutes
4 degrees		15 minutes
7 degrees		13 minutes
10 degrees		10 minutes
16 degrees		7 minutes
23 degrees		3 minutes
32 degrees		1 minute

A table of the starres refraction or rising, by occasion of vapours or exhalations.

The fixed starres being high	shine	higher then they are indeed.
0 degree		30 minutes
1 degree		22 minutes
2 degrees		15 minutes
4 degrees		11 minutes
7 degrees		8 minutes
11 degrees		5 minutes
15 degrees		3 minutes

FIGURE 1.9 Tables for the refraction of the Sun from the *Light of Navigation* (Amsterdam, 1620) by Willem Jansz. Blaeu. National Maritime Museum; PBD8259.

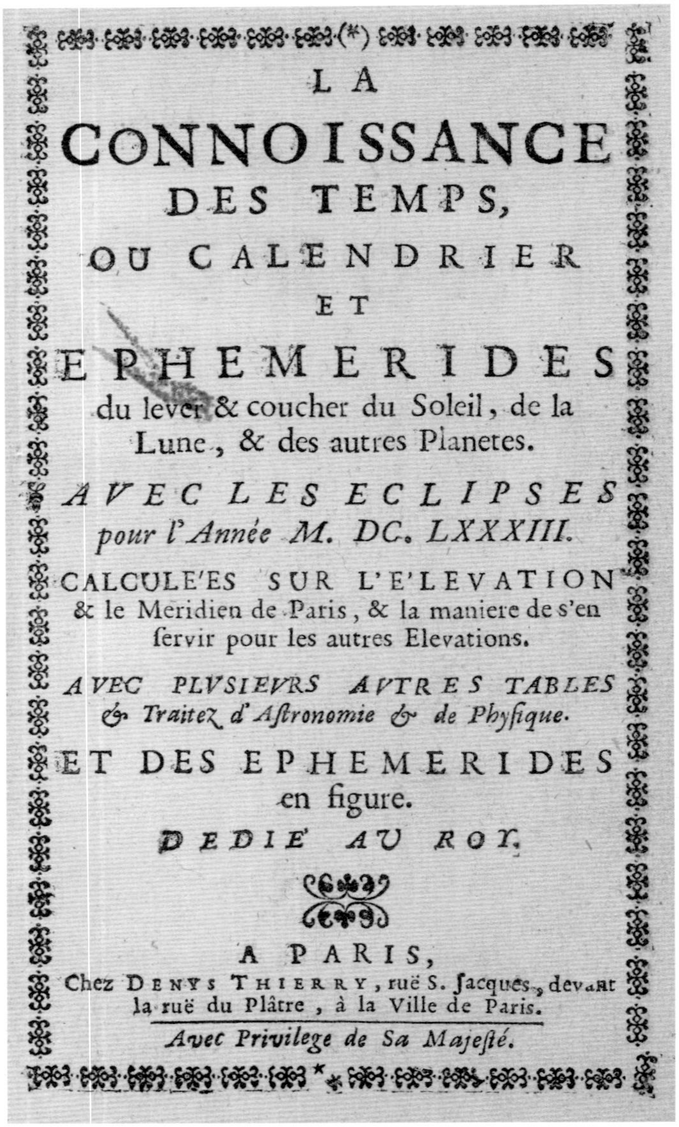

LA
CONNOISSANCE
DES TEMPS,
OU CALENDRIER
ET
EPHEMERIDES
du lever & coucher du Soleil, de la Lune, & des autres Planetes.
AVEC LES ECLIPSES pour l'Année M. DC. LXXXIII.
CALCULE'ES SUR L'E'LEVATION & le Meridien de Paris, & la maniere de s'en ſervir pour les autres Elevations.
AVEC PLUSIEURS AUTRES TABLES & Traitez d'Aſtronomie & de Phyſique.
ET DES EPHEMERIDES en figure.
DEDIE' AU ROY.
A PARIS,
Chez DENYS THIERRY, ruë S. Jacques, devant la ruë du Plâtre, à la Ville de Paris.
Avec Privilege de Sa Majeſté.

FIGURE 1.10 The *Connoissance des temps* for 1683, containing tables of the movement of the Sun, Moon, and planets. Issued annually by the Paris Observatory from 1678, it included tables calculating longitude by lunar distances from 1760. National Maritime Museum; PBD1220.

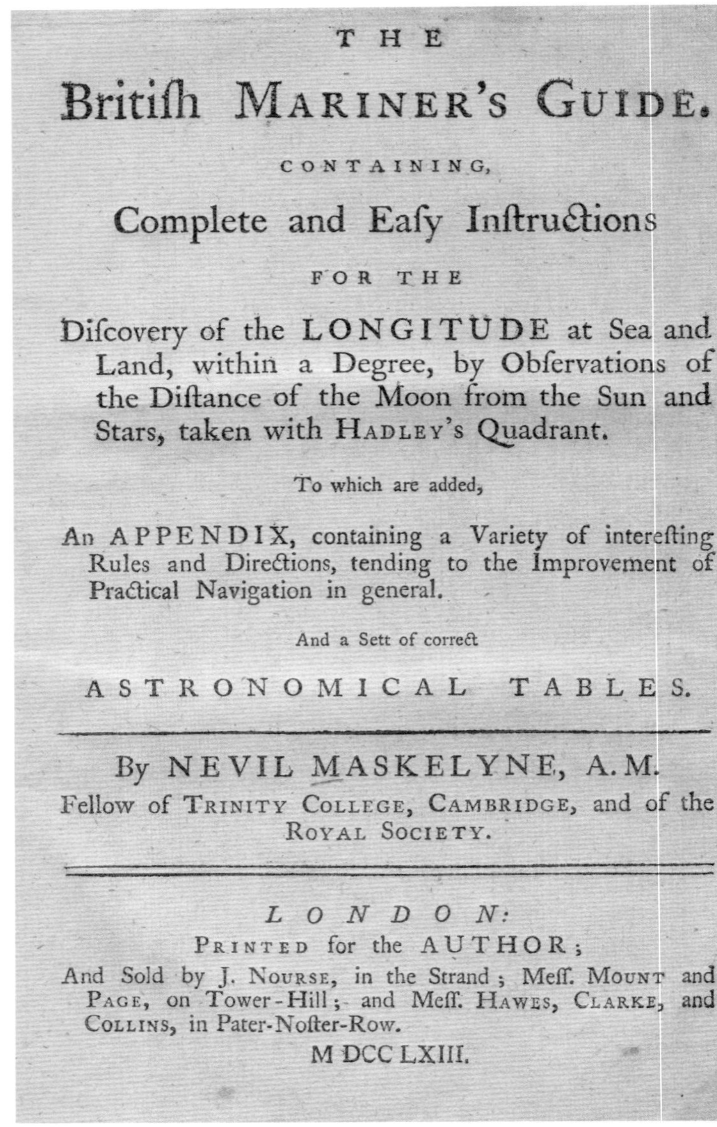

THE
Britiſh MARINER'S GUIDE.
CONTAINING,
Complete and Eaſy Inſtructions
FOR THE
Diſcovery of the LONGITUDE at Sea and Land, within a Degree, by Obſervations of the Diſtance of the Moon from the Sun and Stars, taken with HADLEY's Quadrant.
To which are added,
An APPENDIX, containing a Variety of intereſting Rules and Directions, tending to the Improvement of Practical Navigation in general.
And a Sett of correct
ASTRONOMICAL TABLES.
By NEVIL MASKELYNE, A.M.
Fellow of TRINITY COLLEGE, CAMBRIDGE, and of the ROYAL SOCIETY.
LONDON:
PRINTED for the AUTHOR;
And Sold by J. NOURSE, in the Strand; Meſſ. MOUNT and PAGE, on Tower-Hill; and Meſſ. HAWES, CLARKE, and COLLINS, in Pater-Noſter-Row.
M DCC LXIII.

FIGURE 1.11 *The British Mariner's Guide* containing tables required for calculating the longitude by lunar distances (London, 1763). National Maritime Museum; PBD1847.

of the octant, and its further development and diffusion in the eighteenth century, will be dealt with in the third chapter.

Before the longitude problem at sea was solved, the calculation of latitude from the noon was improved. In 1748 Cornelis Douwes, a prominent teacher of navigation in Amsterdam, developed a method of double altitudes by which latitude could be found without being restricted to the traditional noon observation. The principle had been known since the sixteenth century, but Douwes was the first to provide a practical method. This involved taking two sun altitudes with an octant, one before and the other after the noon passage, the time lapse between the observations being measured with a watch. Seamen in the major seafaring countries in Europe and North America adopted Douwes's method. It was improved in the nineteenth century and remained in use until the 1850s.[29]

Tobias Mayer, an astronomer at the University of Göttingen in Germany, calculated accurate lunar tables in the mid-1750s. At the same time a French astronomer, the Abbé Nicolas-Louis de Lacaille, did this independently, and he was the first to practise the lunar-distance method at sea, when returning to France from the Cape of Good Hope in 1753–54.[30] From 1760 onwards Lacaille's tables appeared in the *Connaissance des temps*, a journal not easily accessible to seamen. At the request of the Admiralty, Captain John Campbell RN and the English astronomer Nevil Maskelyne tried out Mayer's tables at sea, and these tables were published in *The British Mariners' Guide* in 1763. Maskelyne became the driving force behind 'lunars', as the method became known, and he initiated the *Nautical Almanac and Astronomical Ephemeris*. That first appeared in 1766, containing lunar tables for seamen for the years 1767–69. They were based on Mayer's calculations and were for a number of fixed times of each day, calculated for the meridian of Greenwich. They made lunars a practicable and reliable way to find longitude at sea, even though the calculation, called 'clearing the distance', was laborious and required a high degree of numeracy.

During the trials with lunars, the octant's range of measurement often proved too short, because the angular distance between the Moon and stars frequently exceeds the instrument's maximum measurable angle of 90 degrees. Mayer had foreseen this limitation and proposed a reflecting circle that could be used for measuring angles up to 180 degrees. This design proved cumbersome and resulted in the first sextant, made in the second half of the 1750s, with which angles up to 120 degrees could be observed. In turn, the sextant was extended to become a quintant, measuring up to 150 degrees, while in France in the 1770s Mayer's circle was improved into a useful instrument for lunars. A difficulty when manufacturing all such instruments was dividing the circular scale. It was laborious and time-consuming until the late 1760s, when Jesse Ramsden of London invented a circular dividing engine. These developments will be discussed in the fourth chapter.

While the lunar method was being perfected, clockmakers in England and France worked to develop a timekeeper that would perform accurately throughout a voyage. Such a timekeeper should not be influenced by changes in temperature, by gravity, or by the ship's movements. Notable contestants in the process were John Harrison in England, and Pierre Le Roy and Ferdinand Berthoud in France. The two Frenchmen (though Berthoud was born Swiss) made a series of timekeepers, many of which were used at sea. Le Roy deposited his first example at the Académie Royale in 1754, and Berthoud built his first in 1760.[31] It was, however, John Harrison who eventually won the race with his fourth timepiece, completed in 1759.[32]

In the 1780s chronometers, as marine timekeepers were by

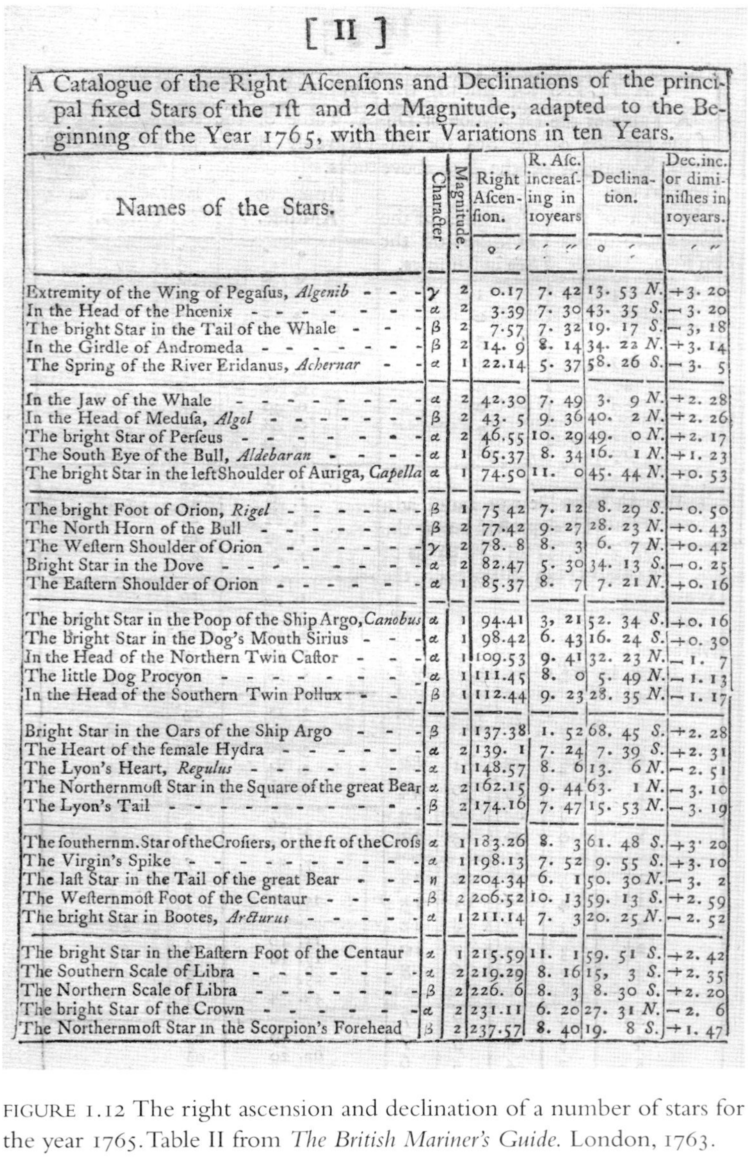

[II]

A Catalogue of the Right Afcenfions and Declinations of the principal fixed Stars of the 1ft and 2d Magnitude, adapted to the Beginning of the Year 1765, with their Variations in ten Years.

Names of the Stars.	Character	Magnitude.	Right Afcenfion. ° ′	R. Afc. increafing in 10years ′ ″	Declination. ° ′	Dec. inc. or diminifhes in 10years. ′ ″
Extremity of the Wing of Pegafus, *Algenib*	γ	2	0.17	7. 42	13. 53 N.	+3. 20
In the Head of the Phœnix	α	2	3.39	7. 30	43. 35 S.	−3. 20
The bright Star in the Tail of the Whale	β	2	7.57	7. 32	19. 17 S.	−3, 18
In the Girdle of Andromeda	β	2	14. 9	8. 14	34. 22 N.	+3. 14
The Spring of the River Eridanus, *Achernar*	α	1	22.14	5. 37	58. 26 S.	−3. 5
In the Jaw of the Whale	α	2	42.30	7. 49	3. 9 N.	+2. 28
In the Head of Medufa, *Algol*	β	2	43. 5	9. 36	40. 2 N.	+2. 26
The bright Star of Perfeus	α	2	46.55	10. 29	49. 0 N.	+2. 17
The South Eye of the Bull, *Aldebaran*	α	1	65.37	8. 34	16. 1 N.	+1. 23
The bright Star in the leftShoulder of Auriga, *Capella*	α	1	74.50	11. 0	45. 44 N.	+0. 53
The bright Foot of Orion, *Rigel*	β	1	75 42	7. 12	8. 29 S.	−0. 50
The North Horn of the Bull	β	2	77.42	9. 27	28. 23 N.	+0. 43
The Weftern Shoulder of Orion	γ	2	78. 8	8. 3	6. 7 N.	+0. 42
Bright Star in the Dove	α	2	82.47	5. 30	34. 13 S.	−0. 25
The Eaftern Shoulder of Orion	α	1	85.37	8. 7	7. 21 N.	+0. 16
The bright Star in the Poop of the Ship Argo, *Canobus*	α	1	94.41	3, 21	52. 34 S.	+0. 16
The bright Star in the Dog's Mouth Sirius	α	1	98.42	6. 43	16. 24 S.	+0. 30
In the Head of the Northern Twin Caftor	α	1	109.53	9. 41	32. 23 N.	−1. 7
The little Dog Procyon	α	1	111.45	8. 0	5. 49 N.	−1. 13
In the Head of the Southern Twin Pollux	β	1	112.44	9. 23	28. 35 N.	−1. 17
Bright Star in the Oars of the Ship Argo	β	1	137.38	1. 52	68. 45 S.	+2. 28
The Heart of the female Hydra	α	2	139. 1	7. 24	7. 39 S.	+2. 31
The Lyon's Heart, *Regulus*	α	1	148.57	8. 6	13. 6 N.	−2. 51
The Northernmoft Star in the Square of the great Bear	α	2	162.15	9. 44	63. 1 N.	−3. 10
The Lyon's Tail	β	2	174.16	7. 47	15. 53 N.	−3. 19
The fouthernm. Star of the Crofiers, or the ft of the Crofs	α	1	183.26	8. 3	61. 48 S.	+3. 20
The Virgin's Spike	α	1	198.13	7. 52	9. 55 S.	+3. 10
The laft Star in the Tail of the great Bear	η	2	204.34	6. 1	50. 30 N.	−3. 2
The Wefternmoft Foot of the Centaur	β	2	206.52	10. 13	59. 13 S.	+2. 59
The bright Star in Bootes, *Arcturus*	α	1	211.14	7. 3	20. 25 N.	−2. 52
The bright Star in the Eaftern Foot of the Centaur	α	1	215.59	11. 1	59. 51 S.	+2. 42
The Southern Scale of Libra	α	2	219.29	8. 16	15, 3 S.	+2. 35
The Northern Scale of Libra	β	2	226. 6	8. 3	8. 30 S.	+2. 20
The bright Star of the Crown	α	2	231.11	6. 20	27. 31 N.	−2. 6
The Northernmoft Star in the Scorpion's Forehead	β	2	237.57	8. 40	19. 8 S.	+1. 47

FIGURE 1.12 The right ascension and declination of a number of stars for the year 1765. Table II from *The British Mariner's Guide*. London, 1763. National Maritime Museum; PBD1847.

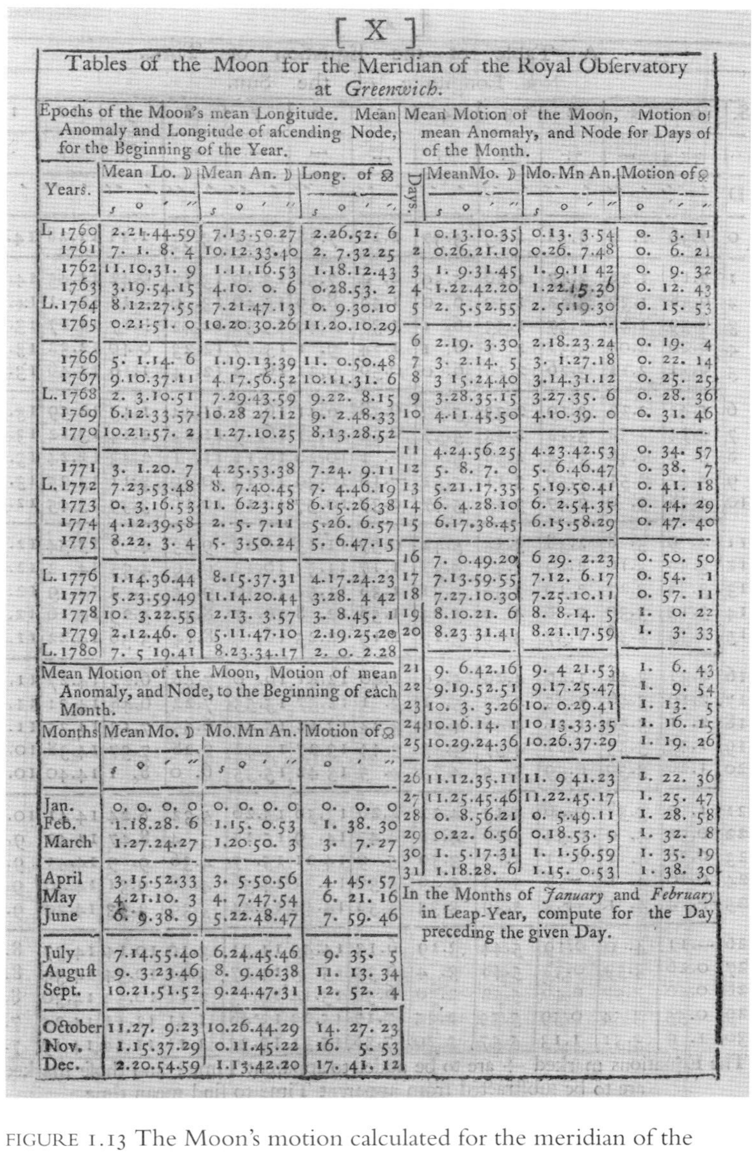

[X]

Tables of the Moon for the Meridian of the Royal Obfervatory at *Greenwich*.

Epochs of the Moon's mean Longitude. Mean Anomaly and Longitude of afcending Node, for the Beginning of the Year.

Years.	Mean Lo. ☽ s ° ′ ″	Mean An. ☽ s ° ′ ″	Long. of ☊ s ° ′ ″
L. 1760	2.21.44.59	7.13.50.27	2.26.52. 6
1761	7. 1. 8. 4	10.12.33.40	2. 7.32.25
1762	11.10.31. 9	1.11.16.53	1.18.12.43
1763	3.19.54.15	4.10. 0. 6	0.28.53. 2
L. 1764	8.12.27.55	7.21.47.13	0. 9.30.10
1765	0.21.51. 0	10.20.30.26	11.20.10.29
1766	5. 1.14. 6	1.19.13.39	11. 0.50.48
1767	9.10.37.11	4.17.56.52	10.11.31. 6
L. 1768	2. 3.10.51	7.29.43.59	9.22. 8.15
1769	6.12.33.57	10.28 27.12	9. 2.48.33
1770	10.21.57. 2	1.27.10.25	8.13.28.52
1771	3. 1.20. 7	4.25.53.38	7.24. 9.11
L. 1772	7.23.53.48	8. 7.40.45	7. 4.46.19
1773	0. 3.16.53	11. 6.23.58	6.15.26.38
1774	4.12.39.58	2. 5. 7.11	5.26. 6.57
1775	8.22. 3. 4	5. 3.50.24	5. 6.47.15
L. 1776	1.14.36.44	8.15.37.31	4.17.24.23
1777	5.23.59.49	11.14.20.44	3.28. 4 42
1778	10. 3.22.55	2.13. 3.57	3. 8.45. 1
1779	2.12.46. 0	5.11.47.10	2.19.25.20
L. 1780	7. 5 19.41	8.23.34.17	2. 0. 2.28

Mean Motion of the Moon, Motion of mean Anomaly, and Node, to the Beginning of each Month.

Months	Mean Mo. ☽ s ° ′ ″	Mo. Mn An. s ° ′ ″	Motion of ☊ ° ′ ″
Jan.	0. 0. 0. 0	0. 0. 0. 0	0. 0. 0
Feb.	1.18.28. 6	1.15. 0.53	1. 38. 30
March	1.27.24.27	1.20.50. 3	3. 7. 27
April	3.15.52.33	3. 5.50.56	4. 45. 57
May	4.21.10. 3	4. 7.47.54	6. 21. 16
June	6. 9.38. 9	5.22.48.47	7. 59. 46
July	7.14.55.40	6.24.45.46	9. 35. 5
Auguft	9. 3.23.46	8. 9.46.38	11. 13. 34
Sept.	10.21.51.52	9.24.47.31	12. 52. 4
October	11.27. 9.23	10.26.44.29	14. 27. 23
Nov.	1.15.37.29	0.11.45.22	16. 5. 53
Dec.	2.20.54.59	1.13.42.20	17. 41. 12

Mean Motion of the Moon, Motion of mean Anomaly, and Node for Days of the Month.

Days.	MeanMo. ☽ s ° ′ ″	Mo. Mn An. s ° ′ ″	Motion of ☊ ° ′ ″
1	0.13.10.35	0.13. 3.54	0. 3. 11
2	0.26.21.10	0.26. 7.48	0. 6. 21
3	1. 9.31.45	1. 9.11 42	0. 9. 32
4	1.22.42.20	1.22.15.36	0. 12. 43
5	2. 5.52.55	2. 5.19.30	0. 15. 53
6	2.19. 3.30	2.18.23.24	0. 19. 4
7	3. 2.14. 5	3. 1.27.18	0. 22. 14
8	3 15.24.40	3.14.31.12	0. 25. 25
9	3.28.35.15	3.27.35. 6	0. 28. 36
10	4.11.45.50	4.10.39. 0	0. 31. 46
11	4.24.56.25	4.23.42.53	0. 34. 57
12	5. 8. 7. 0	5. 6.46.47	0. 38. 7
13	5.21.17.35	5.19.50.41	0. 41. 18
14	6. 4.28.10	6. 2.54.35	0. 44. 29
15	6.17.38.45	6.15.58.29	0. 47. 40
16	7. 0.49.20	6 29. 2.23	0. 50. 50
17	7.13.59.55	7.12. 6.17	0. 54. 1
18	7.27.10.30	7.25.10.11	0. 57. 11
19	8.10.21. 6	8. 8.14. 5	1. 0. 22
20	8.23 31.41	8.21.17.59	1. 3. 33
21	9. 6.42.16	9. 4 21.53	1. 6. 43
22	9.19.52.51	9.17.25.47	1. 9. 54
23	10. 3. 3.26	10. 0.29.41	1. 13. 5
24	10.16.14. 1	10 13.33.35	1. 16. 15
25	10.29.24.36	10.26.37.29	1. 19. 26
26	11.12.35.11	11. 9 41.23	1. 22. 36
27	11.25.45.46	11.22.45.17	1. 25. 47
28	0. 8.56.21	0. 5.49.11	1. 28. 58
29	0.22. 6.56	0.18.53. 5	1. 32. 8
30	1. 5.17.31	1. 1.56.59	1. 35. 19
31	1.18.28. 6	1.15. 0.53	1. 38. 30

In the Months of *January* and *February* in Leap-Year, compute for the Day preceding the given Day.

FIGURE 1.13 The Moon's motion calculated for the meridian of the Royal Observatory Greenwich, for use with lunar distances. Table X from *The British Mariner's Guide*. London, 1763. National Maritime Museum,; PBD1847.

then called, became available in sufficient numbers and at affordable prices, mainly thanks to English makers such as John Arnold, Thomas Mudge, and Thomas Earnshaw.[33] To find longitude by chronometer, the local time on board ship was calculated from an altitude observation of the Sun with an octant, and by using the estimated latitude.[34] The difference between local time and the time at one's meridian of departure according to the chronometer – Greenwich for seamen using the *Nautical Almanac* – provided the longitude.

Despite the increase in the number of chronometers, the lunar method also remained in use during the nineteenth century. It served to check the rate and error of chronometers at sea, a task that in the early twentieth century was taken over by radio signals transmitted from shore stations.[35] For this purpose, and for those who could not afford a chronometer, the annually published *Nautical Almanac* continued to include lunar tables until 1907.

POSITION-LINE NAVIGATION

The first major improvement in navigational techniques in the nineteenth century was the advent of position-line navigation.[36] This is the name for various ways by which, through an altitude measurement of a celestial body, a reading of the chronometer time, and a subsequent calculation, a position line can be constructed on a chart. When a number of stars (usually four or five) lying in different directions are observed at (almost) the same time, the intersection of the subsequent position lines will provide a fix. As an accurate altitude measurement requires a clear horizon, star observations are generally limited to the morning and evening twilight.

The American captain Thomas H. Sumner of Boston is credited with the discovery of position-line navigation, which he found quite accidentally in 1837 as he sailed from Charleston, South Carolina, to Greenock in Scotland. Because of bad weather, Sumner was unable to observe the Sun's altitude for many days. When his dead reckoning told him he was nearing the south coast of Ireland, he was at last able to observe the Sun's altitude one forenoon as it broke briefly through the clouds. Sumner noted the chronometer time and, by using that time and the measured altitude, calculated his longitude three times with three different estimated latitudes. When he plotted the results on his chart, Sumner found that the longitudes lay in one line which, when extended, passed through the Small's lighthouse in south Wales. Sumner drew the conclusion that his Sun altitude had been the same as for an observer in the lighthouse. He changed course to follow the line and, after an hour, indeed sighted the Small's lighthouse. By chance Sumner had discovered a way to find the imaginary circle of equal altitudes and a method by which a navigator, with a chronometer and the altitude of a celestial body, could establish a position line at any time of the day. Sumner published his discovery in *A New and Accurate Method of Finding a Ship's Position at Sea* (Boston, 1843) and the method was internationally recognized as a major improvement in navigation. Two decades later, the French naval officer Aved de Magnac and his countryman the astronomer Yvon Villarceau improved position-line navigation, naming it *la nouvelle navigation astronomique.* Soon thereafter another French naval officer, Marc de Blond de St Hilaire, developed a method by which the zenith distance of the Sun or a star, as measured with an octant or sextant, is compared with the zenith distance calculated by using the dead-reckoned latitude. In principle this method remained in general use until satellite positioning replaced celestial navigation in the twentieth century.[37] During the nineteenth century the sextant gradually replaced the octant and was technically perfected, not least as a result of improvements in the methods and techniques of surveying and hydrographical work. These developments will be discussed in the fifth chapter.

SATELLITE NAVIGATION: THE END OF THE ERA OF THE SEXTANT

In 1964, the Navy Navigation Satellite System (Transit) programme was set up in the United States, heralding satellite navigation and the end of the sextant as a tool of professional seamen. Three years later, satellite navigation became available on a very limited basis, and in 1976 the NAVSTAR Global Positioning System (GPS) became available.[38] Owing to insufficient numbers of satellites, GPS was initially limited to certain periods of the day, and did not cover the entire globe. Since 1996, it has been fully operational and anyone with a receiver can obtain a position instantly and continually, anywhere in the world. This is quicker and far more accurate than calculating a position by celestial navigation and, as GPS receivers have become cheaper, celestial navigation has become obsolete.[39] Today only nostalgic yachtsmen, those interested in historic instruments, and sailors with a dislike for modern technology navigate with a sextant.

1. Quoted in Waters, *The Art of Navigation*, 3.
2. Taylor, *The Haven-Finding Art*, 40.
3. The Amplitude is the angle from the east point where the star rises or the west point where it sets, measured along the horizon northward or southward.
4. See Tibbetts, *Arab Navigation*; Levathes, *When China Ruled the Seas*, 96; Lewis, *We, the Navigators*.
5. Taylor, *The Haven-Finding Art*, 35–64; Cotter, *A History of Nautical Astronomy*, 1–20.
6. Tibbetts, *Arab Navigation*, 290–91.
7. Taylor, *The Haven-Finding Art*, 65-121. For Viking navigation see the more recent Schnall, *Navigation der Wikinger*.
8. Parry, *The Discovery of the Sea*, 165–92.
9. Randles, 'The Emergence of Nautical Astronomy in Portugal', 48–9. The Guards are the stars Kochab and Pherkab, which circulate at a small angular distance around the Pole Star in about 24 hours.
10. The construction, development, and use of altitude-measuring instruments used at sea prior to 1731 are discussed in the next chapter.
11. Randles, 'The Emergence of Nautical Astronomy in Portugal', 53.
12. Randles, 'The Emergence of Nautical Astronomy in Portugal', 49–50. The two volumes are preserved in the Bavarian State Library, Munich, Germany, and in the Evora Public Library, Evora, Portugal, respectively. The tables in the *Regimento* were based on the Alfonsine tables, which had been compiled in the thirteenth century by order of King Alfonso the Wise of Castile.
13. Depending on the date, the value of the Sun's semi-diameter is a little more or less than 16 minutes of arc.
14. Randles, 'The Emergence of Nautical Astronomy in Portugal', 49.
15. Lamb, 'Nautical Scientists and Their Clients in Iberia, 49–51.
16. Crone, *Cornelis Douwes 1712–1773,* 71–7, provides a brief survey of these manuals and their translations. See also Davids, 'Dutch Contributions to the Development of Navigation Technology', 60–1.
17. Waters, *The Art of Navigation*, 47. This method was also called 'running down the altitude'.
18. Randles, 'Portuguese and Spanish Attempts to Measure Longitude in the Sixteenth Century', 402–4.
19. Howse, *Greenwich Time and the Longitude*, 20–3.
20. The Moon is so close to the Earth that its apparent position against the stars is different when viewed from widely separated places on earth: the difference is called parallax.
21. Howse, *Greenwich Time and the Longitude*, 24–5.
22. Howse, *Greenwich Time and the Longitude*, 26–7, 57–9; Davids, *Zeewezen en wetenschap*, 69–74.
23. Cotter, *A History of Nautical Astronomy*, 184–5.
24. Mörzer Bruyns, 'The Importance of the Van Keulens to Navigation', 33–4; Davids, 'Dutch Contributions to the Development of Navigation Technology', 67.
25. Waters, *The Art of Navigation*, 402–13; Davids, *Zeewezen en wetenschap*, 116–17.
26. Cotter, *A History of Nautical Astronomy*, 97–110; Davids, *Zeewezen en wetenschap*, 125–6. The publication referred to in the text appeared as *The Light of Navigation* in 1612, and in 1628 as *Le flambeau de la navigation.*
27. Further corrections are those for parallax, irradiation, and the observer's personal error; see Cotter, *A History of Nautical Astronomy*, 111–22.
28. Forbes, *The Birth of Navigational Science*, 3–5.
29. Cotter, *A History of Nautical Astronomy*, 148–9, 168.
30. Howse, *Greenwich Time and the Longitude*, 68. Lacaille's lunar tables were also used at sea by Alexandre-Guy Pringré in 1761. According to Howse, Lacaille's tables were less precise than those of Mayer.
31. Cardinal, 'Ferdinand Berthoud and Pierre Le Roy', 281–92.
32. Randall, 'The Timekeeper that Won the Longitude Prize', 235–54.
33. Betts, 'Arnold and Earnshaw', 311–28.
34. Cotter, *A History of Nautical Astronomy* 51–2; May, *A History of Marine Navigation*, 171.
35. Chauvenet, *New Method of Correcting Lunar Distance.*
36. For details of the techniques mentioned in the following text, see Vanvaerenbergh and Ifland, *Line of Position Navigation*; Cotter, *A History of Nautical Astronomy* 268–308.
37. In the nineteenth and twentieth centuries various short methods were developed by which the calculation was simplified, but all were based on a celestial altitude and a chronometer reading; see Cotter, *A History of Nautical Astronomy*, 317–34.
38. Logsdon, *The Navstar Global Positioning System*.
39. The European Commission supports its own satellite navigation system, named Galileo, which is expected to be operational before the end of the next decade.

CHAPTER TWO

THE HISTORY AND DEVELOPMENT OF INSTRUMENTS FOR MEASURING ALTITUDE AT SEA BEFORE THE OCTANT

INTRODUCTION

Before scientific instruments were used at sea to measure altitudes, the height of the Pole Star above the horizon was estimated, often simply with the use of hands and fingers. The width of a man's finger held at arm's length covered approximately two degrees; his palm or fist approximately eight. The mast of a ship could be divided in hand's breadths, in such a way that the navigator could regularly check the altitude and thereby his latitude.[1] Arab seamen used this technique, and Chinese pilots followed it in the early fifteenth century.[2] In 1454 an Italian pilot, sailing south along the coast of Africa, estimated the altitude of the Pole Star by the length of a lance. Others did this against the height of a man above the ship's deck.[3]

THE MARINER'S, OR NAUTICAL, QUADRANT

The first scientific instrument to be used in Western navigation for measuring an altitude was the mariner's, or nautical, quadrant, an adaptation of a tool used by astronomers at least as early as the thirteenth century.[4] The mariner's quadrant came into use in the 1460s; it was a quarter of a circle made of wood or brass, and graduated along the arc from 0 to 90 degrees. One edge was fitted with two sight vanes with pinholes; a plumb bob was also suspended from the quadrant's apex by a silk thread, hanging down in contact with the scale when in use. The quadrant was held vertically and the star was viewed through the pinholes; the thread was then clamped against the arc and the altitude read off the scale. Because the quadrant was not practical on board ship, navigators often went on shore for the observation, for wherever possible ships still hugged the coast. At first mariner's quadrants were used to find the observer's latitude in an indirect way. Every degree on the scale represented a number of miles that the vessel had sailed on a northerly or southerly course. For each degree the Pole Star was found to be lower or higher in the sky (taking into account the correction according to the position of the Guards), a certain distance had been covered. As Iberian seamen sailed along the coast of Africa more frequently, it became practice to mark the quadrant with the altitude of the Pole Star at various places so that a particular angle corresponded with, for example, a cape, a promontory, or a river estuary. As the mathematical ability of navigators increased, charts were marked with latitude scales and seamen became accustomed to working with degrees of latitude. The fact that the quadrant is depicted along with other navigational tools on the title page of Waghenaer's *Mariner's Mirror* of 1588 shows that Dutch and English navigators were familiar with the instrument. It is unlikely, however, that it remained in use at sea until the end of the sixteenth century. As already mentioned, the quadrant was not very suitable for use on a moving ship's deck: it could also not be used for sighting the Sun, as the observer would be blinded. Waghenaer did not include a description of the quadrant in *The Mariner's Mirror* because, by that time, more practical tools had superseded it, namely the mariner's astrolabe and the cross-staff.[5]

THE SEA, OR MARINER'S, ASTROLABE

Like the mariner's quadrant, the mariner's astrolabe was an adaptation of an existing astronomer's instrument. The only function of the latter that remained in the mariner's version was the scale of degrees to measure altitude and zenith distance. The earliest recorded use of a mariner's astrolabe dates from 1481, on a voyage down the coast of Africa.[6] Initially the instrument consisted of a large, flat, circular plate of wood or brass sheet, suspended by a hinged ring that allowed it to hang vertically.[7] A sighting alidade was pivoted at its centre so that the pointers at each end moved along the edge of the plate, which bore a scale in degrees. This early version of the sea astrolabe was probably primarily used on shore in the same way as the quadrant. Well before 1500, the mariner's astrolabe came to consist of a smaller, open, wheel-like frame made of cast brass or bronze (for example cat. no. 005; see Plate 1).[8] It was suspended in the same way as the earlier version and had an alidade with two sight vanes with pinholes. The edge of the upper half of the frame was graduated with stamped-in digits, reading from 90 directly below the suspension ring down to zero on either side. When the method of finding the latitude from a noon observation of the Sun was established in the 1480s, a complementary graduation was added for zenith distance. By reading this directly from the instrument, the observer avoided having to subtract the altitude from 90.[9]

In order to steady the instrument in the wind, the frame was made quite heavy, usually weighing approximately two kilograms, often with additional ballast at the bottom. The openings in the frame also helped to steady it by allowing the wind to pass through.

The astrolabe could be used in two ways. For an observation of the Pole Star, it was held up by the suspension ring and the seaman rotated the alidade until he saw the star through both sight vanes, and then read the altitude from the scale. Since observing the Sun in this way would blind him, the seaman would hold the astrolabe in front of his body, directing its plane towards the Sun. The alidade was rotated so that the shadow of the upper sight vane covered the lower completely, and the zenith distance was then read from the scale. Over the years, the scale was refined from one degree to a quarter of a degree.[10]

The mariner's astrolabe remained popular for a long time. It was manufactured in Spain and Portugal, and later also in France, England, and the Netherlands, and remained in use well into the seventeenth century. By the 1660s it was regarded as crude and inaccurate when compared with the cross-staff and backstaff, which had then undergone a number of improvements. At best, the stamped-in scales of a mariner's astrolabe were no finer than 15 minutes of arc, and the pointers of the alidade were vulnerable. Because of its weight an astrolabe was also awkward to use, and far more expensive to produce than the other instruments, which were made of wood.[11] The Dutch East India Company (VOC) stopped providing its ships with mariner's astrolabes after 1670, and in the following thirty years the astrolabe gradually disappeared from the sea altogether; the last recorded use was on a Spanish vessel in 1702.[12] Most examples were probably melted down for their metal; in 1966 only 21 were recorded, mainly preserved in museum collections. Just over two decades later, largely through maritime archaeology, this number had increased to 65.[13]

THE CROSS-STAFF, OR FORE-STAFF

Chronologically, the cross-staff was the third instrument for measuring altitudes at sea and it too was an adaptation of an existing astronomical instrument, the *baculus Jacob*.[14] Its transformation was inspired by the *kamal*, a tool for the measurement of altitudes at sea used by Arab seafarers and seen by Vasco da Gama in 1498. The *kamal* consisted of a set of nine rectangular wooden tablets of different lengths, each threaded through the centre with a cord.[15] For an observation, a tablet was selected according to the estimated altitude of the Pole Star. The cord was held taut by hand at eye level or by the teeth, and the tablet's upper and lower edges were brought into alignment with the celestial body and the horizon beneath it, respectively. Before their great voyages Chinese seafarers used a *liangtianchi*, a vertical ruler, but by the early fifteenth century they were using a *qianxingban* to measure altitudes. This consisted of twelve square pieces of wood of different sizes, one of which was held out at arm's length, similar to the working of the *kamal*.[16]

The cross-staff was introduced at sea around 1515; the earliest reference in a navigational work dates from shortly before 1520.[17] It consisted of a square-section wooden staff about 800 mm long, and one or more wooden vanes or crosses.[18] The sides of the staff were smooth, about 16 mm wide, and bore incised graduated scales marked with stamped-in digits. The vane was shorter than the staff, flat, and approximately 40 mm wide. In the middle it had a square hole just large enough for the staff to slide through, maintaining a firm right angle to the vane.

For an observation, the vane was placed vertically on the staff. The eye end of the staff was then held as near the eye as possible and the other end directed at a point in the sky about halfway between the chosen celestial body and the horizon directly below it. The vane was slid along the staff until its upper edge 'touched' the celestial body and the lower edge the horizon. The vane was then clamped to the staff by a screw, and the altitude or zenith distance read from the scale.

Originally the cross-staff was used for Pole Star observations only, and had one scale of 0–90 degrees. In the second half of the sixteenth century a complementary scale for the zenith distance was added, for observing the Sun. In 1574 William Bourne described, in *A Regiment for the Sea*, how a piece of coloured glass could be placed at the upper end of the vane to prevent the observer being blinded by sunlight. If no coloured glass was available, Bourne advised observing the Sun's upper limb and subtracting the semi-diameter from the measured altitude.[19] In the tropics, the noon altitude of the Sun often exceeds the maximum reliable range of the cross-staff of approximately 40 degrees, so that there the Pole Star was used instead to find the latitude.[20]

During the next hundred or so years a number of improvements made the cross-staff into a reliable and accurate instrument. First, the altitude and zenith distance scales were spread over three sides of the staff, each side representing a section of the entire 0–90 graduation. Michiel Coignet, in *Nieuwe Onder-wijsinghe* of

1580, first described this division.[21] It required three vanes of different lengths: side one of the staff was for angles over 30 degrees, side two for angles between 10 and 30 degrees, and side three for angles under 10 degrees. Spreading the graduation gave more space per degree, and consequently a more refined division was possible. In the 1630s, the fourth side of the staff was given a graduation for altitude and zenith distance, for larger angles. At the end of the sixteenth century, the scales were divided in half degrees, similarly to a mariner's astrolabe. By the 1660s, however, the reading accuracy had been refined to 10 minutes of arc, and forty years later to 2 minutes[22] (see Plate 2).

One of the causes for error was the necessary eye movement, or blinking, between the upper and lower ends of the vane to ascertain if it was in the correct position. This became avoidable after 1595, when the English pilot John Davis proposed a method of using the staff with one's back to the Sun.[23] In this method, the shadow of the upper end of the longest vane was cast on the shortest, the 'horizon vane'. Two different methods of observing in this way became common; that prescribed by Davis, and the 'Dutch fashion', so named by the English, as it was popular with Dutch seamen.[24] The difference was in the way the staff was held; English seamen found the Dutch manner more convenient. Another significant error eliminated in the backward observation method was ocular parallax, first described by Bourne. This was caused because the cross-staff was placed near rather than in the eye – which is physically impossible – so that the lines of sight never fall on the eye's retina. To compensate for ocular parallax, Bourne recommended shortening the staff a little at the eye end; this, however, increases the error.[25]

In the seventeenth century a number of variations of the cross-staff were designed, mostly for the backward observation method, to avoid blinking and ocular parallax.[26] No surviving examples of these variant instruments are recorded and little is known about their construction. An exception is the *spiegelboog* (mirror-staff) patented in 1670 by the Dutchman Joost van Breen. Recent research has shed more light on this instrument, and a reconstruction has been undertaken.[27] It consisted of a staff similar to that of a cross-staff, a fixed transom at the eye end with one or two vanes, and a sliding horizon vane on which a removable mirror was mounted to reflect the image of the Sun or a star in one line with the horizon. Mirror-staffs were soon used on Dutch ships, and were still for sale in Amsterdam in the 1770s, a century after their invention. The *spiegelboog* was the first reflecting instrument for navigation that was actually used at sea.

Refined graduation and backward observation made the cross-staff into a reliable and accurate instrument for altitude measurement of the Sun and stars, and one that was relatively easy to construct and not very expensive. After its development in Spain and Portugal, French and English seafarers adopted the cross-staff, probably around the 1530s. The Dutch were using it in the late 1550s, and through them it spread to northern and eastern Europe.[28] At first, seamen made their own cross-staffs by using tables provided for that purpose in navigational manuals. Towards the end of the sixteenth century they were made professionally in Spain, France, England, and the Netherlands. In the seventeenth and eighteenth centuries, the last-named country became predominant in cross-staff manufacturing.[29] Surviving examples and archival evidence show that there have been a vast number of these instruments.[30] The cross-staff declined after the introduction of the octant but remained sporadically in use at sea until the end of the eighteenth century, a longer period than for any other instrument for measuring altitudes.

THE BACKSTAFF, OR DAVIS QUADRANT

The backstaff was the first instrument designed especially for measuring the noon altitude of the Sun at sea. It was invented by John Davis (*c.*1550–1605), who also proposed the adaptation of the cross-staff for backward observation.[31] He mentioned both in *The Seaman's Secrets*, a successful treatise on navigation that appeared in 1595.[32] This includes an illustration of a backstaff but without a detailed description.[33] A staff is shown with two fixed crosses: the upper – with a sliding shadow vane – was the chord of a circle, and the lower cross was an arc of a circle, to which a sliding sight vane was fitted. Both crosses had a graduation. The vanes were set to form an angle equal to the estimated altitude of the Sun; then the sight vane was set precisely by the user's observation, and the angles read from the crosses were added up.

Others followed with similar instruments. In 1623 Edmund Gunter proposed his own quadrant, an instrument resembling a crossbow for backward observation of the Sun's altitude, which he published in *The Description and use of the Sector*.[34] Another adaptation of Davis's design – in fact a transitional instrument between the original backstaff and the improved form – was the 'plough', – so named because of its resemblance to that agricultural tool. It could also be used as a cross-staff for forward observation and was described in 1680 by John Seller as an instrument ' . . . anciently in use among Mariners . . . '. It had an arc of a circle and two sliding crosses with vanes.[35] No complete working example of the plough is known, but a probably full-size representation forms part of Grinling Gibbons's decorative woodcarving of about 1695 in the Admiralty Boardroom.[36]

Only the backstaff prevailed, and in the next stage of its development it had two fixed crosses that were chords of concentric circles, forming part of a wooden frame. The Dutch named this a *hoekboog*, a term that was translated into English as 'double triangle' in Willem Janszoon Blaeu's *Sea-Mirror* of 1625. The upper chord covered an arc of 65 degrees and the lower chord on arc of 25 degrees. Like the plough, no complete example of a double triangle is known to have survived.[37]

In the middle of the seventeenth century, arcs of similar angles replaced the chords; these are usually referred to as the 60- and 30-degree arcs. From this time on, the backstaff was also known

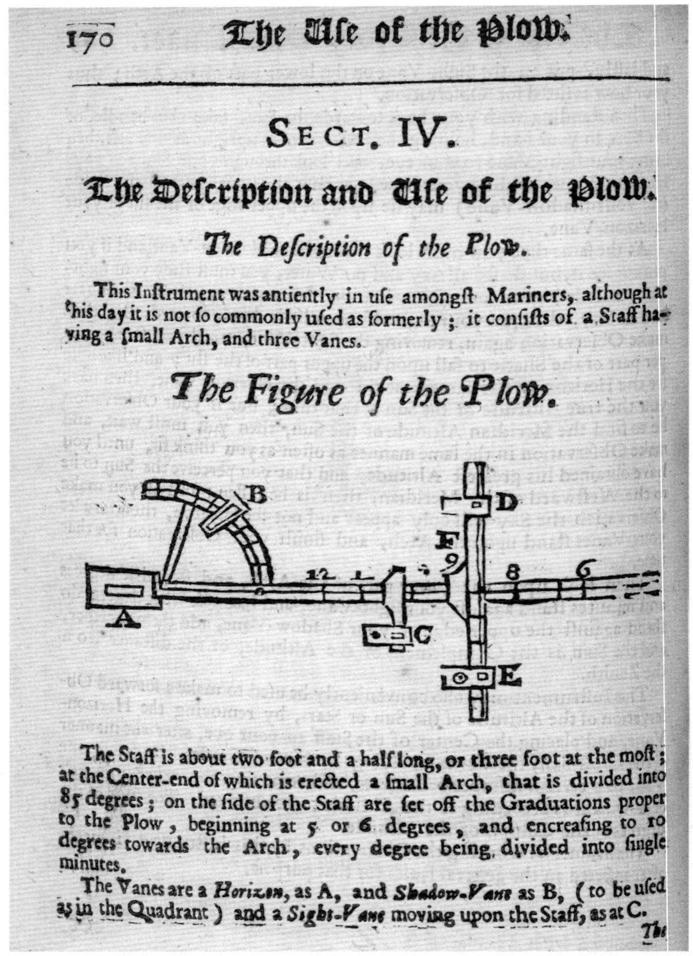

170 The Uſe of the Plow.

SECT. IV.

The Deſcription and Uſe of the Plow.

The Deſcription of the Plow.

This Inſtrument was antiently in uſe amongſt Mariners, although at this day it is not ſo commonly uſed as formerly; it conſiſts of a Staff having a ſmall Arch, and three Vanes.

The Figure of the Plow.

The Staff is about two foot and a half long, or three foot at the moſt; at the Center-end of which is erected a ſmall Arch, that is divided into 85 degrees; on the ſide of the Staff are ſet off the Graduations proper to the Plow, beginning at 5 or 6 degrees, and encreaſing to 10 degrees towards the Arch, every degree being divided into ſingle minutes.

The Vanes are a Horizon, as A, and Shadow-Vane as B, (to be uſed as in the Quadrant) and a Sight-Vane moving upon the Staff, as at C.

The

FIGURE 2.1 (*above*) The 'crossbow' of Edmund Gunter depicted on the lower left of the title page of his *The Description and use of the Sector*, London, 1624. National Maritime Museum; PBE5165.

FIGURE 2.2 (*above, right*) The 'figure of the plough', from *Practical Navigation* by John Seller, London, 1672.National Maritime Museum; PBE1327.

FIGURE 2.3 (*right*) Observing the Sun's altitude with a backstaff, or Davis quadrant. From *Practical Navigation* by John Seller, London, 1672. National Maritime Museum; PBE1327.

as a Davis quadrant, simply a 'quadrant', or an 'English' quadrant. Besides a shadow and a sight vane it also had a horizon vane, with a slit parallel to the horizon, and it was used as follows[38] (see Plate 3). The shadow vane was fastened to the 60-degree arc and the sight vane to the 30-degree arc, so that the sum of the angles on the two arcs was approximately equivalent to the estimated altitude of the Sun. Both vanes had a slightly bent brass pressure plate, by which they remained tightly in place. With his back to the Sun, the observer viewed the horizon through the pinhole in the sight vane and through the slit in the horizon vane. He then slid the sight vane slowly up or down the 30-degree arc until the shadow of the vane fell on the slit of the horizon vane.The sum of the angle read from the 30-degree arc and that found on the 60-degree vane

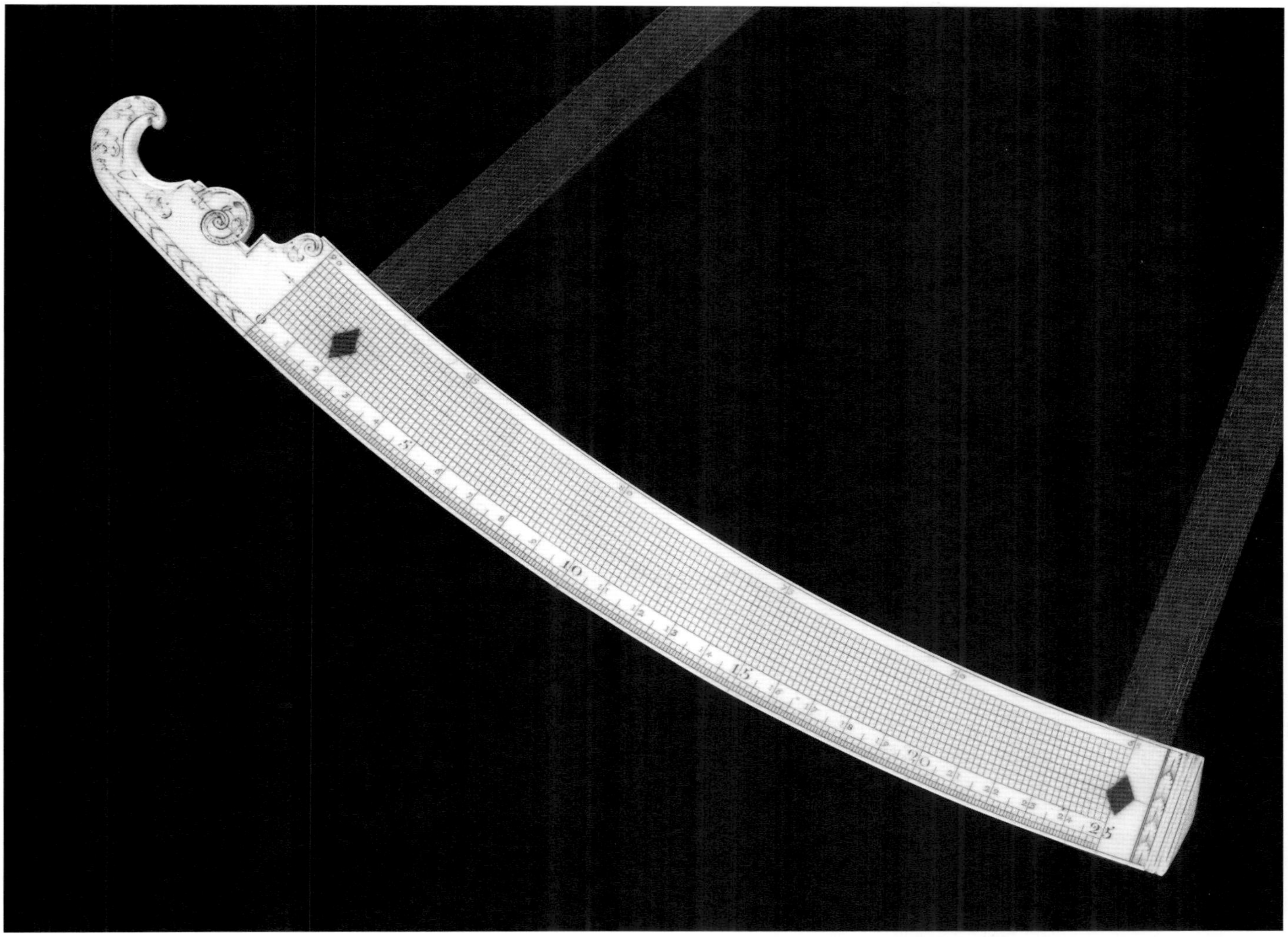

FIGURE 2.4 The thirty-degree arc of a backstaff with transversals (cat. no. 013).

was the altitude. For fine reading, the 30-degree arc had a transversal, or diagonal, scale – the first secondary scale for refined division that appeared on a navigational instrument. This allowed a reading accuracy of up to 1 or 2 minutes along the fiducial edge of the sight vane, the edge at the centre of the vane, in the same line as the pinhole. The reading accuracy depended on the division of the transversal scale, the actual accuracy of an observation on the ability of the observer, and has been estimated to be between 6 and 12 minutes of arc.[39]

Backstaff frames were often made of a hardwood, and the arcs and vanes of pear wood or boxwood. An observation with a backstaff required bright sunlight; a thin layer of cloud rendered the instrument useless. This was overcome around 1670 when Astronomer Royal John Flamsteed proposed a small convex lens be fitted on the shadow vane.[40] Weak rays of sunlight were focused by the lens to cast a bright spot on the horizon vane, and several surviving backstaffs are fitted with such a lens. English backstaffs are often found to have a secondary scale on the 60-degree arc, by which the correction for the Sun's semi-diameter – which is around 16 minutes of arc – was included in the reading. Some also have tables for the Sun's northern and southern declination and for the right ascension, inscribed on the back of the 30-degree arc.

The backstaff remained in use throughout the eighteenth century, with the cross-staff and the octant. Backstaffs were manufactured in Great Britain, Ireland, and the Netherlands and also in North America, where they became especially popular.[41] Many have been preserved, mainly from the eighteenth century.

The reward set by the Longitude Act stimulated the improvement of navigational instruments, although these were not always lasting. Jacob Rowe, a mariner, invented a 'fluid quadrant', which he patented (no. 431 of 1720). He hoped for a reward under the Act, since according to him it could be used for lunar observations.[42] It was a quadrant with an artificial spirit-level horizon to align it with the horizontal plane, independent of the visibility. A telescope was mounted along one edge of the quadrant, and the spirit level was hinged to its apex. Rowe especially recommended his quadrant for use with star observations, and according to him it was given sea trials by order of the Admiralty in 1719. No example is known to have survived.

The next navigational instrument to be fitted with an artificial horizon was the backstaff. In 1728 John Elton, mariner, merchant,

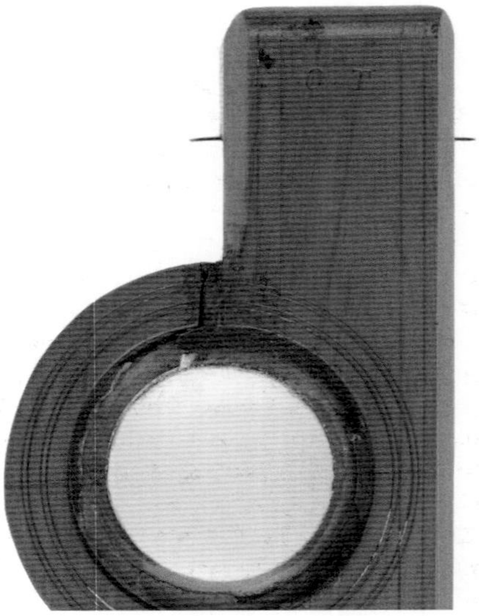

FIGURE 2.5 The shadow vane of a backstaff, with a Flamsteed lens (cat. no. 023).

FIGURE 2.6 The outer edge of the 60-degree arc of a backstaff, showing the scale for correcting for the Sun's semi-diameter (cat. no. 020).

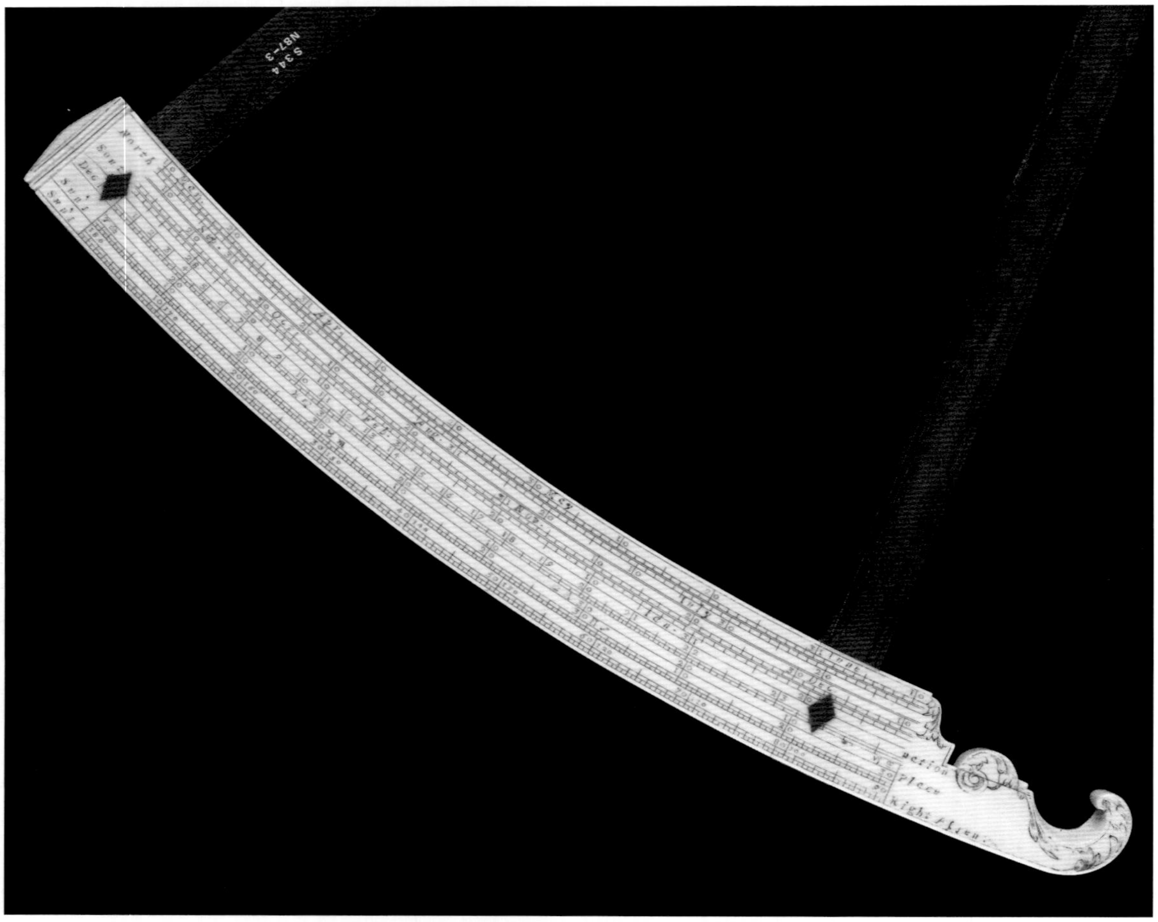

FIGURE 2.7 The back of the 30-degree arc of a backstaff with tables for the Sun's right ascension and declination (cat. no. 013).

and adventurer, patented this horizon (no. 501 of 1728).[43] It consisted of two spirit levels: one to hold the quadrant vertically, and the other to keep the line of sight vertically[44] (see Plate 5). With an extra alidade, a vertically mounted third spirit level, and an oil lantern, Elton's design could also be used to take forward altitudes of stars. This facility was probably not very practical and only one incomplete example of Elton's artificial horizon has survived, on a backstaff by Jonathan Sisson (cat. no. 022). Interestingly, it had a vernier for refined reading instead of the usual transversal scale. In the 1730s a number of artificial horizons were invented, mostly for use with an octant, and they will be dealt with in the following chapter. An exception was an artificial horizon for a backstaff by Charles Leigh, published well after the octant was known. It used fluid levels that were fixed to the instrument frame to align it with the horizon, and a second version worked with a mercury level.[45]

1. Taylor, *The Haven-Finding Art*, 49.
2. Levathes, *When China Ruled the Seas*, 96.
3. Taylor, *The Haven-Finding Art*, 161.
4. Taylor, *The Haven-Finding Art*, 159–62; Waters, *The Art of Navigation*, 46–7.
5. Davids, *Zeewezen en wetenschap*, 66.
6. Stimson, *The Mariner's Astrolabe*, 16.
7. Stimson, *The Mariner's Astrolabe*, 14–15.
8. There were also semicircular astrolabes, whereby the lower half of the frame was left out. These were of Dutch manufacture; see Stimson, *The Mariner's Astrolabe*, 27.
9. Stimson, *The Mariner's Astrolabe*, 36–7.
10. Resulting in an accuracy of 15 nautical miles to the north or south of the latitude, at the best.
11. In 1670 the Dutch East India Company valued a mariner's astrolabe at 13 guilders and a cross-staff at just over 2 guilders; see Schilder and Mörzer Bruyns, 'Navigatie', 180.
12. Mörzer Bruyns, *The Cross-Staff*, 16; Stimson, *The Mariner's Astrolabe*, 42.
13. Waters, *The Sea- or Mariner's Astrolabe*; Stimson, *The Mariner's Astrolabe*. The known number has since increased to about 85 (personal communication from Alan Stimson).
14. The cross-staff is therefore often named 'Jacob's staff'; see Mörzer Bruyns, *The Cross-Staff*, 23.
15. Waters, *The Art of Navigation*, 53.
16. Levathes, *When China Ruled the Seas*, 96.
17. For the following information, see Mörzer Bruyns, *The Cross-Staff*, 25–9, 32.
18. The staff was often made of ebony or lignum vitae, hardwoods that do not warp and can easily be marked. The vanes were generally made of the softer pear wood.
19. By observing the upper limb the observer ensured that the Sun was obscured by the vane. When the lower limb was observed, the semi-diameter had to be added.
20. Mörzer Bruyns, *The Cross-Staff*, 43.
21. Waters, *The Art of Navigation*, 204.
22. Mörzer Bruyns, *The Cross-Staff*, 41–3.
23. Mörzer Bruyns, *The Cross-Staff*, 25.
24. Mörzer Bruyns, *The Cross-Staff*, 25–7.
25. Mörzer Bruyns, *The Cross-Staff*, 28.
26. For a listing, see Mörzer Bruyns, *The Cross-Staff*, 24–5; Hilster, 'The *Spiegelboog*', 6.
27. Hilster, 'The *Spiegelboog*'.
28. Mörzer Bruyns, *The Cross-Staff*, 32–3; Davids, 'Dutch Contributions to the Development of Navigational Technology in the 17th Century'.
29. Mörzer Bruyns, *The Cross-Staff*, 31–2.
30. Mörzer Bruyns, *The Cross-Staff*, 44–5; Mörzer Bruyns, 'The Cross-Staff Ten Years Later', 20, documents a total of about 120 examples.
31. Michael Hicks, 'Davis [Davys], John (*c.* 1550–1605), *Explorer*'. *Oxford Dictionary of National Biography* <http://www.oxforddnb.com> Davis was an explorer and expert mathematician, who sailed to the Arctic, where the Davis Strait is named after him. He discovered the Falkland Islands and in 1598 sailed as a pilot with the Dutch to the East Indies.
32. To distinguish the backstaff from the cross-staff which could also be used for backward observations, Henry Gellibrand named the latter the 'fore-staff' in *An Epitome of Navigation*, which probably first appeared about 1634, however, no copy of that date appears to survive.
33. Waters, *The Art of Navigation*, 201–9; Cotter, *A History of Nautical Astronomy*, 70–3.
34. One 'Gunter's sea quadrant' or crossbow is known to survive, in Stokloster Castle, Sweden; see Losman and Sigurdsson, 'Äldre vetenskapliga instrument på Skokloster', 99.
35. Seller, *Practical Navigation*, 165–7; May, *A History of Marine Navigation*, 130.
36. Stimson, 'The Old Admiralty Boardroom Carvings', 128.
37. What are probably some pieces of an example of the latter have been preserved; see Hilster, 'The *Spiegelboog*', 8.
38. Bennett, 'Catadioptrics and Commerce in Eighteenth-Century London', 250.
39. May, *A History of Marine Navigation*, 23.
40. Mörzer Bruyns, *The Cross-Staff*, 27.
41. After 1731 the VOC ceased to provide its ships with backstaffs, as the improved cross-staff was regarded as adequate; see Mörzer Bruyns, *The Cross-Staff*, 16. For the popularity of the backstaff in America, see Warner, 'Davis' Quadrants in America'.
42. See Rowe, *Navigation Improved* for the full description.
43. E.I. Carlyle, 'Elton, John (d. 1751) *Merchant and Adventurer*'. *Oxford Dictionary of National Biography* <http://www.oxforddnb.com/>
44. Bennett, 'Catadioptrics and Commerce in Eighteenth-Century London', 265–6; Ifland, *Taking the Stars*, 91–4.
45. 'A Description of a Water-Level Fix'd to Davis's Quadrant, Whereby an Observation May Be Taken at Sea, in Thick and Hazy Weather, without Seeing the Horizon', and 'The Description and Use of an Apparatus Added as an Improvement to Davis's Quadrant, Consisting of a Mercurial Level, for taking the Co-Altitude of Sun or Star at Sea, without the Usual Assistance of the Sensible Horizon, Which Frequently is Obscured'. *Philosophical Transactions*, vol. 40 (1737–38) no. 451, 413–17, 417–24.

CHAPTER THREE

THE INVENTION OF THE OCTANT, AND ITS DEVELOPMENT AND DIFFUSION

The invention of the double-reflecting quadrant, which soon became known as an octant, for measuring the altitude of celestial bodies at sea, had a rather long prelude. As early as 1665, five years before Joost van Breen's mirror staff, Robert Hooke, secretary and curator of experiments of the Royal Society, designed a reflecting instrument with one speculum and tested it at St Paul's Cathedral in London.[1] It consisted of a triangular frame, of which one side was the index arm which revolved around a pivot and was fitted longitudinally with a telescope. A speculum was fixed perpendicularly on the other arm at the pivot, and the third arm of the frame had a graduation. The observer held the instrument so that he viewed the horizon through the telescope and at the same time observed the reflected image of the celestial body. The lower end of the index arm then indicated the altitude on the graduated scale.

Sir Isaac Newton was the next to design a reflecting instrument for altitude measurement, and he was the first to apply two specula.[2] Newton's quadrant consisted of a brass plate in the form of a sector of one-eighth of a circle, with a radius of two to three feet and an arc of 45 degrees. A telescope was fitted longitudinally to one edge of the sector with a small speculum rigidly mounted at the objective end, perpendicular to the frame. An index arm was pivoted at the centre of the sector and a slightly larger speculum was mounted rigidly on the arm at a right angle, at the pivot. The other end of the arm had a fiducial edge in the centreline of the pivot, index glass, and arm, and the arm was moved along the graduated scale from which the altitude was read off. An observer held the sector with its apex in the direction of the celestial body and looked through the telescope, past the small speculum towards the horizon. He then moved the index arm until the light

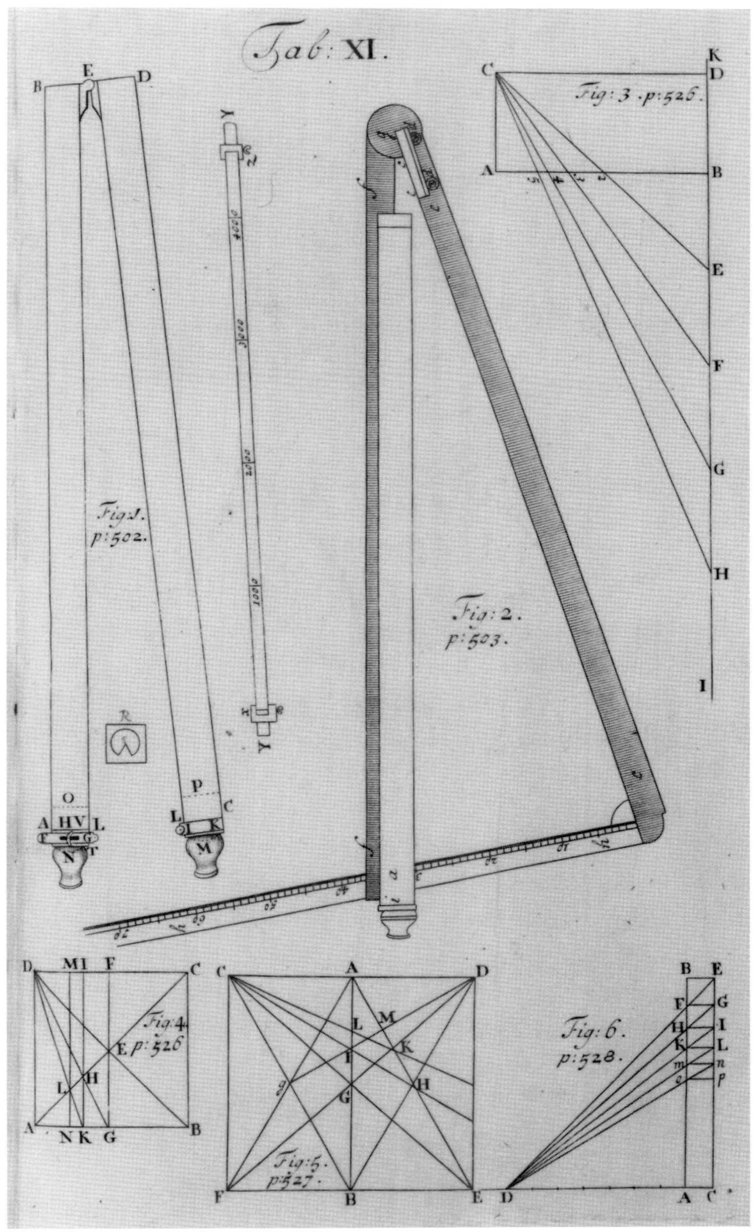

FIGURE 3.1 Robert Hooke's design for a reflecting quadrant. From Richard Waller (ed.), *The Posthumous Works of Robert Hooke*. London, 1705. Courtesy of the Museum of the History of Science, Oxford.

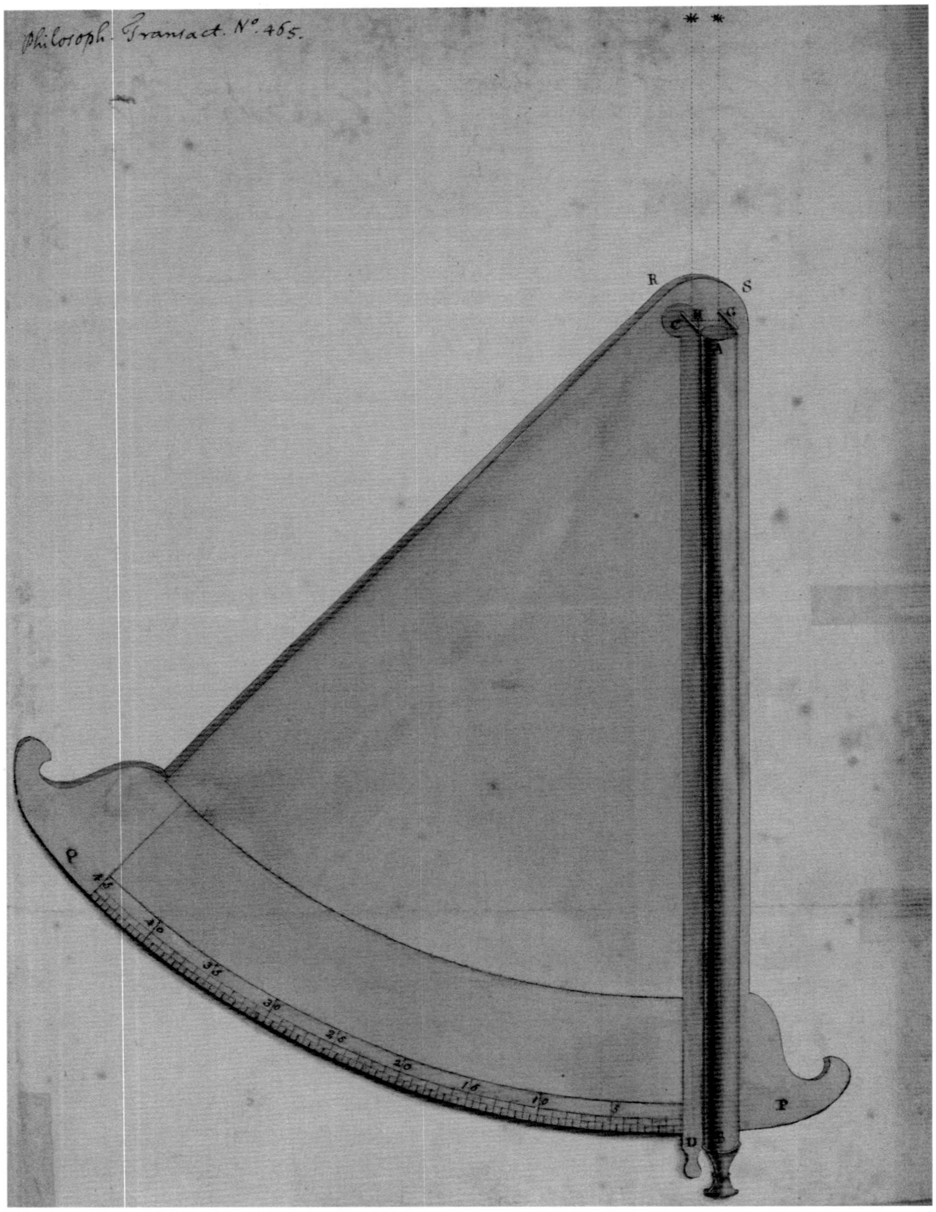

FIGURE 3.2 Isaac Newton's design for a reflecting quadrant. Drawing from 'a true copy of a paper found in the handwriting of Sir Isaac Newton, among the papers of the late Dr Halley, containing a description of an instrument for observing the distance off the Moon from the fixt stars at sea'. Taken from Royal Society manuscripts Letters and Papers/i/120. Courtesy of the Royal Society of London.

from the celestial body that was reflected by both specula became visible through the telescope. The reflected image was then accurately brought in line with the horizon, the index arm clamped, and the altitude read from the scale. Because the light is reflected twice, the measured angle is halved and the graduated scale runs to 90 although the arc of the instrument is 45 degrees.[3]

Apparently Newton showed his design at the Royal Society in 1699 and also communicated it to Edmond Halley (1682–1744), who in 1719 became the second Astronomer Royal. After his death, a drawing and description of Newton's instrument was discovered among Halley's papers. It was titled 'An Instrument for observing the distance of the Moon from the fixt Stars at Sea'. It was read before the Society nine months later and published in the *Philosophical Transactions*.[4]

Halley also designed a reflecting instrument for taking altitudes at sea, and described it to the Society in 1692. The quadrants of Hooke and Newton were constructed but have apparently not survived; Halley's was probably never made. Hooke's design was examined by the Society in 1670, who ordered that an example be made for sea trial.[5] Nothing came of these reflecting instruments; at the end of the seventeenth century, navigators were still content with the cross-staff and backstaff. It was not until after the Longitude Act of 1714, which was prompted by a fairly recent navigational disaster, that new navigational instruments and methods were developed and employed at sea.[6]

HADLEY'S QUADRANT, OR OCTANT

John Hadley (1682–1744) (Plate 6) presented his double-reflecting quadrant at a Royal Society meeting on 13 May 1731.[7] He offered two designs that worked in the same manner, the difference being in the way that the instruments were constructed and held by the observer. The first was very similar to that of Newton, which, a week after the presentation, led Halley to observe that it had already been proposed by Newton. This led to some confusion and a subsequent search in the Society's minutes, and to a notice that Newton had in 1699 exhibited 'an improvement of an old instrument', without further specification.[8] At a later meeting, held in December 1731, the elderly Halley declared that Hadley's design differed from that of Newton and was indeed an original invention. This was incorrect, but as Halley was the only person who knew about Newton's quadrant, Hadley's originality was not challenged.

Hadley published his two designs in the *Philosophical Transactions*, including illustrations, and here used the word 'octant' to indicate that the frame was one-eighth of a circle. He included a description of the principle of catoptrics on which the invention was based: the divergence and convergence of a ray of light on a polished surface.[9] Both of Hadley's designs consisted of a wooden frame. They had a dark glass (a shade or light filter) fitted in a brass mount that was hinged to the frame and could be placed between the large speculum and the eye, for protection from strong sunlight.[10] (The large speculum is now called the index glass and the small one the horizon glass, the terms henceforward used here.) Hadley states no size for the frame, but on the basis of surviving early examples the radius may have been about two feet.

The principal difference between the two designs was that with the first the octant's apex was held towards the celestial body, as proposed by Newton, and with the second the frame had to be held vertically, with its apex at the top. Here the telescope and glasses were mounted at a different angle, making the instrument easier to handle on a vessel, and therefore the second design prevailed. It was used as follows.

For an observation of the Sun, the observer set the index arm at zero on the graduated scale, so that the glasses were parallel, and placed the shade in position. He then looked through the pinhole in the sight vane towards the Sun. While its reflected image was kept in sight, the index arm was moved along the limb, slowly bringing the frame into a vertical position, until the horizon became visible directly. The lower limb of the Sun's reflection was brought onto the horizon and the octant was tilted slightly to the left and right (swung), 'skimming' the Sun's limb over the horizon. The index arm was then clamped and the altitude read off the scale. Hadley also made a provision to measure angles up to 180 degrees by adding a second horizon glass and a second sight vane. These were used by holding the frame with the index glass facing the observer, with his back towards the Sun. This back observation was used when land or low cloud obscured the

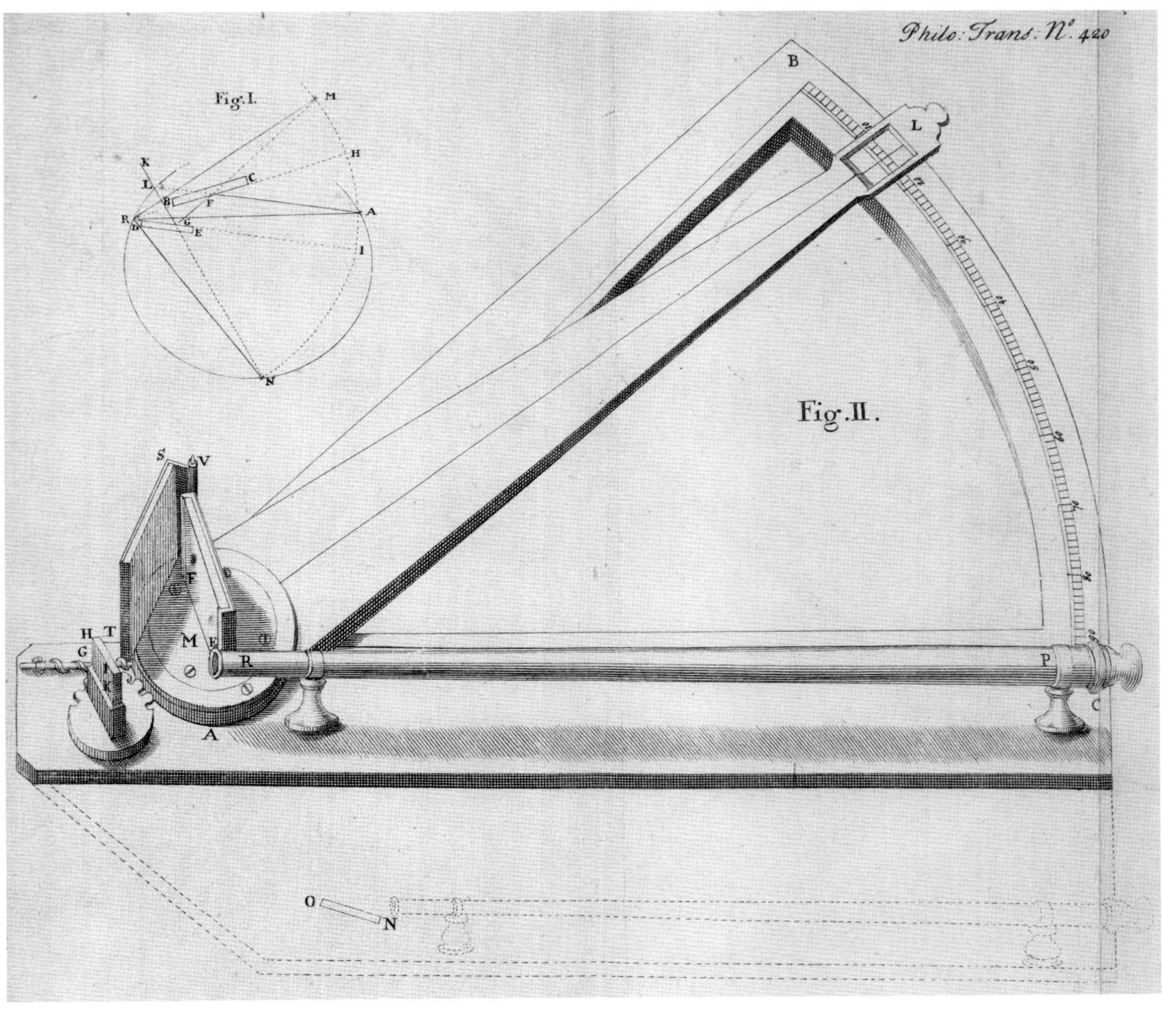

FIGURE 3.3 John Hadley's first octant design. From *Philosophical Transactions* volume 37, 1731/1732, no. 420, 'The Description of a New Instrument for Taking Angles'. Courtesy of the Royal Society of London.

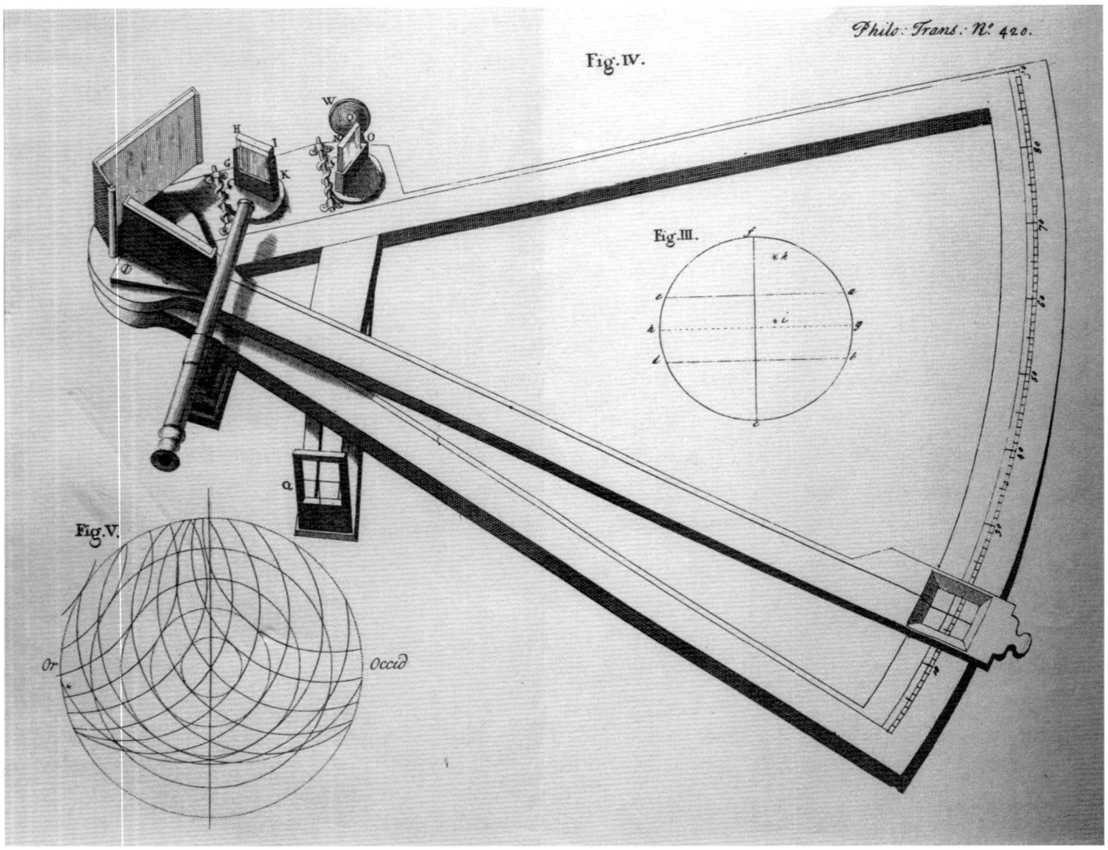

FIGURE 3.4 John Hadley's prevailing octant design. From *Philosophical Transactions* volume 37, 1731/1732, no. 420, 'The Description of a New Instrument for Taking Angles'. Courtesy of The Royal Society of London.

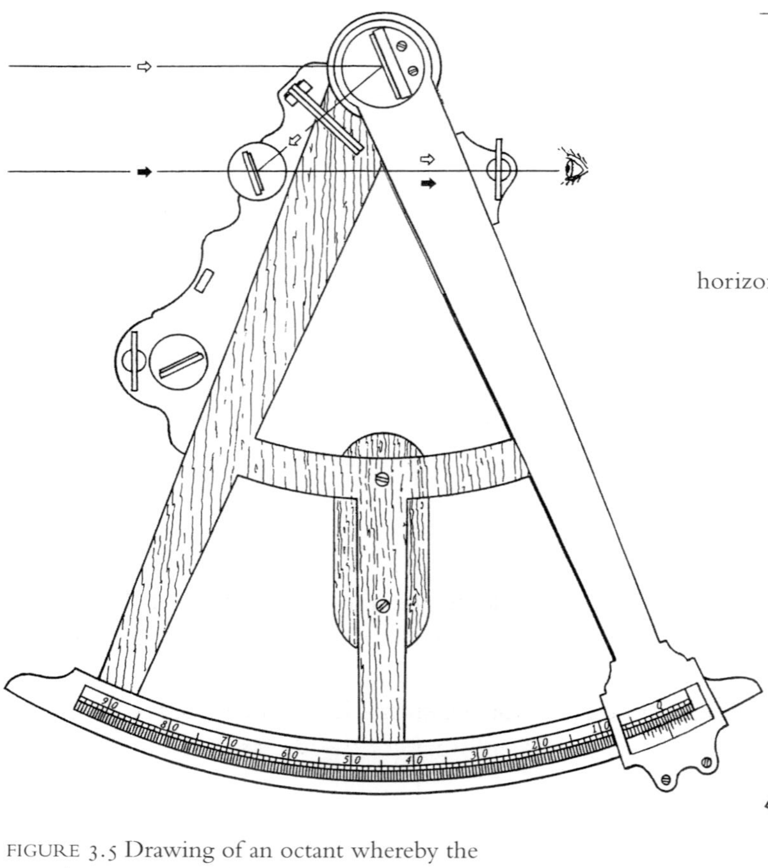

FIGURE 3.5 Drawing of an octant whereby the two glasses are parallel and the index arm is aligned with the zero on the scale. Author's collection, drawing by N. Kamer.

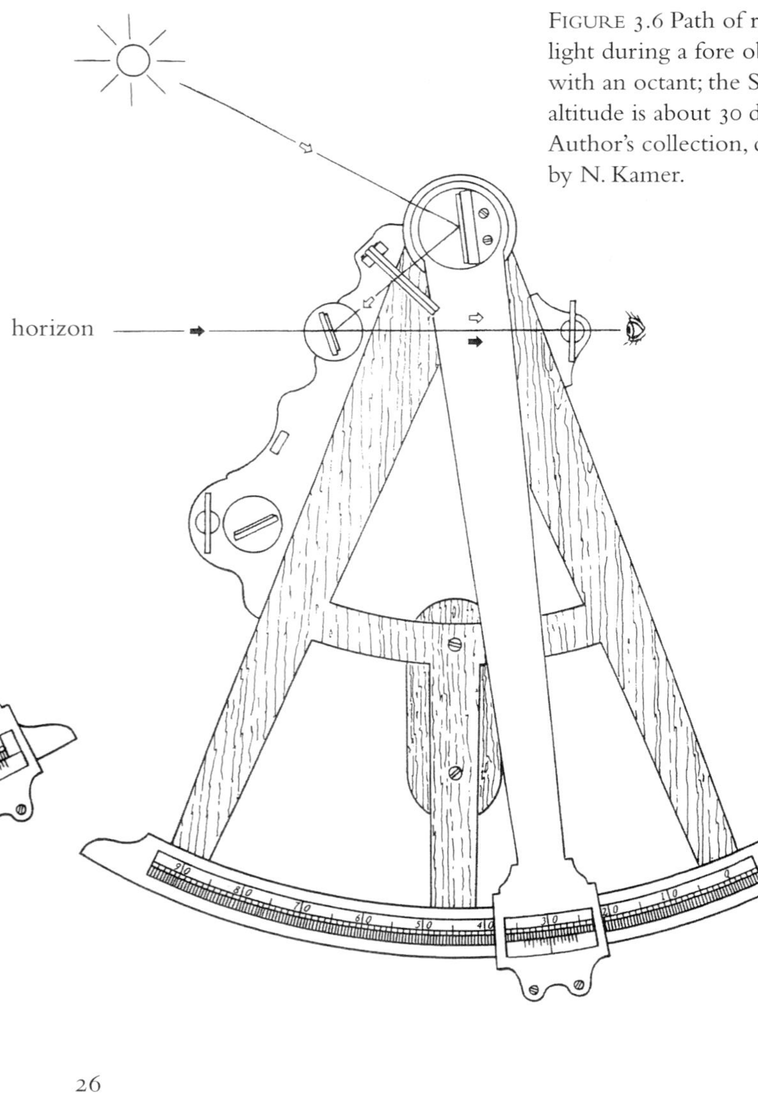

FIGURE 3.6 Path of ray of light during a fore observation with an octant; the Sun's altitude is about 30 degrees. Author's collection, drawing by N. Kamer.

horizon directly below the celestial body and for larger angles, such as those involved in taking lunar observations.[11]

Through the intermediary of Halley, the Admiralty agreed to have Hadley's invention tested at sea, which was done over three days in August 1732 in the Admiralty's yacht *Chatham*. On board were Hadley, his brother Henry, the later Astronomer Royal James Bradley, and James Young, a representative of the Admiralty, practising with two octants. The Hadley brothers and Bradley took the observations in generally poor weather conditions.[12] The instruments they used were the wooden example made by Hadley to show to the Royal Society the previous year, and a brass one made by the London instrument maker Jonathan Sisson (1690?–1747).[13] Sisson was an obvious person to make the second octant. He had been an independent craftsman since 1722, after an apprenticeship to George Graham, a London instrument maker and clockmaker who was renowned for his accurate division of scales.[14] Later Graham employed him on the 8-foot mural quadrant for the Royal Observatory, and in 1729 Sisson was appointed mathematical instrument maker to the Prince of Wales.

Despite the poor weather, the sea trials were a success and the octant proved both handy and reliable. Apart from altitude measurements of celestial bodies above the horizon, two-thirds taken by back observation, angles were also taken between celestial bodies, thus anticipating the utility of the octant for lunar distances.[15]

The advantages of the octant over the cross-staff and backstaff were evident.[16] It was compact and therefore practical on the unsteady deck of a vessel, and less sensitive to the influence of wind. Provided the horizon was visible, it could be used with equal accuracy for both Sun and star observations. Star observations were impossible with a backstaff, and because of the unavoidable 'blinking', very difficult to take with a cross-staff. Adjusting the focus of the eye between the horizon vane and the horizon, as with a backstaff and also a possible cause of error, was no longer necessary. Altitudes up to almost 90 degrees could be measured with an octant, which were impossible with a quadrant or staff. This was especially useful in low latitudes, where the Sun's noon altitude is often high. But the octant could also be used to measure the altitude of the Sun and the Pole Star when they were low, in high northern latitudes and near the equator, respectively. Finally, the older instruments did not offer back observation up to 180 degrees.

OTHER EARLY REFLECTING QUADRANTS

Before continuing on the development of Hadley's octant, other inventions must be discussed. The first that should be mentioned was by Thomas Godfrey (1704–49), a glazier and self-taught astronomer in Philadelphia, who is credited with the near-simultaneous invention of the octant.[17] Godfrey's principle was

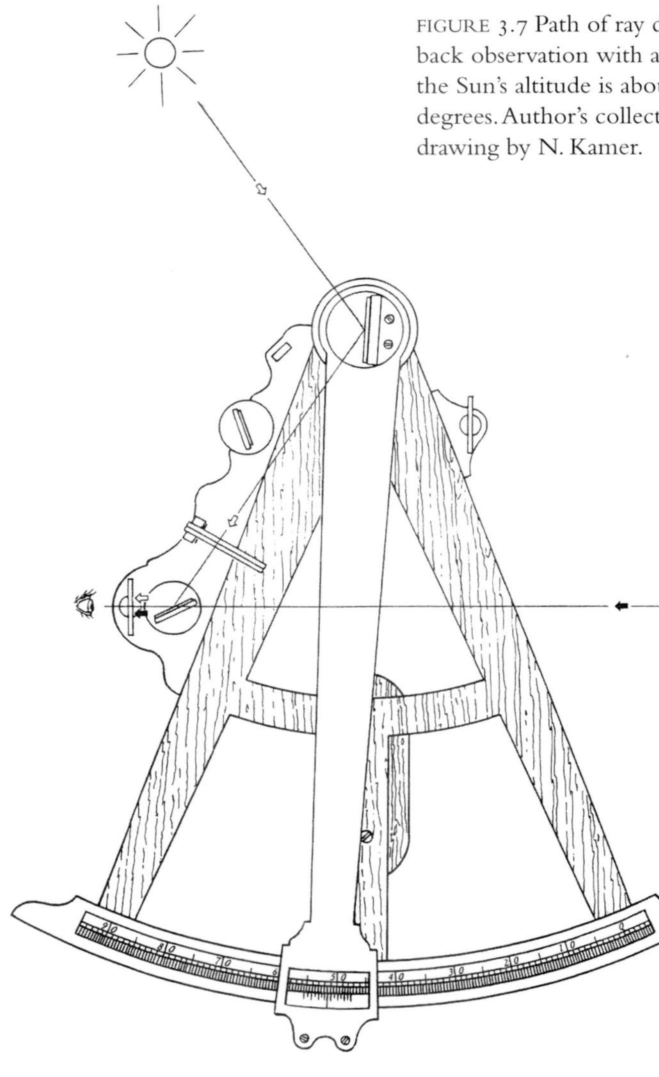

FIGURE 3.7 Path of ray during a back observation with an octant; the Sun's altitude is about 50 degrees. Author's collection, drawing by N. Kamer.

the same as that of Newton, and of Hadley's first design. It was tried out at sea on voyages to Jamaica and Newfoundland in the sloop *Trueman*, in 1730–1.[18] Godfrey was acquainted with Benjamin Franklin, for whom he calculated ephemerides for an almanac that Franklin printed. He showed his reflecting octant to James Logan, Chief Justice of Pennsylvania, asking him to transmit it to the proper authorities in the hope of a reward under the Longitude Act. It was not until Logan saw Hadley's description in the *Philosophical Transactions* that he realized the significance of Godfrey's invention, and described it in a letter to Halley in London.[19] An investigation by the Royal Society, held in January 1733, led to the conclusion that Hadley's and Godfrey's inventions were made independently at almost the same time. Logan published the Philadelphian octant in the *Philosophical Transactions* but devoted more attention to Godfrey's other inventions: a modified backstaff designed for measuring lunar distances, and an entirely new instrument for measuring altitudes at sea that he named the 'mariner's bow'.[20]

Another near-simultaneous invention of the octant took place in France in 1733. On 14 February that year, the astronomer Jean-Paul Grandjean de Fouchy (1707–88), later the permanent secretary of the Académie Royale and a prominent figure in that

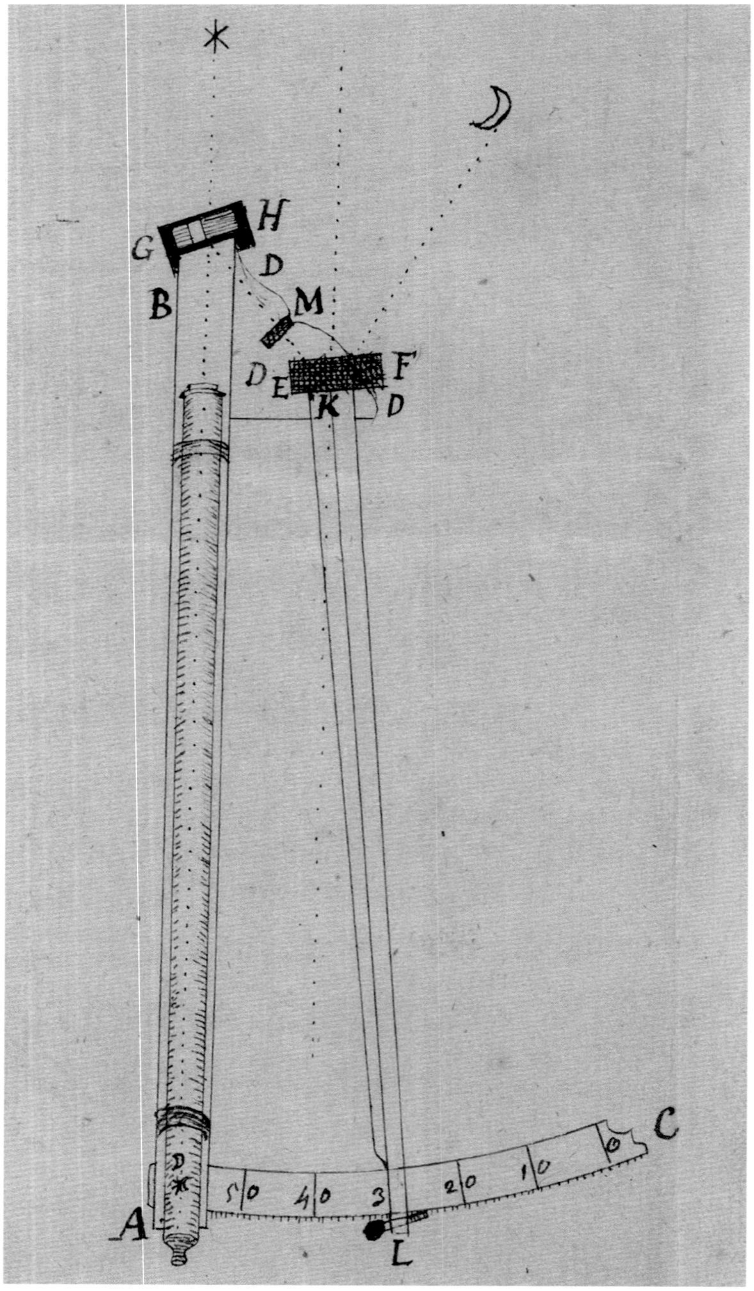

FIGURE 3.8 Thomas Godfrey's octant design. From a letter from James Logan, Pennsylvania, 25 May 1732, to Edmond Halley. Royal Society Archive EL/L6/59. Courtesy of the Royal Society of London.

institution's politics, described a reflecting instrument for use at sea that promised to replace the cross-staff and backstaff. Nothing much is known about this invention, which according to Jean-Hyacinthe de Magellan was similar to Newton's design.[21] Magellan wrote that in 1740 Grandjean de Fouchy presented a second octant design before the Académie, including a system for reading fractional values of the divisions on the scale. It is possible that in his first presentation Grandjean de Fouchy was not discussing his own invention, but in his capacity as secretary of the Académie was referring to that of Hadley. The Frenchman could easily have heard about the octant from fellow academians and French Fellows of the Royal Society, or from instrument makers in London who were working for French clients. Fontenelle, for instance, a previous secretary of the Académie, had been elected a Fellow of the Royal Society shortly before Grandjean de Fouchy's announcement. In the early 1730s the Académie was involved in the preparation of expeditions to Peru and Lapland to measure the linear distance of a degree of latitude, and for these measurements scientific instruments were ordered from London manufacturers.[22] It was hoped to establish the true shape of the Earth to settle the controversy between the Cartesian natural philosophy current in France and the Newtonian gravitational theory. Fontenelle had encouraged this, and Grandjean de Fouchy not only played a significant part in promoting the Peru expedition but also was considered for its leadership. Académien Pierre-Louis Moreau de Maupertuis, who had been a Fellow of the Royal Society since 1728, was a follower of Newton and a friend of Voltaire, then living in exile in London. Maupertuis was appointed the leader of the expedition to Lapland and the expedition to Peru was led by Louis Godin, who was elected a Fellow of the Royal Society in 1735. As will be discussed below, the two expeditions were also the two first occasions on which the French used octants that had been purchased in London. Another connection between scholars in London and Paris was that strong promoter of Newtonian physics the French-born Huguenot scientist and Fellow of the Royal Society since 1714 John Theophilus Desaguliers, who showed a high regard for the octant.[23]

No examples of the inventions by Godfrey and Grandjean de Fouchy are known to have survived, something that cannot be said of the 'sea quadrant' by Caleb Smith (*fl.* 1734–45), a shipping insurance broker in London and inventor of compasses and other nautical instruments.[24] Like the quadrants of Hadley and Godfrey, Smith's had a triangular frame with a telescope on one arm, two glasses, and a shade (see Plate 6). The glasses were placed in the line of sight; the index glass was mounted rigidly on the movable index arm, and the smaller horizon glass was fixed rigidly to the frame.[25] As with the Newton–Hadley–Godfrey design, an observer held the quadrant with its apex towards the celestial body and looked through the telescope towards both glasses. When the reflected horizon was seen in the horizon glass, the index arm was moved until the reflected image of the celestial body was viewed in the index glass. The two images were brought in line, the index arm was clamped, and the altitude read from the graduated arc. Smith showed that a prism could replace the horizon glass, and this was the first time such an optical device was used for an instrument for measuring altitudes at sea. If the horizon was indistinct or obscured, a spirit level could be put in place of the horizon glass, its bubble replacing the horizon. There is evidence of successful sea trials of this sea quadrant. In his description of it Smith refers to recommendations by three ship's masters, Christopher Middleton, George Spurrel, and Joseph Harrison. Middleton sailed for the Hudson's Bay Company from London to the Bay in 1735 and regularly observed his latitude with four recently invented instruments. They were a sea quadrant of Caleb

Smith fitted with a prism; an octant; a backstaff with an artificial horizon by John Elton; and a sea quadrant by Caleb Smith and William Ward, with an artificial horizon.[26] Middleton, a proficient navigator with a strong bent for science, was elected a Fellow of the Royal Society in 1737, after publishing the results of his trials in the *Philosophical Transactions*.[27] He praised Caleb Smith's sea quadrant, especially for star altitudes, and had also successfully used the artificial horizon.

Several prominent instrument makers manufactured Smith's quadrant. There are surviving examples by the London makers Thomas Heath and Tycho Wing (both were still offering it for sale in 1765), and John Bird and George Adams senior; by Benjamin Ayres of Amsterdam; by Gedney King of Newport, Rhode Island; and by Nicolas Jacques Baradelle of Paris. In the Netherlands, it was for sale in Jan van Musschenbroek's workshop in Leiden in 1748.[28] The Rhode Island General Assembly commissioned King to make Smith's sea quadrants in 1750, for use in running the boundary with Massachusetts.[29] Baradelle made a sea quadrant that originally belonged to the Académie de Marine, an institution established in 1752 to improve the knowledge of French naval officers, and which also owned a sea quadrant by Bird.[30] Baradelle modified the design in 1752 by replacing the horizon prism with two index glasses. All this shows that Smith's sea quadrant must have enjoyed considerable popularity for about thirty years, and that it spread to France, the Netherlands, and North America.[31]

Two other sea quadrants should be mentioned, albeit briefly as neither was an instrument of reflection. Almost identical in form, both were invented in 1748 by London instrument makers, the first by Benjamin Cole, the second by George Adams. Both employed a shadow or a spot of light concentrated by a lens.[32]

IMPROVEMENTS TO HADLEY'S QUADRANT UP TO THE 1770s

Despite the success of Smith's sea quadrant, Hadley's octant was to prevail as the instrument for altitude measurement at sea and it soon underwent a number of improvements. One of the first was by Hadley himself, who invented an artificial horizon by use of a spirit level fixed to the frame, whereby the reflected bubble replaced the natural horizon. He patented this (no. 550 of 1734) and published an illustrated description in the *Philosophical Transactions*.[33] As far as is known only one octant with this device has been preserved, although slightly modified. Benjamin Cole made it in about 1760 (cat. no. 038).

Hadley wrote that the specula could either be of metal or glass plates backed with silver foil.[34] (As no octant with a metal speculum is known, this must have been unusual.) The surfaces of the glasses should be absolutely flat; the slightest curvature would divert the reflected light and cause an error in the observation. A small deviation was allowed, however, providing the thickest or thinnest edge of the glass was parallel to the instrument's frame. For an accurate observation, the correct position of the glasses was essential. A slight disturbance, not unusual on the deck of an unsteady ship, could render the octant unreliable. In case this happened, the brass mounts in which the glasses were fitted were adjustable. The index glass had to be perpendicular to the plane of the instrument; in this case a deviation is an 'error of perpendicularity'.[35] The horizon glasses too had to be perpendicular to the frame, and for these a deviation is called a 'side error'. When the index arm is aligned with the zero on the scale, the index and horizon glasses must be parallel to each other; a deviation here is called an 'index error'.[36] Hadley's design shows a screw for the side error, and almost all surviving octant and sextant glasses have adjustment facilities.[37]

In the 1770s, various instrument makers designed techniques for adjusting the glasses. In 1760 John Dollond published a brochure dealing with the adjustment of the back horizon glass, and Peter Dollond patented a complicated lever system for its parallel adjustment (no. 1017 of 1772).[38] The back glass was rotated to the position of the horizon glass, the telescope placed in a lower position, and the glass moved to be parallel to the index glass. This system, which must have been expensive to make and vulnerable, is found on an octant by John Bird in the National Maritime Museum (cat. no. 051; see Plate 7). At the same time, Dollond proposed an extended lever for adjusting the fore observation glass from below the telescope bracket, and this simple technique is found on several preserved octants and sextants (for example cat. no. 057). Joshua Lover Martin patented a system of screws and a lever by which the index glass could be adjusted for perpendicularity and for the index error (no. 1088 of 1774, cat. no. 052).[39] The aforementioned Magellan, a prominent Portuguese scholar who moved to London late in 1763, published a treatise on English octants and sextants, in which he proposed a system for adjusting the index error on an octant. Both horizon glasses were mounted on a notched wheel, by which they could be properly aligned with the index glass and secured.[40] Gabriel Wright patented a system by which the large glass could be adjusted for the index error by revolving it round the pivot (no. 1229 of 1779, cat. no. 148).[41]

Probably none of these designs were really successful at sea, and eventually they were replaced by simpler methods. A screw, sometimes in the shape of a detached key, achieved perpendicularity of the index glass and both horizon glasses by minutely tilting these and their brass fittings forward or backward. The index error was adjusted by slightly moving the horizon glass. This was done by a lever and a wing nut or a worm gear, and then clamping the lever with a milled screw. Sometimes it was achieved by a capstan screw and an adjusting pin, by a thumbscrew, or by a square-headed screw with a detached key. Often these devices were on the back of the frame and, to prevent them from being moved accidentally when the instrument was laid on its back, three feet were provided, one near the pivot, and the others at the bottom of the struts or at the ends of the limb. An adjusting pin, key, or spanner was supplied with the instrument, or a screwdriver was fixed in the frame's crossbar and central strut to work the adjustment screws. Sometimes a pencil is found in the crossbar, for noting an altitude in the blank ivory plate that is inlaid on the back of the frame.[42]

Eighteenth- and early nineteenth-century octants often have a small brass stop screwed to the side of the right arm of the frame. The user knew that when the index arm was placed against the stop, it was aligned with the zero on the scale. For a negative reading – for example in the case of a negative index error – the stop was turned downward. The stop simplified use in poor light, and also prevented the index arm from sliding off the limb. In later octants, and in sextants, this was prevented by the construction of the telescope bracket.

Two problems that could not be dealt with by a navigator were faults in scale graduation and a centring error. The former was the result of poor craftsmanship by the manufacturer. The latter occurred when the pivot of the index arm did not coincide with the centre of the frame's arc. This too could be the result of poor craftsmanship; but it could also be a result of wear, or could occur if the instrument fell or received a severe blow, for example in bad weather.[43]

Caleb Smith's sea quadrants were fitted with a telescope, and Hooke, Newton, and Hadley had also proposed this. Nevertheless, all early octants are found with a cheaper and more easily made brass sight vane with a pinhole. Often there are two adjacent pinholes or a narrow slit parallel to the horizon. The left pinhole and the left end of the slit were a little out of the line of the reflected light, thereby weakening the sunlight and preventing the observer from being dazzled.[44] The sight vane remained in use on octants well into the nineteenth century, but then mainly on cheaper examples. Elaborate and therefore expensive octants, and sextants, were provided with one or more telescopes that had to be fixed into the telescope bracket. This consisted of a threaded collar mounted on a rising-piece, a device by means of which the telescope in its collar may be raised from or lowered to the plane of the frame with a milled screw on the back of the frame. The rising-piece provided the same facility as the left pinhole in a sight vane; it was used to regulate the light in the observer's eye. Once a telescope was fixed it had to be parallel to the plane of the instrument, and any deviation is known as a 'collimation error'.[45] The collar usually consisted of two parts connected by adjusting screws, by which the axis of the telescope could be moved parallel to the frame and corrected for collimation error.

Early octants had features similar to cross-staffs and backstaffs. The wing nuts to adjust the glasses were identical to those found on cross-staffs to clamp the vanes to the staff.[46] Sight vanes were made of pear wood or boxwood just like those on a backstaff, although smaller, and the boxwood or pear-wood limb of an octant was graduated for altitude and zenith distance with transversals similarly to the 30-degree arc of a backstaff. An octant's transversal scale was read along a fiducial edge – often lined with a thin piece of ivory for visual contrast – to two minutes of arc. In the 1740s the reading accuracy increased to one minute, by

FIGURE 3.9 (*left*) The back of an octant, showing the ivory pencil head fitted in the crossbar and the blank ivory plate inlaid in the frame for noting an altitude (cat. no. 047).

FIGURE 3.10 (*above*) The fiducial edge at the end of an index arm, with an ivory lining for visual contrast (cat. no. 037).

1760 to half a minute, and by 1790 to a second of arc or less.[47] Other improvements in the 1740s were a brass index arm, instead of wood, and scales engraved in brass or ivory, materials that are less subject than wood to damage and changes in temperature and humidity. Brass scales were screwed on the front of the frame's wooden limb, and ivory scales were inlaid. Frames were increasingly made of ebony, as this does not easily warp. The number of hinged shades increased from two to three, in a range from light to dark glass. For the same reason as for the glasses, the fronts and backs of the shades had to be parallel, and one colour refracted more than the other.[48] If an octant was fitted for back observation, a set of socket shades was provided that could be placed according to choice between the index glass and either the fore horizon or back horizon glass.

An adjustment device was also introduced, by means of which a slow, controlled motion of the index arm along the limb could be achieved. It eliminated sudden starts when one was moving the arm, which made it particularly difficult to align the reflection of a star with the horizon. The earliest example of such a device on an octant is from around 1749 by Benjamin Ayres, an English instrument maker in Amsterdam who was apprenticed to Sisson and married his sister Anne. It consists of a rack at the bottom of the brass scale, along which a pinion moves the index arm. Ayres probably copied this from Sisson, who had used it on a universal quadrant and a theodolite.[49] Less than ten years later, John Bird was the first to apply a tangent screw for adjusting the index arm on a navigational instrument: a reflecting circle. Bird had used this assembly for his observatory instruments for the first time in 1745, and afterwards also on his octants and sextants. It was situated on a clamp on the instrument's limb, which was connected to the left or right of the index arm.[50] Another slow-motion adjustment was designed by Cornelis Douwes, in which the index arm worked along catgut in a groove at the bottom of the limb.[51] Neither the rack and pinion nor Douwes' catgut were continued on octants, probably because they were slow in use, as they did not allow the index arm to be disengaged from the limb. Moreover, catgut was vulnerable. The tangent screw prevailed, and after Bird's examples it is found from about 1770 onwards on sextants, and a decade later on octants, and remained in use for approximately the next 150 years. The device is situated at the end of the index arm, on the front or back, or at the bottom, and consists of a tangent screw with a short pitch. The screw turns in a

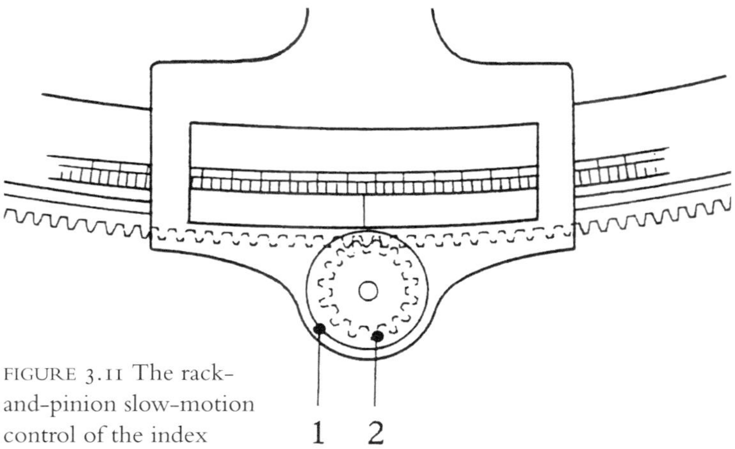

FIGURE 3.11 The rack-and-pinion slow-motion control of the index arm found on octants by Benjamin Ayres. Author's collection, drawing by N. Kamer.

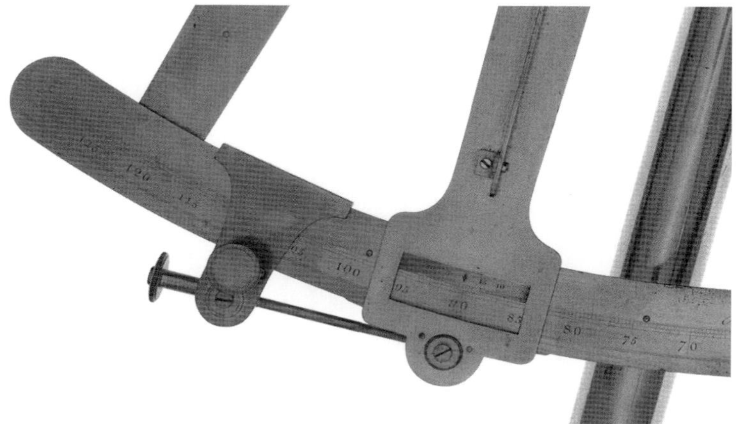

FIGURE 3.12 The tangent screw and clamping screw found on octants and sextants by John Bird (cat. no. 133).

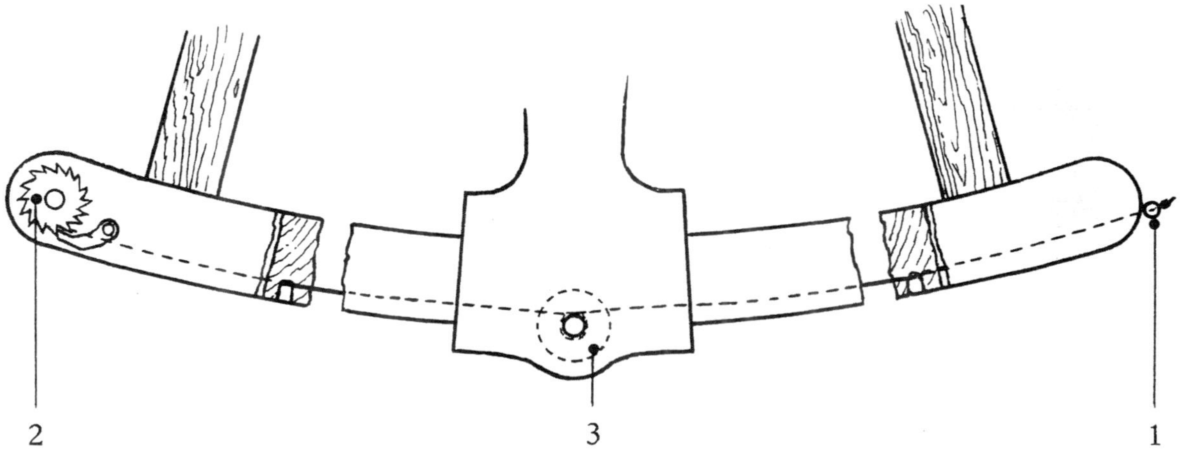

FIGURE 3.13 The catgut and groove slow-motion control of the index arm designed by Cornelis Douwes. The ratchet wheel on the left is to tighten the catgut. Author's collection, drawing by N. Kamer.

ring that is fastened to a clamp. Halfway along its length, the screw runs through a threaded collar that is fastened to the index arm. When the Sun's reflected image is near the horizon the clamping screw is tightened, locking the index arm, which can then be moved slowly and gradually by turning the tangent screw in the threaded collar. The arm is moved forward or backward along the scale until the Sun's lower limb is aligned with the horizon.

Hadley's design does not show a vernier; the earliest known octant with such a secondary scale dates from about 1743.[52] More often on early octants, a transversal scale is found on the boxwood limb. Octants with transversals remained for sale throughout the eighteenth century, but this pattern started to decrease after about 1760 in favour of the vernier. Until the 1770s the zero of an octant's vernier was situated at its centre, usually in a range of 20–0–20 or 15–0–15, depending on the division of the main scale.[53] The zero was in the same position as along the fiducial edge of a backstaff and on an octant with a transversal scale. This arrangement was difficult to read and therefore prone to errors, and this is probably the reason why sextants usually have the zero at the right-hand end of the vernier scale. Octants followed this practice after about 1780. The main scale of a vernier instrument was extended beyond the actual angle limit of the instrument, to allow for complete overlap of the vernier.

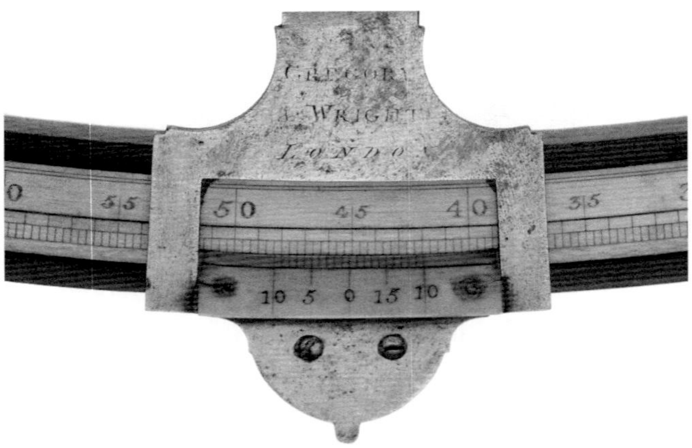

FIGURE 3.14 An octant's vernier with the zero at the centre (cat. no. 065).

FIGURE 3.15 An octant's vernier with the zero at the right (cat. no. 070).

INTRODUCTION AND EARLY DIFFUSION OF THE OCTANT IN BRITAIN

The first description of the octant aimed at the seamen who were its future users appeared in 1734, and was probably written by John Hadley's brother George.[54] The second edition of *A Description of a New Instrument* appeared in 1738 and listed the names of the instrument makers from whom octants could be obtained. Besides Jonathan Sisson, these were Thomas Wright, Joseph Jackson, John Gilbert, John Fowler, Edmond Culpepper, Charles Digby, and William Coleman. All were from the Strand and Fleet Street in London, or from the shipping area around the Tower of London and in Wapping. With the exception of Sisson, no octant by any of these manufacturers is known to survive, although it is certain that Jackson and Gilbert were making them.[55] Gilbert was a manufacturer of backstaffs who advertised the sale of octants in 1738.[56] The oldest octant now on record is by Sisson.[57] Stuck on the inside of one of its struts is a piece of paper containing the handwritten text 'J Sisson Single N° 8. 19–28. June 1735 Geo: Hadle[y]'.[58] 'Single' indicates that it was made for fore observation only, and the note suggests George Hadley's further involvement with the introduction of his brother's invention. A second octant, not signed, has an identical piece of paper in the same place, with the manuscript note 'N.249 Sept. 22 1744 G. Hadley'.[59] Here the word 'single' is omitted because it is fitted for back observation. The figures 19 and 249 on the two notes may suggest the number of octants produced in that period of nine years.

Recent research has given insight into the diffusion of the octant in Britain, and surviving examples and printed evidence suggest that it was not widely used before the 1750s.[60] An early record of simultaneous use of a backstaff and an octant is by Robert Lambert, master of HMS *Gloucester* in 1740–44.[61] In 1750 officers of the East India Company were using octants, perhaps made by George Adams senior, who is known to have supplied some of the mates of Company ships.[62] During the 1750s, the instrument was promoted in navigational manuals. The Master and Wardens of Trinity House recommended equipping apprentices from the Mathematical School of Christ's Hospital with an octant when they went to sea, instead of a cross-staff and backstaff. That decade saw the rise of a number of prolific manufacturers of octants. Besides Adams they were John Bird, John Dollond senior, Benjamin Cole, and his sometime associate John Goater in Wapping, who also made backstaffs. A reason why the diffusion was perhaps slow at first was the high price of the octant compared with a backstaff; in 1755, the difference was £2 8s.[63] As the octant's popularity increased it became cheaper, while the price of a cross-staff and a backstaff remained stable. A 1768 trade list for Benjamin Cole junior's shop shows that the cheapest octant cost £1 14s and the cheapest backstaff £0 12s, a difference of just over £1.[64]

THE INTRODUCTION OF THE OCTANT IN FRANCE

France was the first country outside Britain to adopt the octant. It was employed on the expedition of the Académie Royale under Louis Godin that in 1735 sailed to Peru from La Rochelle in *Le Portefaix*.[65] At the end of 1734 Godin had travelled to London to buy various scientific instruments for this expedition, including an octant. En route in the Atlantic to Martinique, Pierre Bouguer took daily noon sights of the Sun.[66] He was a former teacher of navigation at Le Croisic in Brittany, and France's foremost expert in navigation. In his unpublished journal of the voyage, preserved in the Paris Observatoire, Bouguer testified to the superiority of the octant over the cross-staff. The manuscript shows the index glass placed lower on the index arm, which according to Bouguer facilitated the observation but also reduced its precision.

The second use of the octant by the French was during the expedition to Lapland that sailed in *Le Prudent* from Dunkirk under Moreau de Maupertuis in 1736. Maupertuis had invited the Swedish scholar Anders Celsius to join the expedition, and en route to Dunkirk Celsius visited London, where Graham and Sisson were working on a zenith sector for the expedition. Celsius was elected a Fellow of the Royal Society and met Hadley. In January 1736 they practised together with an octant and the Swede purchased one. It was used in the Gulf of Bothnia to take sights of the Sun, where the *messieurs* used the *instrument Anglois*, as they called it, and the captain and mate of *Le Prudent* a cross-staff. The results were in favour of the octant and were published in the report of the voyage.[67]

The first printed description of the octant in French appeared in 1739, written by J.-B. d'Après de Mannevillette, a lieutenant of the Compagnie des Indes, and the main French propagandist of Hadley's invention.[68] François de Brémond, who was elected a Fellow of the Royal Society that year, delivered a French translation of Hadley's description in 1741.

The booklet by d'Après de Mannevillette contained an advertisement by Pierre Lemaire, a scientific instrument maker in Paris, with an illustration of his octant. He was the first to manufacture octants in his country, starting in 1739, when Lemaire changed the name of his workshop to 'Au nouveau quartier Anglois', the most-used French name for an octant.[69] The striking similarity between an early octant by Lemaire and Sisson's oldest known example suggests contact between the two men.[70] Other French octant makers include Dardenne of Nantes, from whom one example is recorded from the 1740s' and also Baradelle, whose name has already been mentioned, and Joseph Roux. The Roux

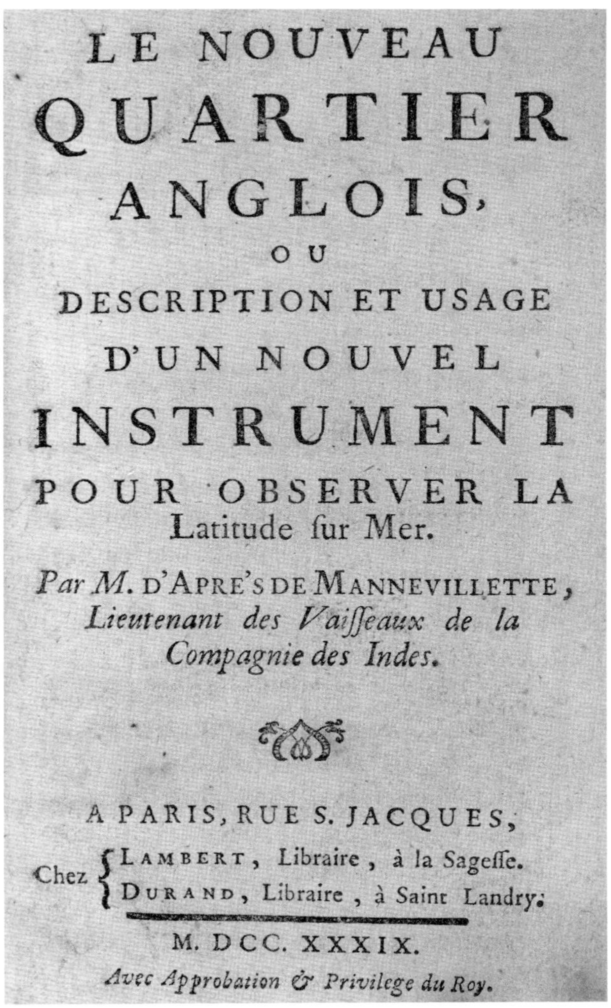
LE NOUVEAU
QUARTIER
ANGLOIS,
OU
DESCRIPTION ET USAGE
D'UN NOUVEL
INSTRUMENT
POUR OBSERVER LA
Latitude ſur Mer.
Par M. D'APRE'S DE MANNEVILLETTE,
Lieutenant des Vaiſſeaux de la Compagnie des Indes.

A PARIS, RUE S. JACQUES,
Chez { LAMBERT, Libraire, à la Sageſſe.
DURAND, Libraire, à Saint Landry.
M. DCC. XXXIX.
Avec Approbation & Privilege du Roy.

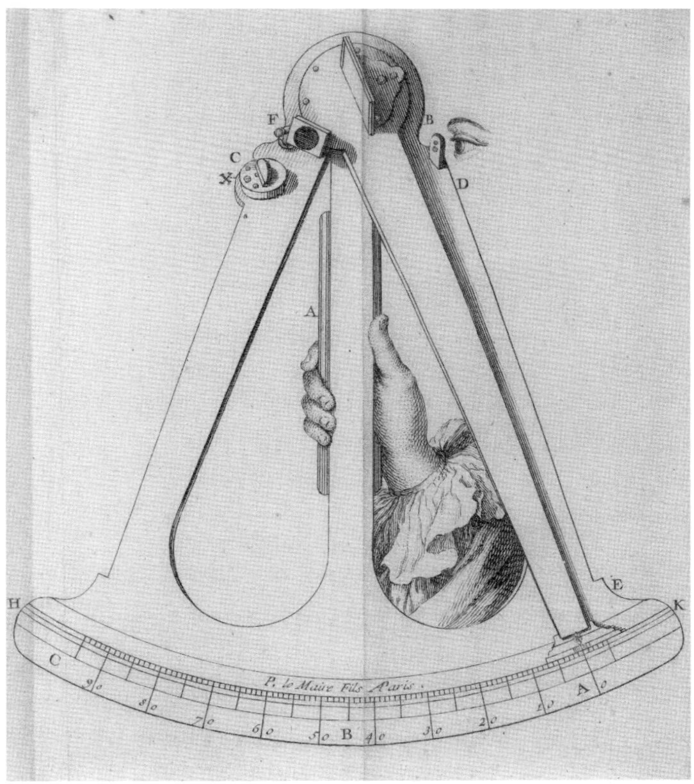

FIGURE 3.16 (*left*) The first description of the octant in French, by J.-B. d'Après de Mannevillette, an officer in the French East India Company. Paris, 1739. National Maritime Museum; PBE4606.

FIGURE 3.17 (*above*) An octant by Pierre Lemaire, the first French manufacturer to make such instruments, about 1739. From the French description of the octant, by d'Après de Mannevillette. National Maritime Museum; PBE4606.

family were instrument makers, hydrographers, and marine artists in Marseilles between around 1760 and 1790.[71]

THE INTRODUCTION OF THE OCTANT IN THE DUTCH REPUBLIC

The third country where the octant was used was the Dutch Republic, and there the acceptance of Hadley's invention was probably faster than in Britain and France. Octants were for sale in Musschenbroek's workshop in 1735, the same year that Frenchmen used it for the first time.[72] The following year an illustrated Dutch translation of George Hadley's brochure of 1734 appeared, in *Uitgeleeze Natuurkundige Verhandelingen over de Natuurkunde*, a scholarly journal inspired by the *Philosophical Transactions*. The earliest recorded use of an octant by a Dutchman dates from 1737, when Lieutenant George de Visscher, commander of the *Teijlingen* of the Amsterdam Admiralty, used one on a voyage from Amsterdam to the West Indies.[73]

Willem Jacob's Gravesande was responsible for the introduction of the octant in the Dutch Republic. He had spent a year as a diplomat in London and met Newton and Desaguliers, whose lectures he attended. Gravesande was elected a Fellow of the Royal Society and became the main Dutch follower of Newtonian physics. After returning to the Republic he was appointed professor of mathematics and astronomy at the University of Leiden, and for his lectures used demonstration apparatus that was provided by Musschenbroek. Being a Fellow of the Royal Society, Gravesande was among the first Dutchmen to know about Hadley's octant, and he encouraged Musschenbroek to include these in his range.[74] The fact that it was first used on a naval vessel was due to the Anglophile Admiral Hendrik Lijnslager, who was a relative of Visscher.

The first octants for ships of the VOC were purchased from Musschenbroek, in 1741, and three years later four were bought from Benjamin Ayres. In 1747 it was decided that each Company vessel was to be supplied with an octant, together with printed instructions for use, which were written by Cornelis Douwes. These developments led to a demand for octants in the Republic, and the main manufacturers were Johannes van Keulen and his son Gerard in Amsterdam, and Jacobus Kley in Rotterdam.[75]

THE MANUFACTURE OF OCTANTS IN NORTH AMERICA

Although the octant was simultaneously invented in Philadelphia, little is known about the instrument's early years in Britain's American colonies. Joseph Halsey junior of Boston advertised in 1738 that he made Hadley's quadrants and also backstaffs, but no examples by him seem to have survived. After about 1730 Anthony Lamb made backstaffs in his workshop 'At the Sign of the Compass and Quadrant' in New York.[76] Claims that Lamb made Godfrey's first octant have not been substantiated by research. As in Britain, it appears that the octant was not widely used in America before the 1750s.[77] As far as is known, James Keappock of Philadelphia and John Dupee of Boston made the oldest preserved American octants, in 1753 and 1755, respectively.[78] Before the War of Independence (1775–83), American seamen mainly used octants imported from England.[79] When this market fell away as a result of the war, it became apparent that American octants had technical shortcomings.

THE MANUFACTURE OF OCTANTS IN DENMARK AND SWEDEN

The only eighteenth-century Danish instrument maker known to have made octants and sextants was Jeppe Smith (1759–1821) of Copenhagen, a few specimens by whom survive.[80] It is likely that when Smith was active, at the end of the century, he could not compete with the quality and quantity of products that were being offered by the English makers who had begun to dominate the market. When Christian Carl Lous, the most distinguished teacher of navigation in Copenhagen, offered the Board of Longitude the draft and description of his newly invented instrument for measuring lunar distances in 1788, he added that it would have to be made by an English instrument maker.[81] The number of English-made octants preserved in Danish collections confirms their dominance in that country.[82]

In Sweden, Daniel Ekström (1711–55) was a promising scientific instrument maker who, with Celsius, was involved in founding the Royal Swedish Academy of Sciences. Ekström was in London from 1736 to 1741, as an agent of Celsius, to purchase scientific instruments for the observatory at Uppsala. During this time he attended lectures by Desaguliers and worked with Graham. Ekström's sojourn probably inspired him to build his only known octant, around 1745.[83] He died young and it is unlikely that there was other significant octant production in Sweden, as a lack of qualified scientific instrument makers rendered the country dependent on the British market.[84]

THE MANUFACTURE OF OCTANTS IN GERMANY

George Friedrich Brander of Augsburg (1713–83) was an accomplished German scientific instrument maker, who in 1775 associated himself with his son-in-law Christoph Caspar Höschel.[85] From 1764 onward Brander made octants and sextants, but surviving examples date from 1777 and later. The absence of shades and the provenance of these instruments indicate that they were for meant for land surveying. Because of the distance of Augsburg from the coast, it seems unlikely that seamen could easily have purchased them. This cannot be said of the octants that were made on the Frisian island of Föhr, a whaling community on the coast of Schleswig-Holstein, by a small group of clockmakers, who also taught navigation. Surviving octants date from between about 1760 and 1790, and some have highly decorative brass frames. The most prolific manufacturer was Jens Nickels(en) (1721–85), who is better known under his Dutch name Jan Corneliszoon (cat. no. 053; see Plate 10). Centuries of whaling ties between the Dutch Republic and Föhr led island seamen to translate their Frisian names into Dutch, and octants from the island were especially popular among Dutch and German whalers. Despite the confined market, Corneliszoon's production figures rank high when compared to contemporary sales in Amsterdam. Technically, however, the octants from Föhr were far from perfect when compared with examples that could be purchased in London and Amsterdam. For example, transversal scales were applied well into the 1780s, while elsewhere these had long been replaced by vernier scales.

1. Bennett, 'Catadioptrics and Commerce in Eighteenth-Century London', 252; Jardine, 'Monuments and Microscopes', 304; Cotter, *A History of the Navigator's Sextant*, 104–6
2. Cotter, *A History of the Navigator's Sextant*, 106–9; Bennett, 'Catadioptrics and Commerce in Eighteenth-Century London', 252–4.
3. For the mathematical proof, see Cotter, *A History of the Navigator's Sextant*, 114–15.
4. 'A True Copy of a Paper Found in the Hand Writing of Sir Isaac Newton, among the Papers of the Late Dr Halley, Containing a Description of an Instrument for Observing the Moon's Distance from the Fixt Stars at Sea'. *Philosophical Transactions* vol. 42 (1742–43) no. 465, 155–6.
5. Bennett, 'Catadioptrics and Commerce in Eighteenth-Century London', 252.
6. Howse, *Greenwich Time and the Longitude*, 53–4.
7. Gloria Clifton, 'Hadley, John (1682–1744)'. In: *Oxford Dictionary of National Biography* <http://www.oxforddnb.com>
8. Cotter, *A History of the Navigator's Sextant*, 111; Bennett, 'Catadioptrics and Commerce in Eighteenth-Century London', 262–3.
9. Bennett, 'Catadioptrics and Commerce in Eighteenth-Century London', 259.
10. John Hadley, 'The Description of a new Instrument for taking Angles'.
11. Hadley, 'The Description of a new Instrument', 151–2; Cotter, *A History of the Navigator's Sextant*, 115; Maskelyne, 'Remarks on the Hadley's Quadrant', 51. With the back-observation the upper limb of the Sun or Moon was observed and the semi-diameter had to be added to the altitude. Maskelyne called the regular observation 'fore-observation'. See also Nichols, 'Hadley's Quadrant: The Back Observation'.
12. Cotter, *A History of the Navigator's Sextant*, 119.
13. Derek Howse, 'Sisson, Jonathan (1690?–1747)'. In: *Oxford Dictionary of National Biography*. <http://www.oxforddnb.com>.
14. Jeremy Lancelotte Evans, 'Graham, George (*c.* 1673–1751)'. In: *Oxford Dictionary of National Biography* <http://www.oxforddnb.com/>
15. Forbes, *The Birth of Navigational Science*, 4.
16. Cotter, *A History of the Navigator's Sextant*, 132–3; Mörzer Bruyns, *Schip Recht door Zee*, 40.
17. Bedini, *Thinkers and Tinkers*, 118–20; Cotter, *A History of the Navigator's Sextant*, 122–3; Warner, 'American Octants and Sextants', 97–8.
18. Howse, *Greenwich Time and the Longitude*, 63–4.
19. Letter from Logan to Halley, 25 May 1732, Royal Society Archive, EL/L6/59.
20. 'An Account of Mr. Thomas Godfrey's Improvement of Davis's Quadrant, transferred to the Mariner's-Bow, communicated to the Royal Society, by Mr. J. Logan'. *Philosophical Transactions* vol. 38 (1733–34) no. 435, 441–50; Cotter, *A History of the Navigator's Sextant*, 122–7.
21. Magellan, *Descriptions des octants et sextants anglois*, 2; Fauque, 'The Introduction of the Octant in Eighteenth-Century France', 95; Cotter, *A History of the Navigator's Sextant*, 129–30.
22. Hoare, *The Quest for the True Figure of the Earth*, 48, 69, 111.
23. Clifton, 'The Adoption of the Octant in the British Isles', 90.
24. Cotter, *A History of the Navigator's Sextant*, 127–79; Talbot, 'Caleb Smith's Sea Quadrants'.
25. Although the horizon glass was smaller, the heights of the two glasses above the frame were equal; see Maskelyne, 'Remarks on the Hadley's Quadrant', 59.
26. Bennett, 'Catadioptrics and Commerce in Eighteenth-Century London', 269.
27. Glyndwr Williams, 'Middleton, Christopher (d. 1770) *Explorer in Canada*'. In: *Oxford Dictionary of National Biography*. <http://www.oxforddnb. com> 'Observations Made of the Latitude, Variation of the Magnetic Needle, and Weather, by Capt. Christopher Middleton, in a Voyage from London to Hudson's-Bay, Anno 1735'. *Philosophical Transactions* vol. 39 (1735–36) no. 442, 270–80.
28. Talbot, 'Caleb Smith's Sea Quadrants', 5–6; Warner, 'American Octants and Sextants', 102–3; Mörzer Bruyns, 'Octants and Sextants in the Musée national de la Marine in Paris', 42–3; Clercq, *At the Sign of the Oriental Lamp*, 250; Taylor and Richey, *The Geometrical Seaman*, 109–10.
29. Warner, 'American Octants and Sextants', 102–3.
30. Mörzer Bruyns, 'Octants and Sextants in the Musée national de la Marine in Paris', 42–3.
31. Magellan, *Description des Octants et Sextants Anglois*, 156, was still referring to Smith's quadrant in 1775.
32. Cotter, *A History of the Navigator's Sextant*, 130.
33. 'A Spirit Level to be fixed to a Quadrant for taking a Meridonial Altitude at Sea when the Horizon is not visible'. *Philosophical Transactions* vol. 38 (1733–34) no. 430, 167–72. This included Hadley's description of his octant, which thereby was included in the patent, see Bennett, 'Catadioptrics and Commerce in Eighteenth-Century London', 269.
34. Hadley, 'The Description of a new Instrument', 154.
35. This is now called the First adjustment. For this and the following adjustments and how to test an instrument for them, see Watts, *The Sextant Simplified*, 49–55.
36. The side error is called the Second adjustment, and the index error the Third adjustment.
37. Very cheap instruments sometimes lack these facilities.
38. [Dollond], *The Construction and Use of the Sea Quadrant*; Dollond and Dollond, *Directions for using the new and improved Hadley's Quadrant*; Bennett, *The Divided Circle*, 136.
39. Ifland, *Taking the Stars*, 71–2, 84–5.
40. Magellan, *Description des octants et sextants anglois*, 12–13; see also Ifland, *Taking the Stars*, 86–9.
41. Wright, *The description and use of a new constructed quadrant, sextant and quintant*, 5–8. It is also found on an octant by Gilbert, Wright, and Hooke in the ScML, inv. no. 1928–925, on a sextant by Gilbert and Wright in the MSM, inv. no. 1951–394, and on a quintant by Gilbert in the ScML, inv. no. 1923–475. For Magellan, see R.W. Home, 'Magellan, Jean Hyacinthe de (1722–1790) *natural philosopher*'. In: *Oxford Dictionary of National Biography* <http://www.oxforddnb.com>

42. Cittert, *Astrolabes*, 43.
43. Watts, *The Sextant Simplified* 48–9.
44. Crone, 'Pieter Holm en zijn octant', 5–6. For example, this pinhole could be used when observing a bright star for which the shade was too dark.
45. Called the Fourth adjustment, see Watts, *The Sextant Simplified*, 57–8.
46. Mörzer Bruyns, *Schip Recht door Zee*, 37.
47. Daumas, *Scientific Instruments*, 182; John Adams, 'On the subdivision of a nonius by means of which the divisions of a quadrant . . . may be read of to a second or less'. In: *Transactions of the Society for the Encouragement of Arts, Manufactures and Commerce* 4 (1790) 173–8. The actual accuracy of a measurement, of course, depended on the quality of the scale and the experience of the observer.
48. James Horsburgh found errors introduced by coloured glass when observing with a Ramsden sextant in 1798; see Horsburgh and Mackenzie, *A Treatise on Marine surveying, supplement*, 151.
49. Mörzer Bruyns, *Schip Recht door Zee*, 112, 138, 206, 209; Leopold, 'Some Notes on Benjamin Ayres', 396. The universal quadrant and theodolite are in the NMM (NAV1443 and NAV1451). The device is also found on an octant by Jacobus Kley of Rotterdam, of around 1754, and on one by Sisson of *c.* 1760; see Cittert, *Astrolabes*, 43. It is also found on an octant dated 1769, by John Williamson of Ulverstone; here the rack is on the top of the brass scale. This instrument was offered for sale at Phillips, London, on 12 December 1995.
50. Moskowitz, 'The World's First Sextants', 30–1; Mörzer Bruyns, *Schip Recht door Zee*, 138, 212–13; Jan van Wijk of Amsterdam copied it on some of his octants.
51. It is found on octants by Jan van Wijk and Johannes van Keulen of around 1760; see Mörzer Bruyns, *Schip Recht door Zee*, 138, 210–11.
52. Mörzer Bruyns, *Schip Recht door Zee*, 205; this octant is by Jan van Musschenbroek.
53. At this time a degree on an octant or sextant scale was divided into three parts (20 minutes of arc) or four parts (15 minutes of arc).
54. Anita McConnell, 'Hadley, George (1685–1768)'. In: *Oxford Dictionary of National Biography*. <http://www.oxforddnb.com> For the scholarly world, a Latin version was published at the same time.
55. Clifton, *Directory*, 62, 73, 84, 104, 112, 149, 253, 306; Culpepper died in 1737, and besides this reference nothing is known about Coleman. Clifton, 'The London Mathematical Instrument Makers and the British Navy', 27.
56. Clifton, 'The Adoption of the Octant in the British Isles', 91.
57. Signed 'J. Sisson in the Strand London', in the CHSI, inv. no. 1998–1–1663.
58. Mörzer Bruyns, *Schip Recht door Zee*, 41; Clifton, 'The Adoption of the Octant in the British Isles', 90–1.
59. In the NSA, inv. no. B.0021(02).
60. Clifton, 'The Adoption of the Octant in the British Isles', 90–4.
61. May, *A History of Marine Navigation*, 32.
62. Warner, 'John Bird and the Origin of the Sextant', 4; Cotter, *A History of the Navigator's Sextant*, 144; Millburn, *Adams of Fleet Street*, 18–20.
63. Clifton, 'The Adoption of the Octant in the British Isles', 93.
64. Taylor and Richey, *The Geometrical Seaman*, 110.
65. Fauque, 'The Introduction of the Octant in Eighteenth-Century France'; Mörzer Bruyns, *Schip Recht door Zee*, 42–3.
66. Hoare, *The Quest for the True Figure of the Earth*, 121. The sights were reported to the Académie by a letter from Godin in 1739, but little notice was taken.
67. Outhier, *Journal d'un voyage au nord*, 11–13, 21–2.
68. d'Après de Mannevillette, *Le nouveau quartier Anglois ou description et usage d'un nouvel instrument*. Paris, 1739
69. Daumas, *Scientific Instruments*, 182, 259; Chapuis, *A la mer comme au ciel*, 42. Lacaille called it a *quartier de reflection*.
70. Mörzer Bruyns, 'Octants and Sextants in the Musée national de la Marine in Paris', 41–2.
71. Brewington, *Navigating Instruments*, 11.138. An example is preserved in the MMNN, inv. no. 1986.46.01, and one in MSM, inv. no. 1959–1138. Marcelin, *Dictionnaire des fabricants*.
72. Clercq, *At the Sign of the Oriental Lamp*, 241. For the rest of this paragraph see also especially Mörzer Bruyns, *Schip Recht door Zee*, 51–66.
73. Davids, *Zeewezen en wetenschap*, 229.
74. It is uncertain if Musschenbroek made octants or imported them from England, perhaps from Sisson with whom he had contact; see Mörzer Bruyns, *Schip Recht door Zee*, 61.
75. Mörzer Bruyns, *Schip Recht door Zee*, 116–21.
76. Bedini, 'At the Sign of the Compass and Quadrant', 20, 40–5.
77. Warner, 'American Octants and Sextants', 86, 100.
78. Warner, 'American Octants and Sextants', 102; Brewington, *The Peabody Collection of Navigating Instruments*, 9.
79. Bedini, *Thinkers and Tinkers*, 338–9.
80. Andersen, 'Jeppe Smith (1759–1821): A Danish Instrument-Maker', 408.
81. ROG BOL papers 14/6, p.135.
82. Andersen, *Historic Scientific Instruments in Denmark*; Thirslund, 'Samlingen af nautiske instrumenter', 156–57.
83. Pipping, *The Chamber of Physics*, 41–2, 60–1, 118.
84. Sinnerstad, 'Astronomy and the First Observatory', 54.
85. For this paragraph see Mörzer Bruyns, 'Octants from the Friesian Island of Föhr'.

CHAPTER FOUR

THE SEXTANT AND FURTHER IMPROVEMENTS IN THE EIGHTEENTH CENTURY

THE SEXTANT

The sextant differs from the octant in having an extended arc that makes it possible to measure angles up to about 120 degrees. While in astronomy a sextant refers to an instrument with an arc of 60 degrees, the first use of the word for an instrument to measure altitudes of celestial bodies at sea appears to date from 1735.[1] Made by Thomas Heath of London, this had a brass frame and an artificial pendulum horizon. No example is known to survive, and it is generally accepted that John Bird built the first marine sextant, based, like the octant, on the reflecting principle. Bird was an accomplished scientific instrument maker who had worked for Jonathan Sisson and Graham and in 1745 established his own workshop in London.[2] Captain (later Vice Admiral) John Campbell RN had been ordered by the Admiralty to test the usefulness of Hadley's octant for lunars and did so in the *Essex* in February and early March 1757, while cruising off Cape Finisterre. For these trials he also used a circular reflecting instrument made of brass that was built after Mayer's design and could measure angles up to 180 degrees.[3] Bird made it based on a wooden example, which had been passed on to him by the Admiralty. However, Campbell found it too cumbersome for effective use and, together with Bird, he then devised the sextant. The first example was made of brass and had a radius of 16 inches, twice the diameter of Mayer's circle.[4] Campbell tested it on the *Royal George* while cruising off Ushant, in September 1758 and from July to September 1759.

A small number of sextants by Bird survive; three have frames of wood, the others being of brass.[5] The large sextants (with radii of between about 13 and 18 inches) have the previously mentioned tangent screw, and some include a sight-tube or a fixed telescope. Bird's sextants have crossed reinforcing bars on the backs of the frames to make them more rigid. To bear the weight, these sextants had a supporting pole that could be placed in the observer's belt (cat. no. 133).

FIGURE 4.1 Portrait of John Bird with a beam compass that was used for dividing circular scales prior to the invention of the circular dividing engine. Engraving by N. Green after a drawing by Lewis, published in London, 1776. National Maritime Museum; PAF3435.

The transition from the octant to the sextant, developed by Bird and Campbell, led to a number of improvements that made the sextant into a sophisticated navigational instrument. Henceforth brass was generally used for sextant frames and the graduation was engraved in the brass limb.[6] Brass is more rigid than wood and not subject to deformation by humidity. Because a brass surface can cause inconvenient reflection, parts of the frame near the glasses were lacquered black. Sextants were also given a wooden handle on the back of the frame to make them easier to hold during an observation. A direct-image telescope of low power permanently replaced the sight vane; a second telescope with parallel cross-wires was used to check if the axis of the telescope was parallel to the instrument's plane, as a check against collimation error.[7] The number of index shades increased to four or five, and hinged horizon shades – usually two or three – were fitted to eliminate the reflection of sunlight or moonlight from the sea.[8] Separate shaded eyepieces were also supplied to be screwed on the ocular end of the telescopes and the sight-tube. The tangent screw became a standard part of the sextant, as did a magnifier, often on a swivelling arm, to read the vernier. Towards the end of the century the graduation was often engraved on a silver arc that was inlaid in the brass limb. Silver allows finer engraving but discolours by oxidation, and because of its softness also wears quickly through polishing. To remedy this problem, instrument makers sometimes used gold, platinum or palladium as a material for their inlaid arcs.[9]

Nevil Maskelyne (Plate 8) published an article on the octant and sextant, obviously aimed at seamen as it appeared in the *Nautical Almanac*.[10] He recommended Peter Dollond's design for adjusting the back horizon glass, described in the previous chapter.[11] Maskelyne also gave instructions on how to adjust the axis of the telescope for collimation error. He dealt with the problem of the lack of parallelism of the front and back of the glasses, and warned navigators not to rely on octant makers placing the thickest or the thinnest edge of the glass nearest the instrument's plane, as Hadley had advised. He added that the grinding of glass was not perfect either. Maskelyne provided a device to deal with lack of parallelism when taking a sight of the Sun or the Moon. The back of the upper part of the index glass had to be roughened and blackened; the lower part remained silvered as usual, and could, at the user's choice be covered by a hinged brass flap: this was called the Maskelyne flap[12] (for example cat. no. 136). When taking a sight, the observer turned the flap to cover the silvered part, and with the rising-piece raised the telescope from the instrument's plane and centred it on the blackened part of the index glass. Sunlight and moonlight were sufficiently strong to be reflected by the upper surface of the 'black mirror' (and, according to Maskelyne, images would appear much more distinct). Because the reflection came from the front of the glass, rather than from the silvered back, any lack of parallelism was avoided. For star observations, the telescope was lowered back towards the frame and the flap opened to bring the lower silvered part of the index glass back into play.

While the octant remained in use to measure altitudes of celestial bodies, the sextant was primarily used for lunar distances. It did not have a back horizon glass, as its extended arc allowed back sights to be taken of altitudes higher than sixty degrees. Lunars were practised at sea after Mayer's tables appeared in *The British Mariner's Guide*. John Horsley, fourth mate of the East India Company ship *Glatton*, for example, used a quadrant (i.e. octant) by Bird to find the vessel's longitude by lunars on a voyage from Britain to Asia in 1763.[13] Application of the lunar method increased after the tables appeared regularly in the *Nautical Almanac*. James Cook and his astronomers used sextants in the Pacific, in combination with the first issue of the *Nautical Almanac*. On the voyage in *Endeavour* (1768–71), there was a 15-inch brass sextant by Jesse Ramsden for the astronomer, and Cook too had one by this maker. There were two 15-inch brass sextants for the astronomers on the second voyage, in *Resolution* and *Adventure* (1772–5), one by Ramsden and the other by Peter Dollond. Cook used a wooden-frame sextant by Bird and there is mention of a sextant by Edward Nairne.[14] On the third voyage, in *Resolution* and *Discovery* (1776–80), there were four sextants, including an 18-inch wooden example by Bird said to have been Cook's own. Those by Ramsden included the example that was given the number D.33 by the Hydrographic Office in 1828 and which now is in the National Maritime Museum (cat. no. 135; see Plate 13).[15]

In the last decades of the eighteenth century, Edward Troughton was the most innovative manufacturer of sextants in Britain.[16] He patented a 'pillar and plate frame for astronomical instruments' (no. 1644 of 1788), consisting of two parallel frames of thin brass plate that were connected by a series of pillars. By this construction, the instrument was light but at the same time rigid (for example cat. no. 145). Ramsden adopted this, and at about the same time introduced the 'bridge sextant', in which a frame (bridge) covered the fittings and protected these against damage

FIGURE 4.2 A Maskelyne flap fitted on the index glass of a sextant (cat. no. 136).

by rough handling[17] (for example cat. no. 151). Ramsden's apprentice Matthew Berge also built bridge sextants, as did Berge's apprentice Nathaniel Worthington, but this construction is not found after the 1830s. Ramsden's lattice frame, however, was used throughout the nineteenth century and was renamed the triangle-pattern frame.[18] Troughton's pillar frame was also adopted by a larger group of manufacturers and was still available in the 1870s.[19] In 1816 Jan Marten Kleman (1758–1845), a prolific mathematical instrument maker in Amsterdam, advertised that he made sextants after Troughton's design.[20]

THE QUINTANT

The next development of Hadley's invention was the quintant (one-fifth of a circle, and therefore sometimes called a pentant). As with the sextant, lunar distances that occasionally required the measurement of large angles inspired the arc's extension.[21] With the quintant, angles up to about 150 degrees could be measured. A difficulty when measuring a wide angle is that the view in the index glass shrinks to a narrow slot and may even disappear. The moment at which this happens depends on the angle of the horizon glass to the axis of the telescope. For sextants this is usually twenty degrees, and for quintants it is around twelve degrees, limiting the maximum angle that could still be read to around 155 degrees.[22] In practice, this meant that quintants could not always be read to their maximum measuring angle.

Gabriel Wright was probably the first to use the word 'quintant', in a description that appeared in 1779.[23] A quintant made by him is preserved; it has a scale extended to 160 degrees, divided down to 20-minute intervals, and the vernier can be read to 1 minute of arc.[24] Apparently Wright's idea did not appeal, because other surviving quintants are considerably younger. In the National Maritime Museum collection, the oldest preserved example dates from about 1810 and was made by Berge[25] (cat. no. 275).

Hydrographical work is the basis for sea charts and, among other skills, it requires the accurate measurement of angles, in the horizontal plane, between two or three points on shore of which the position is known. On land this is done with a theodolite placed on a tripod, an arrangement that is not practical on a moving vessel. Hydrographers used an octant, a sextant, or a quintant, by holding these horizontally.[26] An advantage was that, unlike with a compass, the angle did not have to be corrected for magnetic variation and could be measured to a greater accuracy. A position was even more accurate if two angles between three points were measured simultaneously. For this process two observers using two instruments were engaged, but this involved two potential instrumental errors. Around 1784 Graeme Spence designed the first 'double sextant', which enabled one observer to measure two angles between three points at once.[27] A double sextant had two sets of index glasses, scales, verniers, and index arms. Spence was an assistant to his uncle Murdoch Mackenzie, a famous hydrographer, and they used the double sextant in conjunction with an improved station pointer.[28] Although the Admiralty rewarded Spence's invention and Troughton built an example, the development of double sextants was not taken further until the nineteenth century, and will be dealt with in the next chapter.

THE REFLECTING CIRCLE

The brass circle that Campbell used without success to measure lunar distances has already been mentioned. It was consequently ignored in favour of the sextant, but this did not mean the end of Mayer's idea; rather, the reverse. The optical principle of Mayer's reflecting circle was the same as that of an octant.[29] He designed a circular instrument with an index arm pivoted at the centre of the circle. The arm carried the rigidly mounted index glass at its centre, and a vernier. A second arm, also pivoted at the centre of the circle, carried a longitudinally fitted telescope and the horizon glass; this is called the telescope arm. Both arms could be clamped independently in any position on the limb, on which a scale of 360 degrees was divided into 720 parts. For an observation, the index arm was clamped at zero and the telescope arm moved so that both glasses were parallel, and then also clamped.[30] Through the telescope the observer saw the Sun (or the star) directly and also its double-reflected image. While the direct image of the Sun was kept in sight, the index arm was freed and moved until the double-reflected image of the Moon became visible and its limb was brought against the limb of the Sun (or the star, depending on the required measurement). The index arm was clamped and the angular distance read from the scale by use of the vernier.[31] An important advantage of Mayer's circle was that the observation could be repeated, thereby reducing any error. After the first observation, the index arm remained in place and the telescope arm was moved further along the limb until the glasses were once again parallel, and then clamped. Now the index arm was freed and moved until the star (or Sun) and the Moon once again coincided. The reading was then twice the angular distance, and had to be halved. This procedure could be repeated several times, and the final reading had to be divided by the number of repetitions, thereby reducing the errors each time. Because the apparent movement of the Moon is relatively fast, the moment of each repetition was noted by a deck watch showing local time, and averaged after the final reading.

Mayer's circle was improved by the French mathematician Jean-Charles, Chevalier de Borda (1733–99), and described by him in *Description et usage du cercle de reflection*, published in 1787.[32] Borda extended the telescope arm across the frame, and fitted the horizon glass at its end. The telescope was also fixed at some distance from the index glass (for example cat. no. 298; see Plate 21). With the Borda circle, as the instrument became known, the image of the celestial body on the right was sighted directly and

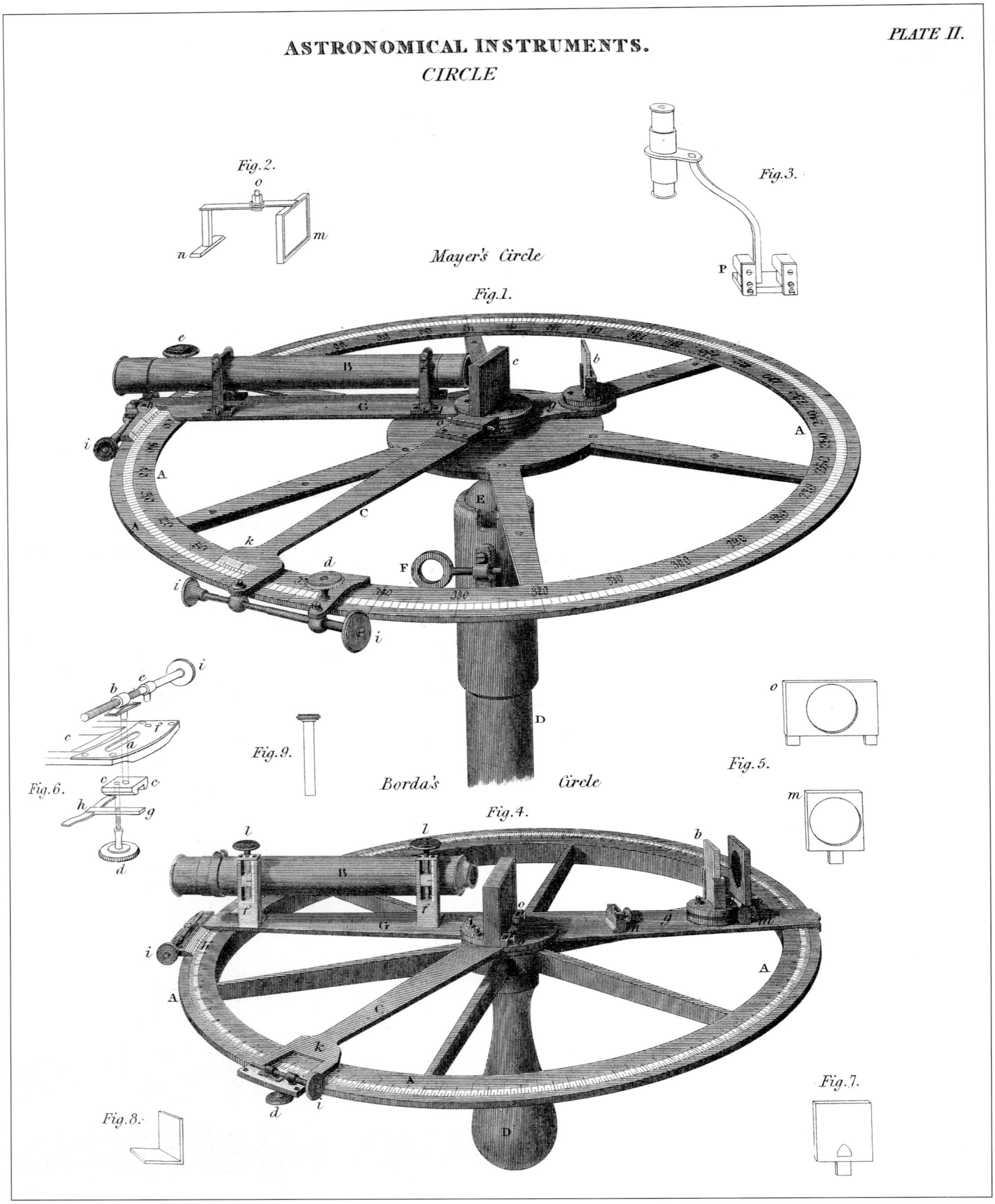

FIGURE 4.3 Tobias Mayer's reflecting circle (above) and that of Borda (below). On the latter the telescope arm has been extended across the frame and the horizon glass fitted at its end. From Abraham Rees, *The Cyclopaedia; or, Universal Dictionary of Arts, Sciences, and Literature*. Volume 1, London, 1820. National Maritime Museum.

the one on the left was seen by double reflection, and the two were brought into coincidence by moving the index arm. The circle was then turned upside down, kept in the same plane, and the observation was repeated in reverse order. The celestial body on the left was sighted directly and that on the right by double reflection. The difference in readings on the index arm was twice the distance and, as the observation was repeated more often, the final reading had to be divided by twice the number of double observations. Another way was to move the index and telescope arms alternately, which Borda called 'the crossed observation'. He eliminated the necessity to set the glasses parallel before each measurement, thereby saving time, and making his circle far more efficient then Mayer's.

Reflecting circles had one handle that was attached to the centre, which made them difficult to hold when the instrument had to be turned upside down to repeat the observation. This was solved by Joseph de Mendoza y Ríos, a Spanish astronomer exiled in London, in a modification which he described in 1801.[33] He proposed a detached U-shaped bracket that had to be mounted on the frame below its handle. The bracket, with a second handle at the other end, was fitted around the frame so that there were handles on both sides of the circle (for example cat. no. 302). A third handle could be screwed into the second one so that the instrument could be held with ease in almost any position.[34] According to Mendoza, Troughton was the first to adopt his U-bracket on his reflecting circles. Mendoza also introduced what he called the 'flying nonius', to simplify the reading of the circle. The circle's scale was divided into 360 degrees rather that the traditional 720 parts, and the arms carried no verniers. The nonius was stretched along the inside of the limb and could be made to slide round – hence 'flying' – by clamping it to either the index or the telescope arm. This flying nonius must have been difficult to construct and vulnerable, especially at sea, which probably is the reason why this modification by Mendoza was not broadly adopted.[35]

Often, the telescope arm of reflecting circles has an attached semicircular brass finder arc. This carries a scale division, usually by one degree, and two sliding brass stops. The arc serves to reposition the index arm relative to the line of sight through the telescope after each observation.[36] In 1796 Edward Troughton designed a reflecting circle with three rigidly connected index arms separated by 120 degrees, pivoted at the centre and moving around the limb (for example cat. no. 303). Each arm had a vernier. The telescope and glasses were rigidly fixed to the other side of the instrument's frame. A measurement between two celestial bodies gave three readings that had to be averaged to find the actual distance. Troughton's circle could not repeat; its main advantage was that it reduced errors of scale and eccentricity.

THE DIVIDING ENGINE

Probably the most difficult part of the production of any circular instrument for measuring angles was the graduated scale. A slight inaccuracy could lead to a considerable error in an observation. Graham and Sisson mastered the art of dividing, but it was John Bird who brought it to great perfection. Manually dividing, as these makers did, involved bisecting computed chords of a circle of known radius (that of the octant or sextant frame), and marking these off on the instrument's scale with a beam-compass.[37] The larger the radius of the octant, the easier it was to work on and the more accurate the division. At the request of the Commissioners of Longitude, Bird published his method in 1767 and was awarded £500.[38]

In 1766, Jesse Ramsden constructed his circular dividing engine, by means of which the time to divide an octant or sextant scale was considerably reduced.[39] The radius of octant and sextant frames could also be reduced, which made them easier to handle, and the accuracy of division was improved twelvefold.[40] Ramsden's second dividing engine was an improvement on his first and was operational by 1775. It consisted of a horizontal brass wheel pivoted on a wooden stand. The wheel's diameter was 40

FIGURE 4.4 Portrait of Jesse Ramsden with his successful dividing engine. Engraving by John Jones, London 1791, after an oil painting by R. Home. National Maritime Museum; PAG6438.

FIGURE 4.5 Jesse Ramsden's successful dividing engine. From Abraham Rees, *The Cyclopaedia; or, Universal Dictionary of Arts, Sciences, and Literature*. Volume 2, London, 1820. National Maritime Museum.

inches, and Ramsden cut 2160 brass notches in its circumference. He engaged an endless screw with a ratchet mechanism with the notches, and a treadle which turned the screw and moved the wheel round.[41] Ramsden chose the ratio between the pitch of the worm screw and the number of notches in the brass wheel so that one movement of the treadle turned the wheel through an angular distance of precisely 10 minutes of arc. To divide a scale, the octant or sextant's frame was fastened on top of the brass wheel with the pivots of the wheel and instrument coinciding. A cutter frame was mounted rigidly between the pivot and the wooden stand, at the edge of the wheel. With each turn of the screw, the wheel and octant or sextant moved 10 minutes of arc; the cutter was then lowered onto the scale and a 10-minute line of arc marked on the brass or silver scale of the instrument.[42]

At the request of the Board of Longitude, Bird examined a sextant divided on the engine and reported favourably, after which Ramsden was awarded £300 by the Commissioners. In addition the Board purchased the engine for £315, although it remained under the inventor's control. Ramsden was further required to publish his invention so that it could be shared with others. He did so in 1777 and was also obliged to divide octant and sextant scales for other instrument makers.[43] Edward Troughton's elder brother John soon built his own dividing engine after Ramsden's design, completing it in 1778.[44] Although Troughton did not have an obligation to divide scales for other makers, his dividing engine was used by some. Gabriel Wright mentions having his instruments divided on the engines of Ramsden and Troughton, as did John Hamilton Moore in 1785.[45] Ramsden had the centres of octant and sextant scales marked by a small stamped foul anchor, flanked by his initials I and R, to signify that the scale had been divided on his engine and as a guarantee of quality.[46] At the end of this chapter is a list of known examples of surviving octants that were divided on Ramsden's engine and on those of others. Ramsden's initiative of marking his scales also led others to do the same, and this became a tradition that continued until the middle of the nineteenth century. The marks include a foul anchor flanked by the initials E and P, probably for Edward Pritchard, an apprentice of Ramsden who inherited his master's small engine. Pritchard was apparently allowed to mark his own initials on octants by Ramsden and others (cat. nos. 057, 058, and 059).[47] The Admiralty, which used the foul anchor symbol, probably inspired Ramsden's choice of scale mark. Although the Board of Longitude owned the engine, the Admiralty had paid the bill. It was not the first time that an anchor was used on an octant scale; a foul anchor mark is found on an octant from the 1750s made by either Benjamin Cole or George Adams.[48] Besides John Troughton, other London instrument makers who soon built their own dividing engines were Peter Dollond (perhaps in the 1770s)[49] and John Stancliffe, a former foreman to Ramsden (in 1788).[50] These were followed by engines made by Edward Troughton in 1793, John Miller of Edinburgh, also in 1793,[51] W.T. Parsons,[52] Samuel Rhee (or Réhé) in 1803, James Allan in 1809,[53] Thomas Jones in 1811, Andrew Ross in 1831, and William Simms in 1843.[54] By about 1830 there were around twelve dividing engines in London.[55]

Around 1775, Ramsden sold his circular dividing engine of 1766 to Baron Bochart de Saron, a French aristocrat with an interest in science.[56] In 1794 the prominent Paris scientific instrument maker Etienne Lenoir (1744–1832) was entrusted with converting it to divide into 400 *grads*, the new post-Revolution, metric measure. Shortly before that, François-Antoine Jecker (1765–1834), who worked for Ramsden from 1786 to 1794, built a straight-line dividing engine, apparently a simplification of his master's design, but he admitted that he could not afford to make the more difficult parts.[57] Another circular dividing engine outside Britain was that by Gerard Hulst van Keulen (1733–1801) of Amsterdam, which he made after his own design, patented in 1792. Ten years before, Hulst van Keulen had his octant and sextant scales divided on an English engine. This is suggested by the inscription *divided cum privilegio* found on several examples, and one octant also has a foul anchor mark on the scale.[58] According to Jonathan Cuthbertson (1744–1806), an English octant and sextant maker who was active in Rotterdam, several instruments could be divided on an engine in the same time in which one was divided manually.[59] Despite the considerable investment of building a dividing engine (it took Troughton three years), and the increased demand for sextants due to the popularization of lunar distances, these instruments became a little cheaper.[60] In the early 1790s Cuthbertson charged 200 guilders for a manually divided sextant, while one by Hulst van Keulen that was divided on his engine cost 150 guilders. In 1792 George Adams junior was making use of Ramsden's dividing engine and charged 15 guineas for a brass sextant, about the same price as Hulst van Keulen.[61]

ARTIFICIAL HORIZONS

There are two basic types of artificial horizon: firstly, those mounted on altitude-measuring instruments and replacing the actual horizon, and secondly, detached examples that were placed on the ground and reflected the image of the celestial body being observed. The mounted type was meant for use at sea, while detached examples were more practical on land. The instrument-mounted inventions of Elton, Hadley, and Smith have been mentioned in the previous chapter, and their development will be

FIGURE 4.6 (*above, left*) A scale mark indicating that the scale was divided on Jesse Ramsden's dividing engine.

FIGURE 4.7 (*above, centre*) The scale mark of Edward Pritchard, an apprentice of Jesse Ramsden.

FIGURE 4.8 (*above, right*) A foul anchor scale mark found on an octant by Cole or Adams from the 1750s, in the collection of the Mariners' Museum, Newport News, Virginia; 2003.03.68.

dealt with further in that which follows. Around 1738, George Adams senior was possibly the first to invent a detached example, a container with a pool of liquid – mercury, for example – that provided a reflective surface.[62] Five years later, the mariner John Serson designed a whirling speculum, or horizontal top, that was accelerated by a spring-loaded bridge mechanism. It was tried out at sea in 1743 but reported to be impractical: Adams published a historical account and description of this instrument in 1786.[63] For use on land, in hydrography and surveying, the detached type using the reflected image of a celestial body prevailed. Towards the end of the eighteenth century this consisted of a rectangular wooden trough about 150 by 100 mm, and about 20 mm deep. Mercury was stored in a wooden or metal bottle and poured into the trough when an observation was taken. A roof-shaped cover with two glass sides at an angle of 90 degrees was placed over the trough to prevent the wind disturbing the surface of the mercury. Sometimes a piece of glass was floated on the mercury to provide a steady surface. The roof and trough were so placed that the sunlight entered through one glass side and was reflected through the other. Another much-used type was the artificial glass horizon. This consisted of a glass mirror (often circular) fitted in a brass case with three or four adjustable feet (see Plate 22). A spirit level was supplied to level the glass. The Board of Longitude provided examples of mercury trough and glass horizons for a voyage to the north-west coast of America in 1791: Troughton made the glass one.[64] Matthew Flinders took a wooden trough and five-and-a-half pounds of mercury in a bottle on his voyage to circumnavigate Australia in HMS *Investigator* in 1801.[65]

FIGURE 4.9 Portrait of John Adams of Edmonton at the age of 56. Adams was a teacher of mathematics and navigation, and is depicted with an artificial glass horizon, a sextant, and a telescope. Engraving by J.T. Smith, published by N. Smith, London, 1795. National Maritime Museum; PAD3019.

THE MARKET

Towards 1780 octants, sextants, quintants, and reflecting circles became firmly established for taking altitudes and lunar distances at sea. At the same time, backstaffs were still being manufactured, especially in the Dutch Republic and in North America, countries where the octant had been in use for three decades.[66] A substantial number of Americans continued to use backstaffs, even after 1767, when the *Nautical Almanac* with lunar tables was available. The popularity of backstaffs with Dutch seamen was not the result of conservatism, but of their price in relation to the octant. In the 1720s, a Swede by the name of Pieter Holm (1696–1776) settled in Amsterdam and ran a nautical school named 'Schip Recht door Zee' (freely translatable as 'ship on a steady course'). He taught navigation and sold a variety of instruments, including backstaffs and octants, the latter made by Jacobus Kley (1716–91) of Rotterdam. Holm kept a manuscript ledger of his business, which has been preserved. It lists his transactions with the names of the buyers, often mentioning their rank and sailing destination. Between 1748 and 1770 he sold 68 octants and 75 backstaffs. Each came in two price categories; the expensive models were more sophisticated, with brass parts and a brass or ivory scale[67] (cat. no. 034). Analysis of the ledger shows that officers in the higher income bracket – masters and chief mates – generally purchased an octant, and that they usually preferred one in the higher price category. Second and third mates more often bought a backstaff: for them, the price of an octant would have been a considerable financial sacrifice. But even so, the lower ranks too preferred the more expensive type of backstaff.

Preserved trade catalogues and price lists give an idea of how much a British seaman had to pay for an octant or a sextant. In 1766 George Adams senior sold octants for between two and four guineas. Two years later, Benjamin Cole junior's prices ranged between £1 14s (for an octant with a transversal scale) and 2 to 3½ guineas. Cole's all-brass-frame octant cost between 3½ and 6 guineas.[68] Henry Pyefinch, an optical and mathematical instrument maker in London, offered octants for between £1 10s and £3 3s in 1771.[69] His brass sextants came in three sizes: small (size not specified), 12-inch, and 18-inch, and cost between £2 12s 6d and 10 guineas. In 1780 George Adams junior charged between 3 and 4½ guineas for octants, the most expensive being of 'best construction'. For a sextant with a wooden frame, Adams charged only £6 16s 6d.[70] That year, Benjamin Martin was asking between £2 15s and £8 18s for an octant, about the same price range as Gabriel Wright in 1779.[71] In 1794 William and Samuel Jones of Holborn, London, were selling octants with wooden scales for £1 11s 6d; their ebony-frame octants with brass fittings cost £8 18s 6d apiece. A brass sextant by Jones 'for taking distances accurately, in order to determine the longitude at sea' cost £12 12s.[72] Possibly because of Ramsden's reputation, his sextants were a little more expensive; in 1787 he charged the Board of Longitude £13 7s for a brass example.[73] The lists show that ebony frames and ivory arcs were more expensive than mahogany frames and brass arcs. Extras such as a telescope, additional shades, and a fitted box to store the instrument also pushed up the price.

Towards the end of the eighteenth century, British octant and sextant makers dominated the market. The reasons were the increased application of the lunar method due to the availability of lunar tables; the dividing engine,which made it possible to increase production; and the inability of instrument makers on the continent to meet the demand for sextants. Ramsden's output before building his engine has been estimated at a hundred sextants: thereafter, he made almost a thousand.[74] Stimson's estimate, based on the serial numbers that Ramsden gave his instruments, attributes about 1450 sextants to him altogether, with an average annual output of about 38.[75] Between 1780, when already using his dividing engine, and his death in 1800, Ramsden's annual average output increased to about 43.[76] Stimson's research has also shown that Troughton made about 480 sextants and 50 reflecting circles.

The most significant production of octants and sextants outside England during the eighteenth century was in France and the Netherlands. Lemaire had ceased to work around 1760, and after him the most prominent manufacturer of this kind of instrument was Lenoir. Although there were several circular dividing engines in France at the end of the eighteenth century, the manufacturers could probably not meet the demand, and a lack of sextants in their own country drove French officers to use British examples.[77]

In the Netherlands, Kley died without a successor, and the two remaining firms in the country that made octants and sextants were those of Hulst van Keulen and Cuthbertson. Because of the political situation after the French Revolution, and Hulst van Keulen's age, his firm went into decline and eventually could not meet the local demand.[78] Although Kleman started his workshop in Amsterdam in 1781, his heyday was after 1814, when the period of French domination had ended and Dutch trade and shipping revived. After Kley's death, Cuthbertson was the only Rotterdam scientific instrument maker of significance and was therefore busy building other apparatus, and no octant or sextant by him is recorded.[79]

The most prominent octant and sextant manufacturers serving the English market in the 1790s were Ramsden, Troughton, Dollond, and Nairne and Blunt. Ramsden's sextants remained in use for years. When Arthur Philip sailed in HMS *Sirius*, taking the First Fleet to Botany Bay in 1787, the Board of Longitude gave him 'an old 14 or 15 inch sextant' by Ramsden. It was returned, and fourteen years later the same instrument was issued to Flinders in HMS *Investigator*.[80]

Judging by the number of surviving octants, the firm of Spencer Browning and Rust in London appears to have been successful. William Spencer and Samuel Browning united their businesses in 1778. Ebenezer Rust joined them six years later, and the firm continued to manufacture octants for approximately the next sixty years.[81] Examples found in museums outside the British

Isles and associated with foreign seamen are evidence of the firm's international orientation.[82] Two octants have *à Londres* added to the makers' name, so were obviously made for retail on the French market[83] (cat. no. 068; see Plate 12). An unknown Hamburg sea captain had his portrait painted in Cadiz, south-west Spain, in 1796, picturing him with his octant, signed Spencer Browning and Rust on the ivory plate in the cross bar (Plate 11). An octant by them was also used on Swedish ships (cat. no. 157). A scale mark with the initials SBR is often found on their octants and also on those by associated firms such as Spencer Browning and Co, Spencer Barrett and Co, and Spencer and Co. They delivered divided instruments to others, as the mark is also found on octants with the names of manufacturers and retailers in England, Scotland, and the United States.[84]

FIGURE 4.11 The scale mark of Spencer Browning and Rust.

FIGURE 4.10 Signature on an octant by Spencer Browning and Rust for the French market (cat. no. 068).

TABLE 4.1 SCALE MARKS FOUND ON OCTANTS, SEXTANTS, AND QUINTANTS TO DECEMBER 2007

	Octant/sextant/quintant		Name found	Date	Collection; inv. no. /cat. no.
	**Jesse Ramsden**				
	O	I↓R	Josua Lover Martin, London	c.1774	NMM NAV1258/052
	S	I↓R	Jesse Ramsden, London	c.1780	NMM NAV1179/137
	S	I↓R	Nairne and Blunt, London	c.1780	NMM NAV1110/136
	O	I↓R	Nairne and Blunt, London	c.1780	MMNN 98.39.23
	O	I↓R	not signed, English	c.1780	MMNN NQ29
	O	I↓R	George Adams, London	c.1780	MMH (Stimson)[85]
	O	I↓R	not signed, English	c.1785	MSM 1958.1122
	S	I↓R	George Adams, London	c.1790	HNC 79-B-3
	O	I↓R	not signed, English	c.1790	MSM 1941.244
	O	I↓R	not signed, English	c.1790	MSM 1972.325
	O	I↓R	John Hamilton Moore, London	c.1790	Stimson
	O	I↓R	Watkins, Charing Cross, London	c.1800	Private, Belgium (Stimson)
	O	I↓R	Watkins, Charing Cross, London	c.1800	ScML 1918–102
	O	I↓R	J. and L. Hardy, London	c.1800	Stimson
	O	I↓R	William Cary, London	c.1800	Private, New York (Stimson)
	O	I↓?	not signed, English	c.1800	MMNN NQ17
	O	I↓R	not signed, English	c.1800	NMS T.2004.512 (ROE)
	O	I↓R	Matthew Berge, London	c.1813	Tain HTRA301.1[86]
	O	I↓R	not signed, English	c.1820	MSM 1946.1335
	O	I↓R	not signed, English	c.1820	MSM 1947.1688
	O	I↓R	George Stebbing, Portsmouth	c.1840	NMS T.2004.529 (ROE)
	Q	I↓R	Nathaniel Worthington, London	c.1840	NMM NAV1214/277

Octant/sextant/quintant		Name found	Date	Collection; inv. no. /cat. no.
Probably Edward Pritchard				
O	E↓P	Jesse Ramsden, London	c.1780	NMM NAV1353/057
O	E↓P	A. Chevalier, Guernsey	c.1780	NMM NAV1320/058[87]
O	E↓P	not signed, English	c.1800	MSM 1940.155
O	E↓P	not signed, English	c.1800	NMM NAV1352/075
O	E↓P	Dring and Fage, London	c.1800	PEM M3477
O	E↓P	George Richardson,London	c.1815	Private (Stimson)
O	E↓P	Benjamin Wood, Liverpool	c.1820	PEM M11831
O	?↓P	not signed, English	c.1820	NSA S.3428
O	E↓?	William Cary, London	c.1825	NMM NAV1267/089
O	E↓P	Thomas Jones, London	c.1840	Stimson
Probably James Allan junior				
O	J↓A	Richard Patten, New York	c.1825	MSM 1968.167
O	J↓A	Joseph Fairey, London	c.1825	QMM 07:05
O	J↓A	Robert Brettell Bate, London	c.1830	PEM M2780
Anchor, no initials				
O	↓	not signed, English	c.1780	MSM 1955.1193
O	↓	William Heather, London	c.1800	PEM M27198
O	↓	Robert Banks, London	c.1800	NMM NAV1343/070
O	↓	not signed, English	c.1800	MMNN NQ24
O	↓	J. Collwell, Liverpool (Holliwell?)	?	Stimson
O	↓	Richard Barry, London	c.1805	Stimson
O	↓	William Cook, London	c.1810	NMM NAV1288/078
O	↓	William Harris, London	c.1810	NMM NAV1301/079
O	↓	not signed, English	c.1810	NMM NAV1333/083
O	↓	M. Wardell and Sons	c.1820	Stimson
O	↓	Isaac/George Bradford, London	c.1820	NMM NAV1310/087[88]
O	↓	William Cary, London	c.1825	Stimson
O	↓	Benjamin Brown, Bristol	c.1831	Stimson
O	↓	G.W. Peters and Zn, Rotterdam	c.1840	NSA S.0228
Anchor, with initials				
O	G↓N	not signed	?	MMB (Stimson)
O	I↓D	E.G.W. Blunt, New York	c.1815	Stimson
O	R↓A	Richard Patten, New York	c.1825?	Stimson
O	R↓G	Dring and Fage, London	c.1830	Stimson
Foul anchor, no initials				
O	↓	G. Hulst van Keulen, Amsterdam	c.1786	NSA 1998.2017
O	↓	not signed, English	c.1790	PEM M10658
O	↓	William Heather, London	c.1795	MSM 1935.106
S	↓	not signed, English	c.1795	MSM 1951.3366
S	↓	not signed, English	c.1797	PEM M3343
O	↓	illegible, English	c.1800	PEM M23595
O	↓	Alexander Wellington, London	c.1800	MSS 706[89]
O	↓	Wilkinson and Sons, London	c.1800	Private, Canada (Stimson)
O	↓	Gilbert and Co., London	c.1800	MMNN NQ49
O	↓	Gilbert and Co., London	c.1800	YPM YPM60001
O	↓	Thomas Ripley and Son, London	c.1800	NMM NAV1285/071
O	↓	Thomas Ripley and Son, London	c.1800	MSN 4999
O	↓	not signed, English	c.1800	PEM M1675
O	↓	not signed, English	c.1800	PEM M8859
O	↓	not signed, English	c.1800	NMM NAV1272/076
O	↓	William Holliwell, Liverpool	c.1810	NBWM 00.1.2
O	↓	J and L. Hardy, London	c.1810	PEM M2359
O	↓	Gilkerson and Co., London	c.1815	CHSI 5301
O	↓	Richard Barry, London	c.1815	MSM 1939.933
O	↓	Thomas Jones, Liverpool	c.1818	PEM M8732
O	↓	William Holliwell, Liverpool	c.1820	MSM 1968.180
O	↓	Isaac/George Bradford, London	c.1820	MBL V10411

Octant/sextant/quintant		Name found	Date	Collection; inv. no. /cat. no.
O	↓	Alexander Cameron, Dundee	*c.*1820	NMS T.1980.183
O	↓	not signed, English	*c.*1820	CHSI 5302
O	↓	not signed, English	*c.*1830	PEM M3276
O	↓	Benjamin Brown, Bristol	*c.*1831	MSM 1932.56
S	↓	William Parnell, London	*c.*1835	NMM NAV1151/175
O	↓	not signed, English	*c.*1840	NSA S.3344 (01)
O	↓	Arbon and Krap, Rotterdam	*c.*1850	NSA A.3950(32)
O	↓	not signed, English	*c.*1850	MSM 1968.107
O	↓	Thomas Jones, Liverpool	*c.*1850	PEM M3003
O	↓	Henry Hughes, London	*c.*1850	PEM M775
 **Foul anchor, various initials, others marks**				
O	G↓N	not signed, English	*c.*1785	MSM 1951.393
O	G↓N	not signed, English	*c.*1790	PEM M2536
O	R↓G	Dring and Fage, London	1802	NMS T.1981.52
O	[fleur-de-lis]	Thomas Ripley and Son, London	*c.*1800	ScML 1939-38[90]
O	?↓?	Benjamin Leja, Altona	*c.*1820	MHG[91]
O	IID	Benjamin Messer, London	*c.*1820	Trade (Stimson)
O	M£	J.W. Norie and Co., London	*c.*1830	PEM M723
O	B&J	Blatchford and Imray, London	*c.*1840	Trade (Stimson)
O	W.H	not signed, English	*c.*1840	NMS T.1980.182
O	SP	not signed, English	*c.*1850	NSA S.3428
FW **Initials FW only**				
O	FW	Benjamin Wood, Liverpool	*c.*1830	Private, London (Stimson)
O	FW	J.W. Norie and Co., London	*c.*1830	PEM M3735
O	FW	Somalvico and Co., London	*c.*1830	Museu Marítimo, Macao
O	FW	Henry Hughes, London	*c.*1835	Stimson
S	FW	James Grant, Hull	*c.*1845	MHG
O	FW	George Christian, Liverpool	*c.*1850	Private, Kent (Stimson)
O	FW	Brokovski, Liverpool	*c.*1840	MMH
S	FW	not signed, English	*c.*1850	NMM NAV1160/188

NOTES

1. Strahan *et al.*, *Description and Use of a New Astronomical Instrument*, 6, 11, 12.
2. Anita McConnell, 'Bird, John (1709–1776)'. In: *Oxford Dictionary of National Biography*. <http://www.oxforddnb.com/>
3. Letter from Dr Bradley to John Cleveland, secretary of the Admiralty, Greenwich, 14 April 1760, in: Mayer (ed.), *Tabulae motuum solis et lunae novae et correctae*, cxi-cxv; see also Charnock, *Biographica Navalis*, 34–7. Campbell took a mean of thirty observations on six different days and nights.
4. According to Cotter, *A History of the Navigator's Sextant*, 147, the brass circle by Bird had a diameter of 16 inches. See also Forbes, *The Birth of Navigational Science*, 5; Stimson, 'Board of Longitude Instruments', 95–6.
5. Moskowitz, 'The World's First Sextants'. A brass sextant by Joseph Jackson (d. 1760) is recorded, see Warner, 'John Bird and the Origin of the sextant', 1–2, 7, and Clifton, *Directory*, 149. Jackson was mentioned as among the first manufacturers of Hadley's octants.
6. Some sextants have wooden frames, in the style of an octant, for example those by Edward Nairne, George Adams, and John and Peter Dollond. Others made octants with a brass frame, such as Benjamin Cole and John Urings, and Jean Baptiste Magnié in Dunkirk.
7. Maskelyne, 'Remarks on the Hadley's Quadrant', 52–3. A sight-tube could also be used for this check; see Cotter, *A History of the Navigator's Sextant*, 144.
8. A small brass stud was fitted to every other index shade frame, and sometimes also on horizon shades, to facilitate selecting the appropriate shade.
9. Chaldecott, 'Platinum and Palladium', 93–4; platinum is harder than silver, cheaper than gold and does not require polishing.
10. Maskelyne, 'Remarks on the Hadley's Quadrant'. In slightly different wording, it also appeared as 'Remarks on the Hadley's Quadrant, tending principally to remove the Difficulties which have hitherto attended the Use of the Back-observation, and to obviate the Errors that might arise from a Want of Parallelism in the two Surfaces of the Index-Glass,'. *Philosophical Transactions* vol. 62 (1772) 99–122.
11. Dollond had described this device to Maskelyne in a letter read before the Royal Society on 29 March 1772, *Philosophical Transactions* vol. 62 (1772) 95–8.
12. The lower part of the glass is the half nearest to the plane of the frame; the upper part is the other half. A Maskelyne flap is found on sextants by Bird, Ramsden, Troughton, and by Nairne and Blunt in the NMM and on a sextant by Dollond in the CHSI, inv. no. 0063.
13. *Philosophical Transactions*, vol. 54 (1764) 329–32.
14. Howse, 'The Principal Scientific Instruments Taken on Captain Cook's Voyages of Exploration', 120–3.
15. Stimson, 'Board of Longitude instruments', 111–14.
16. Anita McConnell, 'Troughton, Edward (1753–1835)'. In: *Oxford Dictionary of National Biography* <http://www.oxforddnb.com>.
17. Ifland, *Taking the Stars*, 42–3.
18. Bennett, *The Divided Circle*, 180; NN, *Heath & Co., Ltd. Patentees & Manufacturers of Marine Instruments*, 7.
19. Besides Troughton, there were Troughton and Simms, Thomas Jones, Parkinson and Frodsham, Edward Smith, Worthington and Allan, and E. and E. Emanuel: these frames came with 12, 18, 20, or 24 pillars. Lecky, *'Wrinkles'*, 45, in 1891, still recommended pillar frames over those cast in one piece.
20. [Mörzer Bruyns], 'Instrumenten', 18–19, records three preserved examples by Kleman.
21. Ifland, *Taking the Stars*, 29.
22. With thanks to George Huxtable, communicated by him on 'rete', the Internet discussion group on scientific instruments, on 18 November 2006.
23. Wright, *The description and use of a new constructed quadrant, sextant and quintant*, 27–8.
24. In the ScML, inv. no. 1923–475; it is fitted with Wright's patent index glass adjustment, discussed in the previous chapter.
25. Magellan designed a 'double sextant', with a scale up to 180 degrees, that was made by Ramsden and was tried out during an Arctic expedition in 1773; see Magellan, *Description des octants et sextants Anglois*, 111–16, and Betts, 'John Hyacinth de Magellan (1722–1790) Part 1', 516.
26. Ritchie, *The Admiralty Chart*, 24–31.
27. NN, 'The Surveys of Graeme Spence'.
28. Ritchie, *The Admiralty Chart*, 26–7. A station pointer consisted of a graduated circle, two movable arms, and one fixed arm and was used in surveying to establish a position in a chart; see Lecky, *'Wrinkles'*, 97–112.
29. Cotter, *A History of the Navigator's Sextant*, 145–9; Forbes, *The Birth of Navigational Science*, 4–5.
30. As with a sextant when the index arm is at zero.
31. The semi-diameter of the Sun and Moon had to be taken into account.
32. Bennett, *The Divided Circle*, 135.
33. Mendoza Rios, *On an Improved Reflecting Circle*, 8–9; Eduardo L. Ortiz, 'Mendoza y Ríos, Joseph de (1761–1816) *Astronomer*'. In: *Oxford Dictionary of National Biography* <http://www.oxforddnb.com/>.
34. This U-shaped handle, albeit without the third handle, was found on one sextant by William Parnell (worked 1811–39), NZNMM, inv. no. 1992.204.7. Clifton, *Directory*, 209–10.
35. Mendoza Rios, *On an Improved Reflecting Circle*, 8–11; Daumas, *Scientific Instruments*, 185–6.
36. Ifland, *Taking the Stars*, 32.
37. Brooks, 'The Circular Dividing Engine', 108.
38. Bird, *The Method of dividing Astronomical Instruments*; Chapman, *Dividing the Circle*, 71–6; Moskowitz, 'The World's First Sextants', 26.
39. Allan Chapman, 'Ramsden, Jesse (1735–1800)'. In: *Oxford Dictionary of National Biography* <http://www.oxforddnb.com/>.
40. Chapman, *Dividing the Circle*, 135.
41. Brooks, 'The Circular Dividing Engine', 102; Chapman, *Dividing the Circle*, 130–3; McConnell, *Jesse Ramsden*, 41–8.
42. Turning the screw less than one rotation allowed a finer division. A vernier was divided in a similar way to a scale.
43. Ramsden, *Description of an engine for dividing mathematical instruments*.
44. Brooks, 'The Circular Dividing Engine', 118.
45. Wright, *The Description and use of a new Constructed octant, sextant and quintant*, 28. Troughton's engine is preserved in the ScML; Moore, *The Seaman's Complete Daily Assistant*.
46. Stimson, 'Board of Longitude Instruments', 111–14.
47. McConnell, *Jesse Ramsden*, 46. A son of Pritchard later sold the engine to Dollond, who passed it on to Parsons; it is now in the ScML inv. no. 1930–366.
48. Preserved in the MMNN, 2003.03.68.
49. Chapman, *Dividing the Circle*, 133–4.
50. Warner, 'American Octants and Sextants', 110; Bennett, *The Divided Circle*, 156.
51. Clarke *et al.*, *Brass & Glass*, 30.
52. In 1819 W. Parsons was paid 15 shillings to divide a circle to a ½ degree; ROG BOL papers 14/19[150]. Parsons may have been using the small engine made by Ramsden and which he had bequeathed to Edward Pritchard and later sold to Parsons; see McConnell, *Jesse Ramsden*, 46.
53. In 1820 Allan was paid £100 for an improvement to the dividing engine; ROG BOL papers 14/19[160]. As he died in 1816, the prize must have gone to his son and heir James; see McConnell, *Jesse Ramsden*, 260–3.
54. McConnell, *Jesse Ramsden*, 260–70; Brooks, 'The Circular Dividing Engine', 119–20, 127.
55. Stimson, 'Board of Longitude instruments', 101.
56. Daumas, *Scientific Instruments*, 183–4, 203–4. This engine is preserved in the Conservatoire des Arts et Metiers in Paris, see Stimson, 'Board of Longitude Instruments', 104.
57. McConnell, 'Ramsden's Workshop', 44.
58. Mörzer Bruyns, *Schip Recht door Zee*, 123, 216.
59. Cuthbertson, *Verhandeling over het algemeen in gebruyk Octant van Hadley*, 20. Cuthbertson himself had no engine.
60. Warner, 'John Bird and the Origin of the Sextant', 8; Mörzer Bruyns, *Schip Recht door Zee*, 134–5.
61. Mörzer Bruyns, *Schip Recht door Zee*, 135.
62. Ifland, *Taking the Stars*, 94.
63. In *Transactions of the Society for the Encouragement of Arts, Manufactures and*

Commerce 4 (1786) 204–13; Taylor, *The Mathematical Practitioners*, 189, 303, 190; Ifland, *Taking the Stars*, 96–7. A Serson whirling horizon is preserved in the ScML, inv. no. 1927–1853, see Morton and Wess, *Public & Private Science*, 340–1.

64. ROG BOL papers 14/13[155]. In 1794 William and Samuel Jones were offering such artificial horizons for sale for £1 16s; ScML Library, class mark O. B. JON; see note 72 below.
65. ROG BOL papers 14/13[171].
66. In 1777 they were still for sale in Amsterdam, see Mörzer Bruyns, 'The Importance of the Van Keulens to Navigation', 60–1; and in America in the 1780s by Clark Elliot, James Ham, and Benjamin King; see Warner, 'Davis' Quadrants in America', 25, 30, 35, 38.
67. Mörzer Bruyns, *Schip Recht door Zee*, 97, 132–5, 147. Expensive octants had a brass index arm and brass or ivory scales; cheap ones had a wooden index arm and scale.
68. Warner, 'John Bird and the Origin of the Sextant', 8; Taylor and Richey, *The Geometrical Seaman*, 110.
69. Warner, 'American Octants and Sextants', 88.
70. Taylor and Richey, *The Geometrical Seaman*, 111.
71. Millburn, *Retailer of the Sciences*, 46–7. Martin's expensive example was on a wooden stand and could also be used on land; Wright, *The description and use of a new constructed octant, sextant and quintant*.
72. Jones's catalogue is bound in *Description and use of a new portable orrery* (London, 1794). ScML Library, class mark O.B.JON.
73. ROG BOL 14/6, p.109.
74. Warner, 'John Bird and the Origin of the Sextant', 7.
75. Stimson, 'The Influence of the Royal Observatory at Greenwich', 128–30.
76. McConnell, *Jesse Ramsden*, 46.
77. Daumas, *Scientific Instruments*, 203; Chapuis, *A la mer comme au ciel*, 57. It did not help that French instrument makers such as Lenoir and Jecker apparently preferred to build reflecting circles.
78. Mörzer Bruyns, *Schip Recht door Zee*, 105, 131, 239. One reflecting circle by Van Keulen is recorded; it is dated 1798.
79. English manufacturers in the Dutch market were Adams, Dollond, Nairne (and Nairne and Blunt), Troughton, and Spencer Browning and Rust; see Mörzer Bruyns, *Schip Recht door Zee*, 103.
80. ROG BOL papers 14/6, p.103, and 14/1328; see also Stimson, 'The Influence of the Royal Observatory at Greenwich', 128. For the voyage, Philip also received a 'new improved 10 inch' sextant by Ramsden.
81. Clifton, *Directory*, 261.
82. Mörzer Bruyns, 'The Navigating Instruments of the National Maritime Museum', 26.
83. Another example is in the Musée d'Histoire des Sciences in Geneva, inv. no. 1677.
84. The names are Alexander, Yarmouth; Jonathan Thompson, London; Charles Jones, Liverpool; Alexander Adie, Edinburgh; Joseph Fairey, London; David Heron, Glasgow; William Parnell, London, and Henry Duren, New York. For details, see Mörzer Bruyns, 'The Navigating Instruments of the National Maritime Museum', 26.
85. In this table, 'Stimson' refers to Stimson, 'Board of Longitude Instruments', 112–14.
86. At Tain Royal Academy (formerly Royal Tain Academy), Tain, Scotland, founded in 1813, where navigation was taught. It is assumed that this octant was presented to the Academy when it was founded.
87. There is an octant by A. Chevalier of about same date, but without a scale mark, in the MMH.
88. An octant signed Bradford from about the same date, without a scale mark, is in the JUM, inv. no. MUJ 4087.87/V.
89. Miniati, *Museo di Storia della Scienza*, 228, no. 25.
90. Other octants by Ripley have no scale mark; see cat. no. 069 and NMAH inv. no. PH 290534.
91. Benjamin Leja (1798–1870) was an optician in Altona, Hamburg, see 'Kurzmeldungen', in *Judische Gemeinde in Hamburg*, 3 (2007) 13.

CHAPTER FIVE

INNOVATIONS AND PRODUCTION IN THE NINETEENTH AND TWENTIETH CENTURIES

INTRODUCTION

At the beginning of the nineteenth century the octant, sextant, and quintant had reached a basic shape that remained static for many decades. By the 1820s most – if not all – new octants and sextants were divided on an engine, and their radius was reduced to about six or eight inches.[1] Variations of the sextant and quintant were developed in the first half of the century for hydrographical surveying, and in the second half improvements in navigation brought changes to the marine version. Both of these topics will be dealt with below.

Octant frames continued to be made of ebony with a brass index arm and brass fittings, but the technology of the instrument hardly changed until it became obsolete around 1900. Following the sextant, octants were more often fitted with horizon shades, and after the 1840s the back-observation facility is rarely found.[2] They were still for sale in London at the end of the century. In 1891 the firm of Joseph Levi and Co., which ran an optical and photographic warehouse in London, issued a trade catalogue including their navigational and surveying instruments. Besides brass sextants (and one with an ebony frame), they sold ebony octants in various price ranges.[3] Generally an octant came in a fitted oak box, usually of a keystone pattern, which often had a stepped lid, the higher section of which covered the glasses and sight vane. Cheap octants had a fixed sight vane with one or two pinholes: more expensive examples had a threaded telescope bracket for shipping a detached telescope or a sight-tube. The box also contained a pin or key for adjusting the glasses. Squire Lecky, author of the popular manual *'Wrinkles' in Practical Navigation*, wrote that a navigator who took pride in his work should possess both a first-class sextant and a good example of its hardier and more humble relative, the octant. This was reserved for everyday work in the open ocean, while the sextant was for more sophisticated observations.[4] Double ownership as general practice cannot be confirmed by donations from former seamen to the National Maritime Museum; most often, either an octant or a sextant was presented.[5] Although it is unlikely that the manufacture of traditional ebony-framed octants survived into the twentieth century, existing examples very probably remained in use at sea for a considerable time after 1900.[6]

During the nineteenth century, reflecting circles were mainly manufactured in Britain, France, Germany, and the United States. In Germany, the firm of Pistor and Martins in Berlin modified the reflecting circle in 1843 by replacing the horizon glass with a prism[7] (cat. no. 312). A later improvement was made to the circle by Captain Charles Hervey Townshend (1833–1904), who in 1888 patented his design for measuring the angles between three objects simultaneously.[8] Being a navigator, Townshend probably designed his circle for lunars, but a double observation involving two stars must have been difficult on board a ship. Stackpole and Brother in New York built Townshend's circles, of which several are known to survive, some with an aluminium frame.[9] In the 1890s, T. Cooke and Sons in York and A. Hurlimann in Paris brought out modifications of the reflecting circle, both with two separate optical systems.[10]

By the end of the century the lunar method, and with it the reflecting circle, fell into disuse.[11] The ship's chronometer had become more common, so that observing lunars became superfluous. Sometimes the lunar method was used for checking the rate of the chronometer during a voyage but it was easier to use time balls, which were installed in major ports in the second half of the century. From the turn of the twentieth century, radio time signals were transmitted, and after 1907 lunar tables no longer appeared in the *Nautical Almanac*.[12]

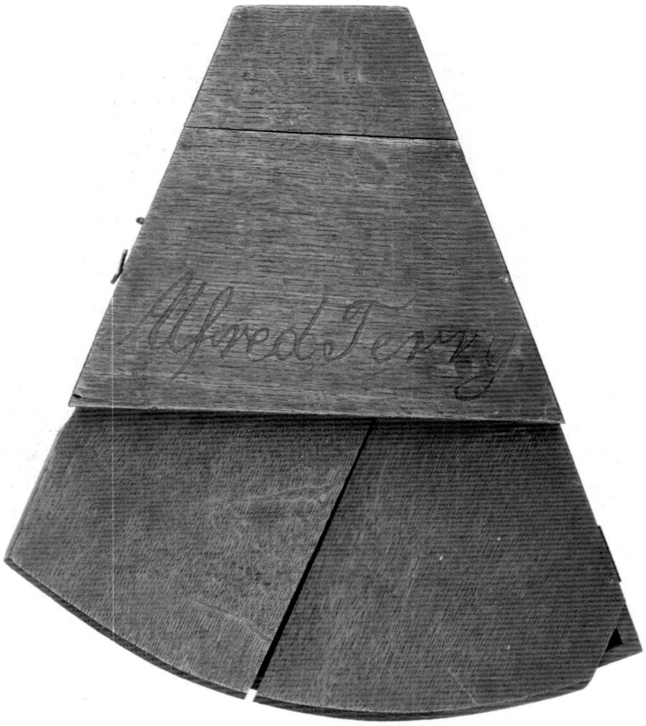

FIGURE 5.1 A stepped keystone oak octant box (cat. no. 096).

FIGURE 5.2 A typical inside of a sextant box of the 1830s (cat. no. 172).

Early in the nineteenth century, manufacturers turned to casting sextants and quintant frames in bronze, rather than assembling them from brass plate. Later, tin bronzes became available, which are easier to cast, allow fine and intricate castings to be made, and, unlike brass, are resistant to salt-water corrosion.[13] At first, like octants, sextants too were kept in keystone boxes but, as manufacturers added more detached parts, that shape became impractical. Consequently, a square box – which was also easier to store – often made of mahogany, rather than oak, became more common. The parts usually included a magnifying glass, one or more shaded eyepieces in various colours (or sometimes rotatable in the eyepiece) that could be screwed on to the ocular end of a telescope, a screwdriver, an adjusting pin and key, and a piece of chamois leather for cleaning the glasses. The box also contained a number of Galilean and Keplerian telescopes that could be fixed in the telescope bracket.[14] The standard issue was a low-powered Galilean telescope, which gave an erect image and was used for observing the Sun and stars. The magnification used to enlarge the Sun's image was between two and four times: the image of a star cannot be magnified. By the 1860s, this telescope often had a large objective lens. The wider field made it easier to locate a star and increased its illumination and that of the horizon: this telescope was called a 'star finder'. The second telescope was a high-powered Keplerian telescope that gave an inverted image and magnified nine or ten times. At its focus, it usually had two parallel cross-wires to check the instrument for collimation error. This Keplerian telescope was also employed with an artificial horizon on shore and came with an extra drawtube of a lower power (generally six times), with four cross-wires at its focus. This tube gave a larger image than the star finder and it was recommended for taking the meridian altitude, because the Sun's limb was clearly defined.[15]

Besides telescopes, almost all sextant and quintant boxes contained a sight-tube with no lens. Its main purpose was to keep the eye parallel to the instrument's plane when measuring shore angles, for example the angular height of a lighthouse, with which the distance in miles between the ship and the object was found from a distance table.[16] For this observation, twentieth-century sextants were sometimes equipped with a prismatic telescope, usually giving an erect image. A shade of frosted white glass was often attached to the magnifier at the end of a swivelling arm, to diffuse the light on the vernier. In the twentieth century expensive sextants were fitted with an electric battery-powered light that was attached at the end of the index arm to allow reading of the vernier in the dark: the battery was fixed in the instrument's handle.

However carefully a seaman treated his sextant, it could still be damaged. The mounts of the glasses, the shades, and the telescope bracket could be moved, and the instrument could be affected by the salty environment. Errors could not always be remedied at sea, and seamen were anyway strongly advised not to touch the adjusting screws.[17] A conscientious navigator returned his sextant to the manufacturer between voyages to have it examined, repaired, and cleaned. Sextant makers had the apparatus for this, and John Browning of London made such a testing device (cat. no. 342). Although manufacturers continued to offer this service, professional institutions provided independent quality control, including an examination of the scale, in the second half of the nineteenth century. This involved a special collimator, and the outcome was a series of corrections for intervals of fifteen degrees on the scale. The corrections that had to be applied to an

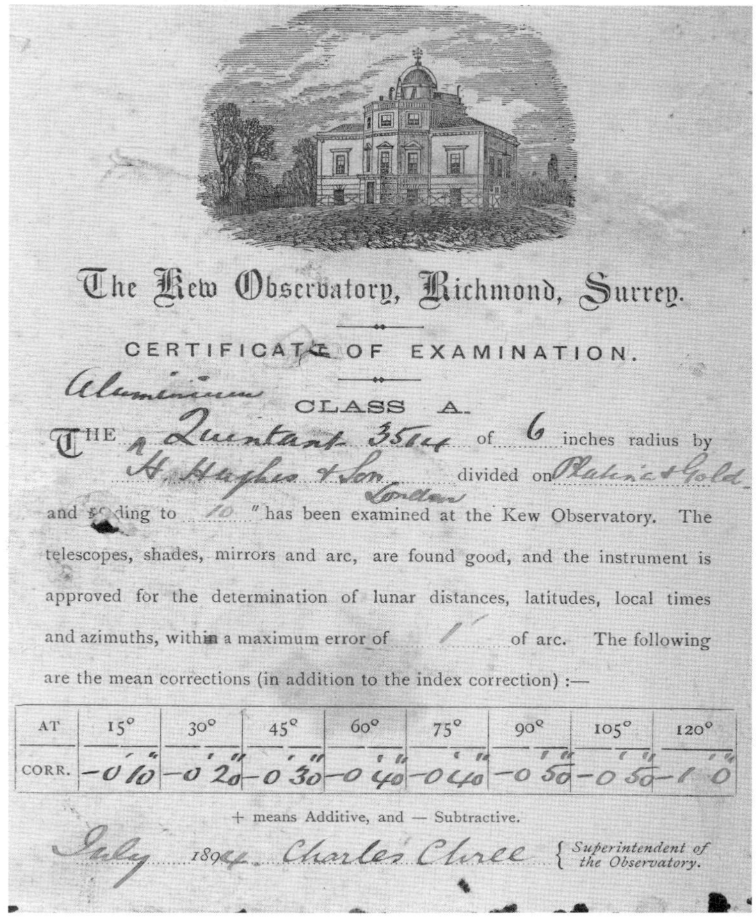

The Kew Observatory, Richmond, Surrey.

CERTIFICATE OF EXAMINATION.

Aluminium CLASS A.

THE Quintant 3564 of 6 inches radius by H. Hughes & Son London divided on Platina & Gold and reading to 10" has been examined at the Kew Observatory. The telescopes, shades, mirrors and arc, are found good, and the instrument is approved for the determination of lunar distances, latitudes, local times and azimuths, within a maximum error of 1' of arc. The following are the mean corrections (in addition to the index correction) :—

AT	15°	30°	45°	60°	75°	90°	105°	120°
CORR.	−0 10	−0 20	−0 30	−0 40	−0 40	−0 50	−0 50	−1 0

+ means Additive, and — Subtractive.

July 1894. Charles Chree {Superintendent of the Observatory.

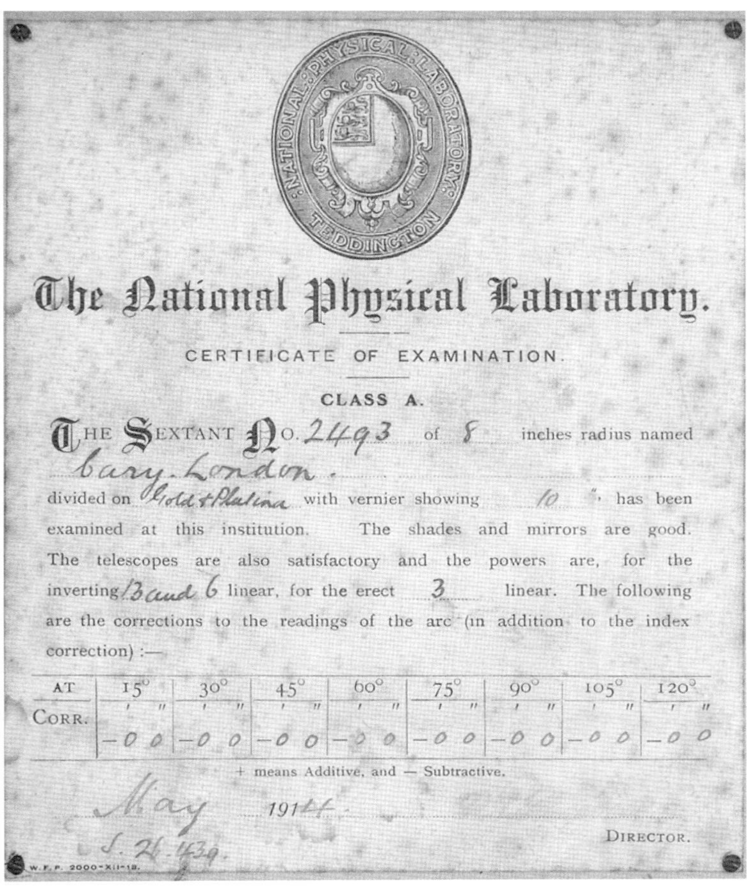

NATIONAL PHYSICAL LABORATORY TEDDINGTON

The National Physical Laboratory.

CERTIFICATE OF EXAMINATION.

CLASS A.

THE SEXTANT No. 2493 of 8 inches radius named Cary. London. divided on Gold & Platina with vernier showing 10" has been examined at this institution. The shades and mirrors are good. The telescopes are also satisfactory and the powers are, for the inverting 3 and 6 linear, for the erect 3 linear. The following are the corrections to the readings of the arc (in addition to the index correction) :—

AT	15°	30°	45°	60°	75°	90°	105°	120°
CORR.	−0 0	−0 0	−0 0	−0 0	−0 0	−0 0	−0 0	−0 0

+ means Additive, and — Subtractive.

May 1914. DIRECTOR.

5. 21. 439

W.F.P. 2000-XII-13.

FIGURE 5.3 (*above, left*) Collimator for sextant testing, with a sextant placed in position for examination. The collimator is not signed but was probably made in Germany. It was acquired by the Amsterdam branch of the Koninklijk Nederlands Meteorologisch Instituut in 1892, where sextant testing commenced in 1881. Courtesy of the Nederlands Scheepvaartmuseum, Amsterdam; RB.0748(25).

FIGURE 5.4 (*above, right*) Certificate of examination from Kew Observatory, Richmond, Surrey (cat. 291).

FIGURE 5.5 (*left*) Certificate of examination from the National Physical Laboratory (cat. no. 214).

observation were noted on a certificate of examination that also included comments on the parallelism of the shades and glasses, and on the quality of the telescopes. Kew Observatory at Richmond in Surrey started verification of meteorological and magnetic instruments in 1851.[18] Seven years later, the verification of surveying and optical instruments for expeditions was taken in hand, including a number of sextants. In the following decades sextant testing was a regular duty of the Observatory, which in 1895 became the government-funded National Physical Laboratory.[19] Every seaman could submit his sextant for testing for a fee of five shillings, not including the carriage of the instrument there and back.[20] Private testing continued too, by such specialists as J. Coombes, optician and scientific instrument manufacturer at 'The Observatory' in Fore Street, Devonport, who also provided

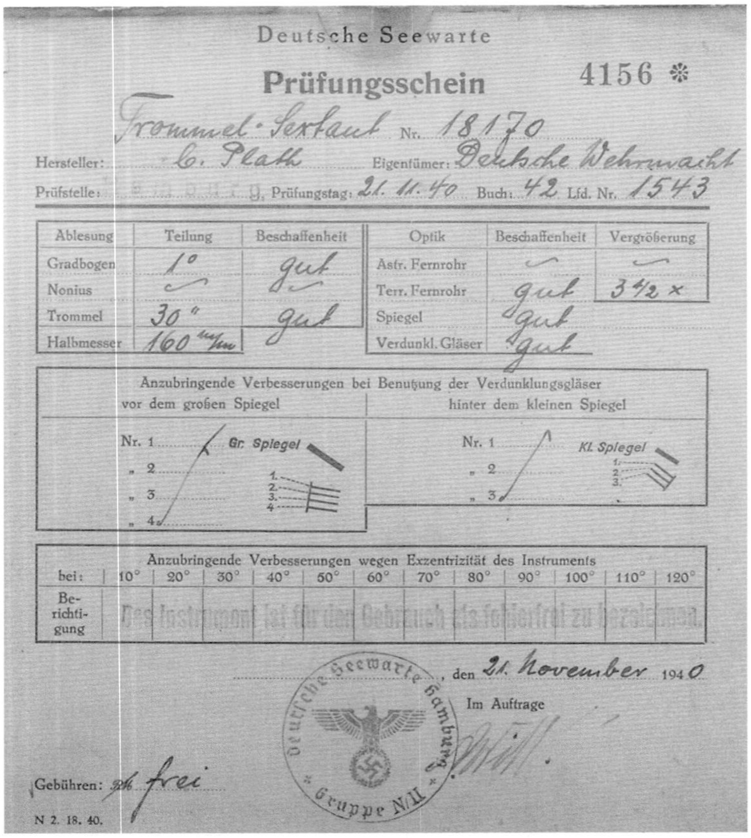

Deutsche Seewarte

Prüfungsschein 4156 *

Trommel-Sextant Nr. 18170

Hersteller: C. Plath Eigentümer: Deutsche Wehrmacht

Prüfstelle: Prüfungstag: 21. 11. 40 Buch: 42 Lfd. Nr. 1543

Ablesung	Teilung	Beschaffenheit	Optik	Beschaffenheit	Vergrößerung
Gradbogen	1°	gut	Astr. Fernrohr	–	–
Nonius	–		Terr. Fernrohr	gut	3½ x
Trommel	30"	gut	Spiegel	gut	
Halbmesser	160 mm		Verdunkl. Gläser	gut	

Anzubringende Verbesserungen bei Benutzung der Verdunklungsgläser

vor dem großen Spiegel — hinter dem kleinen Spiegel

Anzubringende Verbesserungen wegen Exzentrizität des Instruments

bei: 10° 20° 30° 40° 50° 60° 70° 80° 90° 100° 110° 120°

Berichtigung: Das Instrument ist für den Gebrauch als fehlerfrei zu bezeichnen.

..., den 21. November 1940

Im Auftrage

Gebühren: frei

N 2. 18. 40.

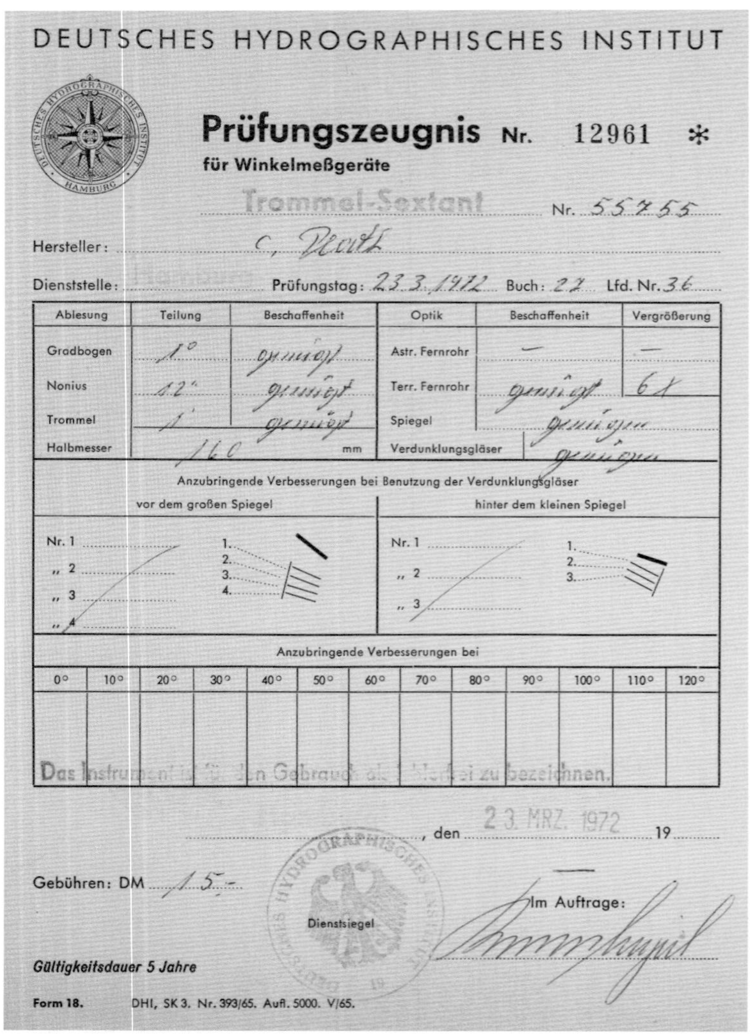

DEUTSCHES HYDROGRAPHISCHES INSTITUT

Prüfungszeugnis Nr. 12961 *

für Winkelmeßgeräte

Trommel-Sextant Nr. 55755

Hersteller: C. Plath

Dienststelle: Hamburg Prüfungstag: 23.3.1972 Buch: 22 Lfd. Nr. 36

Ablesung	Teilung	Beschaffenheit	Optik	Beschaffenheit	Vergrößerung
Gradbogen	1°	genügt	Astr. Fernrohr	—	—
Nonius	12"	genügt	Terr. Fernrohr	genügt	6 x
Trommel	1'	genügt	Spiegel	genügen	
Halbmesser	160 mm		Verdunklungsgläser	genügen	

Anzubringende Verbesserungen bei Benutzung der Verdunklungsgläser

vor dem großen Spiegel — hinter dem kleinen Spiegel

Anzubringende Verbesserungen bei

0° 10° 20° 30° 40° 50° 60° 70° 80° 90° 100° 110° 120°

Das Instrument ist für den Gebrauch als fehlerfrei zu bezeichnen.

..., den 23 MRZ. 1972

Gebühren: DM 15.–

Im Auftrage:

Dienstsiegel

Gültigkeitsdauer 5 Jahre

Form 18. DHI, SK 3. Nr. 393/65. Aufl. 5000. V/65.

certificates of examination, albeit only for index error[21] (cat. no. 207). The Deutsche Seewarte (German Sea-watch) was founded in Hamburg, Germany, in 1874 and started testing octants and sextants the following year.[22] It became part of the Imperial German Navy, but during World War I could not handle all the new sextants being made. This prevented Carl Plath of Hamburg, the main supplier of German sextants, from delivering his instruments in time. So he was given permission by the German naval authorities to test them on his own apparatus.[23] In 1945, the Seewarte was renamed the Deutsches Hydrographisches Institut but continued sextant testing. Branches of the Koninklijk Nederlands Meteorologisch Instituut (Royal Netherlands Meteorological Institute) that were established in Amsterdam and Rotterdam in 1881 and 1889, respectively, were responsible for sextant testing in the Netherlands.[24] As with Kew, the German and Dutch institutions welcomed foreign seamen bringing their sextants for examination.

SOURCES OF INNOVATIONS IN THE SEXTANT

Hydrographic work, and to a somewhat lesser extent expeditions such as those to the polar regions, encouraged innovation. After the Napoleonic Wars ended in 1815, unemployed Royal Navy officers were obliged to seek work elsewhere, and even those not relegated to half pay needed constructive new goals. Polar exploration, especially the combined search for the magnetic pole and a North-West Passage became important. The Admiralty deployed other officers in hydrographic work, and the Navy's systematic approach to the charting of the world's seas, which started in the 1830s, promoted innovations in both hydrography and surveying. In the second half of the century, the growth of steam propulsion brought about significant changes in navigation and navigational instruments. The number of steamships increased, they went faster than sailing ships and at a constant speed, and they did not depend on the force or direction of the wind. Ideally, steamers headed for their destinations on a direct course, leading to regular schedules and heavily frequented shipping routes. The increased risk of collisions brought home the need for accurate calculation of a ship's position. This led to new techniques and methods such as those developed by Sumner, Magnac, Villarceau, and St. Hilaire described in the first chapter, and these required accurate measurements. By the end of the nineteenth century, position-finding at sea with a sextant, a chronometer, a nautical almanac and tables, and a sea chart had reached a standard that

FIGURE 5.6 (*left, above*) Certificate from examination of the Deutsche Seewarte, Hamburg (cat. no. 244).

FIGURE 5.7 (*left, below*) Certificate from the Deutsches Hydrographisches Institut, Hamburg (cat. no. 271).

remained broadly unchanged until the 1970s. There were of course significant developments, such as radio direction finding, radar, and hyperbolic navigation systems such as Decca and Loran, but none of these drove the sextant from the ship's bridge. Seamen relied on the old handwork, astronomical observation, a calculation, and a construction on the chart until the advent of satellite navigation.

Developments in aviation in the twentieth century also led to new requirements and for this branch of navigation the marine sextant underwent a number of significant modifications. Adaptations were necessary owing to the increased speed of aeroplanes and their flying altitude. The traditional navigator's sextant developed a series of offshoots, often complicated and ingeniously designed instruments, which are discussed below.

HYDROGRAPHY, SURVEYING, AND POLAR TRAVEL

The double sextant that Graeme Spence invented around 1784 was mentioned in the previous chapter. The increase in hydrographical activity in the nineteenth century encouraged further development of such instruments. Probably the next was the 'Improved Sextant' by Ebenezer Hoppe of London, in around 1804 (cat. no. 160). It had two scales and could, theoretically, be used to measure an angle of up to 360 degrees by moving the index glass.[25] William Fitzwilliam Owen (1774–1857) was a prominent hydrographer who designed a double-reflecting circle for use in surveying, which was manufactured in 1821 by order of the Admiralty.[26] David Rowland designed a new version of a double sextant, by means of which one observer could measure two angles between three points at once (patent no. 6528 of 1833).[27] In principle, this was probably similar to the double sounding sextant (actually a quintant) by Frederick William Beechey of about 1847, of which some examples are preserved (cat. nos. 185 and 279).[28]

The sounding sextant (or quintant) probably came into use in the 1830s, and was so named as it was used in combination with depth sounding. In order to record the water depth, a ship's boat was rowed along a track near a coastline. The depth was sounded at regular intervals with a lead and line and, at the same instant, the hydrographer, with a sounding sextant and also in the boat, measured the horizontal angle between two points on the shore whose positions were accurately known. When the entire track had been covered, the positions of the soundings were plotted on the chart by use of the measured angles and a station pointer. A sounding sextant differed from the mariner's version in several ways. Because it was not used for measuring the altitude of a celestial body, it lacked shades. The glasses were also larger than on a regular sextant, to make it easier to locate the designated points on the shore (for example cat. nos. 294 and 297, see plate 20).[29] Sometimes the handle was hinged or detached and could be screwed into the back of the frame at a right angle to the instrument's plane, to make it easier to handle during the observation. The geographical position of points on shore was established by celestial observations with a sextant or quintant and an artificial horizon. The latter was of the mercurial, black-glass type discussed in the previous chapter, with which the angle between a celestial body and its reflected image (that is, twice the altitude) was measured.[30] As the natural horizon was not used, error caused by dip and terrestrial refraction was avoided and any other error in the measurement was halved. A disadvantage was that the angle could exceed the range of a sextant and even a quintant. Artificial reflecting horizons were therefore often used in combination with a surveyor's sextant or a box sextant, the range of which sometimes exceeded that of the marine version. The double box sextant designed by Captain Christopher George RN (1809–87) in about 1845, for example, measures up to 156 degrees (cat. no. 182). Box sextants (also called pocket-sextants or snuffbox sextants) were first made in the last decades of the eighteenth century, and often came in a leather carrying case with a strap. Sir William Gell (1777–1836), a classical archaeologist, used a box sextant while travelling in Greece, Asia Minor, and Italy. Box and pocket sextants were too compact to be practical on a ship, but ideal on polar expeditions; for example, one was used on HMS *Assistance* by Sir Allen Young in the Arctic in the 1850s (cat. no. 189). This type was still for sale at the beginning of the twentieth century, and in 1971 Francis Barker and Son of Kent revived the box sextant, aiming it at sailors of yachts and small craft (cat. no. 274).

FIGURE 5.8 Inscription on the lid of a box sextant that belonged to Sir William Gell (1777–1836) who used this instrument in Asia Minor (cat. no. 139).

A very basic and robust instrument for measuring horizontal angles between points on shore was the 'Paget Angle Sextant' (for example cat. no. 220), which was designed around 1892 and brought onto the market by Henry Hughes and Son in London.[31] It had no shades or telescope, and no tangent or clamping screw, and the index arm was moved by hand. The scale was in black digits, clearly marked on a white ivorine overlay, and read to 30 minutes with the naked eye by use of a single indicating line at the end of the index arm. The Paget Sextant was light and compact – but not as compact as a box sextant – and it had a wooden handle that was screwed into the frame at a right angle, so that it was easy to hold both vertically and horizontally. It was used in combination with a station pointer to obtain a fix. Around 1875, a combination of a sextant and a station pointer was invented by the Austro-Hungarian naval officer Konstantin Edler von Pott (1842–1908), and brought onto the market by the firm of H. and F. Müller, located in the port of Trieste on the Adriatic (cat. no. 213).

Sextants and quintants for polar expeditions can often be recognized by pieces of chamois leather that protect those parts of the frame which, during an observation, will touch the observer's skin. It is difficult to handle a sextant while wearing gloves, but touching metal in polar conditions is not without risk to the skin. Therefore the clamp and tangent screw, and the section of the frame nearest the observer's cheek, were covered with leather (cat. no. 288). An innovation in sextant manufacturing that was probably inspired by polar expeditions was the use of aluminium for the frame.[32] Aluminium weighs about one-fifth the weight of brass, an advantage on any expedition. William Ford Stanley (1829–1909) patented the use of aluminium for nautical and other instruments (no. 3092 of 1861). Despite this, nineteenth-century sextants are rarely made of this metal, probably since it corroded easily, and a heavier instrument was preferred.[33] When corrosion-resistant alloys became available, aluminium was reintroduced and used more generally for sextants (for example cat. no. 291).

DEALING WITH A POOR HORIZON AT SEA

At sea, the horizon can be obscured by fog, haze, or land; at night, it can be difficult to distinguish or completely invisible. These circumstances led to a variety of artificial horizons, of which a small selection is preserved in the National Maritime Museum.[34] One of these, patented by Robert Benton Roxby (no. 4695 of 1822), was an artificial pendulum horizon attached to a sextant. Above the pendulum was an indicator that, when the sextant was positioned correctly, became visible in front of the horizon glass (cat. no. 169). It was examined by the Board of Longitude, which dismissed it as unworthy of further attention.[35] Judging by positive reports published in *The Nautical Magazine & Naval Chronicle*, the

FIGURE 5.9 Inscription on an artificial horizon designed by Robert Benton Roxby, dating from 1822 or later (cat. no. 169).

FIGURE 5.10 Inscription on an artificial horizon designed by A.B. Becher of 1834 or later (cat. no. 150).

artificial horizon invented by Lieutenant Alexander Bridport Becher RN (1796–1876) enjoyed some success[36] (see Plate 14). This too consisted of a pendulum, with a cross-wire on top suspended in such a way that when the instrument was held vertically, the wire replaced the natural horizon. To damp its movement, the pendulum was submerged in a reservoir containing almond oil. Later in the century, other artificial horizons were developed on the principle of the 'floating wire'. Among these were Murray's patent of 1893 and a similar design by Newey in 1895 (cat. nos. 333 and 336). Both consisted of an eye tube containing a balanced or gimballed cross-wire that replaced the natural horizon and had to be fixed in the telescope bracket.

In France, Admiral Georges Ernest Fleuriais (1840–95) invented an artificial horizon using a gyroscope in 1886. This was based on the principle of a fast-spinning gyroscope that maintains its position regardless of external influence. The gyroscope was fitted in an airtight casing with two glass-covered openings on opposite sides. A glass piece through the body of the gyroscope had a series of horizontal lines that appeared to the observer as a constant line of reference when the instrument was running. The spinning gyroscope was attached to the sextant so that the lines replaced the horizon. In the early versions a handle and gear propelled the gyroscope; in later years this was replaced by an air pump (cat. nos. 221 and 331). According to a Dutch report, Fleuriais's horizon – although classified among the best of its kind – never became popular with Dutch seamen, and by 1915 was in disuse. The urge to own such an instrument had been greater than the urge actually to use it.[37]

In 1894, the Hamburg instrument maker Georg Butenshön brought an artificial horizon onto the market that was favourably reported on, but it is not represented in the National Maritime Museum collection.[38] It consisted of a quadrant with a telescope rigidly fixed to one side and a graduation along the arc. The celestial body was viewed directly through the telescope and its image was aligned with a spirit level that was fixed to an index arm. The bubble was reflected into the line of sight by a polished metal mirror inside the telescope. To facilitate nighttime observations, later versions of this instrument included a battery-powered light. Butenshön's horizon was reported to be simple and reliable.

Even a visible horizon was not necessarily reliable. Owing to terrestrial refraction, the dip could amount to 2.5 minutes of arc in areas where there was a considerable difference between the sea and air temperatures; for example in the Polar regions, the Red Sea, the Mediterranean, and the Persian Gulf and along the south coast of Australia.[39] The value could be measured with a dip sector. Henry Raper wrote that Edward Sabine used a Wollaston's dip sector in the Arctic in 1822, and that Captain George's double box sextant (cat. no. 182) could also be used as a dip sector.[40] About another dip sector, Raper wrote that it was inconvenient and little used.[41] Around 1906 Dr Carl Pulfrich (1858–1927), a physicist with Carl Zeiss in Jena, designed an instrument to measure dip. It consisted of a telescope mounted on a metal casing

FIGURE 5.11 An advertisement for Heath and Co's 'Hezzanith' artificial horizon (cat. no. 335). Heath's patent greatest-angle clamp is fitted on the arc of the elegant-pattern sextant frame. From Heath and Co's sextant price list of about 1900. National Maritime Museum; PBD8367.

that contained two prisms (cat. no. 345).[42] Through the telescope, the observer saw the reflected horizons to his left and right, at an angle of 180 degrees. The prisms were adjustable by a micrometer, by means of which the two horizons were aligned and the dip was read from a scale. A similar invention around that time was the 'Blish Prism'. This was invented by Commander John Bell Blish USN (1860–1921) and worked in the same way as Pulfrich's instrument, but was attached to a sextant.[43]

Taking a sight of a star requires a lot more practice and skill than sighting the Sun. A fix by the stars – usually four or five – was preferably taken in the morning or evening twilight when the horizon became clearly visible and the stars could still be seen.[44] Twilight in the tropics is short, and high seas causing an uneven horizon could further frustrate the observation. The French merchant Captain Pierre Michel Albert Laurent (1815–88) took out British and American patents in 1859 and thereafter for an invention to take a star sight with a poorly defined horizon.[45] He applied a special lens that elongated the star's image from a small point of light into a thin horizontal line, which was easier to align with the horizon. Laurent called this a 'star glass' (*stellar lenticular*), although this lens had been developed in the first half of the century to remedy astigmatism in eyes. At first, Laurent's invention, whereby the use of an elongating glass for correcting astigmatic sight was reversed, was fitted on special night sextants lacking shades and with a large star-finder telescope. In later years, the star glass was found on regular sextants along with the set of index shades. Prince George (later King George V, 1865–1936) owned a Laurent night sextant and the United States Navy purchased at least one example[46] (cat. no. 198).

Another device for facilitating star observations with a poorly defined horizon was the Wollaston prism. This split unpolarized light into two orthogonal, linearly polarized outgoing beams.[47] Although its principles had been known since the early nineteenth century, this prism was probably not applied to sextants until shortly after 1914.[48] When the prism was placed between the eye and the star, the observer saw two images of the star, one above the other, about a quarter of a degree apart. The horizon was then brought in between the two images and the altitude taken.[49] Like the star glass, it was mounted in a special frame, and hinged along with the set of index shades.

ARTIFICIAL HORIZONS IN THE TWENTIETH CENTURY

Aviation brought about significant changes in navigational instruments, and challenges to instrument makers. As the speed and altitude of aircraft increased in the 1920s, it was no longer practical to use marine navigation. One of the problems was that the horizon was so far below the observer that the dip became a huge correction.[50] Admiral Fleuriais's sextant with a gyroscopic horizon was tried out during a balloon flight in 1907, and Butenshön's horizon was also used in such craft. Even Serson's

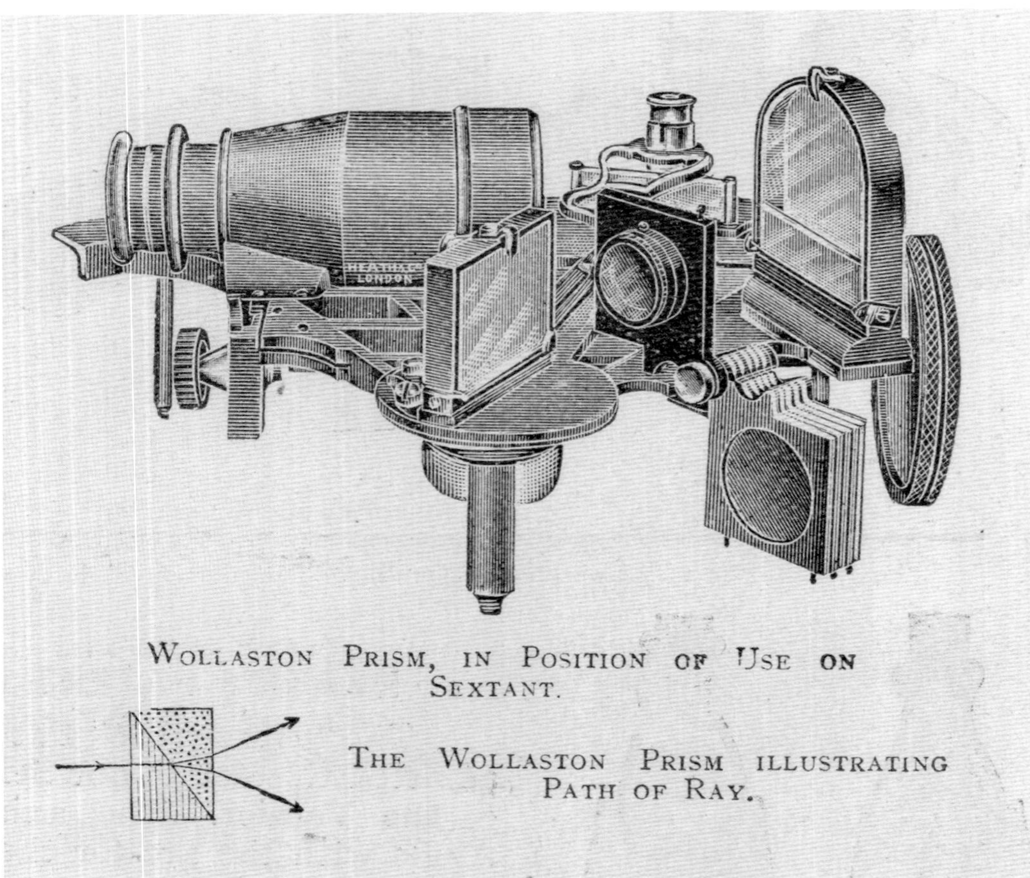

FIGURE 5.12 Wollaston prism in position between the two glasses, and its ray path. From an advertisement of Heath and Co in *The Nautical Almanac* for 1931. National Maritime Museum.

whirling speculum was revived, at least in theory. In 1906 the Frenchman Louis Favé designed a spirit-level sextant, which was produced by Albert Lepetit in Paris, the successor to the firm of Hurlimann, Ponthus, and Therrode. Favé's design was a success; it was improved by Lionel Burton Booth of the Royal Aircraft Establishment at Farnborough and used until World War II. Here, the bubble in the level replaced the natural horizon. A Portuguese naval officer, Carlos Viegas Gago Couthino (1869–1959), designed a spirit-level horizon, and used it on a flight from the Cape Verde Islands to Rio de Janeiro in 1919. Seven years later, Carl Plath began manufacturing Couthino's design commercially. Also in 1919, Lieutenant-Commander Richard Evelyn Byrd USN (1888–1957, later Rear Admiral) made another important flight, from Nova Scotia via the Azores and Lisbon to England, navigating with a bubble sextant of his own design manufactured by Brandis and Sons of Brooklyn, New York. Byrd's was the first in a long series of aeronautical sextants that were subsequently developed in the United States.[51]

World War II stimulated the further development of artificial horizons. Around 1938 Henry Hughes and Son Ltd of London became involved with bubble sextants for aircraft, and they marketed two types. The Mark IX and Mark IXA became available in about 1940 and 1941, respectively, and were used throughout the war.[52] For security reasons, the initials A.M. for 'Air Ministry' replaced the manufacturer's name on the examples produced during the war (cat. nos. 245 and 247).

An important contribution to the development of the aeronautical sextant was the *SOLDLibellensextant* (bubble sextant) by Plath, conceived in 1939 (for example cat. no. 238).[53] It had only one glass, and optical parts to view the artificial bubble horizon. The *Libellensextant* was mainly used in aero-navigation but in 1943 a new version, the *Soldsextant*, was put into production.[54] Like that of Fleuriais, it had a gyroscope for the artificial horizon, driven by an air stream from a hand-operated air pump. The *Soldsextant* was specially designed for use on U-boats, which surfaced only at night, safe under cover of darkness. A *Soldsextant* could be used for direct observations, viewing the celestial body through shades and indirectly via the glass. During the war Plath produced almost 21 000 sextants, of which more than half were built on the SOLD principle. For the same reason as in Britain, they do not carry the manufacturer's name, but the initials J.B., probably for Johannes Boysen, Carl Plath's grandson-in-law, who became a partner in the firm in 1937 (for example cat. no. 250). After the war, the Admiralty Research Laboratory in Teddington and the United States Naval Observatory in Washington took possession of some German gyro-sextants taken from U-boats, and had these tested.[55] One of the conclusions was that they could supplement the traditional sextant but would not replace it, a prophecy that came true. Plath's gyroscopic sextant had been a wartime necessity.

After the war, bubble sextants by Hughes and Son remained in use on aircraft. In addition both Hughes and Heath and Co. developed new bubble horizons for the traditional marine sextant (cat. no. 263). As with all artificial horizons, these required much practice in use and were expensive.[56]

IMPROVEMENTS IN SEXTANT TECHNOLOGY

A significant improvement to the sextant was the introduction of the micrometer, a device by which small subdivisions of linear and angular scales can be measured. It was far easier to use and read than a vernier, especially as the refinement of scales increased. On sextants this could be to 10 seconds, one-sixth of a minute of arc. Such vernier lines are difficult to distinguish in twilight, even with the aid of a magnifier. The micrometer was invented in the mid seventeenth century and soon used with telescopes. The earliest example on a marine sextant is found on one by Ramsden

FIGURE 5.13 Witch on a broomstick holding a Hughes bubble sextant. Frontispiece by Alabaster from *History of Air Navigation* by Arthur J. Hughes (London, 1946). National Maritime Museum; PBE6602.

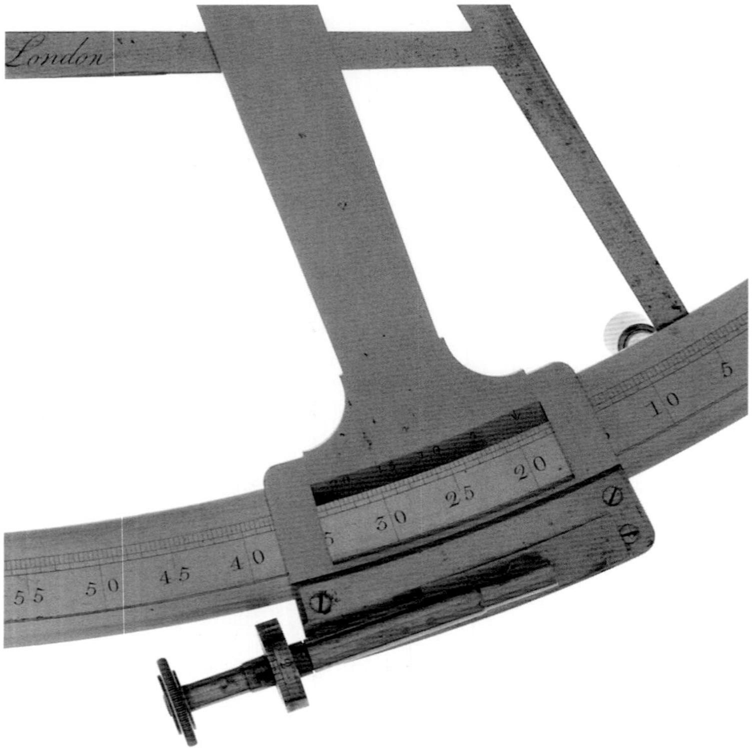

FIGURE 5.14 The micrometer on a sextant by Jesse Ramsden of about 1783 (cat. no. 140).

from around 1783 (cat. no. 140).[57] Because there is no locked-in relation between the micrometer and the arc, the observed angle had to be read in a roundabout way. After the micrometer was read, the Sun or star was approximately aligned with the horizon. The celestial body was then aligned precisely by use of the micrometer, and the micrometer was read again. Then the index arm was moved to the next full degree by turning the micrometer and counting the number of turns and the final part turn. The micrometer was read again, and the shift of the index arm provided the number of arc minutes to be added to the full degrees (or subtracted if the arm had to be moved anticlockwise). Another way was to set the micrometer to zero, align the Sun or star approximately, and then set the index arm to the nearest full degree on the scale. Then the celestial body was aligned with the horizon precisely and the micrometer read. Both methods had to be applied after each observation, which was especially impractical when taking several star sights in the brief twilight. Perhaps that is why the Ramsden sextant also has a vernier.

The earliest printed reference yet found to a micrometer in connection with sextants dates from 1833, when Lieutenant Owen Stanley RN (1811–50) proposed to fit one for finding the mean of a number of altitude measurements.[58] The next reference is from the Deutsche Seewarte. Out of 2353 sextants tested by that institution in the period 1871–1901, two were reported to have a *micrometerschraube* (micrometer screw). One was on a sextant by B. Bromander of Hamburg, by whom no instrument is on record. The other was on an example made by William Gerrard of Liverpool (worked 1862–90), of whose instruments a total of eight were tested.[59] These micrometers may well have been similar to the one found on a sextant by Cary of London, dated around 1896, which also has a vernier reading to 10 seconds of arc.[60] The micrometer has divisions from 1 to 40 in units of 1, which means that it could be read to 1.5 minutes of arc, considerably less than the instrument's vernier. Because there is no locked-in position between the micrometer and arc, it has the same disadvantage as the instrnment made by Ramsden.

The breakthrough in micrometer technology for sextants must be attributed to Carl Plath. But before this happened, Plath invented an indicator sextant to facilitate easier reading.[61] The indicator rotated on a large circular scale about the size of a pocket watch. It was mounted at the bottom of the index arm and was easily readable with the naked eye. The indicator was geared to the sextant's limb by means of a toothed rack in such a way that when the index arm was moved through one degree, the indicator turned a full circle on the scale. This construction proved vulnerable and apparently did not attract much custom.[62]

Plath probably constructed his first *Trommelsextant* (drum sextant) with a micrometer drum around 1906, by order of the German Navy.[63] The micrometer found on Pulfrich's instrument for measuring the dip of the horizon (cat. no. 345) may well have inspired him. Although made by Zeiss, Pulfrich's instrument was advertised in Plath's trade catalogue of 1907.

The micrometer drum replaced the knob of the tangent screw and the pitch of the screw – which engaged in a toothed rack along the bottom of the limb – was chosen so that a full-

FIGURE 5.15 A micrometer on a sextant by Cary of London, of about 1896. Courtesy of the National Museums Scotland, Edinburgh; T1987.

circle turn of the drum moved the index arm one degree along the scale. The advantage of Plath's micrometer was that it was synchronized to the scale and did not have to be set to zero before an observation. If the index arm was engaged while the micrometer was at 30 minutes, for example, the indicating line at the bottom of the arm would point halfway between two degree markings on the scale. With a micrometer drum, the degrees are read from the scale by use of the indicating line, and minutes or half minutes are read from the drum. The screw was conically shaped and could be turned continuously without coming to an end, by which means another potential problem was remedied. If the traditional tangent screw was turned in one direction over a long period of time, it would come to its end and be blocked. The observer had to check for this regularly and, if necessary, turn the tangent screw back to regain sufficient thread on either side of its centre. Rewinding the screw took some time and if this had been neglected, the opportunity for an observation might be lost. On English-made sextants, the micrometer drum (Heath and Co. called it the 'Hezzanith' Rapid Reader) was not introduced until the 1920s, when manufacturing problems were overcome.[64]

George Wilson Heath designed another significant improvement, the quick-release device, first found on a British sounding sextant from around 1900.[65] It replaced the clamping screw for securing the index arm to the sextant's limb. Plath's drum sextant came with a springloaded trigger that, when squeezed towards the observer, disengaged the endless tangent screw from the toothed rack so that the index arm could be moved freely along the limb. Heath and Co. patented an automatic clamp (patent no. 17 840 of 1909), the 'Hezzanith' Endless Tangent Screw Automatic Clamp, which had to be squeezed sideways to free the index arm from the limb.

The Patent Greatest Angle Clamp by Heath and Co. was a clip that could be mounted on the limb just below the expected maximum noon altitude of the Sun.[66] As the observer followed the rising of the Sun prior to culmination, the index arm pushed the clip further along the limb. If the observer missed the moment of culmination and brought the index arm back, the greatest altitude could still be found by moving the arm back against the clamp (see Fig. 5.11). Heath's 'Hezzanith' Centring Error Detector enabled the navigator to check if his sextant had such an error. As explained in the previous chapter, this could not be corrected at sea, and the navigator would have to find another sextant until his was repaired. Finally, Heath's Patent Definition Shade was a small metal screen fixed to the horizon glass to prevent the eye

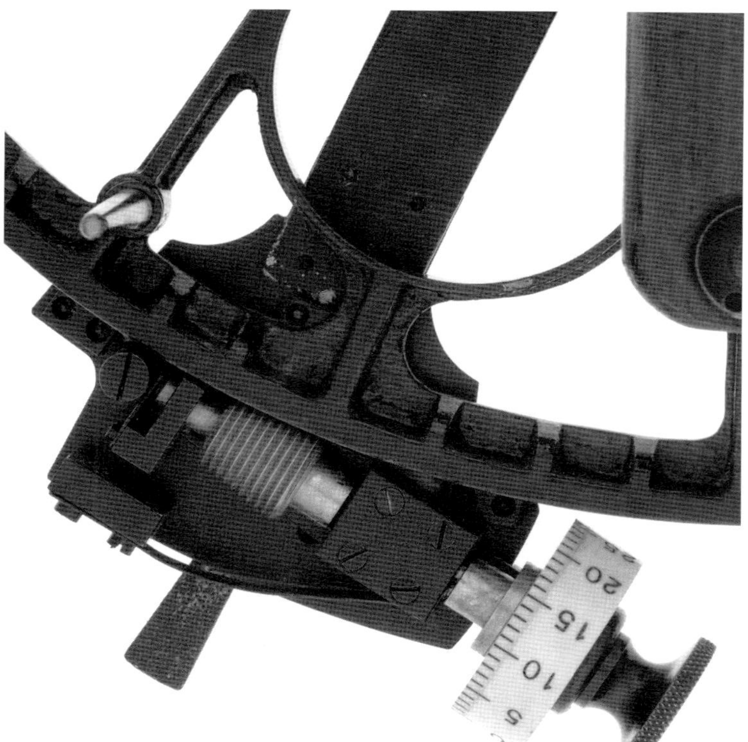

FIGURE 5.16 Carl Plath's drum micrometer and quick-release clamp on a quintant of about 1917 (cat. no. 292).

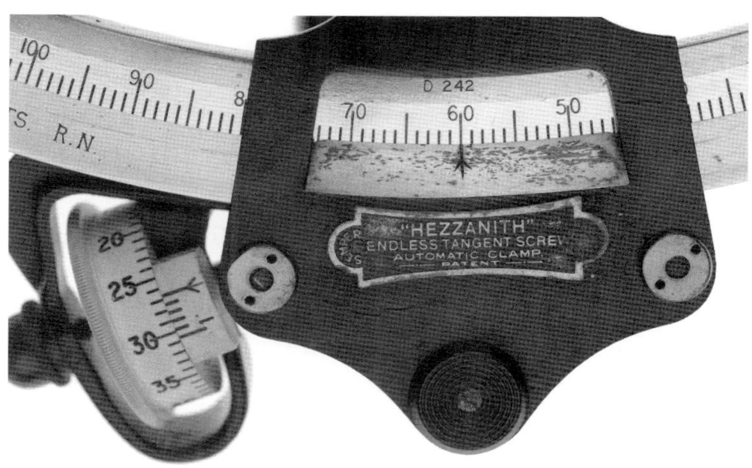

FIGURE 5.17 A Heath and Co Ltd micrometer, endless tangent screw, and automatic clamp on a sextant of 1934 (cat. no. 231).

FIGURE 5.18 Photograph showing three officers taking noon sight of the Sun on the bridge of a merchant vessel in the 1920s. National Maritime Museum; N47907.

being dazzled during an observation in bright sunlight, not unlike the principle of a lens hood on a camera.[67]

SEXTANT MANUFACTURING IN EUROPE AND NORTH AMERICA

Throughout the nineteenth century the British octant- and sextant-making industry continued to dominate, although production continued in France and the Netherlands, and was joined by Germany and the United States. In Britain, the demand for octants and sextants was maintained during the long French wars of 1793–1815, and many a prominent nautical instrument maker – or his firm – survived this period of war and political turmoil. Ramsden died in 1800 but his assistant Matthew Berge carried on the business with Ramsden's successful dividing engine. When Berge died in 1819, there was some competition for that engine. Thomas Jones, a one-time apprentice to Ramsden, asked the Board of Longitude for permission to take it over. So did Nathaniel Worthington, who had been Berge's assistant.[68] Jones was lent a linear dividing engine of Ramsden's, but the circular one went to Worthington, who in 1821 associated with James Allan, an instrument maker who had previously built his own dividing engine.[69] Thomas Jones also built an engine and became a prolific octant and sextant manufacturer in his own right. He delivered work to naval officers, including Lord Hay and William Fitzwilliam Owen (cat. no. 172). William Cary (*c.*1759–1825) was another apprentice of Ramsden who became a prolific manufacturer of sextants, especially box sextants (cat. no. 182). After Cary's death his business was continued until almost the end of the century, and in the 1860s the firm produced the artificial horizon designed by Captain George already mentioned. Spencer Browning and Rust went on to manufacture octants and had connections in the United States. Several octants show the additional name of Henry Duren of New York (cat. no. 120). Duren's name and that of one of his later associates are also found on an octant by William Holliwell of Liverpool and on one by Heath. An octant by Spencer and Co bears the name of Duren and Costigan, and John Crichton of London made an octant for Duren and Alder. Finally, an octant by Spencer Browning and Rust was retailed by William Heaton in Newcastle upon Tyne and thereafter repaired or sold by Henry Duren, as all three names are found on the instrument.[70] George Whitbread was another London instrument maker who delivered to foreign retailers; in the 1850s, the firm of Van Keulen in Amsterdam sold his octants.[71] The workshops of Spencer Browning and Rust, and Whitbread continued for a decade or so into the second half of the century and then faded away.[72] Meanwhile other London firms emerged, such as that of Elliott Brothers and that of Mrs Janet Taylor, who exhibited navigational instruments at the Great Exhibition of 1851[73] (cat. nos. 179, 281; see Plate 19). Members of the Royal family owned quintants by both Elliot Brothers and Taylor. In Scotland, the instrument-making firms of James Whyte and Co. and Duncan McGregor and Co. emerged (cat. no. 201). They sold sextants and in due course opened branches in more lucrative shipping areas. The Whyte firm was associated with Lord Kelvin and later included Bottomly and Baird, establishing branches in London, Montreal, Newcastle upon Tyne, Boston, and New York.[74] McGregor and Co. branched out to Liverpool and London.[75]

At the end of the century two sextant-making firms became prominent, and remained prolific until after World War II: Heath and Co. of Crayford, and later of New Eltham, using the trademark 'Hezzanith', and Henry Hughes and Son of Fenchurch Street, London, trading as 'Husun'. Both had a wide range of attractive frame patterns with fancy names sold at reasonable prices.[76] In a trade catalogue of around 1900 Heath and Co. offered 37 different frame patterns with names such as like 'tulip', 'lunar', 'heart', and 'spectacle'. (The various frame patterns are illustrated in Table 5.1 at the end of this chapter.) Depending on how simple or complicated the cast frame was, the type of scale (inlaid silver, ivory, or brass), and the number of accessories, the price of a Heath sextant ranged from £3 5s to £20.[77] Health and Co. made sextants which were retailed elsewhere and inscribed on the limb with both names, with words such as 'Made for M.L. Baudouin à Dunkerque by'; such sextants were also made for G. and T. Davison in Newcastle-on-Tyne and for S.D. Neill Ltd,

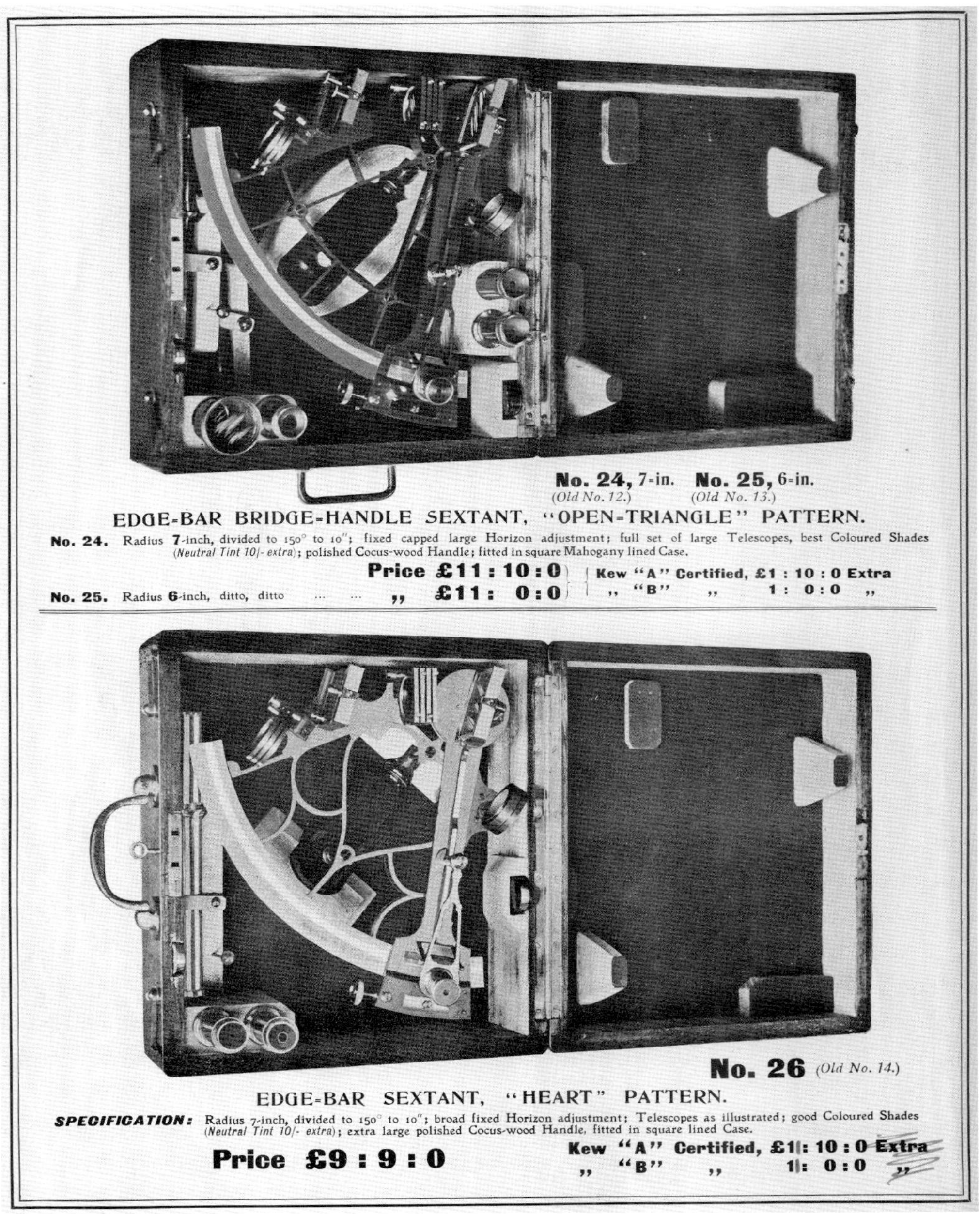

FIGURE 5.19 Page from Heath and Co Ltd's sextant price list of about 1900, showing two of the cheapest available brass sextants. National Maritime Museum; PBD8367.

Belfast. Heath and Co. also retailed sextants through the Liverpool Sailors' Home.[78] A recorded Hughes and Son sextant includes the name of B.C. Cousens and Sons of Falmouth, presumably a retailer.[79]

Despite the availability of affordable modern sextants, old ones remained in use for many decades. The report of the Deutsche Seewarte shows that twenty octants and sextants by Spencer Browning and Co., and as many by Troughton and Simms, were tested in the period 1875–1901, presumably because they were still in use at sea. These two were the best represented names of British manufacturers in the report.[80]

In France, François-Antoine Jecker established his workshop in Paris, where, among other instruments, he made reflecting circles and quintants (cat. no. 309).[81] After Jecker, Henry-Prudence Gambey (1787–1847) was certainly the most important early French supplier of such instruments.[82] It was not until 1845, however, when Gambey's former apprentice Edmond Lorieux formed a business with Schwartz, that sextant and quintant production in France increased, probably to a considerable extent. Lorieux's frame pattern unmistakably distinguished his instruments from designs by other makers (cat. no. 285). Apparently it appealed to Heath and Co., because around 1900 they offered 'Lorieux Borda pattern' sextants, closely resembling the French style.[83] Lorieux's business was later associated with Hurliman, Ponthus, and Therrode, and thereafter continued until about 1950 under Albert Lepetit (cat. no. 293). Finally, the French sextant manufacturer Leon Dubas should be mentioned. Between 1861 and 1873 he was a watchmaker, jeweller, and optician in Nantes, and exhibited at his city's national exposition in 1861.[84] Dubas was a witness to Laurent's American patent for the use of the elongating glass on sextants, and the first to manufacture these.

French rule over the Netherlands ended in 1814. Only a few octant- and sextant-making firms had survived the previous period of maritime economic depression that had affected the shipping industry. The Van Keulens carried on until the 1880s but never regained their previous level of production, and octants and sextants were imported from such British manufacturers as Whitbread and Dollond. Kleman became the most prominent Dutch manufacturer of octants, sextants, quintants, and the odd reflecting circle, and his business continued until the 1850s. From 1832, Leendert Johannes Harri was an instrument and compass maker in Amsterdam, and manufactured octants and sextants. After his death the firm continued until about 1988, selling and repairing sextants. After the 1870s these were no longer manufactured on the premises but imported from Britain, the Harri name being engraved at the point of sale.[85]

Plath of Hamburg was the only firm that equalled Heath and Co. and Hughes and Son in quality, innovation, and quantity. Carl Christian Plath (1825–1910) started as an apprentice to Repsold, the Hamburg manufacturers of astronomical and surveying instruments, and then worked for Pistor and Martins. In 1862 he took over the workshop of David Filby (1810–79) in Hamburg, and in the following four decades developed it into a successful navigational-instrument-making business.[86] Plath purchased his first circular dividing engine from Repsold and used it until he built his own in 1904.[87] Plath's sextants came in three frame designs: regular, without a specific name; a bar pattern; and a three-circle pattern. Their radii ranged between about 118 and 180 mm, and the graduation ran to 135 degrees. Plath used the term *halbsextant* (half sextant) for sextants graduated to 119 degrees.[88] After Carl's death, his son Theodor continued the firm and was succeeded by his son-in-law Boysen, who carried on until the 1960s. Pistor and Martins made high-quality reflecting circles, with the German navy and the Royal Netherlands Navy among their clients, until the firm was dissolved in 1873, two years after the death of Carl Otto Albrecht Martins (1816–71) (cat. nos. 312 and 314). After this the Dutch navy purchased sextants from Th. Wegener of Berlin.[89] Another firm that manufactured sextants for the German navy at the beginning of the twentieth century was H. Haecke of Neukölln, near Berlin (cat. no. 218).

Benjamin Stancliffe of Philadelphia (1782–1834) was probably the first American to use a circular dividing engine to make sextants.[90] He had been apprenticed in London to his uncle John Stancliffe, an employee of Ramsden who built his own dividing engine. Benjamin then worked for Edward Troughton and around 1816 emigrated to the United States. A few of his sextants are preserved. A more prolific sextant-manufacturing firm was that of E. and G.W. Blunt in New York. Edmund Blunt (1799–1866) and George William Blunt (1802–78) built their circular dividing engine between 1851 and 1857. A number of their octants and sextants are preserved, and some have a United States Navy serial number.[91] Stackpole and Brother have already been mentioned: William Stackpole (1819–95) and Robert Stackpole (1823–73) came from Ireland in 1833 and started an instrument-making business in Baltimore, before establishing themselves in New York in 1851.[92]

There are American links with French manufacturers. The nautical instrument firm of T.S. and J.D. Negus in New York were sole agents in the United States for Laurent's sextant with an elongating glass, and the United States Coast and Geodetic Survey owned a quintant by Lorieux Lepetit.[93] Captain Townshend from New Haven owned a sextant by Blunt, but also an octant signed by Charlet of Le Havre – though possibly made by Dubas – and a reflecting circle by Gambey.[94]

THE END OF THE SEXTANT FOR PROFESSIONAL SEAFARERS

At the beginning of the twentieth century Henry Hughes and Son, Heath and Co., and Carl Plath were the most prominent manufacturers of sextants, but industrial growth eventually led to mergers. The first was in 1926, on the retirement of George Wilson Heath, when Heath and Co. was acquired by W.F. Stanley and Co. Ltd of in London, a firm that had been manufacturing surveying and other scientific instruments, including a limited selection of sextants, since 1853.[95] The increased demand for navigational instruments during World War II was reflected in the sextant industry. Henry Hughes and Son Ltd merged with Kelvin, White and Hutton in 1941, and six years later became Kelvin Hughes Ltd.[96] In pre-war America firms such as Brandis and Sons, the Kollsman Instrument Company, the Pioneer Instruments Co., and Bausch and Lomb had been involved in with the development and manufacture of sextants for aviation.[97] During the war, Bausch and Lomb and Pioneer could not meet the demand for navigational instruments, so other firms that did not normally manufacture sextants were encouraged to enter the market. One was the Leupold and Stevens Instruments Company of Portland, Oregon, which for a short time made quintants and bubble sextants. Also during the war, other American firms made plastic and sometimes cardboard sextants for use in lifeboats or the rubber lifesaving dinghies of aeroplanes (cat. no. 261).[98] The production of such lightweight, simple, easy-to-use instruments continued after the war, mainly for yachtsmen (cat. no. 131).[99] After 1945 the Freiberger Präzisions-mechanik, a subsidiary of Carl Zeiss located in Freiberg, East Germany (later the German Democratic Republic), became a competitor to the sextant industry in the West. The factory at Freiberg produced good-quality lightweight sextants at prices far below those of firms such as like Plath, Hughes and Son, and Heath and Co.[100]

Soon after the war, when Plath had recovered from penalties imposed by the Allies for collaboration, the firm entered the American market. Weems and Plath was founded in 1953 with the retired Lieutenant-Commander Philip Van Horn Weems USN, an American navigation expert. This firm is still selling navigational equipment in Annapolis, Maryland, including sextants by Tamaya Techniques Inc. of Tokyo, Japan.[101] Cassens and Plath in Bremerhaven is another offshoot of the Hamburg firm that is still in the sextant business.

As electronics were increasingly applied in navigation this was reflected in sextant development. In the 1960s Plath designed the 'NAVISTAR Professional' an advanced computerized sextant.[102] In cooperation with the Admiralty, around 1970, the marine division of Kelvin Hughes designed a digital sextant bearing the old name of Henry Hughes and Son Ltd (cat. no. 270). The movement of the index arm and the micrometer drum was electrically relayed by wire to a remote device for digital reading and recording. Some years later a similar recording sextant was developed by the Scientific Instrument Company of Milwaukee, Wisconsin, and used by the United States Navy.[103]

These innovations could not prevent the gradual downfall of the sextant. As described in the first chapter, the NAVSTAR Global Positioning System (GPS) was initiated in 1976. In the next two decades it was extended and by the 1990s was fully operational. Meanwhile the major sextant-manufacturing firms continued their losing battle. Stanley went on producing sextants and quintants under the 'Hezzanith' trademark but in 1999 the firm went into liquidation. Kelvin Hughes became a division of Smiths Industries Ltd in 1965 but continued to make sextants (cat. no. 272; see Plate 18).[104] On his retirement in 1963, Boysen had sold the Plath firm to Litton International, who later passed it on to the American Litton Group.[105] Since then Plath of Hamburg has been wound up, although a new firm by that name still sells navigational instruments. The only Dutch firm that continued to make sextants until the mid 1990s was the Observator Instrument Co. of Rotterdam.[106] In 2007, besides Tamaya Techniques, already mentioned, only the Davis Instrument Corporation of San Leandro, California, the ChangZou Feitan Gear Company in China, and the Freiberger Company in Germany continue to manufacture professional sextants.[107] However good these may be, they will no longer be found on the bridge of a naval or merchant vessel, but they continue to be used by nostalgic yachtsmen.[108]

5.1 TABLE OF FRAME PATTERNS

Frame patterns as advertised by Heath and Co. in their price list of *c.*1900[109], and of H. Hughes and Son Ltd and Kelvin Hughes

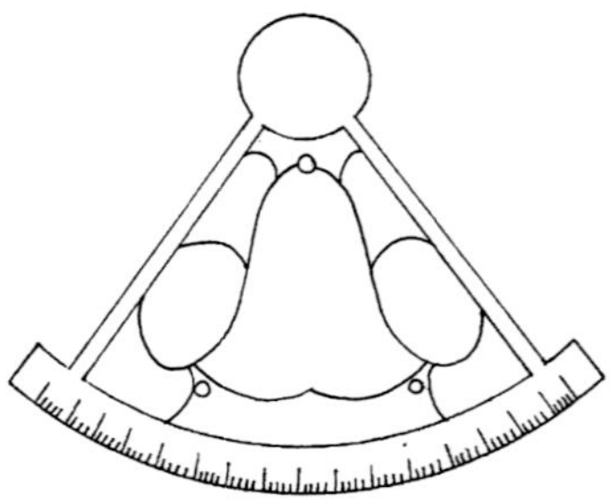

Bell pattern

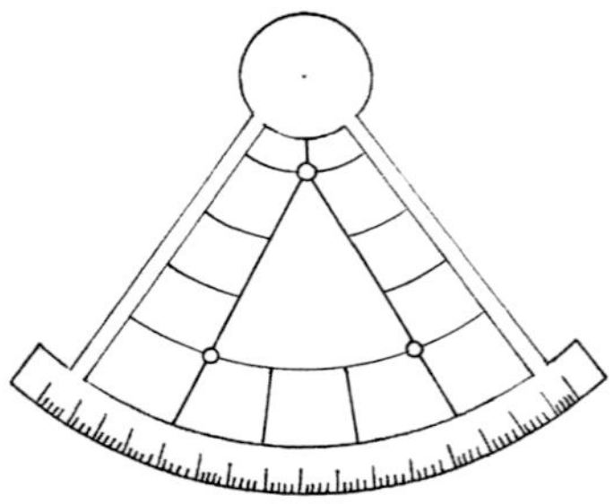

Triangle pattern

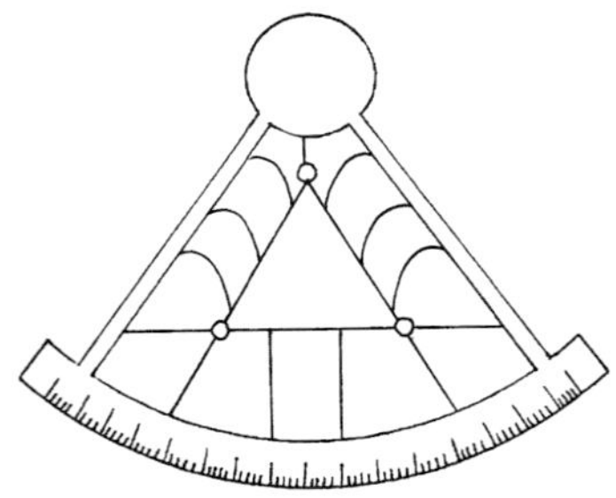

Triangle pattern

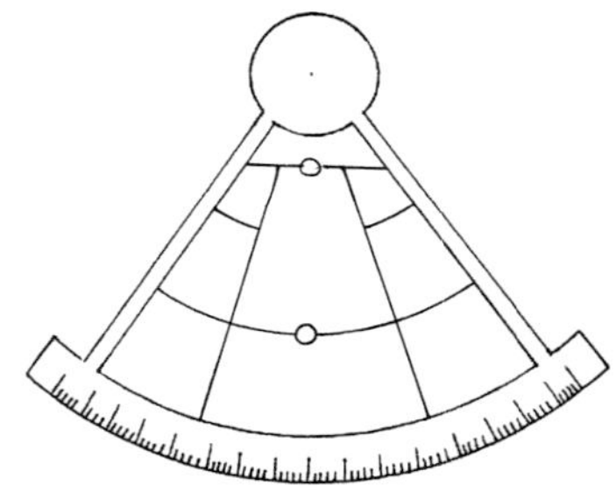

Lorieux-Borda pattern

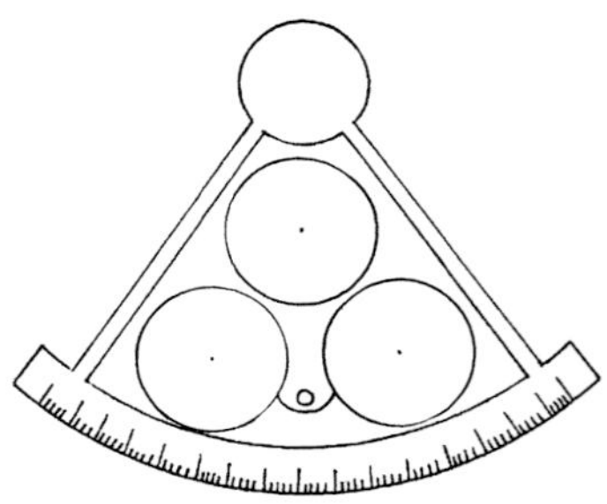

Three-circle pattern

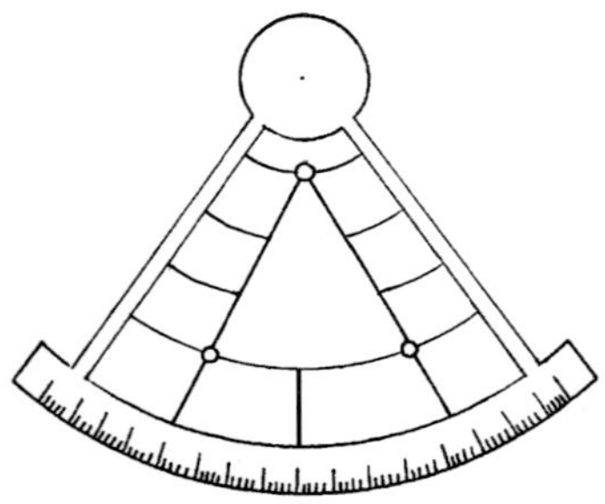

Trinity House pattern

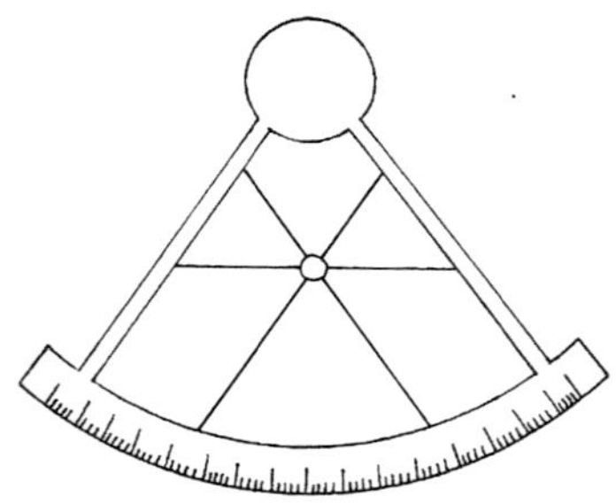

Star pattern

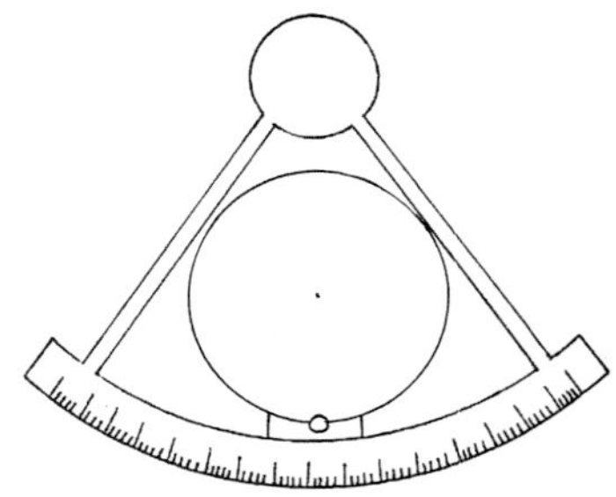

Circle pattern

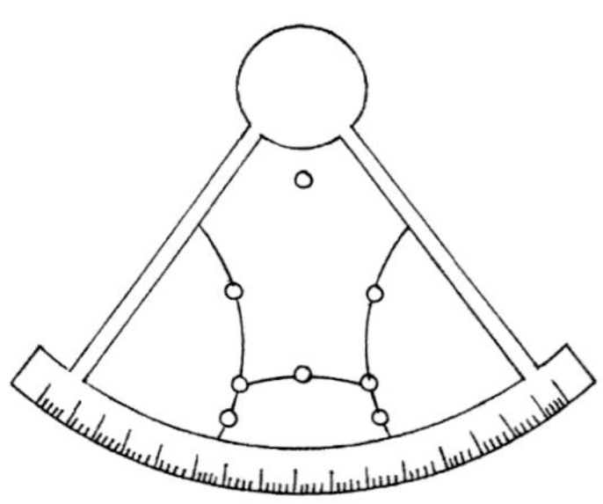

(Double)-curve pattern

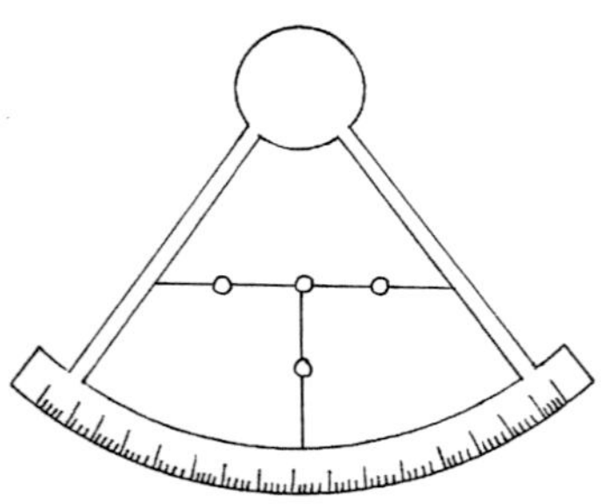

Strait-bar pattern

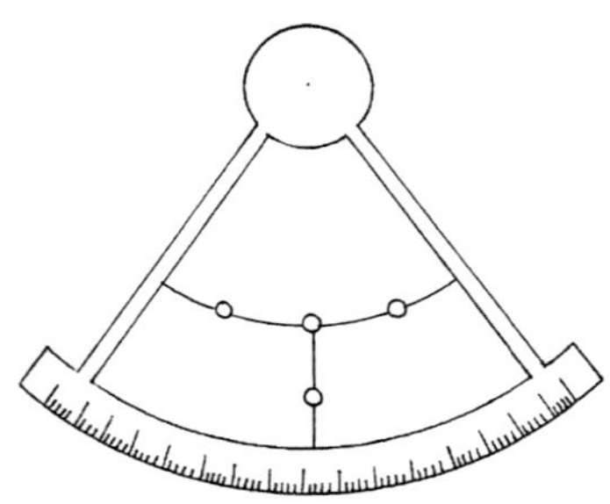

Curved pattern

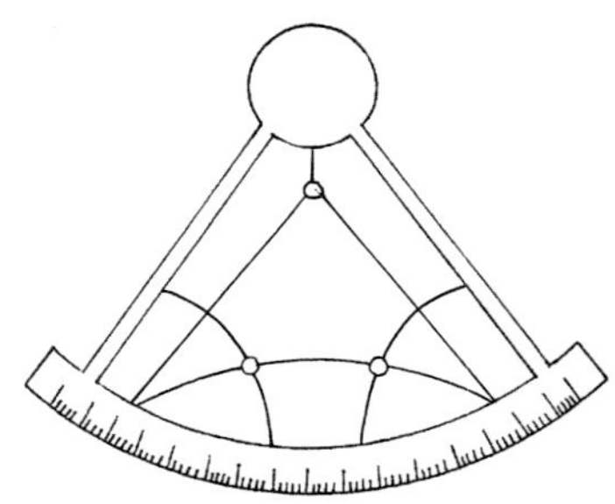

Gothic pattern

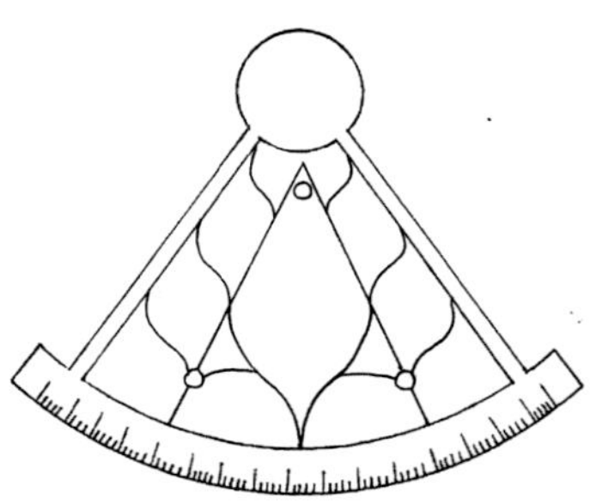

Elegant pattern

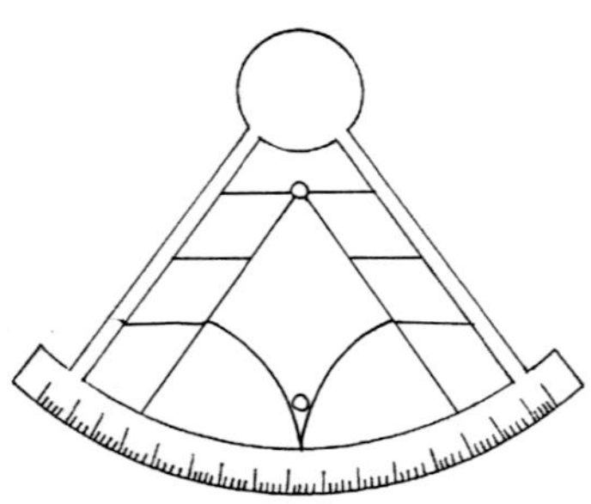

Lunar pattern

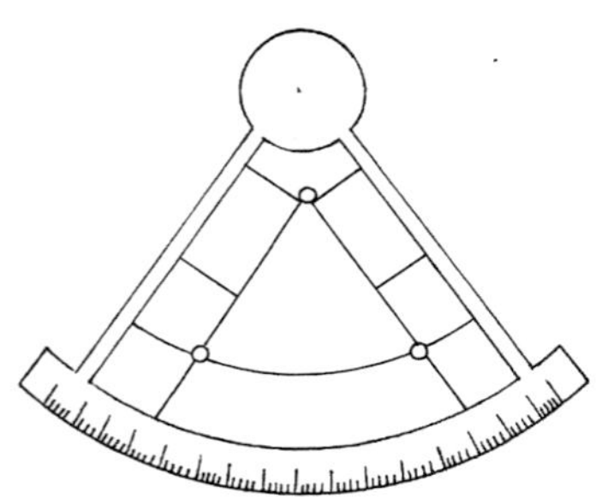

Open Triangle pattern

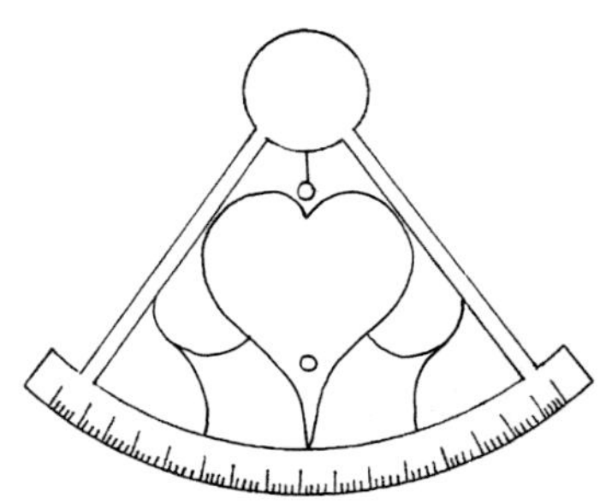

Heart pattern

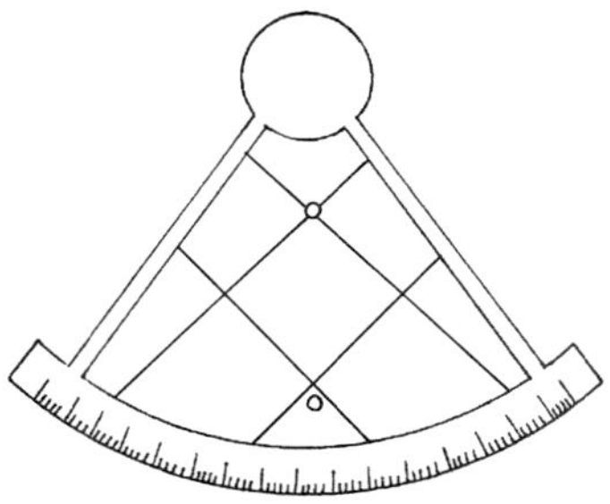
Diamond and Diamond pattern

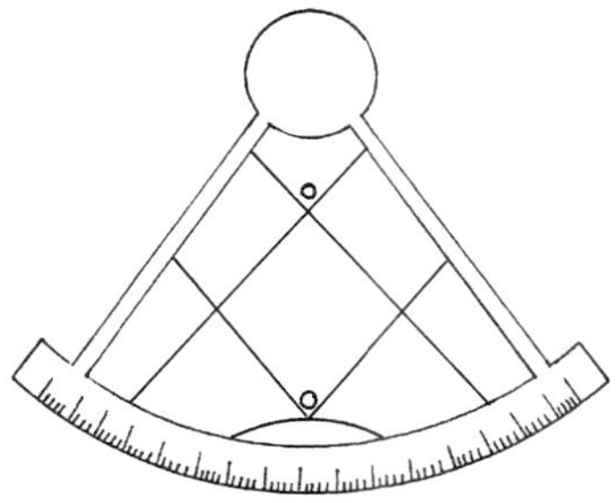
Diamond and Curve-pattern

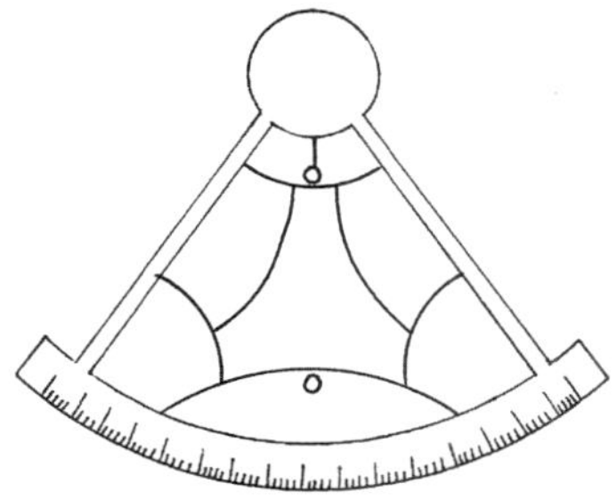
Curve-bar pattern

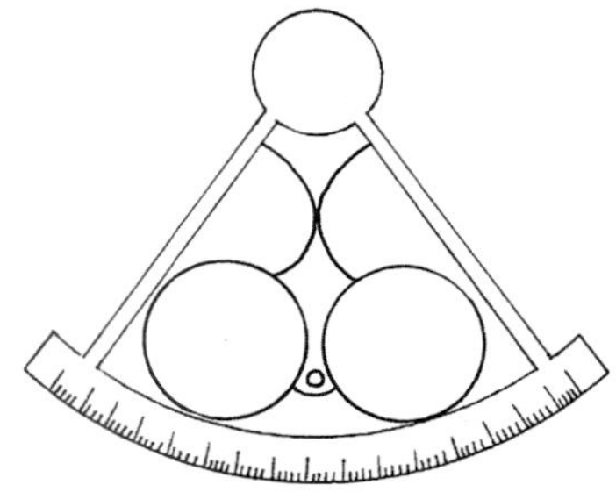
Two-circle pattern

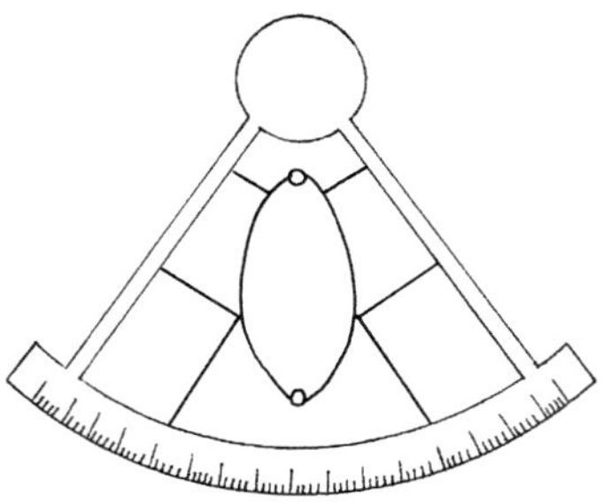
Oval pattern

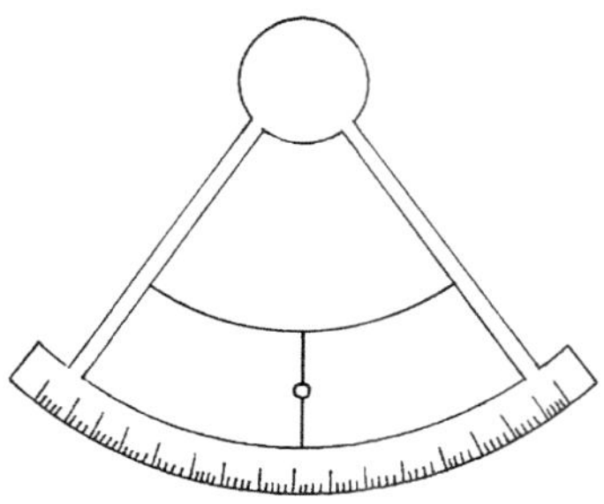
Dollond pattern (as sextant), Plain pattern (as octant))

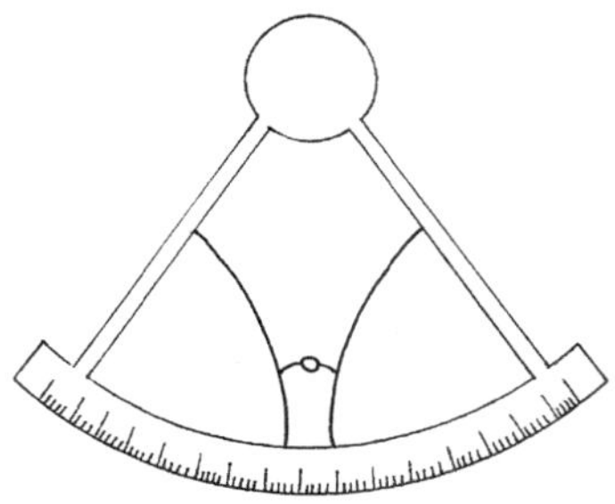
Pillar pattern

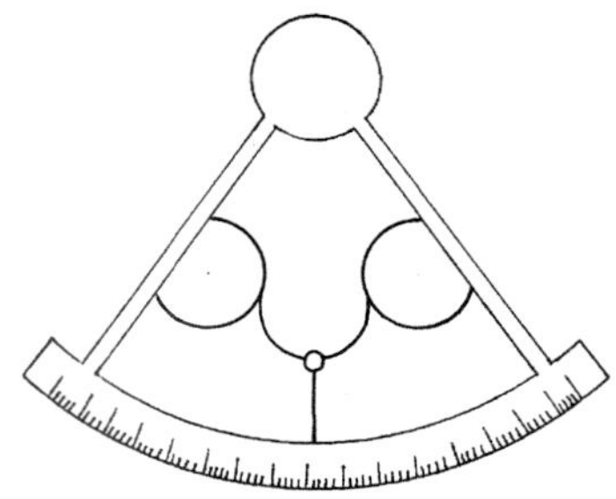
Spectacle pattern

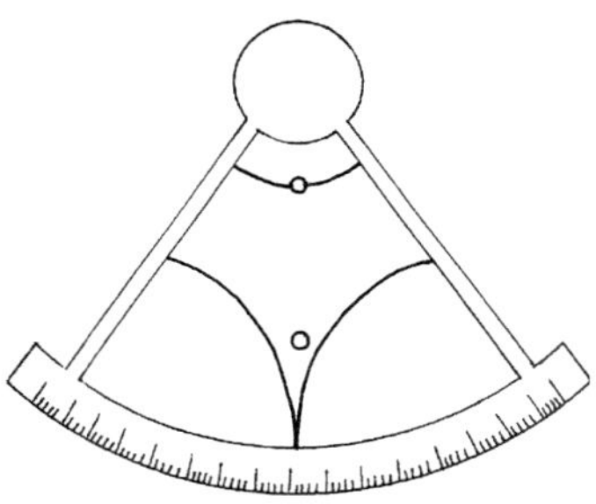
Light pattern

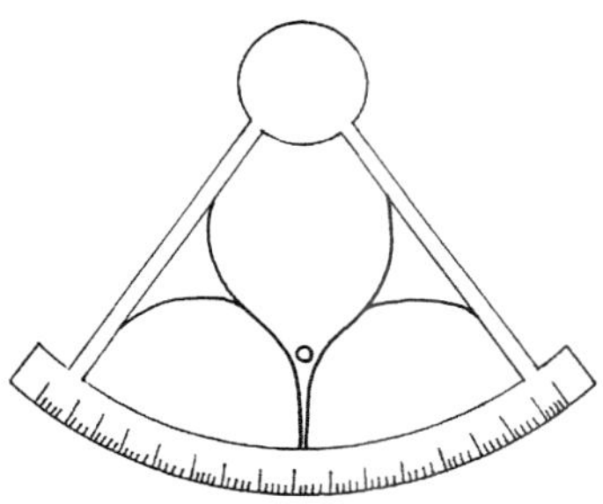
Tulip pattern

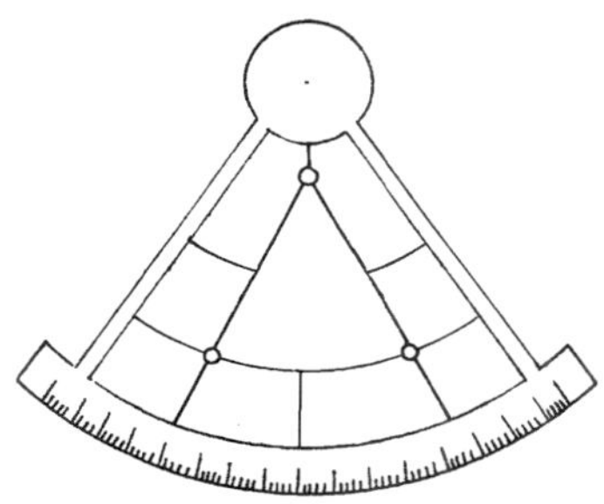
Open triangle pattern

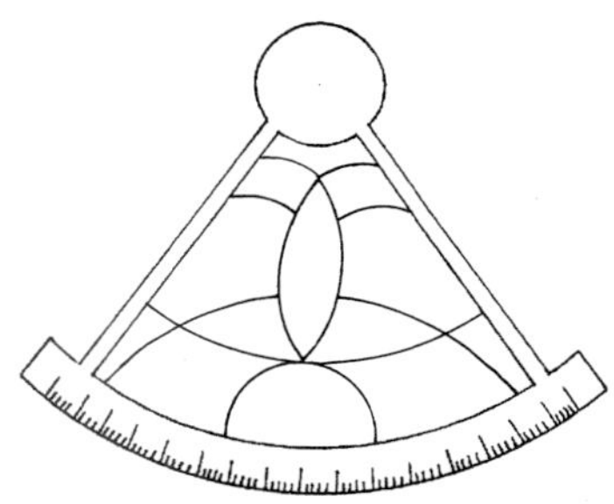
Oval-curve pattern

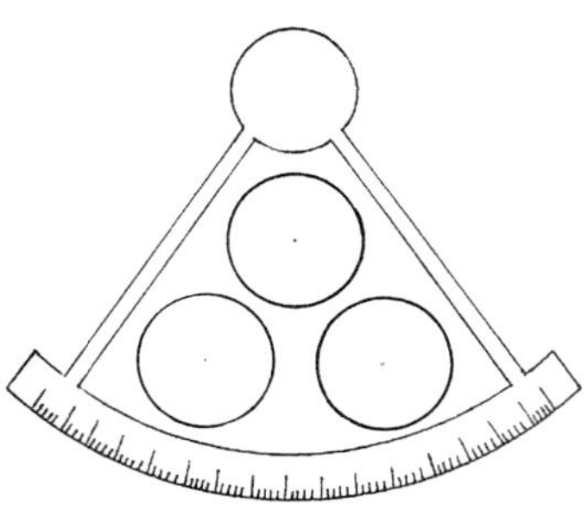
Viking sextant (Hughes, 1912), Britannia cadet sextant, or Mate's sextant (Kelvin Hughes 1961)

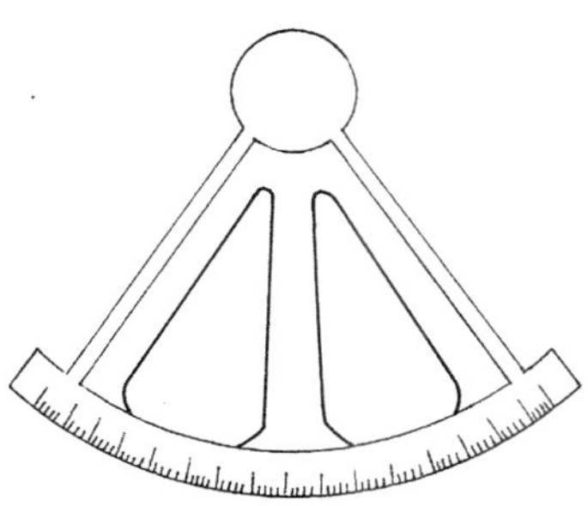
Admiralty or Challenger pattern

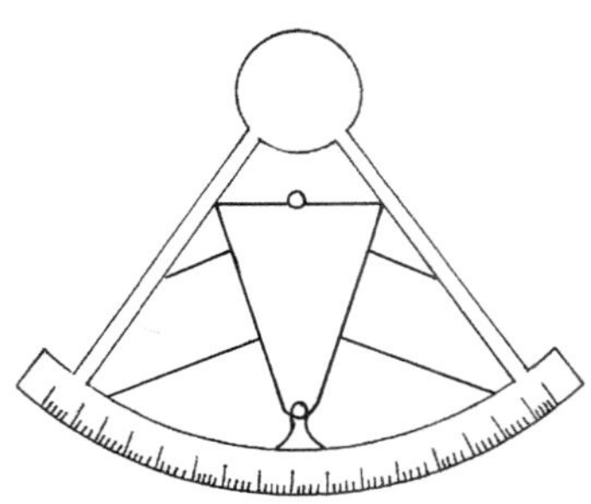
Gothic pattern (Kelvin Hughes 1981)

NOTES

1. May, *A History of Marine Navigation*, 148; Chapman, *Dividing the Circle*, 135.
2. This was not because the chronometer had replaced the lunar method (see Nichols, 'Hadley's Quadrant – The Back Observation'), but because for lunars, that the sextant, quintant, and reflecting circle had replaced the octant.
3. NN, *Joseph Levi & Co Optical & Photographic Warehouse*, 5; the price of an ebony-frame octant was about one-third of Levi's cheapest brass-frame sextant.
4. Lecky, *'Wrinkles'*, 45.
5. An example of double ownership is provided by the American Captain Townshend of New Haven, Connecticut, who had a sextant, an octant, and a reflecting circle; see Preuss and Treworgy, 'Taking a Departure', 29–30.
6. A report on octant and sextant testing in the period 1875–1901 shows that out of 3165 tests, 234 were on octants, many with ebony frames, see Messerschmitt, 'Ergebnisse von Sextantenprüfungen.'
7. Bennett, *The Divided Circle*, 183–4.
8. Preuss and Treworgy, 'Taking a Departure', 32.
9. Warner, 'American Octants and Sextants', 108–9; Tesseract Early Scientific Instruments, Catalogue 82 (Summer 2006) no. 15; McConnell, 'Aluminium and Its Alloys', 618.
10. Ifland, *Taking the Stars*, 137–8.
11. Lecky, *'Wrinkles'*, 279–80.
12. Howse, *Greenwich Time and the Longitude*, 102–5, 154–5.
13. Communicated by Will O'Neill on the Internet discussion group 'Sextants', 29 January 2004.
14. For the following information see Watts, *The Sextant Simplified*, 78–80, and NN, *Katalog über nautische Instrumente*, 27.
15. Binocular telescopes are rarely found; they do not make an observation much easier but introduce an extra risk of collimation error.
16. Lecky, *'Wrinkles'*, 429; Watts, *The Sextant Simplified*, 92–5.
17. Lecky, *'Wrinkles'*, 48.
18. Bryden, 'Quality Control in the Making of Scientific Instruments'. The British Association for the Advancement of Science took over the Observatory from the Crown in 1842.
19. The Laboratory continued to issue certificates until 1967. This task was then taken over by the Admiralty Compass Observatory, which had been issuing certificates of serviceability since 1942; see Fanning, *Steady as She Goes*, 291.
20. Lecky, *'Wrinkles'*, 50.
21. Advertisement in the *Post Office Directory of Plymouth, Devonport and Stonehouse of* 1906–1907, 382.
22. Messerschmitt, 'Ergebnisse von Sextantenprüfungen.' Hughes and Son Ltd and Heath and Co had examination facilities and provided their own certificates until they ceased producing sextants.
23. Jerchow, *From Sextant to Satellite Navigation*, 77. The Deutsche Seewarte was founded as the private Norddeutsche Seewarte.
24. The Koninklijke Paketvaart Maatschappij (Royal Packet Shipping Company) established its own testing facility in Tandjong Priok, the port of Batavia in the Dutch East Indies. In Rotterdam, the navigational instrument firm Observator also examined instruments.
25. NN, *An Explanation of E. Hoppe's Improved Sextant*; Ifland, *Taking the Stars*, 29–31.
26. ROG BOL 14/19[269–270]. The Admiralty paid £42 for the circle to be made.
27. NN, 'Rowland's Double Sextant', 161.
28. NN, 'The Double Sextant', 385–6.
29. Light shades were sometimes included for use against sunlight reflected on the sea.
30. Well into the twentieth century, use of these artificial horizons was a required skill for Royal Naval officers, and included in examinations. Jamieson, 'Simplicity and Precision'; May, *A History of Marine Navigation*, 149.
31. NN, 'A New Angle Sextant', 444, 448, and Ifland, *Taking the Stars*, 131–2. Perhaps Paget was the name of the person who designed this instrument; it was not mentioned in the 1898 and 1909 editions of Wharton, *Hydrographical Surveying*. Around 1892, a Paget compass was designed; see NN, 'The Paget Compass', 446–8.
32. Ifland, *Taking the Stars*, 39–40, 133.
33. An early aluminium sextant was made in Paris in 1857; McConnell, 'Aluminium and Its Alloys', 614. Seamen often preferred a heavier sextant to resist wind buffeting.
34. Jamieson, 'Simplicity and Precision', estimates that from 1720 to 1906, 46 British patents were taken out for artificial horizons.
35. ROG BOL 10[12], June 9, 1823.
36. Becher, 'The Pendulum Marine Artificial Horizon'. Becher invented his first artificial horizon in 1834; see NN, 'Marine Artificial Horizon' (cat. no. 150). He invented a third in 1845; see NN, 'A New Artificial Horizon for Seamen'.
37. Mars, *Zeevaartkundige hoekmeetinstrumenten*, 24. Originally A. Hurlimann in Paris made the sextant and A. Demichel the gyroscope; later examples were manufactured by Hurlimann, Ponthus, and Therrode.
38. Mars, *Zeevaartkundige hoekmeetinstrumenten*, 21–2; Ifland, *Taking the Stars*, 147, 157–8.
39. Mars, *Zeevaartkundige hoekmeetinstrumenten*, 26–8; Lecky, *'Wrinkles'*, 64.
40. Raper, *The Practice of Navigation*, 62, 166; no example of Wollaston's dip sector has been found.
41. Probably the type made by Thomas Jones in about 1845, of which an example is preserved in the NMS, inv. no. 1911.101.
42. Moll, *Kimtiefenmessungen*, 13–18.
43. Hughes, 'The Paget Artificial Horizon', 60–1.
44. Lecky, *'Wrinkles'*, 221–2.
45. Laurent, *Explanation, Adjustments and Use of Laurent's Patent Night Octant*; Mörzer Bruyns, 'The Introduction of the Elongating Glass on Sextants'; Ifland, *Taking the Stars*, 81–2.
46. Mörzer Bruyns, 'The Introduction of the Elongating Glass on Sextants', 77.
47. Invented around 1806 by William Hyde Wollaston (1766–1828), a chemist, physicist, and physiologist.
48. Hughes, 'The Paget Artificial Horizon', 65–6, mentions that it was fitted to a thermometer only. Cat. no. 228 is a sextant with a Wollaston prism by Heath and Co., dated 1921.
49. Watts, *The Sextant Simplified*, 82.
50. Ifland, *Taking the Stars*, 157–89; Dobson, 'Design of Instruments for Navigation of Aircraft', 381–3.
51. Warner, 'Celestial Navigation Aloft'.
52. Hughes, *History of Air Navigation*, 119–29.
53. SOLD would appear to be an acronym but no definite interpretation has been found. Two likely possibilities are Sextant oder Oktant mit Libellenreflektion für Direkte und indirekte Beobachtung, suggested by Dr Wolfgang Köberer, Frankfurt am Main, and Spezial Oktant der Luftwaffe Deutschlands, suggested by Ing. Rudolf Zeiriger, Graz, Austria; both by personal communication.
54. Schaafhausen *et al.*, *C Plath 1862–1962*, 112–15; Jerchow, *From Sextant to Satellite Navigation*, 97–105; NN, *Beschreibung und Bedienungsvorschrift für den Sold-Libellensextanten*; Ifland, *Taking the Stars*, 116–18.
55. Hunt, 'The German Gyro-Sextant'; Ifland, *Taking the Stars*, 117.
56. Watts, *The Sextant Simplified*, 84.
57. A Ramsden inverting land sextant with a micrometer was used on a survey in India in December 1776; see Kochar, 'The Growth of Modern Astronomy in India', 73; Phillimore, *Historical Records*, 200.
58. It was clearly not for finer reading of the scale. See NN, 'Application of a Micrometer-head to the Tangentscrew of the Sextant'.
59. Messerschmitt, 'Ergebnisse von Sextantenprüfungen', 35, 37. The report erroneously states that Gerrard was in London, see Mörzer Bruyns, 'The Navigating Instruments of the National Maritime Museum Greenwich', 27.
60. In the NMS, inv. no. T.1987.147. It has a Kew certificate of examination dated June 1896, and formerly belonged to Leith Navigation School, which was founded in 1846 and became Leith Nautical College in 1903. It may have been an experimental instrument.
61. Schaafhausen *et al.*, *C Plath 1862–1962*, 102–4. An apparently similar indicator was described in Strahan *et al.*, *The Description and Use of a New Astronomical Instrument*, in 1735, but was probably never constructed.

62. Indicator sextants are preserved in the NSA, inv. no. A.5161, in the NMAH, inv. no. 1963–76, and in Copenhagen; see Andersen, *Historic Scientific Instruments*, 42, no. 179. All three are by H. Haecke.
63. The micrometer was temporarily protected by DRGM (Deutsches Reichs Gebrauchs Muster) no. 274505, and almost immediately marketed; see NN, *Katalog über nautische Instrumente*, 9.
64. NN, 'The Micrometer Sextant', 89–90, 192.
65. Schaafhausen *et al.*, *C Plath 1862–1962*, 103–4; Ifland, *Taking the Stars*, 61–3.
66. Ifland, *Taking the Stars*, 66–7.
67. Watts, *The Sextant Simplified*, 79.
68. ROG BOL 14/13[139]; 14/13[140].
69. Ramsden's second engine is preserved in the NMAH; see Stimson, 'Board of Longitude Instruments', 103–4.
70. These octants are preserved in the MSM, inv. nos. 1968.180, 1952.1035, 1983.65, 1968.169, 1959.1011, 1956.820, and in the MMNN, inv. no. NQ0028.
71. Clifton, *Directory*, 295; Mörzer Bruyns, 'De instrumenten van de firma Van Keulen', 69–70.
72. In 1870 Spencer Browning and Rust became John Browning and Co., specializing in magic lanterns.
73. Clifton, 'An Introduction to the History of Elliott Brothers'; Croucher, 'An Exceptional Woman of Science'.
74. The American branch was Kelvin and Wilfred O. White and Co., who around 1942 sold a sextant made by David Shackman and Sons of Chesham, Buckinghamshire, a jeweller who started making sextants at the beginning of World War II; see NMAH inv. no. PH*337006.
75. Clarke *et al.*, *Brass & Glass*, 252–61, 281–85. McGregor and Co. also had an octant made by Spencer Browning and Co.; see NMAH inv. no. EM*331882.
76. Mörzer Bruyns, 'The Navigating Instruments of the National Maritime Museum', 26–8.
77. NN, *Heath & Co., Ltd, Patentees & Manufacturers of Marine Instruments*. There was a wide range of types at less than £6 apiece.
78. Mörzer Bruyns, 'The Navigating Instruments of the National Maritime Museum Greenwich', 27, The Neill sextant is in the MSM, inv. no. 1955.271; a sextant in the MWC&S is signed 'Made for the Sailors' Home Liverpool by Heath and Co Ltd. Crayford, London S.E', inv. no. 1997.4320.122; with thanks to Laureen Sadlier, registrar of the MWC&S.
79. Bauer, *The Sextant Handbook*, 125.
80. Messerschmitt, 'Ergebnisse von Sextantenprüfungen', 42–3. Other English sextants included examples by Ramsden and Berge.
81. Bennett, *Le Citoyen Lenoir*, no. 98; Daumas, *Scientific Instruments*, 279–80. His quintants are rare, there is one in the MNDLM signed 'Jecker à Paris N°, 66', inv. no. 11NA107.
82. Marcelin, *Dictionnaire des fabricants français*.
83. NN, *Heath & Co., Patentees & Manufacturers of Marine Instruments*, 8.
84. Mörzer Bruyns, 'The Introduction of the Elongating Glass', 74–6; Marcelin, *Dictionnaire des fabricants français*.
85. Mörzer Bruyns, 'The Importance of the Van Keulens to Navigation', 68; Mac Lean, *Bijdrage tot de geschiedenis*; Mörzer Bruyns, 'The Navigating Instruments of the National Maritime Museum Greenwich', 28. From about 1900 to 1960 the firm of G. de Koningh in Arnhem manufactured sextants.
86. Filby had worked for the octant manufacturer Heinrich Johann Kosbü, see Schaafhausen *et al.*, *C Plath 1862–1962*, 11, 23; Jerchow, *From Sextant to Satellite Navigation*, 23–31.
87. Schaafhausen *et al.*, *C Plath 1862–1962*, 97–9.
88. NN, *Katalog über nautische Instrumente*, 20.
89. Martins was the son-in-law of Carl Phillipp Heinrich Pistor (1778–1845); see Brachner, 'German Nineteenth-Century Scientific Instrument Makers', 130.
90. Warner, 'American Octants and Sextants', 87–8, 110; Ifland, *Taking the Stars*, 65; see also Miller, 'Circular Dividing Engines in the United States before 1900'.
91. Blunt, *Description of E. & E. W. Blunt's Dividing Engine*, 1–7; Warner, 'American Octants and Sextants', 91–2; Brewington, *The Peabody Museum Collection of Navigating Instruments*, 120; Mörzer Bruyns, 'The Navigating Instruments of the National Maritime Museum', 27.
92. Warner, 'American Octants and Sextants', 108–9.
93. Ifland, *Taking the Stars*, 80; MMNN inv. no. NQ006; MSM inv. no. 1985.68.1. The Negus instruments show much similarity to those by Leon Dubas and Edmond Lorieux.
94. Preus and Treworgy, 'Taking a Departure', 29–30, 32.
95. Allen, *A Century of Scientific Instrument Making*; NN, *Stanley. Great Turnstile, Holborn, London*, 50–1, 54.
96. Clarke *et al.*, *Brass & Glass*, 261.
97. Warner, 'Celestial Navigation Aloft'.
98. Leback, 'World War II: The Lifeboat Sextant'.
99. Watts, *The Sextant Simplified*, 97–8. The Davis Instruments Corporation of San Leandro, California, and the East Berks Boat Company of Wargrave, Berkshire, specialized in plastic sextants.
100. Kandler, 'Nautical Instruments Made in Freiberg'; Ifland, *Taking the Stars*, 69.
101. Jerchow, *From Sextant to Satellite Navigation*, 119.
102. Jerchow, *From Sextant to Satellite Navigation*, 131; Bauer, *The Sextant Handbook*, 159–60.
103. Ifland, *Taking the Stars*, 141.
104. Clarke *et al.*, *Brass & Glass*, 252.
105. Jerchow, *From Sextant to Satellite Navigation*, 122–3. In 1997 it became part of Sperry Marine, which was acquired by Northrup Grumman in 2001.
106. Bauer, *The Sextant Handbook*, 89, 94–5, 112, 166–7.
107. Barlett, 'Choosing & Using a Sextant'.
108. Durham, 'Starman'.
109. Based on NN, *Heath & Co., Ltd. Patentees & Manufacturers of Marine Instruments.*

CHAPTER SIX

COLLECTING AND INTERPRETING NAVIGATION AT GREENWICH

RICHARD DUNN

Having arrived in Greenwich in 1937 and purchased a copy of the catalogue of the recently opened National Maritime Museum, the curious visitor may well have been struck by the Navigation Room, which the plan showed to be on the ground floor of the south-west wing, conveniently opposite the Tea Room.[1] Here,

> The Navigation Screens are covered with manuscript maps and charts, exhibited behind glass, while the desk cases are used for objects which require close examination. Groups of instruments are displayed in wall cases; sometimes on 'staggered' shelves; and sometimes by means of 'cores' and 'Mappin terraces' placed inside the cases and covered with textiles of neutral colour.[2]

From the day the Museum opened to the public that April, navigation and its instruments were central to the collections at Greenwich, although they were displayed separately from the chronological sequence of the other galleries, which primarily featured the art collections. The navigational holdings were displayed in this same gallery for almost half a century.

For a museum about the sea, ships, time, and the stars, it is not surprising that its collection of navigational instruments has been of central importance, or that the history of navigation has formed a significant strand in the Museum's research. The development of both the collection and the subject have always gone hand in hand, so it is appropriate to trace both of these in exploring how the group of instruments described in this catalogue has been assembled and interpreted.

FORMING THE COLLECTION

The Museum was formally established in 1934, but its origins reach as far back as 1824, with the opening of a National Gallery of Naval Art in the Painted Hall of Greenwich Hospital, which, after it became the Royal Naval College in 1872–73 also housed a Naval Museum.[3] These remained the only displays solely concerned with British maritime history until the following century when, almost from the moment of its foundation in 1910, the Society for Nautical Research (SNR) pressed for the establishment of 'a Maritime Museum of National importance and National dignity'.[4] The SNR repeated this call regularly, emphasizing in 1924, for example, that the proposed museum would be 'for the illustration and study of the maritime history of Great Britain in its widest sense'.[5] To support this, the Society also organized occasional displays of historical material, including navigational instruments; a contribution to the British section of the First Dutch Shipping Exhibition in Amsterdam in 1913 was an early instance.[6]

Following the British Government's decision in 1927 to sanction moves towards founding a museum, the Chairman of the SNR, Sir George Hope, emphasized at its annual general meeting that this should be a separate institution. For, he said, the new museum

> cannot conceivably form part of any existing Museum or group of Museums; for the simple reason that it must be based – in respect of knowledge – not upon terra firma, but on that other element which is best understood by those 'who go down to the sea in ships'. To illustrate my meaning, I will cite the case of navigational instruments. One has only to

> look at the arrangement of Nocturnals, Quadrants, Ring Dials, Sextants and Reflecting Circles in our London Museums to realise that in recording the progress of Navigational aids you will be hopelessly at sea without a Navigator.[7]

The SNR's efforts at this point were focused on saving for the nation the Macpherson Collection of maritime prints, drawings, and paintings, for which it had launched a public appeal. This was not doing well, however, until a Scottish shipowner, Sir James Caird (1864–1954), agreed to purchase both the Macpherson Collection and a collection of ship models from the training ship *Mercury* for the new institution. These were held by a temporary board of trustees, replaced by a permanent board established by the National Maritime Museum Act of 1934. The contents of the Naval Museum in the Royal Naval College, as well as other official material, were also transferred to the new Museum, while most of the Greenwich Hospital collection of paintings and other items, hitherto largely displayed in the Painted Hall (which was also in the Royal Naval College), was transferred as a 'permanent loan' in 1936.[8] Each of these founding collections included a handful of navigational instruments.[9] Scientific instruments, in particular those relating to navigation and cartography, therefore not only formed part of these pre-existing collections, but were also of increasing interest to the Museum and its benefactors as they continued to develop the collections.

From the start, the most prominent benefactor was Caird, whose impact on the formation of the collections is impossible to overstate.[10] Caird had become rich as owner of the Scottish Shire Line and spent much of his fortune amassing a large collection of maritime material. As a result, he became an enthusiastic supporter of the SNR and so was involved with the Museum from the start. This and his other support to maritime heritage – notably the earlier preservation of Nelson's *Victory* – earned him a baronetcy and he became a Trustee from 1934. In this role, he soon became a key advocate for the nascent institution – writing in 1935, for instance, that 'now that the nation has shouldered the financial responsibilities of a National Maritime Museum I am certain the shipping community will not be behindhand in contributing its share of the maritime treasures amassed in years gone by'.[11] These were not empty words: by this time he had handed over his own collection at the Trustees' meeting of 8 October 1934, in the symbolic form of a four-volume catalogue, which the Museum's first Director, Geoffrey Callender (1875–1946), considered a 'forerunner of detailed descriptive Catalogues'.[12]

Caird supported the Museum for the rest of his life, initially financing a huge number of additional acquisitions directly, before setting up a trust fund for purchases (the Caird Fund) to continue this. As a result, he was central in defining the range of the Museum's collecting in its early years, aided in particular by Callender. As regards navigational instruments, Caird's generosity allowed purchases from sources worldwide.[13] These included items from the *Exposition des Instruments et Outils d'Autrefois*, held in 1936 at the Musée des Arts Décoratifs (in the Louvre), Paris.[14] Caird also became personally involved in negotiations with and purchases from individual collectors, such as the antiques dealer Nicolas Landau (see Appendix 1). Landau had helped organize the Paris exhibition and negotiated the sale of items from other lenders to the exhibition, such as Georges Bernard, as well as from his own collection.[15] He later also acted as agent in the sale to the Museum of the Barberini collection, which was by then in the possession of a Parisian collector, Charles Michel.[16] An additional player in these negotiations was Frank Maggs, one of the partners of Maggs Brothers Ltd. Although this firm has primarily dealt in antiquarian books and manuscripts since its foundation in about 1853, its involvement with the Museum also included acting as an agent for the acquisition of scientific artefacts, including many navigational instruments, with Frank Maggs becoming a trusted authority in this area.[17] Likewise, both Caird and the Museum, especially in its early years, often made purchases through certain other dealers acting as agents, such as Spink and Son.

Caird's role, therefore, was not just as a financial backer but included developing relationships with dealers and collectors on behalf of the Museum. In this capacity, his most significant contribution to the collection of navigational instruments came through his involvement in the purchase of the Gabb collection.

Having trained as an analytical chemist, George Hugh Gabb (1868–1948) worked for Welford and Son and then as an independent consultant. Having become interested in the history of science by 1885, he began collecting historic instruments and other material, later styling himself as the 'Father of the Scientific Antiquaries'.[18] Indeed, his obituary observed that, 'As one of the first of collectors at a period when there was little interest in old scientific instruments, he was able to make a collection which consisted of really first-class specimens in perfect condition'. The same piece also observed that 'Gabb was a remarkable character – a very typical collector, who would stick at nothing to acquire the instruments he coveted. He was by no means blind to his own qualities in this field, and his amusing conversation was not a little marred by its egocentric character'.[19]

This soon emerged in the negotiations surrounding the Museum's acquisition of his collection in 1937. By that time it contained over 1300 scientific instruments, globes, prints, trade cards, and manuscripts. Negotiating for the Museum, however, Caird was interested only in the maritime material and not the 'scientific' collection. Writing to Gabb at the end of 1936, he complained that these two groups 'are not even complementary to each other . . . To endeavour to sell them together appears to me – as a business man – to make one Collection a serious drag on the other. To do so would, in my view, be like trying to sell a fine elephant and a motor launch together. The buyer – with his chief object the elephant – would be "at sea" with the motor launch, and the motor launch specialist would be "aground" with the elephant.'?[20]

The relationship between Caird and Gabb deteriorated still further when Caird suggested that Gabb was ignorant of the world of serious collecting. Gabb responded abruptly that, 'I can

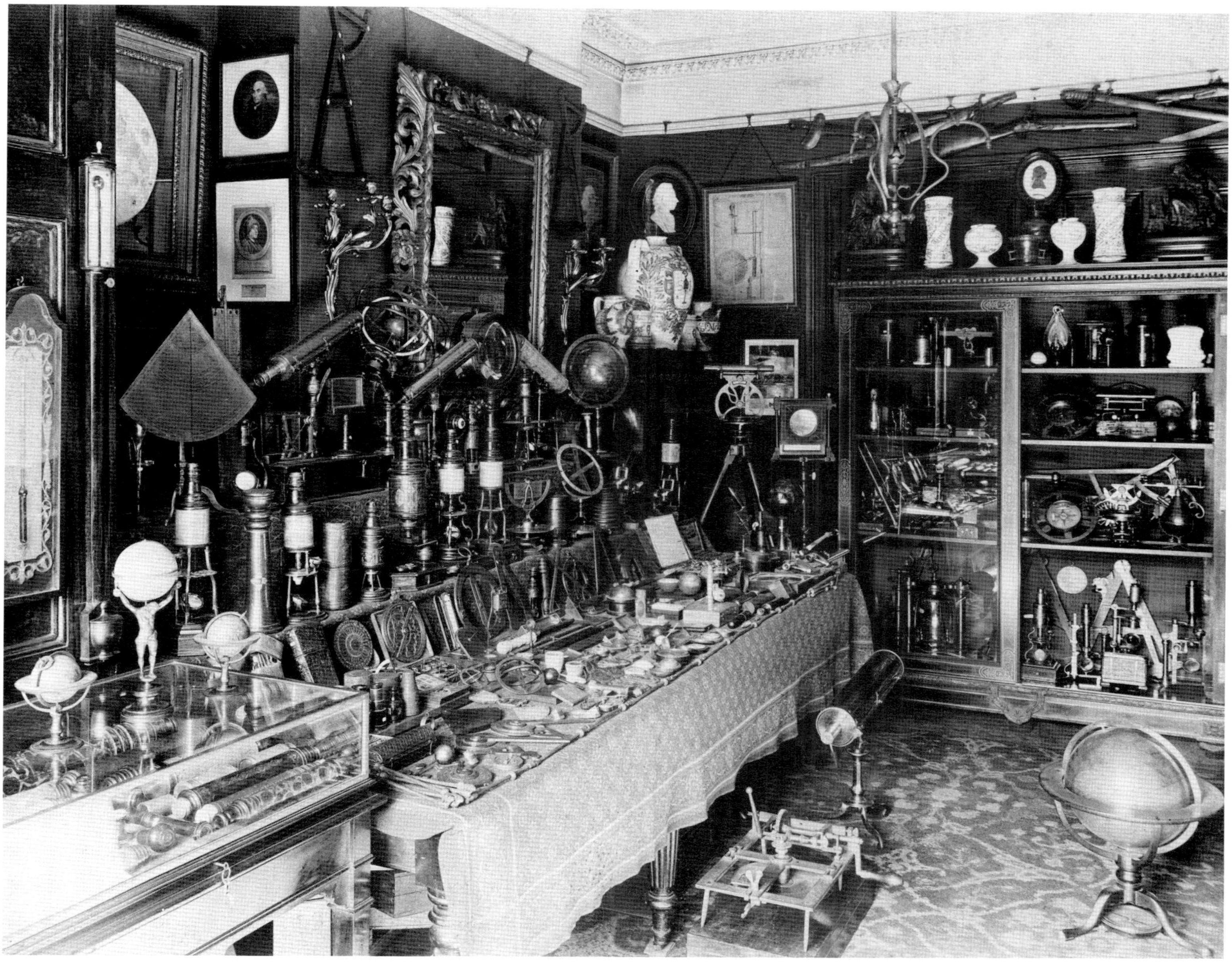

FIGURE 6.1 Part of the collection of George Gabb, displayed at his home in north London in 1911. National Maritime Museum; D8504.

assure you I know more about the world of collecting than all your Johnson-Saints, Maggs and the rest of them rolled into one. I am the Father of them all!'[21] With Gabb threatening to sell to the USA, however, Caird agreed to buy the whole collection for £10 000, although the Museum secured Gabb's agreement that it might subsequently dispose of some parts of it.[22] Yet this tetchy episode did not seem to put Gabb's relationship with the Museum into long-term jeopardy, since the following year he presented an octant (cat. no. 071).

Gabb remains the most significant source for the Museum's collection of scientific instruments as a whole, with twelve items in this catalogue deriving from his collection (see Appendix 1). But for the collection of navigational instruments in particular, a more significant source was a contemporary and acquaintance of Gabb's, Edgar Tarry Adams.

EDGAR TARRY ADAMS, 1852–1926

Edgar's wealth came from the Adams brewery in Halstead, Essex, which his father had purchased in 1876 and which Edgar subsequently took over. Yet even after taking charge of the brewery, Edgar found time to indulge in an astonishing array of interests and activities. These included being a lay preacher, an active member of the local council and the Local Board of Health, an accredited observer for the Royal Meteorological Society (with his own weather station and a column in the local newspaper), founding and becoming captain of the local volunteer fire brigade, and setting up the Halstead Science and Art Club.[23] In later life he also became a lieutenant-commander in the Royal Naval Volunteer Reserve, with an honorary commission awarded by the Admiralty on 23 November 1917. And after buying a holiday

FIGURE 6.2 Edgar Tarry Adams (1852–1926) in his boathouse in Deal, Kent, about 1900. Courtesy of the National Museum of Photography, Film and Television/Science and Society Picture Library.

home in Deal, that house became his base to indulge a passion for painting and photography.

Adams took 'the keenest interest in nautical matters; he . . . sailed a boat since boyhood, and his collection of navigating instruments was very extensive and probably unique'.[24] This collecting, of what his son later described as 'nautical curios', was not confined to instruments but also included other items, such as prints, books, and ship models, many of which have also come to the Museum.[25] Yet, as the list of navigational and other instruments collected by Adams in Appendix 1 shows, he was most unusual in collecting navigational instruments in such a focused way – much more so than Gabb, for example, of whose collection navigational instruments formed only a small part.

Surviving papers from the Adams family reveal something of the nature of Edgar's collecting. One of his main strategies was to keep in regular contact with auctioneers, antique and other dealers, and instrument firms, notably Hughes & Son, whose dealing in second-hand instruments apparently included historic examples. Often Adams's correspondence with retailers discussed very specific requests for items that he wished to add to his collection, including a long-running search for a good-quality back-staff that, surprisingly, seems to have been unsuccessful. Letters from the firms he dealt with also show that Adams preferred to collect instruments that were in working order, and was happy to pay for repairs to bring them to a working state – a very different sensibility from that of many collectors and museums today.

Adams was clearly determined to take advantage of any opportunity to add to his collection. A number of the instrument cases, for example, contain small labels that record how he acquired them. In particular, several show that he encouraged his family to give instruments as birthday and Christmas presents, while others were won as prizes. In many cases, Adams had already purchased the instruments or suggested exactly what the gift or prize should be. For example, Francis Adams noted in his diary for 1908 that the family gave his father a sextant by Ramsden for his birthday (cat. no. 143), although Edgar had 'got it some time ago at Hughes's'.[26]

Adams also kept in contact with other collectors. In 1908, for example, H.E. Head wrote to Adams that 'Mr Hill of the South Kensington Museum [later the Science Museum] tells me that you are a collector of Mathematical Instruments'.[27] This proved to be the beginning of a long-running acquaintanceship in which the two discussed their collections, the possibilities of sale and exchange between them, and the working of specific instruments.

FIGURE 6.3 Instruments from Edgar Tarry Adams's collection, photographed by Adams himself. Francis Adams, Edgar's son, recorded in his diaries helping his father photograph the instruments on a number of occasions. National Maritime Museum; F7654.

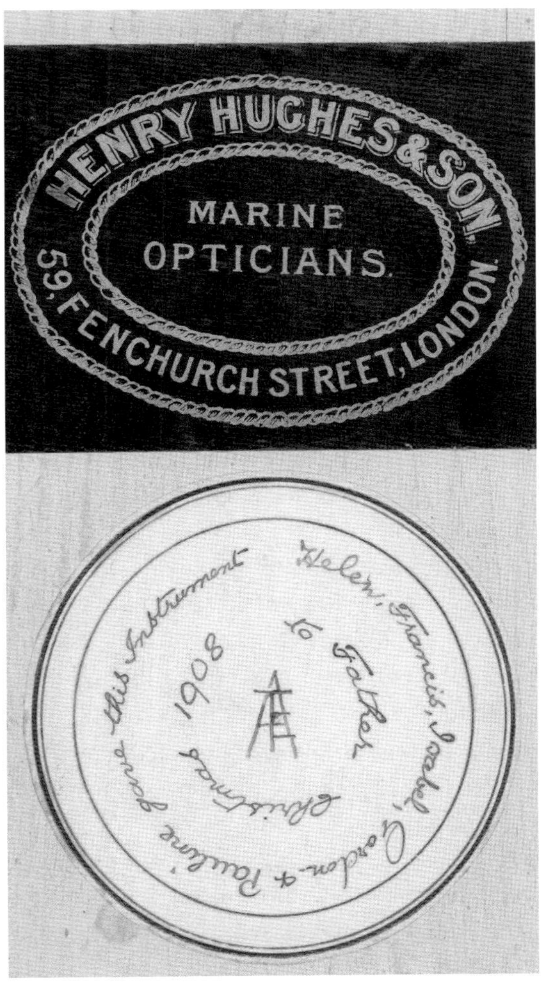

FIGURE 6.4 (*left*) Labels inside the case of an octant from about 1770. The handwritten label records that it was given to Adams by his family for Christmas 1908, which Francis also noted in his diary for that year. The monogram in the centre is made up of Adams's initials. The label from Hughes and Son shows that Adams had bought the instrument through them (cat. no. 047).

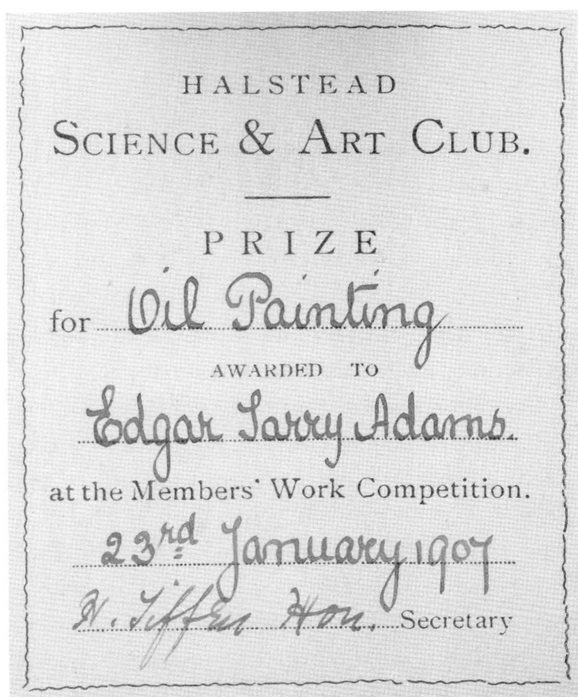

HALSTEAD

SCIENCE & ART CLUB.

PRIZE

for Oil Painting

AWARDED TO

Edgar Tarry Adams.

at the Members' Work Competition.

23rd January 1907

H. Tiffen Hon. Secretary

FIGURE 6.5 (*right*) Label inside the case of an English octant from about 1825. This was one of a number of items that Adams selected as competition prizes (cat. no. 090).

Head's letters to Adams in 1911 include, for instance, ongoing discussion of a circle by Pistor and Martins (probably cat. no. 312), about which Head finally concluded, 'I am rather disappointed in the circle. It is certainly very well graduated . . . but a more hopelessly impractical instrument I have never seen . . . The instrument is evidently intended for surveyors', not sailors, use'.[28] Like Adams, Head was interested in the practical operation of the instruments.

George Gabb first contacted Adams in August 1917, writing that as 'one of the few people who take an interest in early scientific instruments', he had 'heard from two or three sources, that you have a most interesting collection'.[29] The two subsequently kept in touch and met to view their respective collections (or at least to look at photographs of them), and even discussed swapping items. Gabb wrote in 1922, for example, that, 'I never sell anything out of my collection, but if you had something very nice that I wanted we might do an exchange.'[30] Some potential swaps were discussed, but it seems that no mutually satisfactory deal was ever achieved.

Most of Adams's collecting took place between 1906 and his death in 1926, and it seems that he quickly acquired a reputation as a collector with genuine knowledge of navigational instruments and their makers. Letters from W.J. Hill, who worked in the Science Collections at the Science Museum, for example, asked for Adams's advice on a number of matters, such as the working of Beecher's artificial horizon (cat. no. 337). Likewise, when Alexander Hughes (of Hughes & Son) was putting together the *Dictionary of British Scientific Instruments*, Adams was the obvious person to write the introductory piece on navigating instruments.[31]

Sadly, Adams had died by the time the plans for the National Maritime Museum were coming to fruition in the 1930s. Instead, it was his son, Francis Norris Adams (1882–1973), who negotiated the loan of items to the Museum. Francis shared his father's passion for maritime subjects and was a long-standing member of the SNR, serving on its Council from 1930 to 1934. This was a key period in the formation of the Museum and his involvement in the SNR brought him into contact with Geoffrey Callender, with whom he was discussing the idea of a loan of instruments by July 1932. His formal offer in April 1933 was warmly accepted by the Trustees in October.[32] With the loan agreed, the first group of instruments was sent to Greenwich on 12 April 1935 (with the formal paperwork completed in 1936). The loan included ship models and pictures, while many books were given at the same time. Francis initially held on to other items from his father's collection but made additional loans of instruments and further gifts of other material in later years, while the Museum (as well as the Science Museum) also purchased items in the 1950s, a period of financial hardship for Adams. When the loaned items were finally bequeathed to the Museum after Francis's death in 1973, they had been central to the Museum and its displays for over thirty-five years.[33] Indeed, it is hard to overestimate the importance of this group, in terms of both numbers (Adams's collection supplied 74 of the instruments in this catalogue) and quality. The octants he acquired, for example, include some of the earliest in the Museum's possession.

OTHER SOURCES

As well as pre-formed collections such as those of Adams and Gabb, and a few items from other collectors such as Ernest R. Cooper (cat. nos.023 and 043) and W. St John Pym (cat. no. 026), the Museum has also acquired groups of objects that reflect working lives.[34] These include, for example, instruments from the personal collection of John Lort Stokes (1812–85), who served on the *Beagle* as Assistant Surveyor in the period of Darwin's famous voyage (1831–6), subsequently taking command of the ship before going on to carry out survey work in the *Acheron* (cat. nos.168, 174, and 320).[35] Likewise, the collection of instruments presented by Queen Mary in 1936, just before the Museum's opening, included several items from members of the British royal family, including instruments owned by Prince Alfred, Duke of Edinburgh (1844–1900) and King George V (1865–1936, reigned 1910–36), who had both served in the Royal Navy (cat. nos.198, 289, and 327).

In similar vein, a number of instruments have been acquired by transfer from the Admiralty and organizations within it, notably from HMS *Dryad*, the Navy's navigation school, and the museum of the Admiralty Compass Observatory, following its closure. Like the instruments given by the navigational instrument manufacturer Kelvin Hughes, these collections represent not only instruments left over from training or sale stock, but also items collected on account of their historical interest. Other acquisitions, from the Ministry of Defence and from the Royal Greenwich Observatory, based in Cambridge at the time of its closure in 1998, are more obviously instruments once regularly used by those organizations.

COLLECTING POLICY

The collecting of navigational instruments has necessarily been carried out within the remit of the Museum's wider purposes for acquiring artefacts, guided first and foremost by that of providing a 'proper understanding of British maritime development'.[36] By the late 1950s, however, there were significant moves, led in particular by HRH Prince Philip, appointed as a Trustee in 1948, to create a more formalized and proactive collecting policy, which came to fruition in the early 1960s.[37]

By this time, about half of the items in this catalogue (including the Adams material that was then on loan) had entered the Museum, forming part of a rich core of navigational holdings. As a consequence, the acquisition policies created from the 1960s onwards emphasized that filling gaps was the way forward for the navigation collection, which was acknowledged as a key area for the Museum, rather than there being significant areas that were under-represented. Although this subsequently changed slightly to highlight the challenge of representing the twentieth century more fully, the 1974 policy could justly laud the fact that the Museum had 'the world's finest collection of navigation and navigational astronomy material'.[38]

Unsurprisingly, therefore, the rate of acquisition has been slower in recent years, reflecting not only the financially tougher times that museums face, but also the fact that the quality of the collection assembled to date means that 'essential' additions become available much less frequently. Since the 1970s, with the collection already well supplied with examples of the very finest instruments, there has been an increased emphasis on trying to build a more representative collection, not just of the best examples, but also of the more run-of-the-mill instruments that people could afford and used on a daily basis, or with instruments with known histories and personal associations. This was also reflected in moves to keep good contacts not only with dealers but also with the underwater-archaeology community, since by the 1970s this had become the main context in which instruments with specific associations came to light. Efforts to stay in regular contact with underwater archaeologists paid off in acquisitions that included material from the *Stirling Castle*, wrecked in the Great Storm of 1703 (cat. nos.008, 009, 015, 016, and 017), as well as providing much of the information that underpinned Alan Stimson's detailed study of the mariner's astrolabe.[39]

As the above indicates, recent collecting has also reflected the research interests of individual staff. Stimson's research into early serial-numbered instruments, for example, such as those of Ramsden and Troughton, led among other things to the acquisition in 1982 of a sextant believed to have been used on Captain Cook's third voyage to the Pacific (cat. no. 135).[40] Likewise, one of the most recent additions (cat. no. 217) was acquired not only because of its rarity, but also as material evidence of the British scientific instrument trade, which has formed a key part of the work of Gloria Clifton.[41]

EXHIBITING AND RESEARCHING THE HISTORY OF NAVIGATION

Throughout the history of the Museum, including the years before its public opening, navigation and its instruments have been central to the Museum's essential purposes. As early as 1891, for example, Sir Clements Markham (later the first chairman of the SNR) proposed in a talk to the Royal United Service Institution that a collection of instruments, charts, and maps – objects that, Markham felt, had changed how mariners navigated – should be created as a resource for officers at the Royal Naval College. He argued that:

> The officer who uses a sextant to take the sun's altitude would . . . be the better, not only for a knowledge of the history of the gradual development of inventions from early times, from the first rude instrument ever used at sea, down to the construction of the perfected one he holds in his hand; but he would also be the better for an appreciation of the labours and discoveries of the men who . . . were the motives and causes of that development.[42]

Although his idea of forming 'collections which can be inspected . . . during a leisure hour or half hour' received a lukewarm reception at the time, this belief in the efficacy of navigational instruments for telling us about changes in seafaring and in how people have seen the world has been reflected in the ways in which the Museum has sought to research, interpret, and display its collections since its foundation.[43]

This potential was recognized early on by Geoffrey Callender, who by 1936 had requested a 'navigator' to add to the staff, resulting in the appointment of Captain Patrick Maxwell, formerly of the Hydrographic Department of the Royal Navy, as the new Assistant in the Department of Navigation.[44] Maxwell must then have been involved in mounting the original displays in the Navigation Room, which included backstaffs, quadrants and sextants among the instruments on show. According to the *Mariner's Mirror*, the room was arranged slightly differently from the other galleries at the new museum, displaying 'a compromise between strict chronology and individual grouping'. This was done to allow the visitor to gain 'a general view of the broad course of development' of navigation.[45] Other navigational instruments could be seen elsewhere in the Museum; for example, those related to Captain Cook were in the displays dedicated to him.[46]

The Museum's interest in science and its instruments, including those related to navigation, broadened in the 1950s, when the former Royal Observatory buildings in Greenwich Park were transferred to the Museum's care, after the astronomers of the Royal Greenwich Observatory (as the working body was still known) moved to Herstmonceux Castle in Sussex.[47] This resulted in a more explicit interest in purely astronomical instruments, with the 'Old Royal Observatory', as it became formally known from 1967 to 1998, becoming the main focus for the Museum's scientific collections, although displays in the main Museum buildings continued to highlight the role of navigational instruments in maritime history.

During this time, the presence of Commander W.E. May (1899–1989), appointed as Deputy Director in 1951, added to the Museum's expertise in navigation. Having served in the Royal Navy, May had worked at the Admiralty Compass Observatory since 1929, during which time he had written on a number of historical topics, concerning not only the history of the compass but also, to a lesser extent, celestial navigation and its instruments.[48] At the Museum, he continued researching and publishing on a wide range of maritime subjects, notably uniforms, weapons, timekeepers, and shipwrecks, as well as a few further pieces –

largely short notes – on navigational instruments.[49] This latter work culminated in a history of navigation, published five years after his retirement.[50]

Nevertheless, apart from the occasional publication by May, the day-to-day demands of running the Museum hindered attempts to develop an ongoing research agenda. In 1959, for example, the Haylock Report – compiled the previous year as part of an independent enquiry ordered by the Museum Trustees – bemoaned the lack of time available for research on collections that had grown considerably (the Museum had 1300 navigational instruments alone by this time), although it did point out that staff had managed to produce a number of publications.[51] These included a catalogue of the navigation collection, *Instruments of Navigation*. The introduction to this short guide set out its intention, which was to be

> of interest and value to the non-technical visitor and give him an idea of how each object which he sees would have been used. At the same time the catalogue entries themselves have been designed to give sufficient information to the expert who may wish to know to what extent a particular field is represented in the Museum.[52]

In other words, the catalogue was aimed at a range of people of different backgrounds and interests.

These minor successes aside, the difficulty in finding regular time for detailed research was exacerbated in the Navigation Department by an ongoing shortage of staff. By 1957, Henry Hill, the only assistant, was 70 and due for retirement.[53] Attempts to fill the post of Assistant Keeper proved problematic, but were alleviated when Alan Pearsall (1925–2006) moved from the Manuscripts Section to Navigation, only to find that he was sharing his time with his old department.[54] During this same period there were also moves to make significant changes to the existing displays throughout the Museum, which were by then felt to be inadequate.[55]

Things began to improve in 1960, however, with the appointment of Lieutenant-Commander David W. Waters (b. 1911), formerly on the Admiralty historical staff and by that time well known for his seminal work *The Art of Navigation in England in Elizabethan and Early Stuart Times*.[56] On appointment, Waters was the only member of the Navigation Department, although Alan Pearsall gave him valuable assistance. Nevertheless, he was soon allowed to expand his team, with notable additions including Susannah Fisher (as Curator of Hydrography), Lieutenant-Commander Derek Howse (a former specialist navigating officer), and Alan Stimson and Christopher St J.H. Daniel, both former officers in the British merchant navy, within the next few years.[57] All arrived with longstanding seafaring experience, something that had been the norm since the Museum's foundation.[58]

Having successfully increased the number of staff, Waters was able to position the Department more securely, both within the Museum and externally. Internally, Waters brought administrative

FIGURE 6.6 The front cover of *Instruments of Navigation*, the first published catalogue of the navigational-instrument collection, written by H.O. Hill and E.W. Paget-Tomlinson and published in 1956. National Maritime Museum; F7657.

discipline, placing great emphasis on proper cataloguing of the collection, including the many acquisitions that had arrived since the Second World War but which remained poorly documented. He also made good use of his international contacts to promote the Department around the world, putting it firmly on the map as a key academic player in researching and promoting the history of navigation. In general, his approach was to view the subject holistically, insisting that the instruments should be studied, stored, and displayed together with the charts, navigational tables, and manuals that the art of navigation requires. This was also reflected in the special exhibitions that Waters and his department put on, including *Four Steps to Longitude* (1962), *Early British Contributions to the Art of Navigation* (1964), and *The Nautical Almanac Bicentenary* (1967).

Howse also played a significant role in these developments. As an experienced navigator, he soon helped to revitalize the displays in the Navigation Room, which became 'the envy of museums world-wide'.[59] With the added responsibility of the Royal Observatory exercising the Department in the 1960s, however, Howse was soon heavily involved in the astronomical side of the Museum's work, although he remained an important figure for the navigational collections, taking over as Head of the Department of Navigation and Astronomy in 1976.[60]

With more people around, the Department found itself on a firmer footing and staff were able to devote more time to research on the collections. Guided by Howse's more recent managerial experience in the oil industry, the successful cataloguing of the collections led to the compilation and publication of *An Inventory of the Navigation and Astronomy Collections in the National Maritime Museum Greenwich* (1970–3). As Waters said in the preface to what became a three-volume listing, this represented the intermediate stage between the initial internal documentation of the collections and fully published catalogues.[61] Although in some ways not much more than a handlist, with brief introductory notes on the different classes of instruments, the *Inventory* proved invaluable to researchers in subsequent decades. It also reflected a wider trend throughout the Museum that saw the publication of catalogues of many other collection areas.[62]

Alongside the *Inventory*, the staff of the Navigation Department produced a range of papers and other publications exploring aspects of the collections and the history of navigation. These

FIGURE 6.7 (*above*) The Navigation Room at the National Maritime Museum in 1965. National Maritime Museum; A7841–13.

FIGURE 6.8 'The Rival Method', referring to the method of determining longitude by lunars rather than by chronometer. Display case in the Navigation Room at the National Maritime Museum, in 1965. National Maritime Museum; A7840–1.

included a series of Occasional Technical Papers,[63] as well as contributions to the Museum's series of Maritime Monographs and Reports that were produced from 1970 to 1988, with some commissioned from external experts.[64] This was coupled with moves to raise the profile of the Museum among academic researchers by hosting international conferences, including a seminar on 'Fakes and Facsimiles (Scientific Instruments)' in 1981, the second symposium of the Scientific Instrument Commission in 1982, and a conference in 1984 to celebrate the centenary of the adoption of Greenwich as the prime meridian.[65]

RECENT DEVELOPMENTS

In the 1980s and early 1990s, the Department underwent further change as the Museum, like other UK national museums, shifted its priorities owing to financial and political pressures, leading to restructuring and staff changes that greatly altered the nature of research and curatorship at Greenwich.[66] This included a shift towards the recruitment of staff with formal historiographic training rather than of experienced seafarers. The same pressures, and the start of a period of major buildings renovation, also led to reworked displays throughout the Museum, including the dismantling of the Navigation Room in 1985.

As a result, navigation has in recent decades been presented to Museum visitors not as a separate subject but as part of a wider interpretation of Britain's maritime and scientific heritage. This has been typified in displays in and around Neptune Court, where navigational instruments have been used to interpret topics such as the history of oceanic exploration and London's maritime past, as well as in those in the Royal Observatory (redisplayed in 2006), where the story of the discovery of methods for successfully determining longitude remains a key narrative.

Happily, since the mid-1990s, staff at the Museum have again found the time and institutional backing to promote more explicitly research based on the instrument collections, for example through the work of Gloria Clifton, appointed in 1992 as Curator of Navigational Instruments (subsequently becoming Head of the Royal Observatory). Clifton had previously been a

FIGURE 6.9 The *Time and Longitude* gallery in the Royal Observatory, as redisplayed in 2006. The display case discusses the lunar-distance method of longitude determination and includes cat. nos. 133, 140, and 033. National Maritime Museum; F8248–1.

research officer on Project SIMON,[67] and as a result the Museum accepted the task of maintaining the project's database and publishing the *Directory of British Scientific Instrument Makers 1550–1851,* which has become a standard reference work for the field. This renewed emphasis on the study of navigational and other scientific instruments as valuable evidence in the history of science and technology, as well as in the study of Britain's maritime history, informed the decision to produce catalogues such as this one, in particular through the Sackler Research Fellowship in the History of Astronomy and Navigational Sciences, which has run since 1993.[68] The same thinking remains at the heart of the Museum's future research agenda.

Trends in research come and go, just as the accumulation of a museum collection occurs in waves. The collection of sextants, octants and related instruments now held at the Museum has come from a wide variety of sources, and has been researched and interpreted in an equally wide range of contexts. It can only be hoped that such diversity continues to inform the future enrichment of the collections through both acquisition and further contextual research.

NOTES

1. NN, *National Maritime Museum Catalogue.*
2. NN, 'The National Maritime Museum', 275.
3. The Museum was officially opened by King George VI on 27 April 1937, with visitors admitted from 29 April following a private-view day.
4. NN, 'The Society for Nautical Research', 104.
5. NN, 'The National Maritime Museum', 256.
6. NN, 'The Society for Nautical Research', 97.
7. NN,' Annual General Meeting', 202. The same words were repeated three years later in NN, 'The Society for Nautical Research', 105.
8. Littlewood and Butler, *Of Ships and Stars*, passim.
9. Cat. nos. 155 and 171 came from the Royal Naval Museum; cat. nos. 183, 199, 292, 300, and 303 from the Greenwich Hospital Collection.
10. See also Blyzinsky, 'The History of the Collection', esp. 14–16; Dunn, 'Sundials and the Arts', esp. 45–6.
11. Caird, 'A Museum in the Making'.
12. Callender, *The Caird Collection of Maritime Antiquities*, Editor's Foreword.
13. Mason, 'Sir James Caird', for example, states that 'In order to secure for the Navigation Room the instruments necessary to show navigational aids from the earliest times, he purchased large collections from abroad as well as in this country', 224.
14. The museum purchased 319 items at the Paris exhibition, the largest group being sundials. The purchases included cat. nos. 053 and 068. See also NN, *Exposition des instruments et Outils d'Autrefois*; Turner, *Early Scientific Instruments*, 278–9.
15. Forty items were purchased from Landau in 1936, including an octant, a sextant, and a quintant (cat. nos. 056, 195, and 285). A further 41 items were purchased in 1938, of which nine were octants or sextants (cat. nos. 065, 073, 078, 082, 086, 093, 106, 156, and 161).
16. The Barberini collection included a mariner's quadrant of about 1600 (cat. no. 001). For details of Landau, see also Turner, 'Paris Amsterdam, London', 38.
17. Blyzinsky, 'The History of the Collection', 16.
18. Taylor, 'Mr. G. H. Gabb'.
19. Taylor, 'Mr. G. H. Gabb'.
20. Caird papers, NMM 5: letter dated 18 December 1936; quoted in Blyzinsky, 'The History of the Collection', 18.
21. Caird papers, NMM 5: letter dated 14 January 1937; quoted in Blyzinsky, 'The History of the Collection', 18. Johnston-Saint (not Johnson-Saint, as Gabb wrote) was conservator at the Wellcome Museum. Caird was trying to interest him in sharing the purchase of the Gabb collection.
22. Some items from the Gabb collection have since been transferred to the Science Museum, London. See Littlewood and Butler, *Of Ships and Stars*, 128. Cat. no. 005 was recevied as part of the exchange of items in the 1950s.
23. This reflects only a small part of Adams's activities, which also included contributions to many clubs, societies, and charities. See ERO, Acc. A6357.
24. Obituary for Edgar Tarry Adams, *Halstead Gazette*, Friday 30 July 1926, reproduced in Essex Record Office, *The Adams Family – Growing Up in Victorian Times. Further Sources.* See also Essex Record Office, *The Adams Family – Growing Up in Victorian Times. Notes.*
25. F.N. Adams, Letter to Frank Carr, 2 June 1956, NMM file 8/1353.
26. F.N. Adams, personal diary for 1908, entry for 8 December, ERO, Acc. A6357.
27. H.E. Head, letter to Adams, 6 July 1908, ERO, Acc. A6357.
28. H.E. Head, letter to Adams, 27 August 1917, ERO, Acc. A6357.
29. George Gabb, letter to Adams, 20 August 1911, ERO, Acc. A6357.
30. George Gabb, letter to Adams, 25 November 1922, ERO, Acc. A6357.
31. Adams, 'Navigating Instruments'; ERO, Acc. A6357 includes letters from Hughes to Adams in 1918 concerning this article.
32. NMM file 193/1935; F.N. Adams, letter to Callender, 10 April 1953, NMM file 8/1353; James Stanhope, letter to F.N. Adams, 20 October 1933, NMM file 8/1353.
33. NMM files 8/1353 and X74/042. The bequest was formalized in 1974.
34. See, NN, *Catalogue of an Interesting Collection of Scientific Instruments. The Property of W. St. J. Pym, Esq., C.B.E.*. The sale took place on 17 October 1960.
35. NN, *An Inventory of the Navigation and Astronomy Collections in the National Maritime Museum Greenwich*, Vol. III, Section 34.
36. NN, 'The National Maritime Museum', 257.
37. Littlewood and Butler, *Of Ships and Stars*, 171–2.
38. National Maritime Museum, *Museum Collecting Policy*, 1974, NMM file G10/1/01.
39. Stimson, *The Mariner's Astrolabe.*
40. See, for example, Stimson, 'Some Board of Longitude Instruments'.
41. Notably Clifton, *Directory*, as well as numerous other papers.
42. Markham, *On the Advantage of Forming Collections at Greenwich*, 2.
43. Markham, *On the Advantage of Forming Collections at Greenwich*, 15.
44. Littlewood and Butler, *Of Ships and Stars*, 87–8. Interestingly, this reflected Sir George Hope's earlier statements that the Museum needed to be staffed by people with specific seafaring knowledge.
45. NN, 'The National Maritime Museum', 269–70; see also Callender, 'The National Maritime Museum', 51, 55.
46. Callender, 'The National Maritime Museum', 49.
47. The transfer of the Royal Observatory buildings took place between 1953 and 1958.
48. Hitchins and May, *From Lodestone to Gyro-compass*; May, 'Old Altitude Instruments', May, 'Early Reflecting Instruments'; May, 'Position Finding at Sea (1, Latitude; 2, Longitude; 3, Position Lines)'; May, 'The Double Altitude Problem'.
49. May, 'Hadley's Quadrant'; May, 'Mariner's Astrolabes'; see also Munday, 'Commander W. E. May, R.N.'
50. May, *A History of Marine Navigation.* Three years later, he also produced 'How the Chronometer Went to Sea'.
51. Littlewood and Butler, *Of Ships and Stars*, 167–8.
52. Hill and Paget-Tomlinson, *Instruments of Navigation*, v.

53. Commander Henry Oliver Hill worked at the Museum from 1949 to his retirement in 1959. He subsequently gave a number of items to the Museum, including cat. no. 060, which was bequeathed after his death.
54. Littlewood and Butler, *Of Ships and Stars*, 177; NN, *An Inventory of the Navigation and Astronomy Collections in the National Maritime Museum Greenwich*, ii.
55. On 20 June 1957, Commander May produced a report entitled 'A Sense of Direction', following discussions at a number of meetings of the Trustees. Key members of staff commented on the report in the weeks that followed, in what was a key discussion in determining subsequent display strategies at the Museum. See Littlewood and Butler, *Of Ships and Stars*, 171ff.
56. Littlewood and Butler, *Of Ships and Stars*, 177.
57. Littlewood and Butler, *Of Ships and Stars*, 180–1.
58. This tradition continued in subsequent appointments to the Department, including those of Peter Ince (a master mariner), Ken Snowden (from the Royal Navy), and Christopher Terrell (an experienced Royal Navy hydrographer), while other staff, such as Pat Blackett, spent time at sea to add to their knowledge.
59. Waters, 'Derek Howse (1919–98)', 224.
60. Waters, 'Derek Howse (1919–98)', 225; see also Bennett, 'Obituary: Lt-Cdr Derek Howse'. Howse was substantially involved in the full restoration and opening of the Observatory in 1967.
61. NN, *An Inventory of the Navigation and Astronomy Collections in the National Maritime Museum Greenwich*, ii.
62. Knight, 'Afterword. Thirty Years Later', 196.
63. In the context of this catalogue, the most important of these were *The Sextant* and *John Hadley and His Reflecting Quadrant*, both produced by Alan Stimson in 1975.
64. Notably Forbes, *The Birth of Navigational Science*; Taylor, *Navigation in the Days of Captain Cook*; and Waters, *Science and the Techniques of Navigation in the Renaissance*. Likewise, in 1968 the Museum published *Man Is Not Lost* by Donald Sadler, then Superintendent of the Nautical Almanac Office.
65. The proceedings of the 1984 conference were published in *Vistas in Astronomy*, 28 (1985).
66. Knight, 'Afterword. Thirty Years Later', 198.
67. SIMON stood for 'Scientific Instrument-Makers, Observations and Notes'. The project was originally based at Imperial College with funding from the Leverhulme Trust.
68. The first Sackler Research Fellow was Dr Elly Dekker, who wrote *Globes at Greenwich* (1999), the first catalogue in this series.

PART TWO

THE CATALOGUE

EXPLANATION

The catalogue comprises ten sections, by type of instrument, in each of which items are entered in chronological order. Only original instruments are listed. Replicas and copies of mariner's astrolabes, cross-staffs, and backstaffs that the National Maritime Museum owns and uses for educational purposes and as loans to other institutions are not included. All the object descriptions are laid out similarly in the following order.

IDENTIFICATION

The catalogue numbers, from 001 to 347, and the National Maritime Museum inventory number head each description. A concordance between old and new inventory numbers, and catalogue numbers, is provided in Appendix 2. The numbers are followed by the type name for the instrument (for example backstaff, sextant with an artificial horizon, sounding quintant, or night octant). When applicable, this is followed by a specific name such as that of the inventor or manufacturer (for example 'Colby's horizon instrument', 'Newey's horizon finder', or 'Mate sextant').

The difference between an octant, sextant, and quintant has been defined according to the actual range of measurement, rather than by the maximum range of the graduation. There are, for example, instruments on which the scale runs to 125 degrees, suggesting it is a sextant, but which cannot be used to measure an angle over 100 degrees because of the manner in which it was constructed. Such a limitation is indicated as, for example, 'measuring to 86°'. Following this principle, an octant is defined as measuring angles up to 105 degrees, a sextant as measuring between 106 and 135 degrees, and a quintant between 136 and 150 degrees. A reflecting circle is so named when its frame consists of a full circle. One instrument with a range up to 180 degrees is also included in this section eventhough it is only a half-circle (cat.no. 316).

An exception has also been made for box sextants and for twentieth-century gyroscopic and bubble sextants. The actual range of measurement of box sextants varies between 89 and 186 degrees, that of gyroscopic and bubble sextants is usually between 80 and 90 degrees. These instruments are found under 'sextants' because this designation is generally used for that type of instrument.

ORIGIN, MAKER, AND DATE

The town and country of origin of the maker and his or her name (if known) are given; otherwise, this information is given as 'unknown'. 'Possibly' is added when a relevant trade label is present, when initials on the instrument suggest the name, or when the name may be that of a retailer or an owner rather than a maker. 'Probably' is added when documentary evidence (recorded under 'Literature') makes it likely that the person named made the unsigned instrument. A first name in parentheses means that the instrument was probably made by that person rather than, for example, by a son or a brother. The name of the maker is spelled according to modern reference works such as Clifton, *Directory*, Marcelin, *Dictionnaire*, and Brewington, *The Peabody Museum Collection*. If no such reference is known, the spelling found on the instrument is followed. Dating is by the year found on the object, by documentary evidence as recorded under 'Literature', or by the object's provenance. If none of these was available, the object is dated on the basis of its appearance when compared with similar instruments, or on the basis of its technology.

INSCRIPTIONS

These are the inscriptions found on the instrument, where a distinction is made between 'signed' and 'marked'. The former includes the name or initials of the maker and his address or town as found on the instrument. 'Marked' refers to other inscriptions on the object, such as an owner's name, numbers, patents, or trademarks, etc. The inscriptions are given as closely as possible to the original but in almost every case a photograph of the inscription has been added. Inscriptions found on boxes or cases are recorded with the 'description', described below.

GRADUATION

For all instruments with a graduation, the entire range of the scale is given in degrees and minutes of arc, and the material of which the scale was made is also given. The transversal scale on backstaffs is read along the fiducial edge at the centre of the sighting vane. On octants, the transversal scale is read along the fiducial edge at the end of the index arm. The transversal division is usually into ten parts, allowing a reading accuracy of two minutes of arc. The division of verniers is given in minutes or seconds of arc. The material of which the scale and vernier are made is given, and the location of the zero of the vernier is recorded (to the right or left or at the centre), as this can be an indication of the object's date.

DIMENSIONS

All dimensions are given in millimetres. For the nautical quadrants, the dimension given is the radius, and for the mariner's astrolabes, the diameter, thickness, and weight in grams are given, following the practice in Stimson, *The Mariner's Astrolabe*. For cross-staffs, the lengths of the staff and the vanes are given, as in Mörzer Bruyns, *The Cross-Staff*. For backstaffs, the radii of the 30- and 60-degree arcs are given, and also the overall length of the object. For octants with transversal scales, the radius has been measured from the pivot to the outer end of the scale, nearest the edge of the limb. For octants with a vernier scale, sextants, quintants (including the sounding and surveyor's examples), and reflecting circles, the radius is measured from the pivot to the bottom of the graduated scale, which is usually a few millimetres less than the radius of the frame. For these instruments, this dimension is also provided in inches, because, since the eighteenth century, the British have often described these as, for example, 6-, 8-, 10-, 12-, or 18-inch octants or sextants. By providing inches, we allow the original classification of a particular instrument to be more easily determined.

For box sextants, the radius is provided in millimetres and inches, and the diameter in millimetres only. For all other instruments (gyroscopic sextants, bubble sextants, artificial horizons, instruments for measuring the dip of the horizon, etc.), the largest dimensions are given (for example, height × length × width).

DESCRIPTION

This includes the main materials of which the object's body or frame is made; with sextants and quintants, the frame pattern is also named. British instrument makers started the custom of producing frames in various patterns in the nineteenth century, a development dealt with in Chapter 5 in Part 1 The description includes details of vanes, and of the clamping and tangent screws or any other slow-motion device to move the index arm; details of the adjustment of the glasses, the number of shades and their colours, the magnifier for reading the vernier, and the sight vane or telescope bracket are also included. The number and type of telescopes (erect or inverted image, and with or without cross-wires), sight-tubes, and eyepieces are given. The size of a telescope is the closed length in millimetres. Other detachable parts are listed, for example handles, oil bottles, a magnifying glass, an adjusting pin, or a spanner; and missing parts are reported. Details of the shape and material of the original box or case in which the instrument is kept are given, including any labels, certificates, notes, indications of previous ownership, dedications, and inscriptions (for example on the lid). Directly quoted texts are placed between single quotes. The date of a trade label is added in parentheses if it is relevant to the dating of the object. In some instances a remark is made on the condition of the instrument, but the work for this catalogue did not include a full condition survey.

ARTIFICIAL HORIZON

Here the working of the artificial horizons fitted to specific backstaffs, octants, sextants, or quintants is described. Detached artificial horizons are catalogued as separate objects.

PROVENANCE

The provenance details include the names of donors or suppliers, and previous known owners, and the year of inclusion in the collection. In some cases the details are sparse and are based on the Museum's oldest preserved documentation. The names of lenders have not been included as the Museum is undertaking a programme to review long-term loans which are not on display, which is likely to result in such items being either acquired or returned. The catalogue represents the state of the National Maritime Museum collection and the knowledge of its provenance at the date of the completion of this particular survey, in December 2007.

NOTES

Additional information is provided about the instrument, its use, its invention or patent, and its previous owners. Unless stated otherwise, patent numbers refer to British patents, listed in *Patents for Inventions*, of 1875 and 1905.

LITERATURE

The abbreviated titles of publications are listed; the full references can be found in the Bibliography. Publications can refer to the specific instrument and its history, date, invention, construction, or use, and to previous ownership.

1. MARINER'S QUADRANTS

001 NAV1053
Mariner's quadrant

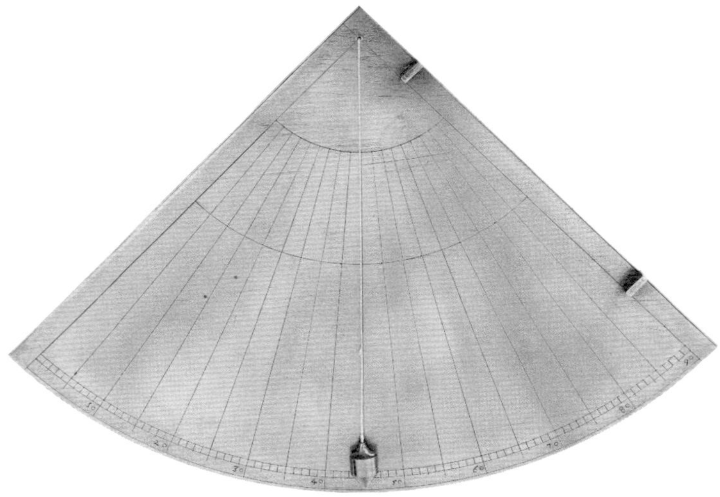

Origin Unknown
Maker Unknown
Date About 1600
Inscriptions None
Graduation Engraved scale along the circular edge of the instrument, from 0° to 90° every 1°. Every 5-degree line has been extended to the apex of the instrument.
Dimensions Radius 175 mm.
Description A polished brass quarter-circle with two sights on the right-hand edge. The back of the instrument is blank apart from a thinly engraved circle with a diameter of 108 mm.
Provenance Presented by Sir James Caird in 1947; purchased as part of the Barberini collection through Maggs Brothers.
Notes The plumb bob is probably a later replacement.

002 NAV1062
Mariner's quadrant

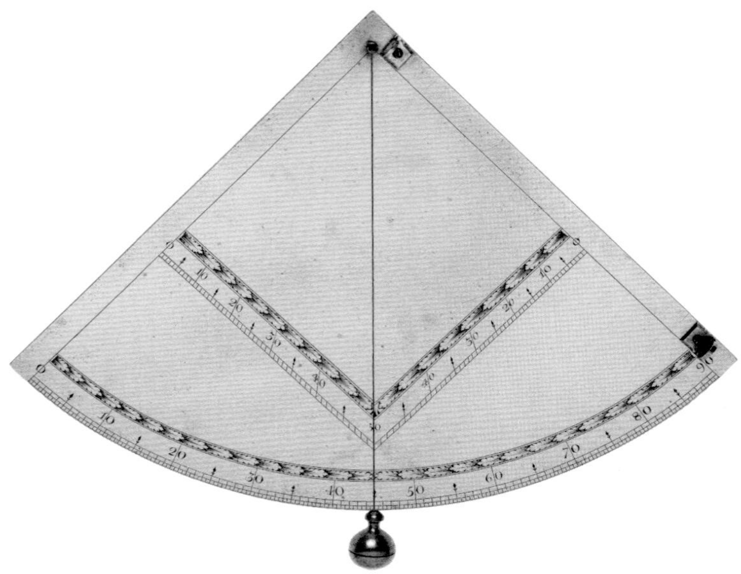

Origin England
Maker Unknown
Date About 1725
Inscriptions None
Graduation Engraved scale along the circular edge of the instrument, from 0° to 90° every 30′; the shadow square runs from 0° to 50° to 0° every 1°.
Dimensions Radius 153 mm.
Description A polished brass quarter-circle with two sights on the right-hand edge. The shadow square is used for taking altitudes during surveying; the back of the instrument is blank.
Provenance Purchased from David Weston Ltd, London, in 1980.
Notes The sights and plumb bob are probably later replacements.

2. MARINER'S ASTROLABES

003 NAV0030
Mariner's astrolabe

Origin Possibly Spain
Maker Unknown
Date Second half of the sixteenth century
Inscriptions There is a small star-like mark on the face of the bottom of the ballast.
Graduation Several digits of the scale are visible, indicating a graduation on the upper half of the frame, on either side of 0° at the top: a division of 90° to 0° to 90° numbered to 5°, but the graduations are obliterated.
Dimensions Diameter 207 mm, thickness at the top 14 mm, thickness at the bottom 28 mm, and weight 3288 g.
Description Cast brass frame, the alidade and suspension ring are missing; part of the axial pin is jammed in the pivot hole. The mater has been made from a single casting with bottom ballast and is wedge-shaped. This is to assist the instrument to hang plumb in the vertical plane during an observation. Badly eroded.
Provenance Found in Mount's Bay, Cornwall, in 1982; purchased from the Receiver of Wrecks, HM Customs and Excise, Customs House, Penzance, Cornwall, in 1989.
Notes A Spanish escudo of about 1520 found at the same site suggests that the astrolabe may have been made in Spain.
Literature Stimson, *The Mariner's Astrolabe*, 140.

004 NAV1801
Mariner's astrolabe

Origin Unknown
Maker Unknown
Date Second half of the sixteenth century
Inscriptions None
Graduation Scales and other marks are obliterated.
Dimensions Diameter 179 mm, thickness at the top 13 mm, thickness at the bottom 19 mm, and weight 1588 g.
Description Cast brass frame, the alidade and suspension ring are missing. The mater has been made from a single casting with bottom ballast and is wedge-shaped. This is to assist the instrument to hang plumb in the vertical plane during an observation. Badly eroded.
Provenance Found on the beach at Telegraph Bay, Alderney, Channel Islands, in 1925, after a storm. Bequest from the late Miss Ethelwyn C. Shiel of Albury, Guildford, Surrey, in 1993.
Notes The instrument's pronounced wedge section might indicate that it is of the second rather than the first half of the sixteenth century.
Literature Stimson, *The Mariner's Astrolabe*, 99.

005 NAV0022
Mariner's astrolabe

Origin Possibly Spain
Maker Unknown
Date About 1588
Inscriptions None
Graduation Engraved scales to 1 degree without digits on the upper half of the frame.
Dimensions Diameter 178 mm, thickness 17 mm, and weight 2270 g.
Description Cast brass frame with a brass alidade and suspension ring. The alidade is fitted on the face of the instrument by an axial pin. The throne is low and moulded. The mater has been made from a single casting, cut out in a wheel shape with a greater weight left in the lower semicircle, to assist the instrument to hang plumb in the vertical plane during an observation. The face of the mater is engraved with circles and decorative lines but the incomplete scale suggests that the instrument was never finished.
Provenance Found in 1845 under a rock on the north side of Combe Hill on Valentia Island, Co. Kerry, Ireland, by a boy working for a local quarry and eventually deposited in the Science Museum, London. Transferred to the National Maritime Museum in 1954, in exchange for part of the Gabb collection.
Notes Valentia is within sight of the Blasket Islands, to the north across Dingle Bay, where three homeward-bound ships of the scattered Spanish Armada sank in 1588. The style suggests Spanish or at least Iberian origin. It was found in the remains of a leather bag, suggesting that it was deliberately concealed, possibly by survivors.
Literature Stimson, *The Mariner's Astrolabe*, 64–5; Waters *The Sea- or Mariner's Astrolabe*; Hill and Paget-Tomlinson, *Instruments of Navigation*, 8.
For image see colour section plate 1.

006 NAV0029
Mariner's astrolabe

Origin Possibly England
Maker Unknown
Date About 1600
Inscriptions None
Graduation Scale and other marks are obliterated.
Dimensions Diameter 180 mm, thickness at the top 11 mm, thickness at the bottom 23 mm, and weight 2495 g.
Description Cast brass frame; the alidade and suspension ring are missing. The mater has been made from a single casting with bottom ballast, and has a pronounced wedge shape. This is to assist the instrument to hang plumb in the vertical plane during an observation. Originally this was an ornate casting, much more so than is usual for a mariner's astrolabe. Badly eroded.
Provenance Found on the seabed near Great Mewstone Rocks in Wembury Bay, Devon, by Mr Wilfred T. Jenkins of the Plymouth Sound branch of the British Sub-Aqua Club, on 29 September 1970. Lent by Mr W.T. Jenkins, Plymouth, Devon in 1971; purchased from him in 1998.
Notes Taking into account the location of the find, this instrument may well have been English, but a more recent discovery of two similarly styled astrolabes off the Iberian coast has deepened the uncertainty surrounding its origin.
Literature Stimson, *The Mariner's Astrolabe*, 114–15, 148–9, 172–3.

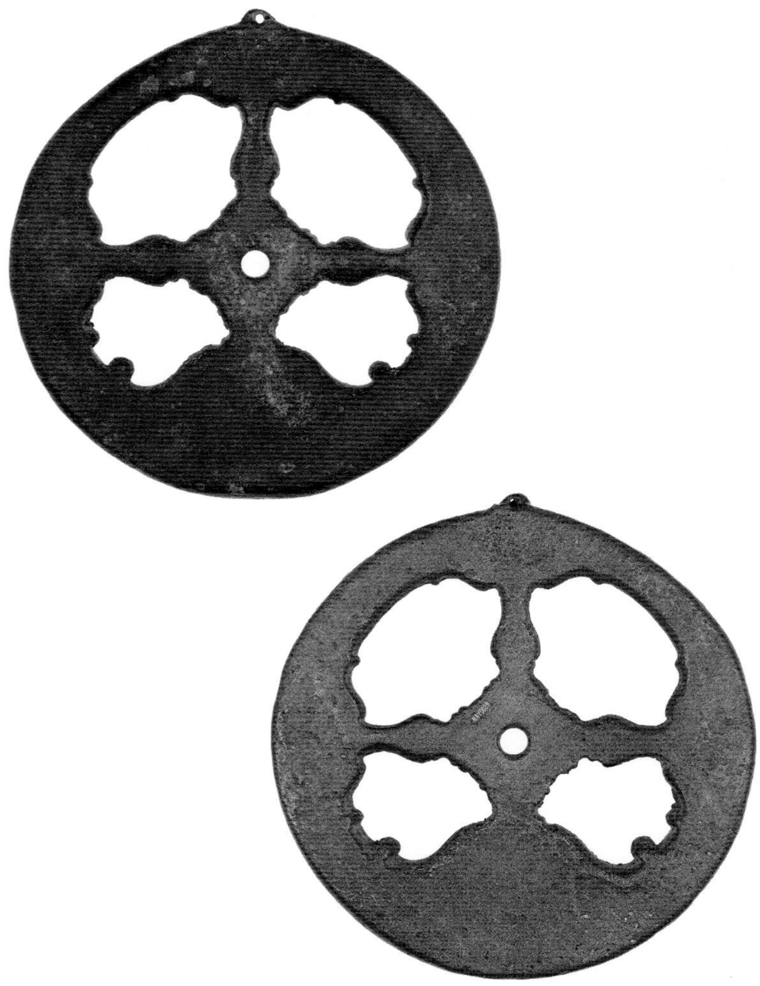

3. CROSS-STAFFS

007 NAV0505
Cross-staff

Origin London, England
Maker Thomas Tuttell
Date About 1700
Inscriptions Signed on side one: Thomas Tuttell Charing + Londini fecit
Signed on side two: Fait par Tho: Tuttell Ingenieur de sa MAIESTÉ BRITANIQUE | pour les Instruments Mathemathiques á LONDRES
Graduation The scales are from 90° to 34°(side one), from 90° to 20° (side two), from 90° to 10° (side three), and from 90° to 4° (side four). All four sides also have scales running in the reverse direction, for the zenith distance. The maximum reading accuracy is 10′ for sides one to three, and 5′ for side four.
Dimensions Length of staff 796 mm, width 14 × 14 mm; the vanes are 506, 292, 136, and 101 mm in length.
Description An ivory staff and four decoratively engraved ivory vanes. The sky end of the staff is pyramid-shaped. The vanes are in two parts, held together by brass pins. Vane 4 has a fixed ivory horizon. The beginnings of the scales on all four sides are marked *Altitude*, and *Complement* for the zenith distance.
Provenance Presented by Sir James Caird in 1938; purchased through Maggs Brothers from the Rufford Abbey sale in Ollerton, Nottinghamshire, by Knight, Frank and Rutley in conjunction with Christie, Manson and Woods, October 1938.
Notes Probably not meant for use on board a ship. Part of a presentation set including a backstaff (cat. no. 011), two Gunter's scales (NAV0108 and NAV0107), and a sector (NAV0166), all fitting in a lyre case. The inscription may point towards an original French owner.
Literature Mörzer Bruyns, *The Cross-Staff*, 54; Clifton, *Directory*, 284; Hill and Paget-Tomlinson, *Instruments of Navigation*, 10.
For image see colour section plate 2.

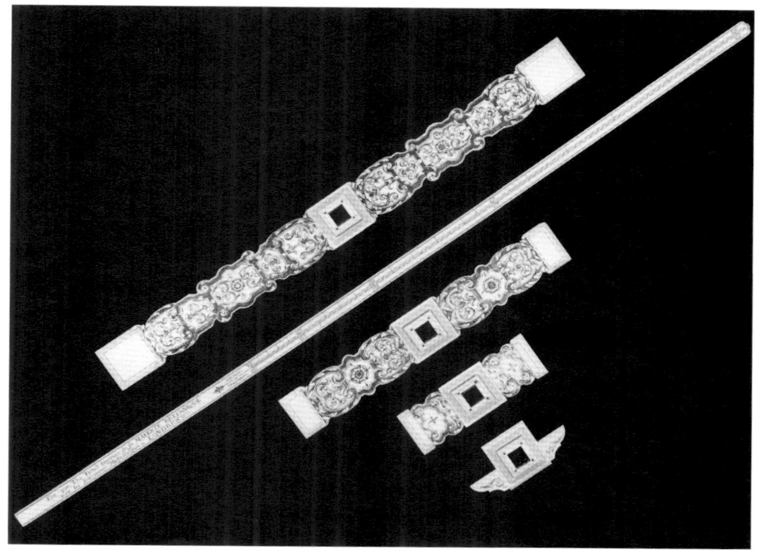

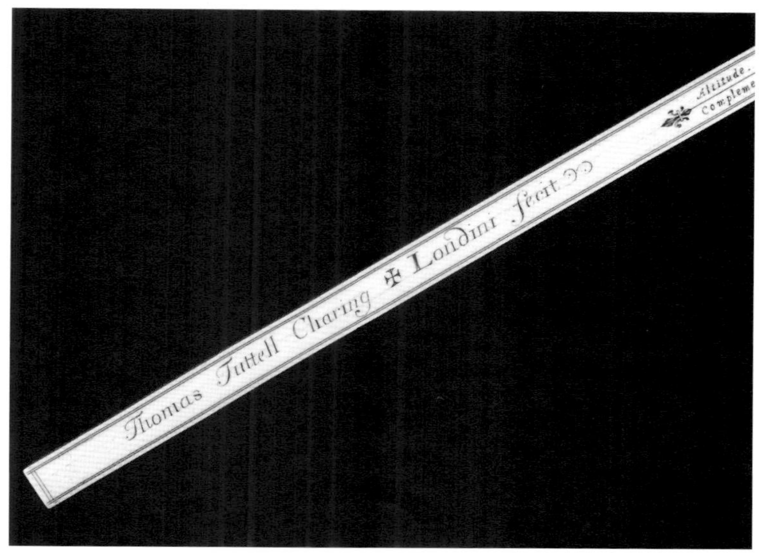

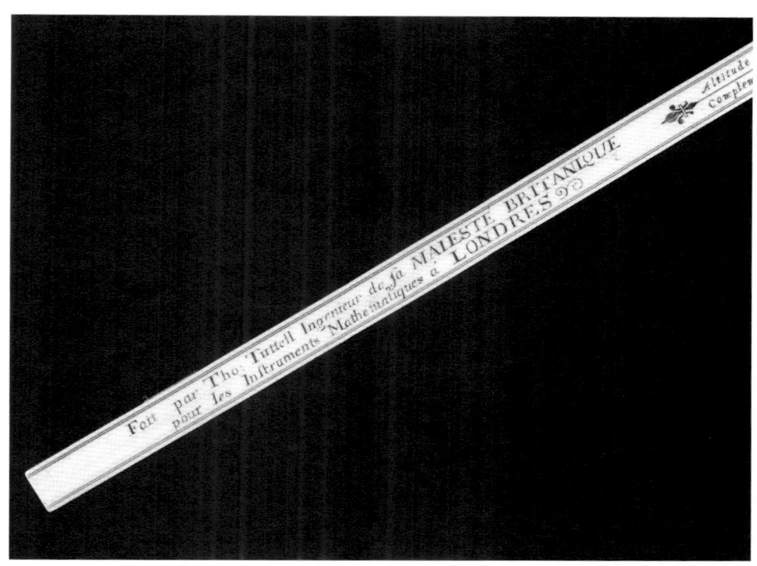

008 NAV0510
Cross-staff

Origin England
Maker Unknown
Date 1703 or earlier
Inscriptions None
Graduation The scales are from 90° to 40° (side one), from 90° to 25° (side two),from 90° to 15° (side three), and from 90° to 10° (side four). All four sides also have scales running in the reverse direction, for the zenith distance. The maximum reading accuracy is 10′ for each side.
Dimensions Length of staff 837 mm (which is less than the original length), width 16 × 16 mm; the vanes are 508, 302, 179, and 118 mm in length.
Description A wooden staff (incomplete) and four vanes. It is possible that either one or three of the vanes do not belong to this staff. The vanes are in two parts, held together by brass pins. The beginnings of the scales are marked *al* for altitude, and *co* for complement, the zenith distance. The vanes are not fitted with clamping screws. The three longest are stamped in 90, 30, and 10 for use with the respective scales, consistent with the English naming of vanes.
Provenance Recovered from the wreck of the *Stirling Castle*, which went aground on the Goodwin Sands off the Kent coast during the Great Storm of 1703. Found in a chest on the orlop deck in the after part of the vessel. The missing parts were subsequently reconstructed in the Museum's workshop in 1983. Purchased from the Receiver of Wrecks, HM Customs and Excise, Customs House, Ramsgate, Kent, in 1982.
Literature Mörzer Bruyns, *The Cross-Staff*, 41, 56.

009 NAV1973

Cross-staff, part of vane

Origin England
Maker Unknown
Date 1703 or earlier
Inscriptions None
Dimensions 62 × 24 mm (length × width); the internal measurements are 16 × 17 mm, which means that the staff to which these vanes belonged was about 16 × 16 mm.
Description Part of a cross-staff vane. A small wedge-shaped brass plate is fitted in the vane; it was pressed against the staff by the tightening screw, to clamp the vane without damaging the staff.
Notes Recovered from the wreck of the *Stirling Castle*, which went aground on the Goodwin Sands off the Kent coast during the Great Storm of 1703. Purchased from the Receiver of Wrecks, HM Customs and Excise, Customs House, Ramsgate, Kent, in 1982.
Literature Mörzer Bruyns, *The Cross-Staff*, 56.

010 ZBA4373

Cross-staff

Origin Amsterdam, Netherlands
Maker Hendrik Mooy
Date 1779
Inscriptions Signed and dated on side one: 1779 | *H * Mooy *
Graduation The scales are from 90° to 38° (side one), from 90° to 28° (side two), from 90° to 13° (side three), and from 90° to 6° (side four). All four sides also have scales running in the reverse direction, for the zenith distance. The maximum reading accuracy is 10′ for each side.
Dimensions Length of staff 746 mm, width 18 × 18 mm; the lengths of the vanes would have been 510 mm (side one), 358 mm (side two), 168 mm (side three), and 68 mm (side four).
Description An ebony staff; the four vanes are missing. The sky end of the staff is pyramid-shaped.
Provenance Purchased at Christie's, London, May 2006, lot no. 202.
Notes Hendrik Mooy (1715–65) was a cross-staff maker at *In de Jonge Lootsman* ('at the sign of the Young Pilot') in Amsterdam. The business was continued by his widow until her death in 1795. Presumably she continued to sell cross-staffs with her husband's name, as the only other recorded cross-staff by Hendrik Mooy, in the Maritime Museum Rotterdam (inv. no. M.483), is also dated after his death.
Literature Mörzer Bruyns, *The Cross-Staff*, 80, 93.

4. BACKSTAFFS

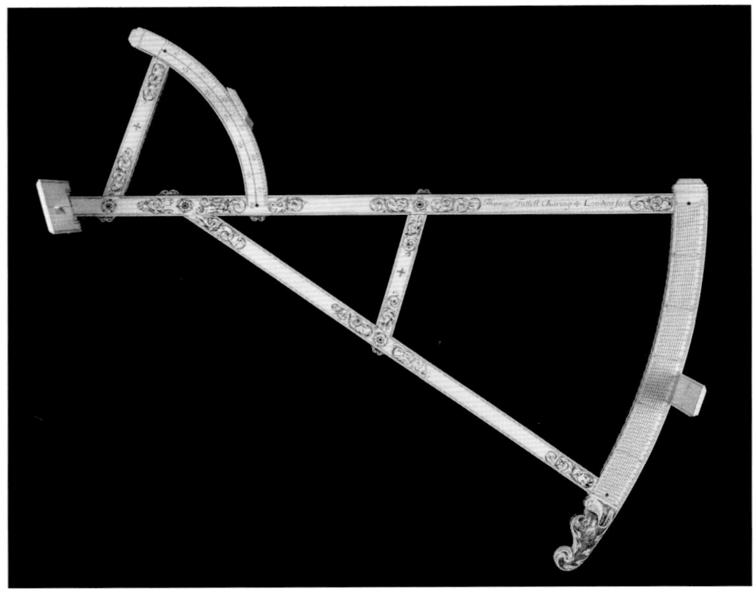

011 NAV0040
Backstaff

Davis quadrant
Origin London, England
Maker Thomas Tuttell
Date About 1700
Inscriptions Signed on the main strut: Thomas Tuttell Charing + Londoni fecit
Signed on the back of the 30-degree arc: Fait par Tho: Tuttell Ingenieur | de sa MAIESTÉ .BRITANIQUE | pour les Instruments.Mathematiques, á | LONDRES
Graduation The transversal scale on the 30-degree arc is from 0° to 30° every 10′, and reads to 1′ along the fiducial edge of the sight vane. The graduation on the 60-degree arc is from 0° to 60 every 1°.
Dimensions The radius of the 30-degree arc is 686 mm and that of the 60-degree arc is 229 mm. The overall length of the instrument is 689 mm.
Description An ivory frame and arcs, with brass rivets; an ivory sight vane, horizon vane, and shadow vane. The sight vane and shadow vane slide in grooves cut in the outer edge of the arcs instead of being clamped against the arcs by the usual brass pressure plates. Floral decorations are found on all parts of the instrument's frame. A decorative mermaid and merman flank the signature in French.
Provenance Presented by Sir James Caird in 1938; purchased through Maggs Brothers from the Rufford Abbey sale at Ollerton, Nottinghamshire, by Knight, Frank and Rutley, in conjunction with Christie, Manson and Woods, October 1938.
Notes Probably not meant for use on board a ship. Part of a presentation set including a cross-staff (cat. no. 007), two Gunter's scales (NAV0108 and NAV0107), and a sector (NAV0166), all fitting in a lyre case. The inscription may point towards an original French owner. The shadow vane is a modern replacement.
Literature Clifton, *Directory*, 284; Hill and Paget-Tomlinson, *Instruments of Navigation*, 11.

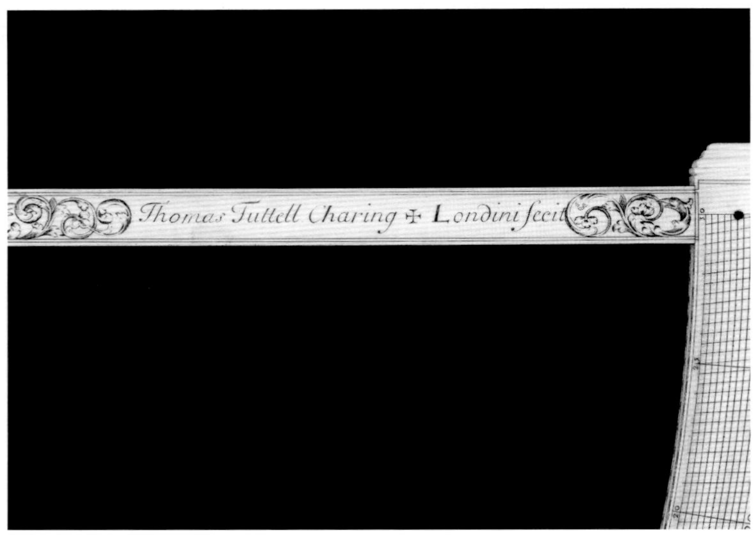

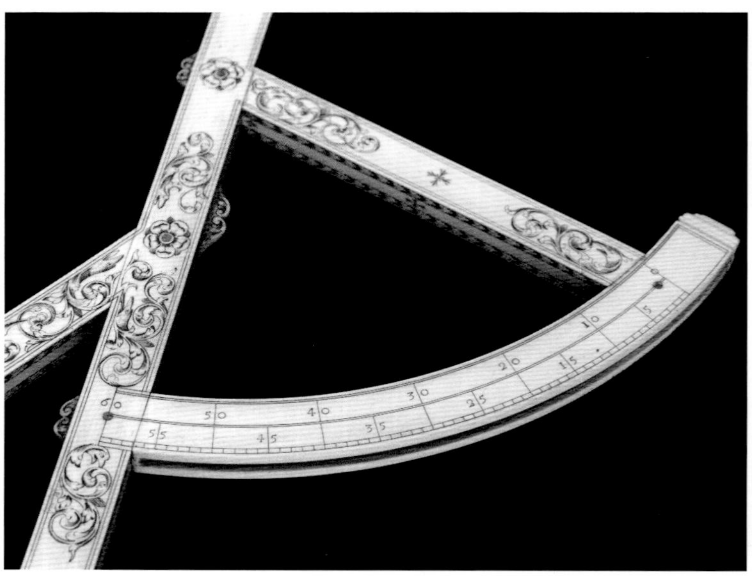

012 NAV0045
Backstaff

Davis quadrant
Origin England
Maker Unknown
Date About 1700
Inscriptions Marked (stamped) on the back of the 60-degree arc and on both vanes: 23
Graduation The transversal scale on the 30-degree arc is from 0° to 25° and 5′ vice versa every 5′, and reads to 1′ along the fiducial edge of the sight vane. The graduation on the 60-degree arc is from 0° to 65° every 1°, and on the outer edge has a scale for correction for the Sun's semi-diameter.
Dimensions The radius of the 30-degree arc is 608 mm and that of the 60-degree arc is 190 mm. The overall length of the instrument is 635 mm.
Description A lignum vitae frame with boxwood arcs, sight vane, and horizon vane. The sight vane has a brass pressure plate for holding it tightly in place on the 60-degree arc, and an inlaid brass pinhole sight. Decorative flowers and fleurs-de-lis are stamped on both arcs.
Provenance Presented by the Royal United Service Institution Museum in 1963, on its closure.
Notes According to the entry in the seventh edition of the R.U.S.I. Museum's *Official Catalogue* (of 1924), the sights were wanting. The replacement shadow vane has a Flamsteed lens (lens missing), and a brass pressure plate.
Literature Wylly, *Official Catalogue*, 176, cat. no. 2657; Hughes, *Official Catalogue*, 141, cat. no. 2657.

013 NAV0048
Backstaff

Davis quadrant
Origin London, England
Maker John Bellinger
Date About 1700
Inscriptions Signed on the inlaid plate: Iohn Bellinger Fecit
Graduation The transversal scale on the 30-degree arc is from 0° to 25° and 5′ vice versa every 5′, and reads to 1′. The graduation on the 60-degree arc is from 0° to 65° every 1°, and on the outer edge has a scale for correction for the Sun's semi-diameter. The back of the 30-degree arc has scales for the Sun's declination and the right ascension.
Dimensions The radius of the 30-degree arc is 612 mm and that of the 60-degree arc is 84 mm. The overall length of the instrument is 635 mm.
Description A laburnum frame with ivory arcs and an inlaid ivory plate on the main strut, with brass rivets capped by decorative diamond-shaped brass caps, and no vanes. Decorations are found on the end of the 30-degree arc (scrollwork and leaves), and stamped fleurs-de-lis in the frame.
Provenance Purchased from Mr T. Mackay, London, in 1987.
Literature Clifton, *Directory*, 26.

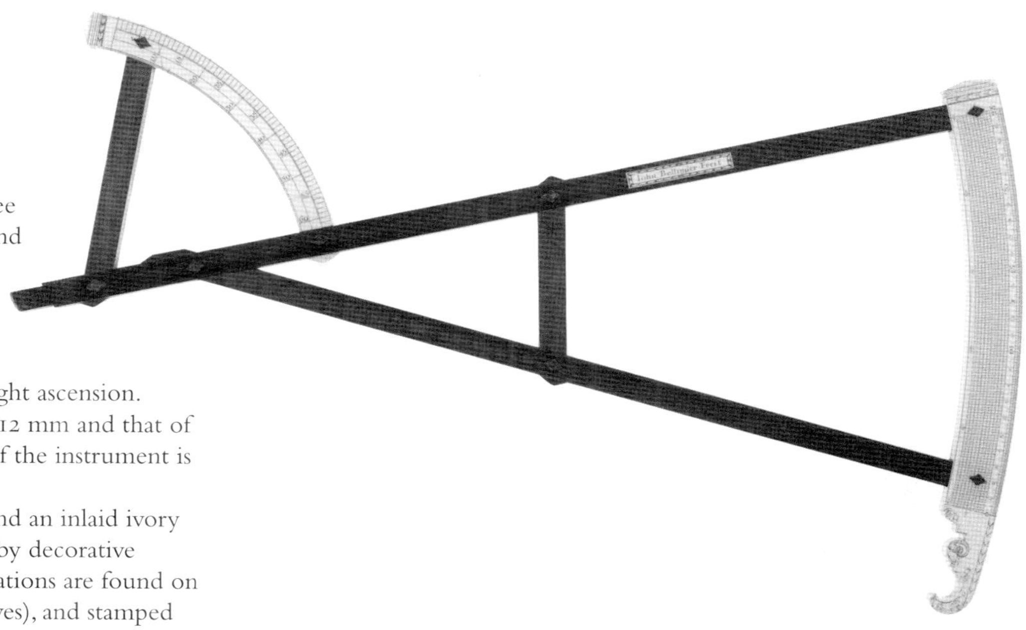

014 NAV0034
Backstaff

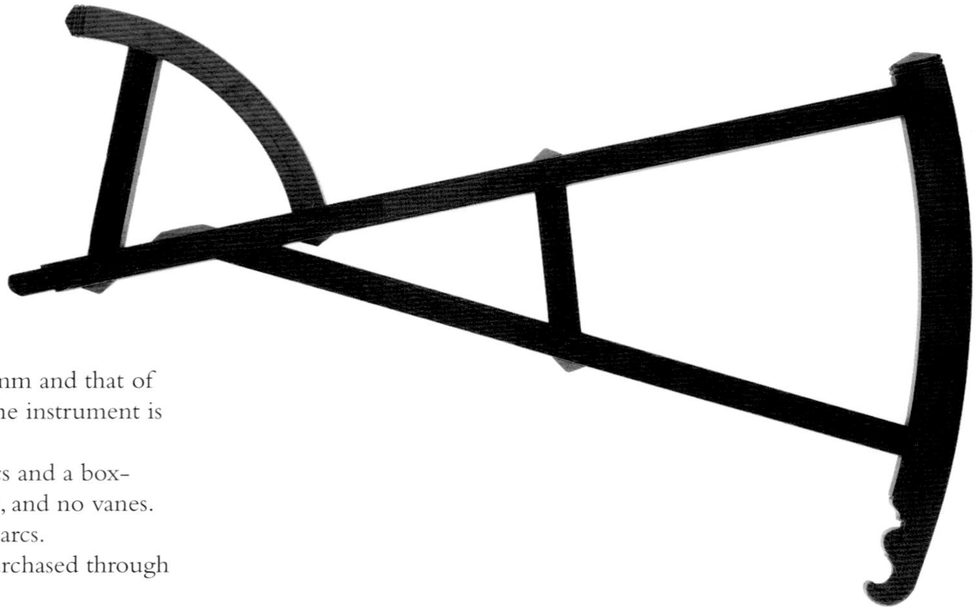

Davis quadrant
Origin England
Maker Unknown
Date First half of the eighteenth century
Inscriptions Marked (stamped) on the back of the 60-degree arc: 31
Graduation The transversal scale on the 30-degree arc is from 0° to 25° and vice versa every $5'$, and reads to $1'$. The graduation on the 60-degree arc is from 0° to 65° every 1°, and on the outer edge has a scale for correcting for the Sun's semi-diameter.
Dimensions The radius of the 30-degree arc is 609 mm and that of the 60-degree arc is 192 mm. The overall length of the instrument is 635 mm.
Description A lignum vitae frame with boxwood arcs and a boxwood inlaid plate on the main strut, with brass rivets, and no vanes. Small decorative stars are stamped at the ends of the arcs.
Provenance Presented by Sir James Caird in 1936; purchased through Maggs Brothers.
Notes Three replacement vanes have been added.
Literature Callender, *The Caird Collection: Supplement*, 91, no. 625.

015 NAV1974
Backstaff, sight vane

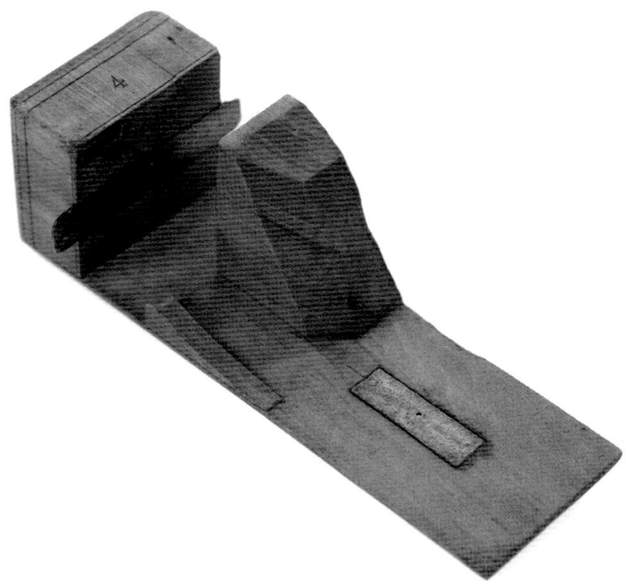

Davis quadrant
Origin England
Maker Unknown
Date 1703 or earlier
Inscriptions Marked (stamped): 4
Dimensions 118 × 37 × 43 mm (length × height × width)
Description Wooden sight vane with an inlaid brass pinhole sight and a brass pressure plate for holding it tightly in place on the 30-degree arc.
Provenance Recovered from the wreck of the *Stirling Castle*, which went aground on the Goodwin Sands off the Kent coast during the Great Storm of 1703. Purchased from the Receiver of Wrecks, HM Customs and Excise, Customs House, Ramsgate, Kent, in 1982.

016 NAV1975
Backstaff, vane part

Davis quadrant
Origin England
Maker Unknown
Date 1703 or earlier
Inscriptions Marked (stamped): R I R
Dimensions 37 × 41 mm (height × width)
Description Part of a wooden sight vane or shadow vane (broken).
Provenance Recovered from the wreck of the *Stirling Castle*, which went aground on the Goodwin Sands off the Kent coast during the Great Storm of 1703. Purchased from the Receiver of Wrecks, HM Customs and Excise, Customs House, Ramsgate, Kent, in 1982.

017 NAV1976
Backstaff, horizon vane part

Davis quadrant
Origin England
Maker Unknown
Date 1703 or earlier
Inscriptions None
Dimensions 53 mm (height)
Description Part of a wooden horizon vane (broken).
Provenance Recovered from the wreck of the *Stirling Castle*, which went aground on the Goodwin Sands off the Kent coast during the Great Storm of 1703. Purchased from the Receiver of Wrecks, HM Customs and Excise, Customs House, Ramsgate, Kent, in 1982.

018 NAV0036
Backstaff

Davis quadrant
Origin London, England
Maker Benjamin Macy
Date 1720
Inscriptions Signed (stamped) on the inlaid plate: Made By Beni. Macy For | M[r] Iames Austin 1720
Graduation The transversal scale on the 30-degree arc is from 0° to 25° and vice versa every 5′, and reads to 1′. The graduation on the 60-degree arc is from 0° to 65° every 1°, and on the outer edge has a scale for correcting for the Sun's semi-diameter. The back of the 30-degree arc has scales for the Sun's declination and the right ascension.
Dimensions The radius of the 30-degree arc is 203 mm and that of the 60-degree arc is 610 mm. The overall length of the instrument is 638 mm.
Description A lignum vitae frame with boxwood arcs and crossbar, and an inlaid boxwood plate on the main strut, with brass rivets; boxwood horizon vane and shadow vane. Decorative fleurs-de-lis are stamped in various places on the frame.
Provenance Presented by Sir James Caird in 1937; purchased as part of the Gabb collection.
Notes The sight vane is a later replacement. The shadow vane may be an old replacement, or taken from another instrument.
Literature Clifton, *Directory*, 176; Hill and Paget-Tomlinson, *Instruments of Navigation*, 11–2.
For image see colour section plate 3.

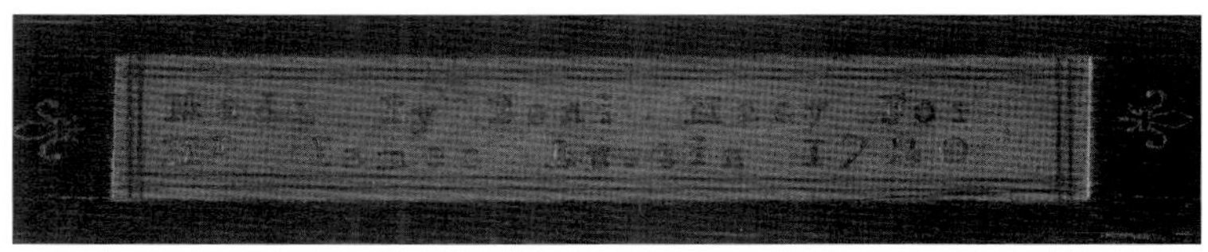

019 NAV0037
Backstaff

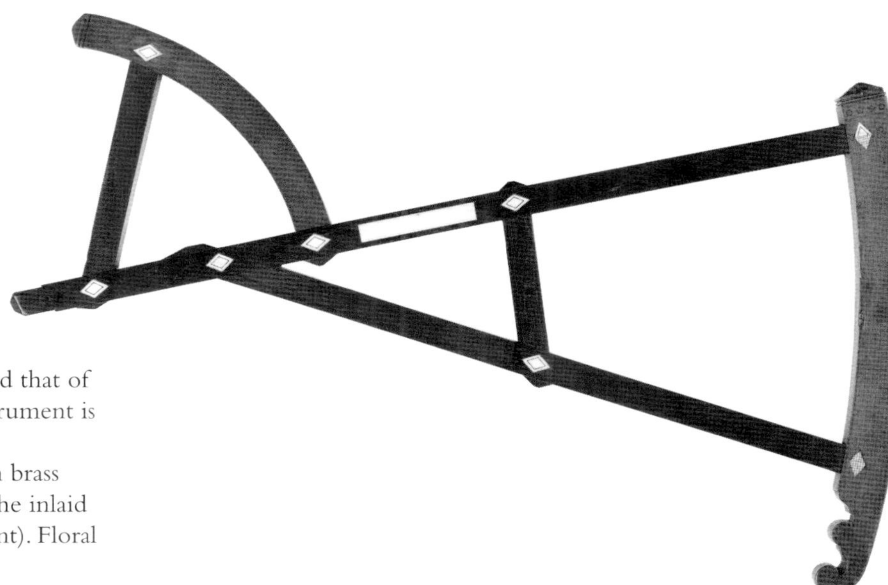

Davis quadrant
Origin England
Maker Unknown
Date About 1720
Inscriptions None
Graduation The transversal scale on the 30-degree arc is from 0° to 25° and vice versa every 5′, and reads to 1′. The graduation on the 60-degree arc is from 0° to 60° every 18, and on the outer edge has a scale for correcting for the Sun's semi-diameter.
Dimensions The radius of the 30-degree arc is 567 mm and that of the 60-degree arc is 190 mm. The overall length of the instrument is 593 mm.
Description A lignum vitae frame with boxwood arcs, with brass rivets capped by decorative ivory triangles, and no vanes. The inlaid plate in the main strut is missing (modern ivory replacement). Floral decorations are stamped on the 30-degree arc.
Provenance Presented by Sir James Caird in 1937; purchased as part of the Gabb collection.
Notes Alan Branch made the replacement vanes in 1963 in the Museum's workshop.

020 NAV0031
Backstaff

Davis quadrant
Origin England, possibly London
Maker Possibly John Patrick
Date About 1720
Inscriptions Marked (stamped) on the inside of the handle on the 30-degree arc: I P [capped by a fleur-de-lis]
Marked (scratched) in the plate in the main strut: R x P 1751.
Graduation The transversal scale on the 30-degree arc is from 0° to 25° every 5′, and reads to 1′ along the fiducial edge of the sight vane. The graduation on the 60-degree arc is from 0° to 65° every 1°, and on the outer edge has a scale for correcting for the Sun's semi-diameter.
Dimensions The radius of the 30-degree arc is 612 mm and that of the 60-degree arc is 190 mm. The overall length of the instrument is 637 mm.
Description A lignum vitae frame, boxwood arcs and an inlaid boxwood plate in the main strut, brass rivets, and a boxwood sight vane. Decorative stars and fleurs-de-lis are stamped at the ends of the arcs.
Provenance Unknown; the typed record indicates that it was acquired in 1936.
Notes The initials I P are possibly for John Patrick of London. The other initials and the date are presumably for an owner. Replacement sight and shadow vanes have been added.
Literature Clifton, *Directory*, 212.

021 NAV0038
Backstaff

Davis quadrant
Origin England, possibly Bristol
Maker Possibly William Wright
Date About 1725
Inscriptions Engraved on the horizon vane: W★W
Graduation The transversal scale on the 30-degree arc is from 0° to 25° every 5′ vice versa every 5′, and reads to 1′. The graduation on the 60-degree arc is from 0° to 65° every 1°, and on the outer edge has a scale for correcting for the Sun's semi-diameter.
Dimensions The radius of the 30-degree arc is 610 mm and that of the 60-degree arc is 198 mm. The overall length of the instrument is 629 mm.
Description A lignum vitae frame with boxwood arcs, the rivets are capped by decorative ivory triangles; a boxwood horizon vane. The sight vane and shadow vanes have brass pressure plates for holding them tightly in place on the arcs; the sight vane has an inlaid copper pinhole sight. The original inlaid plate in the main strut is missing. Decorative stamps are found at the ends of the arcs.
Provenance Presented by Sir James Caird in 1937; purchased as part of the Gabb collection.
Notes The initials are thought to be of William Wright of Bristol. The replacement sight vane and shadow vane, and the inlaid blank ivory plate in the main strut, were made in the Museum's workshop in 1963.
Literature Clifton, *Directory*, 306; Hill and Paget-Tomlinson, *Instruments of Navigation*, 12.

022 NAV0039
Backstaff with artificial horizon

Davis quadrant
Origin London, England
Maker Jonathan Sisson
Date About 1732
Inscriptions Signed (stamped) on the 30-degree arc on the handle: I:Sisson. London.
Marked (stamped) at the end of the 30-degree arc: L M
Marked (stamped) at the other end of the 30-degree arc: In°: Elton. Invent.
Graduation The scale on the 30-degree arc is from -2° to 32° and vice versa every 10′, and reads to 1′ by means of a vernier with an indicating line. There is no 60-degree arc, so that angles larger than 32° cannot be measured.
Dimensions The radius of the 30-degree arc is 619 mm; the overall length of the instrument is 661 mm.
Description A lignum vitae frame with boxwood arcs, brass-capped rivets, and brass fittings. The sight vane and the horizon vane are of brass, with a bubble level perpendicular to the frame. The shadow vane and lens, the bubble level that should be parallel to the frame, the bubble level for the label arm, and the oil lamp are all missing.
Artificial horizon This was designed by John Elton (patent no. 501 of 1728) and consists of two bubble levels that are used to keep the index arm horizontal while the frame of the instrument rotates. As with a regular backstaff, the shadow or a spot of light falls on the horizon vane, and the altitude is read from the scale. A third bubble level mounted on the label arm was used for forward observation.
Provenance Presented by Sir James Caird in 1937; purchased as part of the Gabb collection.
Notes This is the earliest navigational instrument found with a vernier.
Literature E.I. Carlyle, 'Elton, John (d. 1751) *merchant and adventurer*', in *Oxford Dictionary of National Biography*, Oxford, 2004; http://www.oxfordsdnb.com/; Taylor, *The Mathematical Practitioners*, 175–6; Ifland, *Taking the Stars*, 92–3; Clifton, *Directory*, 253; Hill and Paget-Tomlinson, *Instruments of Navigation*, 12.

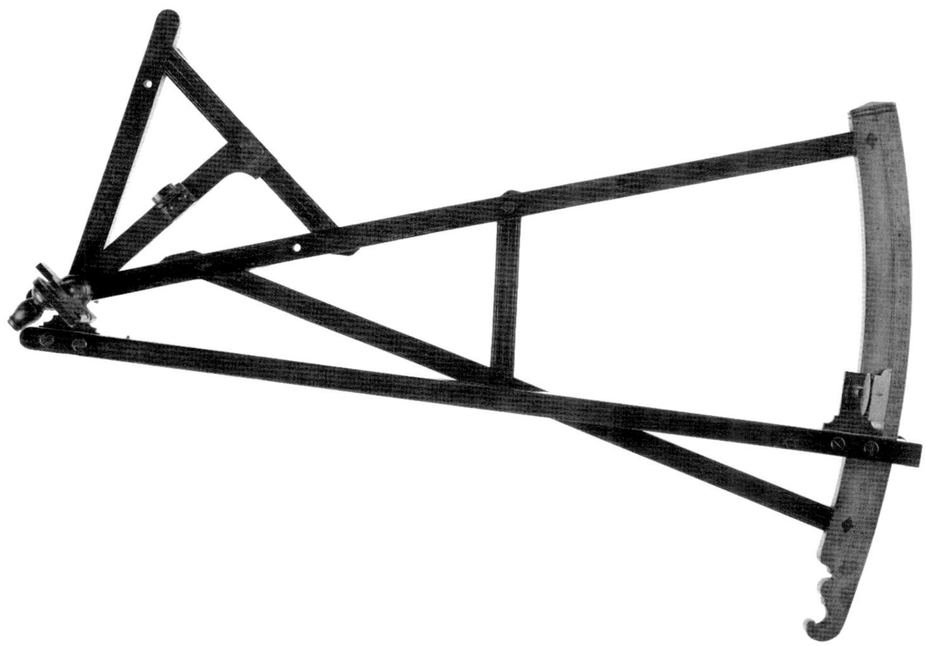

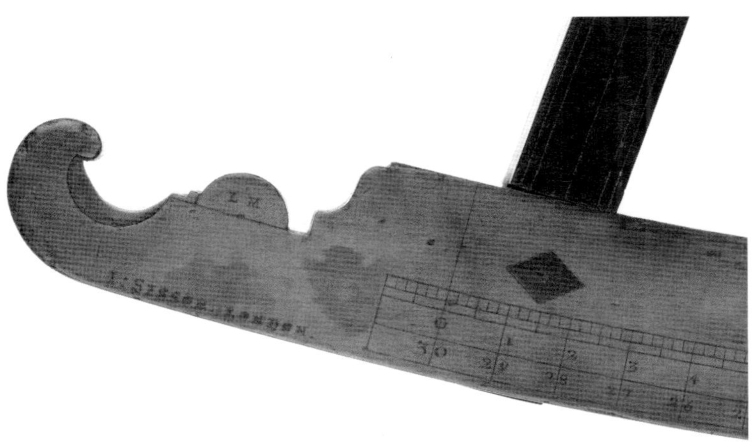

023 NAV0041
Backstaff

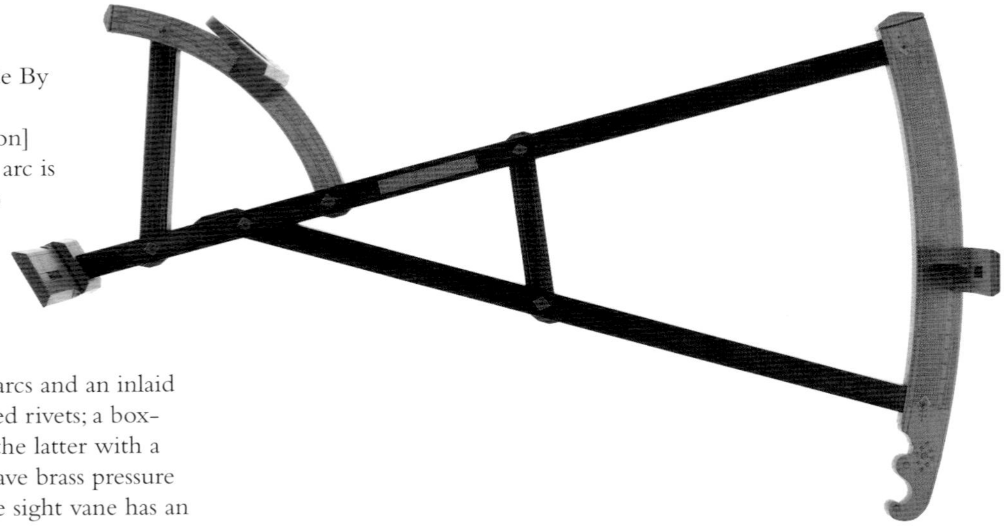

Davis quadrant
Origin London, England
Maker Will Garner
Date 1734
Inscriptions Signed (stamped) on inlaid plate: Made By Will Garner For | Oliver Thompson 1734
Marked on all three vanes: O T [for Oliver Thompson]
Graduation The transversal scale on the 30-degree arc is from 0° to 25° and vice versa every 5′, and reads to ½′. The graduation on the 60-degree arc is from 0° to 65° every 1°.
Dimensions The radius of the 30-degree arc is 606 mm and that of the 60-degree arc is 190 mm. The overall length of the instrument is 635 mm
Description A lignum vitae frame with boxwood arcs and an inlaid boxwood plate on the main strut, with brass-capped rivets; a boxwood sight vane, horizon vane, and shadow vane, the latter with a Flamsteed lens. The sight vane and shadow vane have brass pressure plates to hold them tightly in place on the arcs; the sight vane has an inlaid brass pinhole. Small decorative motifs are stamped on the lower end of the 30-degree arc.
Provenance Presented by Sir James Caird in 1943; purchased through Maggs Brothers, presumably from Ernest R. Cooper, who, in the article cited below, indicated he owned this instrument and also the octant cat. no. 043.
Literature Cooper, 'The Davis Back-Staff or English Quadrant', 59–64; Clifton, *Directory*, 110; Hill and Paget-Tomlinson, *Instruments of Navigation*, 12; Mörzer Bruyns, 'The Navigating Instruments of the National Maritime Museum', 23.

024 NAV0042
Backstaff

Davis quadrant
Origin London, England
Maker Will Garner
Date 1737
Inscriptions Signed (stamped) on the inlaid plate: Made by Will Garner for | Thomas White 1737
Graduation The transversal scale on the 30-degree arc is from 0° to 25° and vice versa every 5′, and reads to 1′. The graduation on the 60-degree arc is from 0° to 65° every 1°.
Dimensions The radius of the 30-degree arc is 201 mm and that of the 60-degree arc is 608 mm. The overall length of the instrument is 635 mm.
Description A lignum vitae frame with boxwood arcs and a boxwood inlaid plate in the main strut, brass rivets, and no vanes. Decorative fleurs-de-lis and stars are stamped on both arcs.
Provenance Purchased from Mr R. J. Emery, Ramsgate, Kent, in 1954.
Notes The three replacement vanes were made in the Museum's workshop in 1954.
Literature Clifton, *Directory*, 110; Hill and Paget-Tomlinson, *Instruments of Navigation*, 12.

025 NAV0032
Backstaff

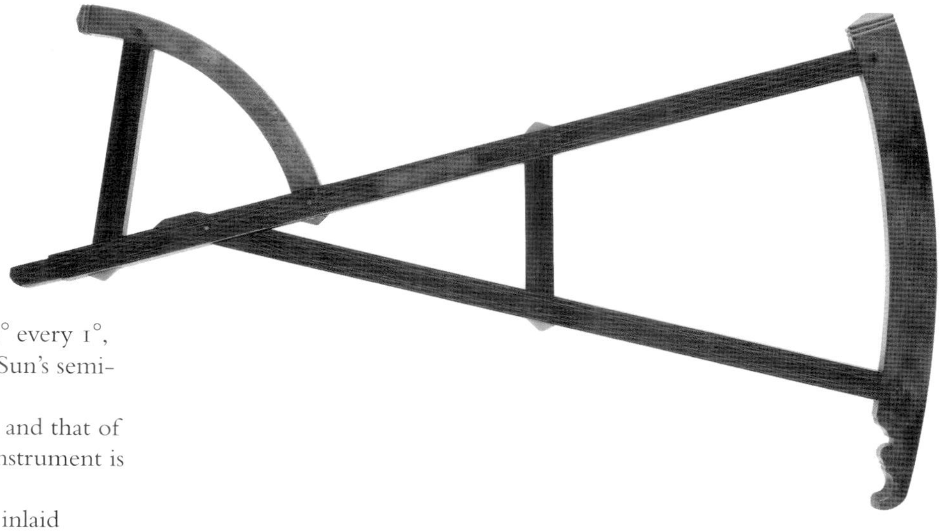

Davis quadrant
Origin London, England
Maker John Gilbert
Date About 1740
Inscriptions Signed (stamped) on the inlaid plate:
I.Gilbert.London
Marked (stamped) on the back of the 60-degree arc: 3
Graduation The transversal scale on the 30-degree arc is from 0° to 25° and vice versa every 10′, and reads to 1′. The graduation on the 60-degree arc is from 0° to 65° every 1°, and on the outer edge has a scale for correcting for the Sun's semi-diameter.
Dimensions The radius of the 30-degree arc is 601 mm and that of the 60-degree arc is 191 mm. The overall length of the instrument is 635 mm.
Description A mahogany frame with boxwood arcs and inlaid boxwood plate in the main strut, brass rivets, and no vanes. Decorative fleurs-de-lis (stamped) are found at the ends of the arcs.
Provenance Unknown; the typed record indicates that it was in the collection before December 1964.
Notes The three replacement vanes were made in the Museum's workshop in 1964.
Literature Clifton, *Directory*, 112; Hill and Paget-Tomlinson, *Instruments of Navigation*, 13.

026 NAV0044
Backstaff

Davis quadrant
Origin England
Maker Unknown
Date About 1740
Inscriptions Marked (stamped) on the back of the 60-degree arc and on the three original vanes: 18
Graduation The transversal scale on the 30-degree arc is from 0° to 25° and vice versa every 5′, and reads to 1′ along the fiducial edge of the sight vane. The graduation on the 60-arc is from 0° to 65° every 1°, and on the outer edge has a scale for correcting for the Sun's semi-diameter.
Dimensions The radius of the 30-degree arc is 611 mm and that of the 60-degree arc is 190 mm. The overall length of the instrument is 631 mm.
Description A mahogany frame with boxwood arcs and an inlaid boxwood plate in the main strut, and brass rivets. Three boxwood vanes: a sight vane, a shadow vane, and a second shadow vane with a Flamsteed lens. This lens and its wooden frame may be an old replacement, or may have been taken from another backstaff. The shadow vanes have brass pressure plates for holding them tightly in place on the 60-degree arc; the sight vane had an inlaid brass pinhole sight. Decorative fleurs-de-lis are stamped in the frame.
Provenance Purchased at Sotheby's, London, 17 October 1960, lot no. 159, with support from the Scott Paine Fund. Formerly in the Pym collection.
Notes The horizon vane is a replacement.

027 NAV0033
Backstaff

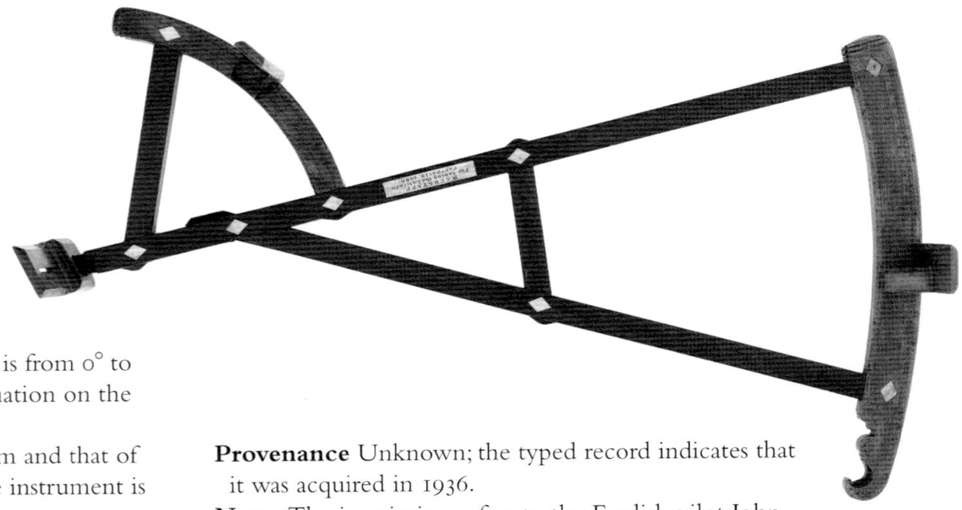

Davis quadrant
Origin Unknown
Maker Unknown
Date About 1750
Inscriptions Marked on the inlaid plate: BACKSTAFF | For taking the Latitude | Cap[t]. DAVIS. 1590.
Marked (stamped) on the main strut near the horizon end: 33
Graduation The transversal scale on the 30-degree arc is from 0° to 25° and vice versa every 5′, and reads to ½′. The graduation on the 60-degree arc is from 0° to 65° every 1°.
Dimensions The radius of the 30-degree arc is 615 mm and that of the 60-degree arc is 216 mm. The overall length of the instrument is 635 mm.
Description A hornbeam frame with boxwood arcs, an inlaid ivory plate on the main strut, rivets capped by decorative ivory triangles, and no vanes. Small decorative stamps, including flowers and fleurs-de-lis, are stamped at the ends of both arcs.
Provenance Unknown; the typed record indicates that it was acquired in 1936.
Notes The inscription refers to the English pilot John Davis of the 1590s but is not original to the instrument. Alan Branch made the set of replacement vanes in 1963 in the Museum's workshop.
Literature Hill and Paget-Tomlinson, *Instruments of Navigation*, 13.

028 NAV0035
Backstaff

Davis quadrant
Origin London, England
Maker John Gilbert
Date 1755
Inscriptions Signed on the inlaid plate: Made by In° Gilbert on Tower Hill | For In° Norman. Aug[t]. 7 : 1755
Marked (stamped) on the back of the 60-degree arc: 1231
Graduation The transversal scale on the 30-degree arc (cracked) is from 0° to 25° and vice versa every 5′, and reads to 1′. The graduation on the 60-degree arc is from 0° to 65° every 1°, and on the outer edge has a scale for correcting for the Sun's semi-diameter.
Dimensions The radius of the 30-degree arc is 601 mm and that of the 60-degree arc is 183 mm. The overall length of the instrument is 639 mm.
Description A lignum vitae frame and boxwood scales, an inlaid ivory plate on the main strut, brass rivets, and no vanes. Decorative stars and fleurs-de-lis are stamped on the frame and handle.
Provenance Presented by Sir James Caird in 1936; purchased through Spink and Son.
Notes The set of replacement vanes was made in the Museum's workshop in 1963; they are stamped 'NMM 1963'.
Literature Callender, *The Caird Collection: Supplement*, 92, no. 641; Hill and Paget-Tomlinson, *Instruments of Navigation*, 13; Clifton, *Directory*, 112.

5. OCTANTS

029 NAV1303
Octant

Origin London, England
Maker Benjamin Cole
Date About 1750
Inscriptions Signed on the left of the limb next to the scale: COLE Maker | Fleet Street | LONDON
Marked on the ivory plate on the crossbar: I*Moss.
Graduation Inlaid boxwood transversal scale from 0° to 90° and vice versa every 20′ (for zenith distance), measuring to 90°. The scale reads to 2′ by use of the fiducial edge at the end of the index arm.
Dimensions Radius 18 inches (457 mm)
Description Pear-wood frame, limb, and index arm; brass fittings and a brass stop for the index arm; inlaid ivory plate on the crossbar. Two socket shades (both red). Index-glass adjustment by a screw; adjustment of both horizon glasses by levers, wing nuts, and milled clamping screws. There is no tangent screw; the clamping screw is on the back of the index arm. The sight vane has two pinholes and a swivelling shutter; the back sight vane has one pinhole. No box.
Provenance Presented by Sir James Caird in 1942; purchased through A.S. Jones, London.
Literature Clifton, *Directory*, 60–1

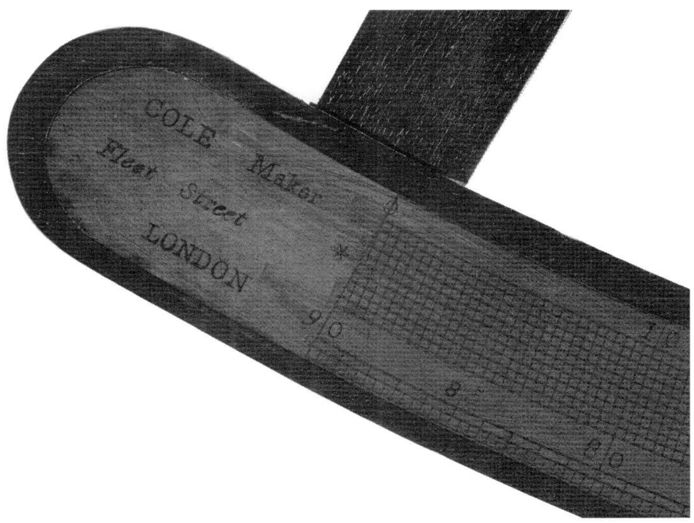

030 NAV1252
Octant

Origin London, England
Maker George Adams (senior or junior)
Date About 1750
Inscriptions Signed on the left of the limb next to the scale: ⋆Made by ⋆ | GEO. ADAMS | in Fleet Street | ⋆LONDON⋆
Marked on the right end of the scale: H⋆GORE⋆
Graduation Boxwood transversal scale from 0° to 90° and vice versa every 20′ (for zenith distance), measuring to 90′. The scale reads to 2′ by use of the fiducial edge at the end of the index arm.
Dimensions Radius 20 inches (508 mm)
Description Mahogany frame and index arm; boxwood limb, brass fittings, and a brass stop for the index arm. There is no tangent screw; the clamping screw (broken, possibly an old replacement) is on the back of the index arm. Two socket shades (red and green). Index-glass adjustment by a screw; adjustment of both horizon glasses by levers, wing nuts, and milled clamping screws. The sight vane has two pinholes and a swivelling shutter; the back sight vane has one pinhole. A small oval label on the back of the frame near the pivot is marked in MS 'H. Hughes & Son' [illegible] '1908' [illegible].
No box.
Provenance Adams collection; lent in 1936, bequeathed in 1974. The Hughes label may suggest that Adams bought this instrument from them in 1908.
Literature Millburn, *Adams of Fleet Street*; Clifton, *Directory*, 2.

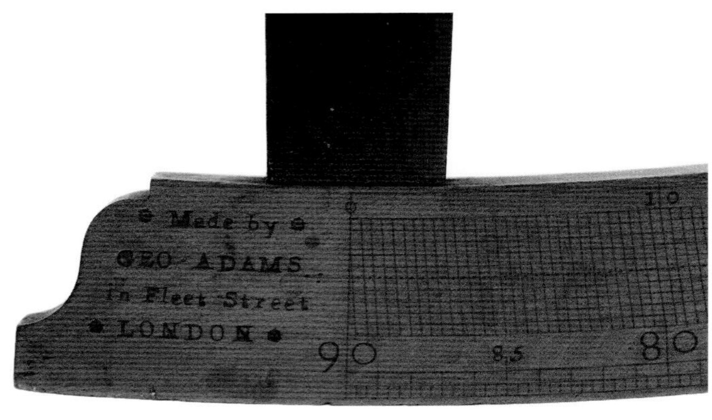

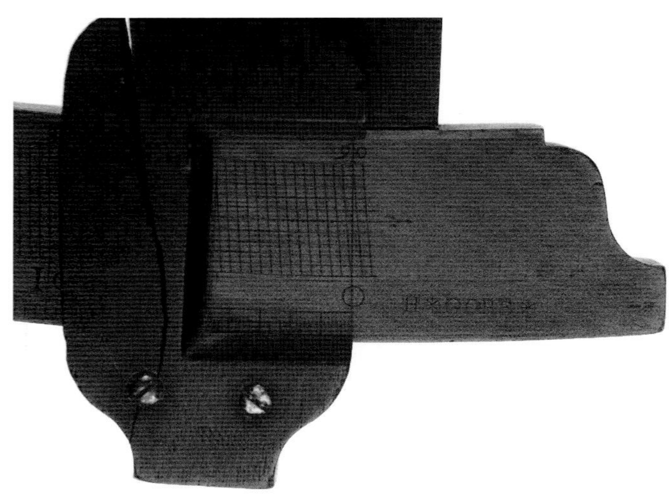

031 NAV1350
Octant

Origin Marseilles, France
Maker Sarrazin
Date About 1750
Inscriptions Signed on the back of the plate: Sarrazin invenit a marseille
Graduation Polished brass scale from 0° to 90° every 20′ on the back of the plate, indicated by a moving pointer. Vernier to 1′, zero at the centre.
Dimensions Radius 3.1 inches front (79 mm) and 2½ inches back (64 mm)
Description A polished brass plate with a detached boxwood handle. The pointer on the front of the instrument is geared to give a 1–4 magnification on the circular scale on the back, which is moved by hand. No shades. The index glass is missing; the glass has been replaced by silver foil; index- and horizon-glass adjustment (with an indicator showing the value) by a screw and a worm screw with a milled knob. The sight vane has two pinholes and a swivelling shutter. Stops to mark the end of the scale are missing. The tip of the pointer is bent, and one screw is missing from the handle. No box.
Provenance Purchased at Sotheby's, London, March 1975, lot no. 98, through Maggs Brothers.
Notes Nothing further is known about Sarrazin or this instrument.
Literature Marcelin, *Dictionnaire des fabricants français*.

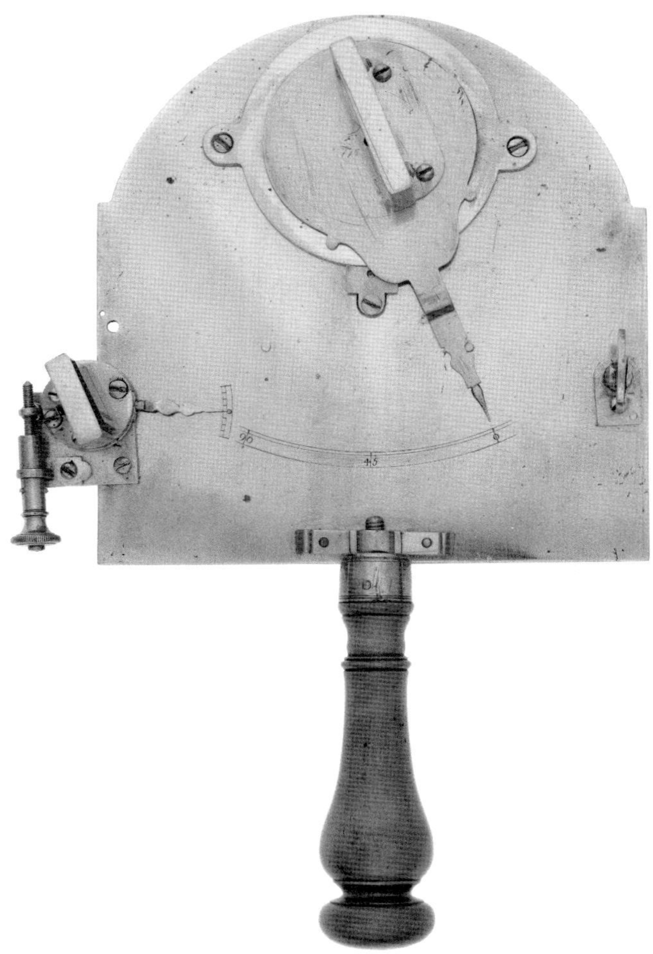

032 NAV1324
Octant

Origin London, England
Maker John Urings
Date About 1750
Inscriptions Signed on the crossbar: I * Urings London.
Marked (scratched) at the centre of the scale: I Bogman
Graduation Silver scale from -5° to 95° and vice versa every 20′ (for zenith distance), measuring to 95°. Silver vernier to 1′, zero at the centre.
Dimensions Radius 17 inches (432 mm)
Description Polished brass frame, silvered brass limb, and a brass stop for the index arm. There is no tangent screw; the clamping screw is on the back of the index arm. Three index shades (pink, red, and green). Index-glass adjustment by a screw; adjustment of both horizon glasses by levers, wing nuts, and milled clamping screws. The sight vane has two pinholes and a swivelling shutter; the back sight vane has one pinhole. A floral decoration is found at each end of the limb. No box.
Provenance Adams collection; lent in 1950, bequeathed in 1974.
Literature Clifton, *Directory*, 285; Hill and Paget-Tomlinson, *Instruments of Navigation*, 17.

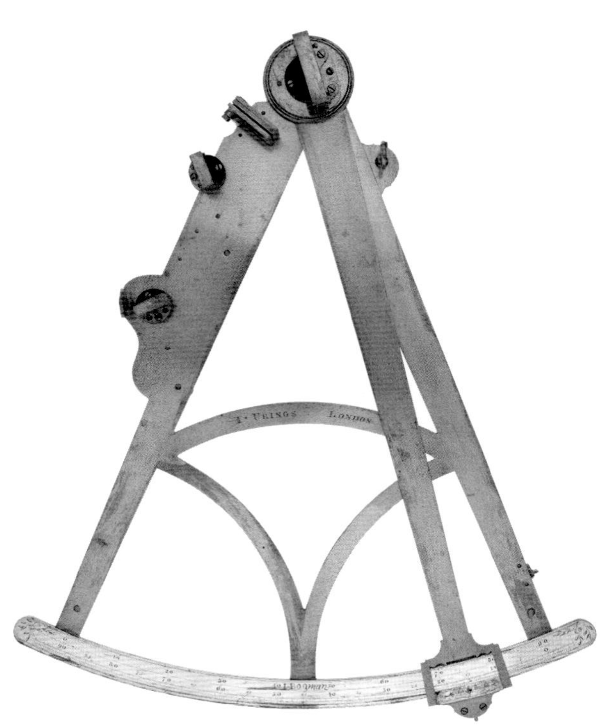

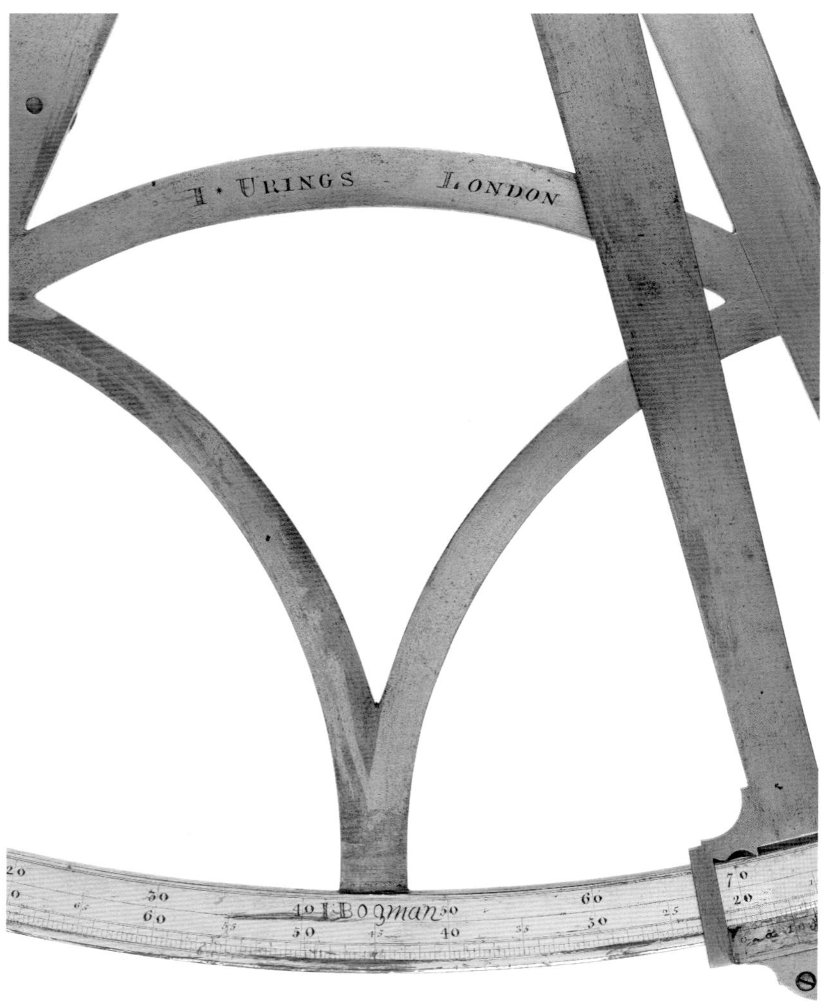

033 NAV1282
Octant

Origin London, England
Maker George Adams (senior)
Date 1753
Inscriptions Signed on the ivory plate on the crossbar: Geo: Adams:. Londini:. | Fecit :1753;
Graduation Boxwood transversal scale from 0° to 90° and vice versa every 20′ (for zenith distance), measuring to 90°. The scale reads to 1′ by use of the fiducial edge of ivory at the end of the index arm.
Dimensions Radius 19¾ inch (502 mm)
Description Mahogany frame and index arm; boxwood limb, brass fittings, and a brass stop for the index arm; an inlaid ivory plate on the crossbar. There is no tangent screw; the clamping screw (missing) was on the back of the index arm. Two socket shades (red and orange). Index-glass adjustment by a screw; adjustment of both horizon glasses by levers, wing nuts, and wing nuts for clamping (one clamping nut is missing). The sight vane has two pinholes and a swivelling shutter (missing); the back sight vane has one pinhole. No box.
Provenance Presented by Sir James Caird in 1937; purchased as part of the Gabb collection.
Notes The earliest dated English octant in the National Maritime Museum collection.
Literature Millburn, *Adams of Fleet Street*, 3, 156; Clifton, *Directory*, 285; Stimson, 'The Influence of the Royal Observatory at Greenwich', 129.

034 NAV1319
Octant

Origin Rotterdam, Netherlands
Maker Jacobus Kley
Date 1753
Inscriptions Marked on the ivory plate on the crossbar: P. Holm, Amsterdam 1753
Graduation Inlaid ivory scale from -5° to 95° and vice versa every 30′ (for zenith distance), measuring to 95°. Ivory vernier to 2′, zero at the centre.
Dimensions Radius 11.85 inches (301 mm)
Description Ebony frame and limb with a black-lacquered brass index arm and fittings. Decorative inlaid diamond-shaped ivory pieces in both struts, and a decorated inlaid ivory plate on the crossbar. There is no tangent screw; the clamping screw is on the back of the index arm. Two index shades (both red). Index-glass adjustment by a screw; adjustment of horizon glass by a lever, wing nut, and milled clamping screw. The sight vane has one pinhole. No box.
Provenance Adams collection; lent in 1950, bequeathed in 1974.
Notes Pieter Holm was a Swedish seaman who ran a nautical school in Amsterdam, where he sold octants that were made by Jacobus Kley of Rotterdam.
Literature Mörzer Bruyns, *Schip Recht door Zee*, 113–16, 208.

035 NAV1253
Octant

Origin England
Maker Unknown
Date About 1755
Inscriptions None
Graduation Boxwood transversal scale from 0° to 90° and vice versa every 20′ (for zenith distance), measuring to 90°. The scale reads to 2′ by use of the fiducial edge at the end of the index arm.
Dimensions Radius 18 inches (457 mm)
Description Mahogany frame and index arm; boxwood limb, brass fittings, and a brass stop for the index arm; an inlaid plate in the crossbar (missing). There is no tangent screw; the clamping screw (broken) is on the back of the index arm. The socket shades are missing. Index-glass adjustment by a screw; adjustment of both horizon glasses by levers, wing nuts, and thumbscrews for clamping. The sight vane and the back sight vane are missing. In a stepped oak box.
Provenance Unknown; the typed record indicates that it was in the collection before September 1963, when it was cleaned.

036 NAV1254
Octant

Origin London, England
Maker Benjamin Martin
Date About 1760
Inscriptions Signed on the lower part of the index arm: B. Martin London
Graduation Inlaid ivory scale from -5° to 95° every 20′, measuring to 888. Ivory vernier to 1′, zero at the centre.
Dimensions Radius 17¾ inches (451 mm)
Description Mahogany frame and limb with a brass index arm and fittings, and a brass stop for the index arm; inlaid ivory plates on the crossbar and on the back of the frame. There is no tangent screw; the clamping screw (missing) is on the back of the index arm. Three socket shades (red, orange, and green). Index-glass adjustment by a screw; adjustment of both horizon glasses by levers, wing nuts, and milled clamping screws. The sight vane has two pinholes and a swivelling shutter; the back sight vane has one pinhole. A brass-capped screwdriver is fitted in the crossbar. In a stepped oak box, containing in the lid a label in MS 'Father from Francis Isabel Gordon Pauline 8th De. 1907'.
Provenance Adams collection; lent in 1936, bequeathed in 1974. The label indicates that his children gave this instrument to Adams at Christmas in 1907.
Literature Clifton, *Directory*, 181; Millburn, *Retailer of the Sciences*.

037 NAV1347
Octant

Origin England
Maker Unknown
Date About 1760
Inscriptions None
Graduation Inlaid boxwood transversal scale from 0° to 90° and vice versa every 20′ (for zenith distance), measuring to 90°. The scale reads to 2′ by use of the fiducial edge (with an ivory lining) at the end of the index arm.
Dimensions Radius 18 inches (457 mm)
Description Mahogany frame, limb, and index arm; brass fittings, and a brass stop for the index arm; an inlaid ivory plate on the crossbar. The bottom of the index arm is of brass, possibly a repair. There is no tangent screw; the clamping screw is on the back of the index arm. Two socket shades (both red). Index-glass adjustment by a screw (missing); adjustment of both horizon glasses by levers, wing nuts, and milled clamping screws (one screw is missing). The sight vane bracket is broken; the back sight vane and the back horizon glass are missing. Two wooden feet are missing from the frame. No box.
Provenance Presented by HMS *Dryad* Museum, Southwick, Hampshire, in 1967.
Notes The instrument resembles octants by Benjamin Cole and John Goater.
Literature Clifton, *Directory*, 60–1.

038 NAV1291
Octant with artificial horizon

Origin London, England
Maker Benjamin Cole
Date About 1760
Inscriptions Signed on the limb: COLE Maker at ye ORRERY Fleet Street LONDON
Graduation Boxwood scale in two sections (joined at 45°) from -5° to 95° and vice versa every 20′ (for zenith distance), measuring to 92°. Ivory vernier (broken) to 1′, zero at the centre.
Dimensions Radius 17¾ inches (451 mm)
Description Mahogany frame, boxwood limb with a brass index arm and fittings, and a brass stop for the index arm; an inlaid ivory plate on the back of the frame. There is no tangent screw; the clamping screw is on the back of the index arm. Three socket shades (two red and one green). Index-glass adjustment by a screw; adjustment of horizon glass by a lever, wing nut, and clamping wing nut. The sight vane has two pinholes and a swivelling shutter; the back sight vane and glass and the lower socket have been removed and replaced by an artificial horizon. In an oak keystone box.
Artificial horizon This consists of a brass frame fitted to the front of the instrument, carrying a slightly curved spirit level. The bubble is reflected to the eye by means by a round lens and a special, adjustable mirror, mounted forward of the horizon glass, which reflects the Sun's image onto the level.
Provenance Presented by Sir James Caird in 1937; purchased as part of the Gabb collection.
Notes John Hadley patented this artificial horizon (no. 550 of 1734); it was improved by Thomas Winter, John Dollond, Daniel Scatliffe, and Henry Gregory (patent no. 752 of 1760).
Literature Clifton, *Directory*, 60–1; Hill and Paget-Tomlinson, *Instruments of Navigation*, 16.

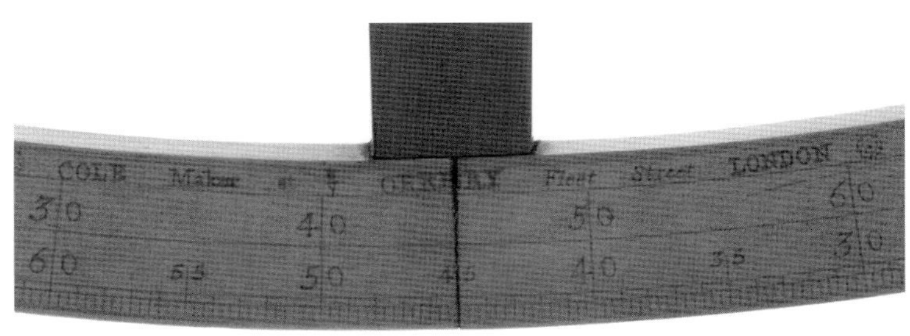

039 NAV1251
Octant

Origin England
Maker Unknown
Date 1761 or earlier
Inscriptions Marked on the brass strip on the crossbar: Jhn Harrison. | Wm Harrison
Marked on a decorative brass plate on the back of the frame: Nov 15th 1761.
Marked on an inlaid brass strip in the limb with decorative engravings at each end: H.M.S. Deptford.
Graduation The inlaid ivory scale has been removed. Ivory vernier to 1′, zero at the centre.
Dimensions Radius 15¼ inches (387 mm)
Description Mahogany frame and limb with a brass index arm and fittings, and a brass stop for the index arm. The inlaid ivory plates on the crossbar and on the back of the frame have been replaced by inlaid brass plates; a brass strip has replaced the scale. There is no tangent screw; the clamping screw is on the back of the index arm. Two socket shades (both red). Index-glass (glass missing) adjustment by a screw; adjustment of both horizon glasses (glass in both missing) by levers, wing nuts (one missing), and clamping screws (both missing). The sight vane has two pinholes and a swivelling shutter; the back sight vane has one pinhole. The lower half of the index arm is decoratively engraved, including a three-mast vessel. The outer ends of the limb have been sawn off and the centre strut, which once contained a pencil or screwdriver, has been removed. Two brass feet are missing from the frame. No box.
Provenance Unknown; departmental documentation indicates that it was in the collection before December 1961.
Notes This instrument was adapted to commemorate the voyage of HMS *Deptford* in 1761–2, in which John Harrison's son William took part to test Harrison's fourth timekeeper, H4, of 1759.
Literature Gould, *The Marine Chronometer*, 79–80; Hill and Paget-Tomlinson, *Instruments of Navigation*, 16.

040 NAV1346
Octant

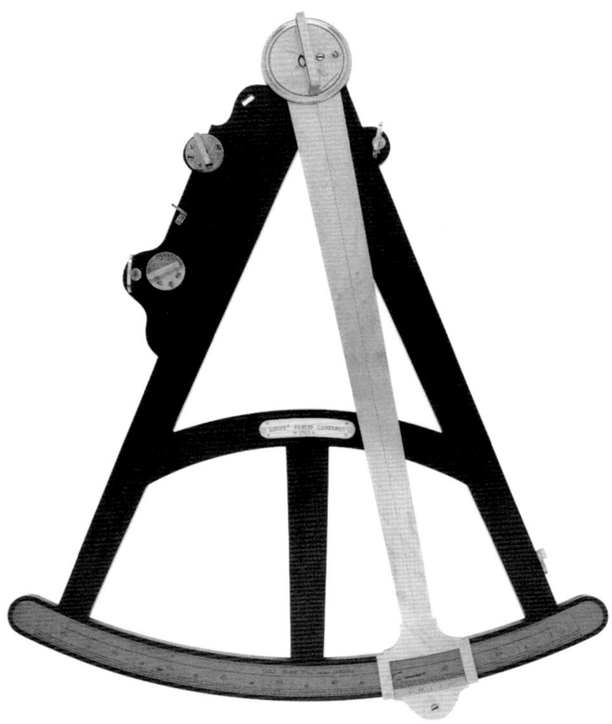

Origin London, England
Maker Benjamin Cole
Date 1761
Inscriptions Signed on the scale: COLE Maker Fleet Street LONDON
Marked on the ivory plate on crossbar: ★Lieu[t]. Philip Carteret ★ | ★1761★
Graduation Inlaid boxwood scale from -5° to 95° every 20′, measuring to 93°. Brass vernier to 1′, zero at the centre.
Dimensions Radius 17.4 inches (442 mm)
Description Mahogany frame and limb with a brass index arm and fittings, and a brass stop for the index arm; an inlaid ivory plate on the crossbar. There is no tangent screw; the clamping screw is on the back of the index arm. The socket shades are missing (the mount is present). Index-glass adjustment by a screw; adjustment of both horizon glasses by levers, wing nuts, and milled clamping screws. The sight vane has two pinholes and a swivelling shutter; the back sight vane has one pinhole (both vanes may be later replacements). No box.

Provenance Lieutenant Philip Carteret (1733–96), the original owner, was in command of HM sloop *Swallow*, sailing from Plymouth under Captain Samuel Wallis in the *Dolphin* in August 1766 to the Pacific, to discover the supposed Southern Continent. They became separated after passing through the Strait of Magellan, Wallis going on to discover Tahiti while Carteret discovered Pitcairn, among other islands, returning in *Swallow* to Spithead in 1769. Purchased from Mme P. Ceresole, Lutry, Vaud, Switzerland, in 1967.
Literature Clifton, *Directory*, 60–1; Elisabeth Baigent, 'Carteret, Philip (1733–1796)' in *Oxford Dictionary of National Biography* Oxford, 2004, <http://www.oxforddnb.com/.>

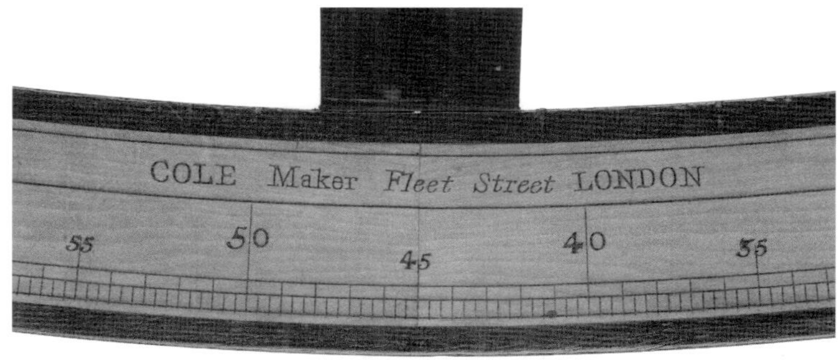

041 NAV1342
Octant

Origin Guernsey, Channel Islands
Maker Possibly Elias De La Rue
Date 1763
Inscriptions Marked on the ivory plate on the crossbar: * Elias: De. Larue: of. Guernsey. * | N° 392 1763
Graduation Inlaid ivory scale in two sections (joined at 45°, damaged) from -5° to 95° every 20′, measuring to 91°. Ivory vernier to 1′, zero at the centre.
Dimensions Radius 17¼ inches (438 mm)
Descriptions Ebony frame and limb with a brass index arm and fittings, and a brass stop for the index arm; an inlaid ivory plate on the crossbar and inlaid decorative ivory diamond-shaped pieces in both struts (one piece is missing). There is no tangent screw; the clamping screw is on the back of the index arm. Two socket shades (both red). Index-glass adjustment by a screw; adjustment of both horizon glasses by levers, wing nuts, and milled clamping screws. The sight vane has two pinholes and a swivelling shutter; the back sight vane has one pinhole. No box.
Provenance Presented by the Royal United Service Institution Museum in 1963, on its closure.
Notes Elias De La Rue may have been an owner; his name is not recorded in Clifton, *Directory*.
Literature Hughes, *Official Catalogue*, 323, no. 7823.

042 NAV1306
Octant

Origin London, England
Maker (John) Goater
Date About 1769
Inscriptions Signed at the centre of the scale: GOATER Maker Wappin LONDON
Marked on the ivory plate on the crossbar: ⋆ Iohn Mason 1769 ⋆
Graduation Inlaid ivory scale from -5° to 95° every 20′, measuring to 91°. Ivory vernier to 1′, zero at the centre.
Dimensions Radius 17¼ inches (438 mm)
Description Ebony frame and limb with index arm two-thirds wood, one-third brass; brass fittings, and a brass stop for the index arm; an inlaid ivory plate on the back of the frame. There is no tangent screw; the clamping screw is missing from the back of the index arm. Three socket shades (two red and one green). Index-glass adjustment by a screw; adjustment of both horizon glasses by levers, wing nuts, and milled clamping screws. The sight vane has two pinholes and a swivelling shutter; the back sight vane has one pinhole. An ivory pencil head (pencil missing) is fitted in the crossbar. A floral decoration, including Masonic symbols, can be found on the brass section of the index arm. One of the feet on the frame is an old replacement. No box.
Provenance Presented by Sir James Caird in 1937; purchased as part of the Gabb collection.
Notes The Museum's record mentions a stepped oak box; this was not found.
Literature Clifton, *Directory*, 114–15; Bennett, 'Catadioptrics and Commerce in Eighteenth-Century London', 248–9.

043 NAV1294
Octant

Origin London, England
Maker Benjamin Cole
Date About 1770
Inscriptions Signed on the ivory plate on the crossbar: ⋆ COLE . Maker . LONDON⋆
Graduation Inlaid ivory scale in two sections (joined at 45°) from -5° to 95° every 20′, measuring to 94°. Brass vernier to 1′, zero at the centre.
Dimensions Radius 18 inches (457 mm)
Description Mahogany frame and limb with a brass index arm and fittings, and a brass stop for the index arm; inlaid ivory plates on the crossbar and on the back of the frame. There is no tangent screw; the clamping screw is on the back of the index arm. Three socket shades (green, orange, and red); the top socket is broken. Index-glass adjustment by a screw (missing); adjustment of both horizon glasses by levers, wing nuts, and milled clamping screws. The sight vane has two pinholes and a swivelling shutter; the back sight vane has one pinhole. In a keystone box with a stepped lid that is painted black on the outside. The lid contains a trade label for Benjamin Cole, next the Globe Tavern in Fleet Street, London. The bottom of the box is marked, in ink, 'Hadley's sea octant otherwise quadrant invented 1731'.

Provenance Presented by Sir James Caird in 1943; purchased through Maggs Brothers, presumably from Ernest R. Cooper, who, in the article cited below,indicated he owned this instrument and also the backstaff cat. no. 023.
Literature Cooper, 'The Davis Back-Staff or English Quadrant', 59–64; Clifton, *Directory*, 60–1.

044 NAV1255
Octant

Origin Glasgow, Scotland
Maker John Gardner
Date About 1770
Inscriptions Signed at the bottom of the index arm: Gardner Glasgow
Graduation Inlaid ivory scale from -5° to 95° and vice versa every 20′ (for zenith distance), measuring to 91°. Ivory vernier to 1′, zero at the centre.
Dimensions Radius 16 inches (406 mm)
Description Mahogany frame and limb with a brass index arm and fittings, and a brass stop for the index arm; an inlaid ivory plate on the back of the frame. There is no tangent screw; the clamping screw is on the back of the index arm. Three socket shades (two red and one green) fastened by a milled screw on the back of the frame. Index glass without adjustment; adjustment of both horizon glasses by levers, wing nuts, and milled clamping screws. The sight vane has two pinholes and a swivelling shutter; the back sight vane has one pinhole. A pencil or screwdriver is missing from the crossbar. Decorative engravings are found on the lower part of the index arm. In a stepped oak box.
Provenance Adams collection; lent in 1936, bequeathed in 1974.
Literature Clifton, *Directory*, 109; Clarke *et al.*, *Brass & Glass*, 164–5.

045 NAV1318
Octant

Origin London, England
Maker (John) Troughton
Date About 1770
Inscriptions Signed at the bottom of the index arm: Troughton London
Graduation Inlaid ivory scale from -5° to 99° every 20′, measuring to 91°. Ivory vernier to 1′, zero at the right.
Dimensions Radius 15¾ inches (400 mm)
Description Ebony frame and limb with a brass index arm and fittings, and a brass stop for the index arm; inlaid ivory plates on the crossbar and on the back of the frame. There is no tangent screw; the clamping screw is on the back of the index arm. Three index shades (red, two missing). Index glass without adjustment; adjustment of horizon glass by a lever, wing nut, and milled clamping screw. The sight vane has two pinholes and a swivelling shutter. A pencil or screwdriver is missing from the crossbar. No box.
Provenance Adams collection; lent in 1950, bequeathed in 1974.
Literature Clifton, *Directory*, 282; McConnell, *Instrument Makers to the World*.

046 NAV1341
Octant

Origin England
Maker Unknown
Date About 1770
Inscriptions None
Graduation Inlaid ivory scale from -5° to 95° every 20′, measuring to 90½°. Ivory vernier to 1′, zero at the centre.
Dimensions Radius 15.6 inches (396 mm)
Description Ebony frame and limb with a brass index arm and fittings, and a brass stop for the index arm; inlaid plates on the crossbar and on the back of the frame are both missing. There is no tangent screw; the clamping screw is on the back of the index arm. The socket shades are missing. Index-glass (glass missing) adjustment by a screw (missing); adjustment of horizon glass (glass missing) and back horizon glass by levers, wing nuts, and milled clamping screws (one missing). The sight vane has two pinholes and a swivelling shutter; the back sight vane is broken. A pencil or screwdriver is missing from the crossbar. No box.
Provenance Presented by the Boy Scouts Association, Formerly on RRS *Discovery*. London, in 1962.

047 NAV1257
Octant

Origin England
Maker Unknown
Date About 1770
Inscriptions None
Graduation Surface-mounted brass scale from -5° to 95° every 20′, measuring to 89°. Brass vernier to 1′, zero at the centre.
Dimensions Radius 18 inches (457 mm)
Description Ebony frame, mahogany limb, a brass index arm and fittings, and a brass stop for the index arm; inlaid ivory plates on the crossbar and on the back of the frame. There is no tangent screw; the clamping screw is on the back of the index arm. Three socket shades (two red and one green). Index-glass adjustment by a screw; adjustment of both horizon glasses by levers, wing nuts, and milled clamping screws. The sight vane has two pinholes and a swivelling shutter; the back sight vane has one pinhole. The sight vane's foot, a foot of the frame, the shutter, and various screws are replacements. An ivory pencil head (pencil missing) is fitted in the crossbar. In a stepped box, painted blue on the outside and cream on the inside, containing in the lid a trade label for Henry Hughes and Son, 59 Fenchurch Street, London and a label in MS 'Helen, Francis, Isabel, Gordon, & Pauline gave this instrument to Father Christmas 1908'.
Provenance Adams collection; lent in 1936, bequeathed in 1974. The presence of the Hughes label may suggest that the instrument was purchased from them and given to Adams by his wife Helen and their children at Christmas 1908.

048 NAV1344
Octant

Origin England
Maker Unknown
Date About 1770
Inscriptions Marked on the ivory plate on the crossbar: James Campbell.
Marked on the index arm: This quadrant belonged to Lieut. James Campbell R.N. who served under Hood, Hotham, Jervis, Nelson & Duncan. Fought in | H.M.S. Dido. 28 at capture of Minerve 44 1795 and at the Battle of St. Vincent 1797 | in temporary command H.M.S. Victor 18 was mentioned in despatches for her well directed fire at the landing at the Texel 1799
Graduation Inlaid ivory scale in two sections (joined at 50°) from -5° to 95° every 20′, measuring to 92°. Ivory vernier to 1′, zero at the centre.
Dimensions Radius 13.6 inches (345 mm)
Description Ebony frame and limb with a brass index arm and fittings, and a brass stop for the index arm; inlaid ivory plate on the crossbar and inlaid decorative ivory strips along all edges of the front of the frame. There is no tangent screw; the clamping screw (missing) is on the back of the index arm. The socket shades are missing. Index-glass adjustment by a screw; adjustment of both horizon glasses by levers, wing nuts (one missing), and milled clamping screws (one missing). The sight vane has two pinholes and a swivelling shutter; the back sight vane has one pinhole and is fastened to the brass back horizon glass mount. A pencil or screwdriver is missing from the crossbar. No box.
Provenance Presented by Commander A.T. Stewart OBE, RN, London, in 1965.
Notes The glasses and sight vanes are unusually small compared with the size of the frame.

049 NAV1357
Octant

Origin England
Maker Unknown
Date About 1770
Inscriptions None
Graduation Inlaid ivory scale from -5° to 95° every 20′, measuring to 91°. Ivory vernier (broken) to 1′, zero at the centre.
Dimensions Radius 16 inches (406 mm)
Description Ebony frame and limb with a brass index arm and fittings, and a brass stop for the index arm; inlaid ivory plates on the crossbar (missing) and on the back of the frame. Remains of a slow-motion device (probably a later addition); the clamping screw is on the back of the index arm. The socket shades are missing. Index-glass (glass missing) adjustment by a screw (missing); adjustment of both horizon glasses by levers, wing nuts, and milled clamping screws. The sight vane has two pinholes and a swivelling shutter; the back sight vane has one pinhole. A pencil or screwdriver is missing from the crossbar. No box.
Provenance Bequeathed by Captain E.L. Davies, Brighton, through Donne, Mileham and Haddock, Solicitors, in 1986.

050 NAV1284
Octant

Origin London, England
Maker Thomas Lake
Date 1770
Inscriptions Signed on the ivory plate on the crossbar: Tho. Lake. Leadenhall Stree[t] | ★London. Fecit.★ | .: For : Martin : Walding : 1770: .
Graduation Inlaid boxwood transversal scale from 0° to 90° and vice versa every 20′ (for zenith distance), measuring to 90°. The scale reads to 2′ by use of the fiducial edge (with an ivory lining) at the end of the index arm.
Dimensions Radius 17.85 inches (453 mm)
Description Mahogany frame, limb, and index arm; brass fittings; a brass stop for the index arm; and an inlaid ivory plate on the crossbar. There is no tangent screw; the clamping screw is on the back of the index arm. Two socket shades (red and orange). Index-glass adjustment by a screw; adjustment of both horizon glasses by levers, wing nuts, and milled clamping screws. The sight vane has two pinholes; the back sight vane has one pinhole. One of the wooden feet is missing from the frame. In an oak keystone box with a stepped lid, partly painted red on the inside and partly grey on the outside. Traces of writing (illegible) on the lid; a carved anchor and a label marked 'Lot 418'.
Provenance Presented by Sir James Caird in 1937; purchased as part of the Gabb collection.
Literature Clifton, *Directory*, 161.

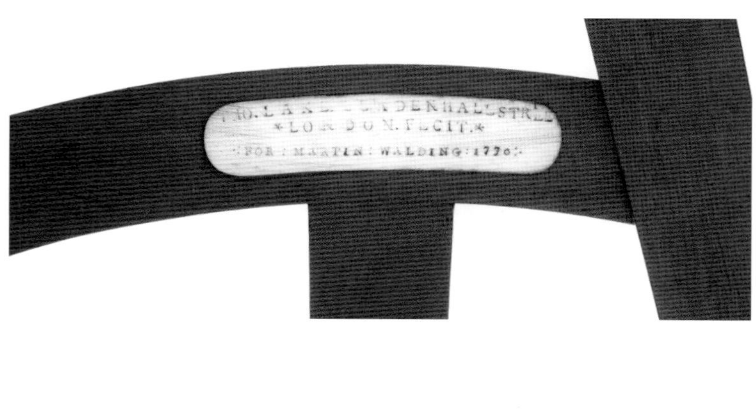

051 NAV 1356
Octant

Origin London, England
Maker John Bird
Date About 1772
Inscriptions Signed on the scale: J. Bird London.
Signed on the back horizon glass adjustment: Dollond London
Graduation Surface-mounted polished brass scale graduated from -4° to 94° every 20′, measuring to 90°. Brass vernier to 1′, zero at the centre.
Dimensions Radius 20 inches (508 mm)
Description Mahogany frame and limb with a brass index arm and fittings; the back of the limb is also brass-covered, and there is a wooden pad-shaped handle on the back of the frame. The clamping screw and tangent screw (both missing) are on a separate shoe, fitted on the limb to the right of the index arm. Horizon glass (glass missing), back horizon glass (glass missing) but with modern replacements. Three socket shades (all red). Index-glass (lower half not silvered) adjustment by a screw; index glass fitted with a Maskelyne flap. Horizon-glass adjustment by a milled screw and a lever, wing nut, and clamping screw (missing); back-horizon-glass adjustment by a 133 mm lever and two milled wheels, operated from below the telescope bracket. Threaded telescope brackets for fore and back observation, both fitted for correcting collimation error; perpendicular adjustment on both by rising-piece and milled knobs (the forward knob is missing). The telescope is missing. Part of the frame near the lower shade socket is broken. In a wooden keystone box with a stepped lid (repaired).
Provenance Purchased at Sotheby's, London, October 1985, lot no. 89.
Notes Peter Dollond patented the back-horizon-glass adjustment (no. 1017 of 1772). By means of a second lever (249 mm), this glass could be swivelled through 90° and brought into the same position as the horizon glass. The telescope was then placed in the lower position, and the back glass was adjusted and reversed to its position for back observation. A sextant by Dollond with the same adjustment device is in the Whipple Museum of the History of Science in Cambridge, inv. no. Wh.1292.
Literature Maskelyne, 'Remarks on the Hadley's Quadrant', 57–8; Dollond, *Directions for Using the New Improved Hadley's Quadrant*; Ifland, *Taking the Stars*, 84–5; Clifton, *Directory*, 30, 227; Warner, 'John Bird and the Origin of the Sextant'; Moskowitz, 'The World's First Sextants'; Bennett, *The Divided Circle*, 136.
For image see colour section plate 7.

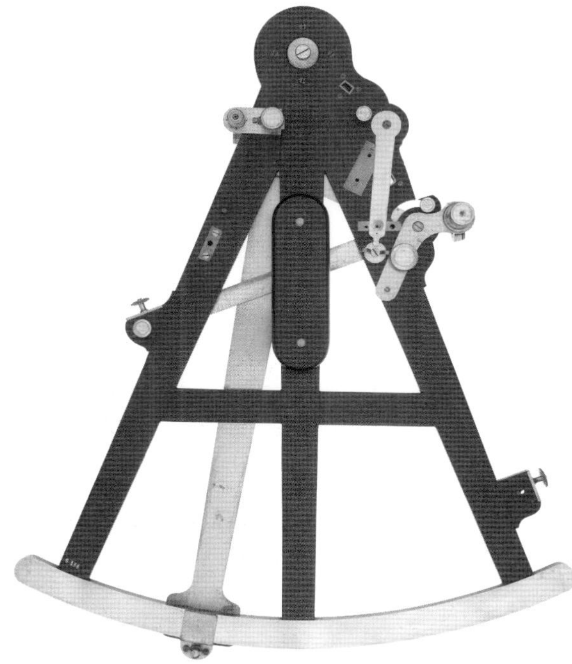

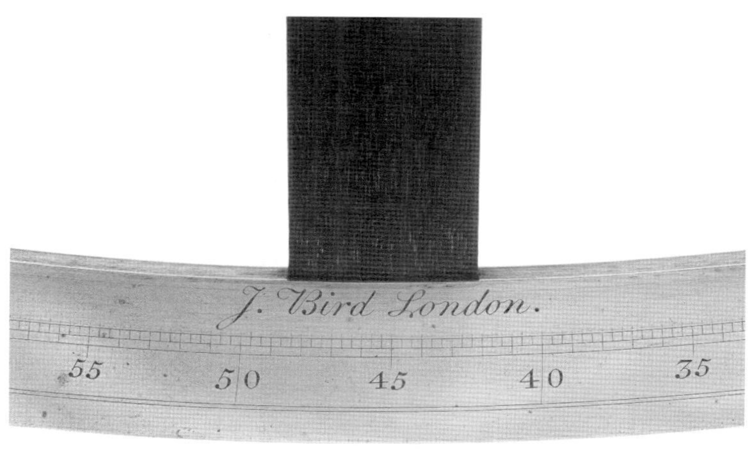

052 NAV1258
Octant

Origin London, England
Maker Joshua Lover Martin
Date About 1774
Inscriptions Signed on the index-glass adjustment arm: Martin Inv.t. et fecit. Londini
Marked (stamped) at the centre of the scale (at 48°): I [foul anchor] R
Graduation Inlaid ivory scale from -2° to 99° every 20′, measuring to 91°. Ivory vernier to 1′, zero at the right.
Dimensions Radius 15.6 inches (396 mm)
Description Ebony frame and limb with a brass index arm and fittings; a brass stop for the index arm; inlaid ivory plates on the crossbar and on the back of the frame (latter missing). There is no tangent screw; the clamping screw is on the back of the index arm. Two socket shades (green and red). The index-glass mount is hinged, with perpendicular adjustment by an adjacent capped thumbscrew. The mount is fitted on a short arm for parallel adjustment by a capped thumbscrew. The adjustment of both horizon glasses is by screws. The sight vane has two pinholes and a lever-operated shaded shutter (red); the back sight vane has one pinhole. A pencil or screwdriver is missing from the crossbar. In a polished stepped oak box; cut in the top of the lid: 'T.P.B.'
Provenance Adams collection; lent in 1936, bequeathed in 1974.
Notes Joshua Lover Martin patented this index-glass adjustment (no. 1088 of 1774). An octant by Benjamin Martin (Joshua's father), with this patent adjustment, is in a private collection in the USA. The scale mark is for Jesse Ramsden.
Literature Stimson, 'Some Board of Longitude Instruments', 111; Ifland, *Taking the Stars*, 85, 196; Clifton, *Directory*, 181, 227; Millburn, *Retailer of the Sciences*, 46.

053 NAV1275
Octant

Origin Föhr, Schleswig-Holstein
Maker Jan Corneliszoon
Date 1775
Inscriptions Signed on the horizon glass mounting plate: Jan Cornelisz: v. Vüer fecit Nom: 123
Marked on the index arm: Comd[r] | Roelof O: Meeuw | 1775
Graduation Surface-mounted brass transversal scale 90° to 0° every 1° for zenith distance, measuring to 90°. The scale can be read to 5′ by use of a division on the fiducial edge at the end of the index arm.
Dimensions Radius 12¾ inches (324 mm)
Description Hornbeam frame and limb with a brass index arm and fittings; a brass stop for the index arm. There is no tangent screw; the clamping screw is on the back of the index arm. Two index shades (one blue; the other is missing). Index-glass adjustment by a screw; adjustment of horizon glass by a screw, lever, and clamping screw. The sight vane has one pinhole. Engraved floral decorations are found on the entire index arm and flowers at the ends of the brass scale.
No box.
Provenance Belonged to Roelof Olphertz Meeuw, master of the Amsterdam-owned whale ship *Vrouwe Jacoba* 1769–80; the British took his unnamed ship in 1798. The instrument was exhibited at the Musée des Arts Décoratifs (Louvre), Paris, March–April 1936, from the Augustin Seguin collection. Presented by Sir James Caird in 1936; purchased through Maggs Brothers.
Literature Callender, *The Caird Collection: Supplement*, 89, no. 316; NN, *Exposition des Instruments et Outils d'autrefois*, 38, no. 282; Mörzer Bruyns, 'Octants from the Friesian Island of Föhr'; Hill and Paget-Tomlinson, *Instruments of Navigation*, 17.
For image see colour section plate 10.

054 NAV 1295
Octant

Origin Liverpool, England
Maker John Leverton
Date About 1777
Inscriptions Signed on the index arm: Leverton Liverpool
Marked on the ivory plate on the crossbar: Peter Richards 1777
Graduation Inlaid ivory scale from -5° to 95° and vice versa every 20′ (for zenith distance), measuring to 86½°. Ivory vernier to 1′, zero at the centre.
Dimensions Radius 15.85 inches (403 mm)
Description Ebony frame and limb with a brass index arm and fittings, a brass stop for the index arm; inlaid ivory plates on the crossbar and on the back of the frame. There is no tangent screw; the clamping screw is on the back of the index arm. Three socket shades (orange, green, and red). Index-glass (glass missing) adjustment by a screw; adjustment of both horizon glasses by levers, wing nuts, and milled clamping screws. The sight vane has two pinholes and a swivelling shutter; the back sight vane (broken) has one pinhole. A pencil or screwdriver is missing from the crossbar. Simple decorative marks are found on the ivory plate on the crossbar. No box.
Provenance Belonged to Joseph Battison Williams, Borough Surveyor of Daventry, Northamptonshire, who had no known connection with the sea. Presented by Major F.B. Topham in 1946, in the name of Mrs K. Williams, Uxley, Hertfordshire, from whom he had bought it in 1944. Mrs Williams is presumed to have been a descendent of J.B. Williams.
Literature Clifton, *Directory*, 166; Morrison-Low, *Making Scientific Instruments*, 63, 331; Hill and Paget-Tomlinson, *Instruments of Navigation*, 17.

055 NAV1283
Octant

Origin London, England
Maker John Gilbert
Date 1777
Inscriptions Signed on the ivory plate on the crossbar: Made by In° Gilbert. Tower Hill. London. | For Adam . Glegg. Sep\r. 25: 1777.
Graduation Inlaid ivory scale from -5° to 95° every 20′, measuring to 89½°. Ivory vernier to 1′, zero at the centre (damaged).
Dimensions Radius 14¾ inches (375 mm)
Description Mahogany frame and limb with a brass index arm and fittings; a brass stop for the index arm; inlaid ivory plates on the crossbar and on the back of the frame. There is no tangent screw; the clamping screw is on the back of the index arm. Three socket shades (one missing, one red, and one orange). Index-glass adjustment by a screw; both horizon glasses by levers, wing nuts, and milled clamping screws. The sight vane has two pinholes and a swivelling shutter; the back sight vane has one pinhole. An ivory-capped pencil is fitted in the crossbar. In an oak keystone box with a stepped lid.
Provenance Presented by Sir James Caird in 1937; purchased as part of the Gabb collection.
Literature Clifton, *Directory*, 112.

056 NAV1281
Octant

Origin Bristol, England
Maker Henry Edgeworth
Date 1779 or later
Inscriptions Signed on the centre of the scale: H. Edgworth Bristol
Marked on the ivory plate on the crossbar: P.re Deschamps
Graduation Inlaid ivory scale from -5° to 95° every 20′, measuring to 90°. Ivory vernier to 1°, zero at the centre.
Dimensions Radius 15¼ inches (387 mm)
Description Ebony frame and limb with a brass index arm and fittings; a brass stop for the index arm; decorative diamond-shaped inlaid ivory pieces in both struts; inlaid ivory plates on the crossbar and on the back of the frame. A double-ended tangent screw is on the front of the index arm; the clamping screw is on the back. Three socket shades (one green and two red). Index-glass adjustment by a screw; adjustment of both horizon glasses by levers, wing nuts, and milled clamping screws. The sight vane has two pinholes and a swivelling shutter; the back sight vane has one pinhole. A pencil or screwdriver is missing from the crossbar. Decorative engravings, including flags and a drum, are found on the lower half of the index arm. In an oak keystone box with a stepped lid, painted dark blue on the outside and containing a trade label for Henry Edgeworth, Opposite the Dial, on the Quay, Bristol, including an inventory of instruments sold by him in 1779. There is another trade label for Antoine Roux of Marseilles, who was selling charts and navigational instruments, sails, and flags of all colours.

Provenance Belonged to Pierre Deschamps, of whom nothing is known; at some stage the instrument was cleaned or repaired by Antoine Roux, a member of the Roux family of instrument makers and hydrographers in Marseilles. Presented by Sir James Caird in 1936; purchased as part of the Landau collection through Maggs Brothers.

Literature Callender, *The Caird Collection: Supplement*, 96, no. 18; Bryant, 'A Note on Henry Edgeworth'; Clifton, *Directory*, 92; Morrison-Low, *Making Scientific Instruments*, 53, 55, 320; Marcelin, *Dictionaire des fabricants français*; Brewington, *The Peabody Museum Collection*, 138; Hill and Paget-Tomlinson, *Instruments of Navigation*, 18.

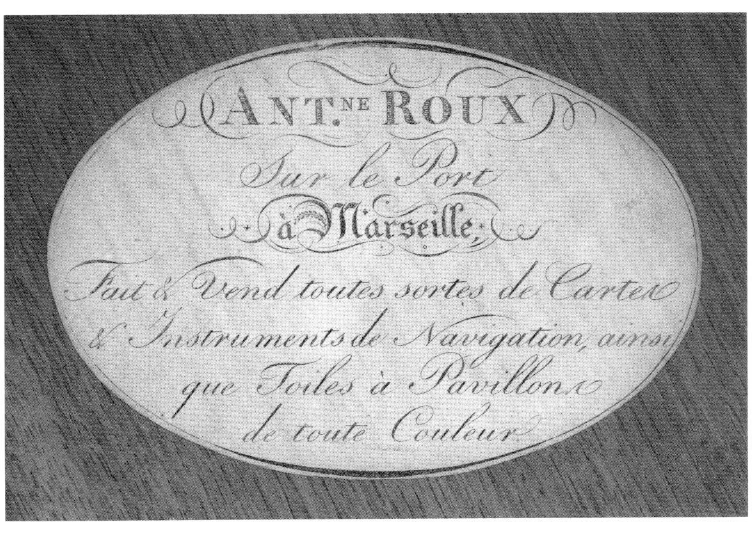

057 NAV1353
Octant

Origin London, England
Maker Jesse Ramsden
Date About 1780
Inscriptions Signed on the ivory plate on the crossbar: Ramsden London.
Marked (stamped) at the centre of the scale (at 55°): E [foul anchor] P
Graduation Inlaid ivory scale from -2° to 108° every 30′, measuring to 90°. Ivory vernier to 1′, zero at the right.
Dimensions Radius 3¾ inches (95 mm)
Description Ebony frame and limb with a brass index arm and fittings; detached wooden handle; an inlaid ivory plate on the crossbar. There are a tangent screw and clamping screw on the back of the index arm. Three index shades (two red and one green). Index-glass adjustment by a screw; adjustment of horizon glass by an extended lever and clamping screw operated from beneath the sight vane. Magnifier (single lens) on a 76-mm swivelling arm. The sight vane has two pinholes and a swivelling shutter. A threaded brass socket on the back of the crossbar for fitting a stand is perhaps a later addition. In a polished mahogany keystone box.
Provenance Purchased from Delehar Antiques, London, in 1978.
Notes Peter Dollond patented the horizon-glass adjustment (no. 1017 of 1772). The scale mark is for Edward Pritchard, an apprentice of Ramsden.
Literature Dollond, *Directions for Using the New Improved Hadley's Quadrant*; Ifland, *Taking the Stars*, 84–5; McConnell, *Jesse Ramsden*, 46, 64, 247–8; Clifton, *Directory*, 224, 227; Stimson, 'Some Board of Longitude Instruments', 113.

058 NAV1320
Octant

Origin Guernsey, Channel Islands
Maker Chevalier
Date About 1780
Inscriptions Signed on the ivory plate on the crossbar: Chevalier. Guernsey
Marked (stamped) at the centre of the scale (at 47°): E [foul anchor] P
Graduation Inlaid ivory scale from -2° to 98° every 20′, measuring to 91°. Ivory vernier to 1′, zero at the right.
Dimensions Radius 13¾ inches (349 mm)
Description Mahogany frame and limb with a brass index arm and fittings; a brass stop for the index arm; inlaid ivory plates on the crossbar and on the back of the frame (the latter is missing). There is no tangent screw; the clamping screw is on the back of the index arm. Three socket shades (two red and one green). Index-glass adjustment by a screw; adjustment of both horizon glasses by levers, wing nuts, and milled clamping screws. The horizon glass and mount are missing. The sight vane has two pinholes and a swivelling shutter; the back sight vane has one pinhole. An ivory pencil head is fitted in the crossbar (pencil missing). No box.
Provenance Adams collection; lent in 1950, bequeathed in 1974.
Notes The scale mark is for Edward Pritchard, an apprentice of Ramsden.
Literature Clifton, *Directory*, 55, 224; McConnell, *Jesse Ramsden*, 46, 64, 247–8; Stimson, 'Some Board of Longitude Instruments', 113.

059 NAV1309
Octant

Origin Liverpool, England
Maker Egerton Smith
Date About 1780
Inscriptions Signed on the index arm: Egerton Smith Liverpool
Graduation Inlaid ivory scale from -2° to 98° every 20′, measuring to 90°. Ivory vernier to 1′, zero at the right.
Dimensions Radius 14¾ inches (375 mm)
Description Ebony frame and limb with a brass index arm and fittings; a brass stop for the index arm; inlaid plates on the crossbar and on the back of the frame (both missing). There is no tangent screw; the clamping screw is on the back of the index arm. Three socket shades (two red and one green), probably not original to this instrument; a second set (two red and one green) is possibly also from another octant. Index-glass adjustment by a screw; adjustment of both horizon glasses by screws and levers. The sight vane has two pinholes and a swivelling shutter; the back sight vane is missing. A pencil or screwdriver is missing from the crossbar. Engraved floral decorations are found around the vernier. In a stepped deal box, painted red on the inside, and containing some illegible pencil notes.
Provenance Purchased from Mrs F.J. Mortimer, London, in 1951.
Literature Clifton, *Directory*, 255; Morrison-Low, *Making Scientific Instruments*, 334.

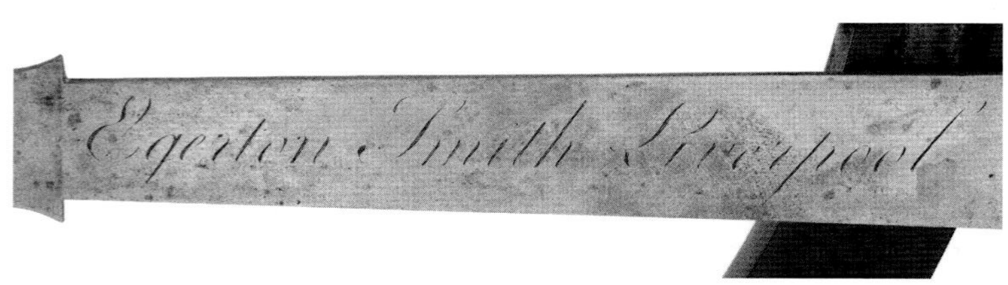

060 NAV1351
Octant

Origin Cork, Ireland
Maker Sweeny
Date About 1780
Inscriptions Signed on the ivory plate on the crossbar: Sweeny Cork Fecit | For Cap[t]. Iohn Leworthy
Graduation Inlaid ivory scale in two sections (joined at 50°) from -3° to 93° every 20′, measuring to 93°. Ivory vernier to 1′, zero at the centre.
Dimensions Radius 15 inches (381 mm)
Description Mahogany frame and limb with a brass index arm and fittings; an inlaid ivory plate on the crossbar. There is no tangent screw; the clamping screw is on the back of the index arm. The shades are missing. Index-glass adjustment by a screw; adjustment of horizon glass by a lever, wing nut, and milled clamping screw. The sight vane is missing. No box.
Provenance Bequeathed by Commander Henry Oliver Hill RN (1887–1974) in 1976. Hill was an assistant keeper in the National Maritime Museum Department of Navigation.
Notes This instrument was made by Edward Nathaniel Sweeney or by his widow.
Literature Clifton, *Directory*, 271; Mollan, *Irish National Inventory*, 416; Littlewood and Butler, *Of Ships and Stars*, 126, 177.

061 NAV1300
Octant

Origin England
Maker Unknown
Date About 1780
Inscriptions None
Graduation Inlaid ivory scale from -5° to 95° every 20′, measuring to 91°. Ivory vernier to 1′, zero at the centre.
Dimensions Radius 15¾ inches (400 mm)
Description Ebony frame, limb, and index arm; the lower end of the arm is of brass, brass fittings, a brass stop for the index arm; inlaid ivory plates on the crossbar and on the back of the frame. There is no tangent screw; the clamping screw is on the back of the index arm. Three socket shades (green, orange, and red). Index-glass adjustment by a screw; adjustment of both horizon glasses by levers, wing nuts, and milled clamping screws. The sight vane has two pinholes and a swivelling shutter; the back sight vane has one pinhole. A pencil or screwdriver is missing from the crossbar. In an oak keystone box with a stepped lid.
Provenance Adams collection; lent in 1946, bequeathed in 1974.

062 NAV1316
Octant

Origin England
Maker Unknown
Date About 1780
Inscriptions None
Graduation Inlaid ivory scale from -2° to 99° every 20′, measuring to 91°. Ivory vernier to 1′, zero at the right.
Dimensions Radius 16 inches (406 mm)
Description Mahogany frame, limb, and index arm; the lower part of the arm is covered by a brass plate and strengthened at the bottom. Brass fittings; a brass stop for the index arm; inlaid ivory plates in the crossbar and on the back of the frame (latter missing). There is no tangent screw; the clamping screw is on the back of the index arm. Three socket shades (red, orange, and green). Index-glass adjustment by a screw; adjustment of both horizon glasses by levers, wing nuts, and milled clamping screws. The sight vane has two pinholes and a swivelling shutter; the back sight vane has one pinhole. A pencil or screwdriver is missing from the crossbar. No box.
Provenance Adams collection; lent in 1950, bequeathed in 1974.

063 NAV1317
Octant

Origin England, possibly Broadstairs, Kent
Maker Possibly John Sandwell
Date 1783
Inscriptions Marked on the ivory plate on the crossbar: * In° Sandwell. of Broadstairs 1783*
Graduation Inlaid ivory scale from -5° to 95° every 20′, measuring to 91°. Ivory vernier to 1′, zero at the centre.
Dimensions Radius 15¾ inches (400 mm)
Description Mahogany frame and limb with a brass index arm and fittings; a brass stop for the index arm; inlaid ivory plates on the crossbar and on the back of the frame. There is no tangent screw; the clamping screw is on the back of the index arm. Three socket shades (two red and one green). Index-glass adjustment by a screw; adjustment of both horizon glasses by levers and clamping screws. The sight vane has two pinholes and a swivelling shutter; the back sight vane has one pinhole. A brass-capped screwdriver (perhaps a replacement) is fitted in the crossbar. Decorative engravings are found on the lower half of the index arm. No box.
Provenance Adams collection; lent in 1950, bequeathed in 1974.
Notes This is the only instrument known bearing the name of John Sandwell, which may suggest he was its owner rather than the maker.
Literature Clifton, *Directory*, 242.

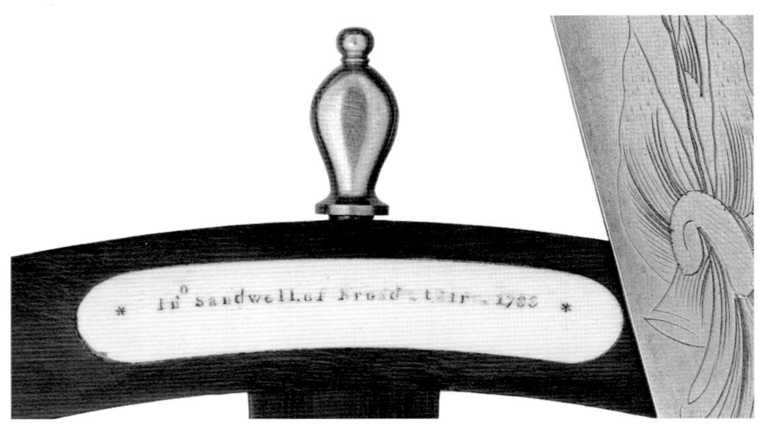

064 NAV 1298
Octant

Origin England
Maker Unknown
Date About 1784
Inscriptions Marked on the ivory plate on the crossbar: *R Maxwell. 1797* [superimposed on erased inscription] Nixon Hill 1784. Marked at the bottom of the index arm: T.G.
Graduation Inlaid ivory scale from -5° to 95° every 20′, measuring to 92°. Ivory vernier to 1′, zero at the centre.
Dimensions Radius 15¾ inches (400 mm)
Description Ebony frame and limb with a brass index arm and fittings; a stop (missing) for the index arm; inlaid ivory plates on the crossbar and on the back of the frame. There is no tangent screw; the clamping screw is on the back of the index arm. Three socket shades (two red and one green). Index-glass adjustment by a screw; adjustment of both horizon glasses by levers, wing nuts, and milled clamping screws. The sight vane has two pinholes and a swivelling shutter; the back sight has one pinhole. A pencil or screwdriver is missing from the crossbar. An engraved decorative floral pattern is found on the lower half of the index arm. No box.
Provenance Adams collection; lent in 1946, bequeathed in 1974.
Notes Presumably both names are of previous owners. Neither is mentioned in Clifton, *Directory*.
Literature Hill and Paget-Tomlinson, *Instruments of Navigation*, 19.

065 NAV1325
Octant

Origin London, England
Maker Henry Gregory and Gabriel Wright
Date About 1785
Inscriptions Signed at the bottom of the index arm: Gregory | & Wright | London
Graduation Inlaid ivory scale from -4° to 94° every 20′, measuring to 90°. Ivory vernier to 1′, zero at the centre.
Dimensions Radius 15¾ inches (400 mm)
Description Mahogany frame and limb with a brass index arm and fittings; a brass stop for the index arm; inlaid ivory plates on the crossbar and on the back of the frame (both loose). There is no tangent screw; there is a clamping screw on the back of the index arm. Three socket shades (one green and two red). Index-glass adjustment by a screw; adjustment of both horizon glasses by levers, wing nuts, and milled clamping screws. The sight vane has two pinholes and a swivelling shutter; the back sight vane has one pinhole. A pencil or screwdriver is missing from the crossbar. In a wooden keystone box (damaged), painted or varnished on the outside, and painted red on the inside.
Provenance Presented by Sir James Caird in 1938; purchased as part of the Landau collection through Maggs Brothers.
Literature Clifton, *Directory*, 120.

066 NAV1256
Octant

Origin Liverpool, England
Maker Egerton Smith
Date About 1785
Inscriptions Signed on the index arm: Egerton Smith Liverpool
Graduation Inlaid ivory scale from -5° to 95° every 20′, measuring to 89½°. Ivory vernier to 1′, zero at the centre.
Dimensions Radius 16 inches (406 mm)
Description Ebony frame and limb with a brass index arm and fittings; a brass stop for the index arm; inlaid ivory plates on the crossbar and on the back of the frame. There is no tangent screw; the clamping screw is on the back of the frame. Three socket shades (red, green, and orange). Index-glass adjustment by a screw; adjustment of both horizon glasses by levers, wing nuts, and milled clamping screws. The sight vane has two pinholes and a swivelling shutter; the back sight vane has one pinhole. A pencil or screwdriver is missing is from the crossbar. A decorative engraved floral pattern is found at the bottom of the index arm. No box.
Provenance Unknown; the typed record indicates that it was in the collection before May 1965, when it was cleaned.
Literature Clifton, *Directory*, 255; Morrison-Low, *Making Scientific Instruments*, 334.

067 NAV 1261
Octant

Origin London, England
Maker Unsigned, but with a scale mark for Spencer Browning and Rust
Date About 1790
Inscriptions Marked (stamped) at the centre of the scale (at 45°): SBR
Marked on the ivory plate on the crossbar: J. Bax
Graduation Inlaid ivory scale from -2° to 99° every 20′, measuring to 91°. Ivory vernier to 1′, zero at the right.
Dimensions Radius 14 inches (356 mm)
Description Ebony frame and limb with a brass index arm and fittings; a brass stop for the index arm; inlaid ivory plates on the crossbar and on the back of the frame. There is no tangent screw; the clamping screw is on the back of the index arm. Three socket shades (two red and one green). Index-glass adjustment by a screw; adjustment of both horizon glasses by levers, wing nuts, and milled clamping screws. The sight vane has two pinholes and a swivelling shutter; the back sight vane has one pinhole. A screwdriver is fitted in the crossbar. In a stepped oak box, containing in the lid a label in MS 'Chapman Light Yarmouth 1906'.
Provenance Adams collection; lent in 1936, bequeathed in 1974. The label may indicate that Adams purchased the instrument in Yarmouth in 1906.
Literature Clifton, *Directory*, 261; Hill and Paget-Tomlinson, *Instruments of Navigation*, 19.

068 NAV1276
Octant

Origin London, England
Maker Spencer Browning and Rust
Date 1791
Inscriptions Signed on the ivory plate on the crossbar: Spencer Browning & Rust à Londres
Marked (stamped) at the centre of the scale (at 48°): SBR.
Marked on the index arm: Prix d'Encouragement donné par Mr. De Gaulle Ingenieur de la Marine, | a Celui de ses Eleves qui aura le mieux repondu a L'Examen Public.
Marked at the bottom of the index arm: Ce Prix | a eté remporte | Le 22 Mars 1791 par | Mr. Montargis
Graduation Inlaid ivory scale from -2° to 99° every 20′, measuring to 91°. Ivory vernier to 1′, zero at the right.
Dimensions Radius 15¾ inches (400 mm)
Description Mahogany frame and limb with a brass index arm and fittings; a brass stop for the index arm; inlaid ivory plates on the crossbar and on the back of the frame (latter missing). There is no tangent screw; the clamping crew is on the back of the index arm. Three socket shades (two red and one green). Index-glass adjustment by a screw; adjustment of horizon glass by lever, wing nut, and a milled clamping screw. The sight vane has two pinholes and a swivelling shutter. An ivory-capped pencil (unused) is fitted in the crossbar. Decorative engravings are found on the index arm. No box.
Provenance This instrument was exhibited at the Musée des Arts Décoratifs (Louvre), Paris, March–April 1936, from the George Bernard collection. Presented by Sir James Caird in 1936; purchased from the Bernard collection.
Literature Callender, *The Caird Collection: Supplement*, 72, no. 72; NN, *Exposition des Instruments et Outils d'autrefoix*, 38, no. 283; Mörzer Bruyns, 'The Navigating Instruments of the National Maritime Museum Greenwich', 24; Clifton, *Directory*, 261.
For image see colour section plate 12.

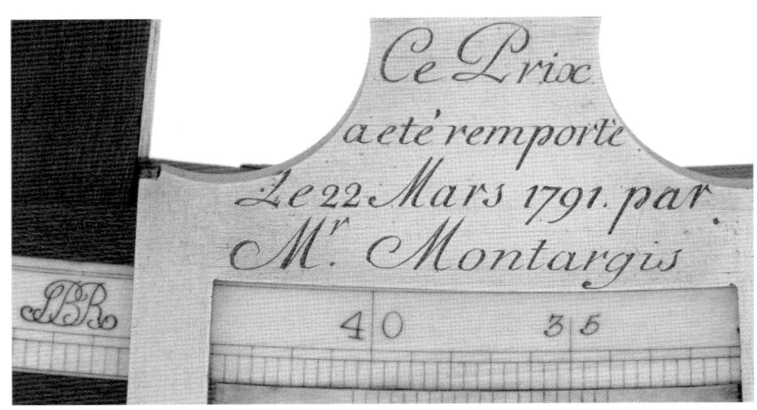

069 NAV1331
Octant

Origin London, England
Maker Ripley and Son
Date 1795
Inscriptions Signed on the ivory plate on the crossbar: Made by RIPLEY & SON Hermitage Bridge London | For Cap[t]. James Henderson March 10th 1795
Graduation Inlaid ivory scale from -2° to 99° every 20′, measuring to 90°. Ivory vernier to 1′, zero at the right.
Dimensions Radius 15¾ inches (400 mm)
Description Ebony frame and limb with a brass index arm and fittings; a brass stop for the index arm is missing and the hole is filled; inlaid ivory plates on the crossbar and on the back of the frame. Tangent screw on top of the index arm; the clamping screw is on the back. The socket shades are missing. Index-glass adjustment by a screw; adjustment of horizon glass by a lever and worm gear; adjustment of back horizon glass by a lever, wing nut, and milled clamping screw. The sight vane has two pinholes and a swivelling shutter; the back sight vane has one pinhole. A pencil or screwdriver is missing from the crossbar. There is an extension on the back of the index arm, which slides through a groove let into the bottom of the limb. The crossbar is cracked. In a stepped oak box (damaged), containing in the lid the remains of a trade label for a London instrument maker, probably Thomas Lorkin.
Provenance Presented by Mrs J. Hochstetler, Hayling Island, Hampshire, in 1960.
Literature Clifton, *Directory*, 233; Wynter and Turner, *Scientific Instruments*, 54; Dunn, 'Touching and Cleaning'.

070 NAV1343
Octant

Origin London, England
Maker Robert Ban(c)ks
Date About 1800
Inscriptions Signed at the bottom of the index arm: Banks London Marked (stamped) at the centre of the scale (50°): [anchor]
Graduation Inlaid ivory scale from -3° to 100° every 20′, measuring to 92°. Ivory vernier to 1′, zero at the right.
Dimensions Radius 11½ inches (292 mm)
Description Ebony frame and limb with a brass index arm and fittings; inlaid ivory plates on the crossbar and on the back of the frame. Tangent screw (knob missing) on the front of the index arm; the clamping screw is on the back. Three socket shades (two red and one green). Index-glass adjustment by a screw; adjustment of horizon glass by a lever, worm gear, and milled clamping screw; adjustment of back horizon glass by a lever, wing nut, and milled clamping screw. The sight vane has two pinholes and a swivelling shutter (missing); the back sight vane is missing. One foot is missing from the frame; a pencil or screwdriver is missing from the crossbar. In a wooden stepped keystone box (damaged).
Provenance Presented by Mr W.O. Gardiner, Charminster, Dorset, in 1965.
Literature Clifton, *Directory*, 16; Stimson, 'Some Board of Longitude Instruments', 113.

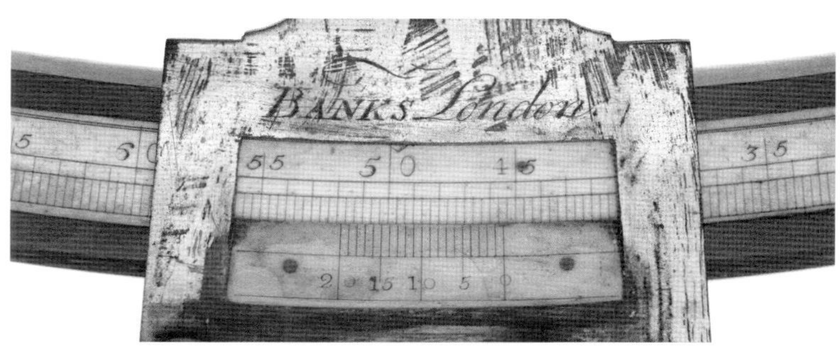

071 NAV1285
Octant

Origin London, England
Maker Thomas Ripley and Son
Date About 1800
Inscriptions Signed on the ivory plate on the crossbar: Made by RIPLEY & SON Hermitage Bridge London
Marked (stamped) at the centre of the scale (at 47°): [foul anchor]
Graduation Inlaid ivory scale from -2° to 94° every 20′, measuring to 89°. Ivory vernier to 1′, zero at the right.
Dimensions Radius 13¾ inches (349 mm)
Description Ebony frame and limb with a brass index arm and fittings; a brass stop for the index arm; inlaid ivory plates on the crossbar and on the back of the frame. There is no tangent screw; the clamping screw is on the back of the index arm. Three socket shades (one orange, and two missing). Index-glass adjustment by a screw; adjustment of both horizon glasses by levers, wing nuts, and milled clamping screws. The sight vane has two pinholes and two swivelling shutters (both missing); the back sight vane has one pinhole. An ivory pencil head is fitted in the crossbar (pencil missing). In a stepped oak box painted brown.
Provenance Presented by Mr George H. Gabb, London, in 1938 (in effect, an addition to the Gabb collection purchased by Caird for the Museum in 1937).
Literature Clifton, *Directory*, 233; Stimson, 'Some Board of Longitude Instruments', 113; Dunn, 'Touching and Cleaning'.

072 NAV1308
Octant

Origin London, England
Maker Spencer Browning and Rust
Date About 1800
Inscriptions Signed on the ivory plate on the crossbar: Spencer Browning & Rust London
Marked (stamped) at the centre of the scale: SBR
Graduation Inlaid ivory scale from -2° to 99° every 20′, measuring to 90°. Ivory vernier to 1′, zero at the right.
Dimensions Radius 8½ inches (216 mm)
Description Ebony frame and limb with a brass index arm and fittings; a brass stop for the index arm; inlaid ivory plates on the crossbar and on the back of the frame. Tangent screw on the front of the index arm; the clamping screw is on the back. Three socket shades (two red and one green). Index-glass adjustment by a screw; adjustment of both horizon glasses by levers, wing nuts, and milled clamping knobs. The sight vane has two pinholes and a swivelling shutter (missing); the back sight vane has one pinhole. A pencil or screwdriver is missing from the crossbar. In an oak keystone box, containing a MS note 'This quadrant was presented to me in October 1844 on my leaving Greenwich school; by Admiral Sir Robert Stopford G.C.B. then Governor of Greenwich Hospital and schools, telling me he had found it a very good instrument. [Signed] George Heather Retired Captain RN September 13 1912'.
Provenance Purchased from Mrs Lindsay Sands, Kempton, Bedford, in 1951.

Notes The reason why Governor Stopford (d. 1847) presented this octant to Heather is not clear, though doubtless it was for some personal merit at the Hospital School. This instrument was old-fashioned by 1844, when Greenwich Hospital was giving sextants (and other instruments) to senior boys nominated to fill naval training vacancies; see cat. nos. 181, 322, and 323. Unlike these official presentations, however, this instrument is not so inscribed. It may therefore have been Stopford's last working instrument, used mainly before he reached flag rank in 1808. In this case his opinion of it, as reported by Heather, is as much a general comment on the type as it is on this instrument specifically.
Literature Clifton, *Directory*, 261.

073 NAV1287
Octant

Origin London, England
Maker Unsigned, but with a scale mark for Spencer Browning and Rust
Date About 1800
Inscriptions Marked (stamped) at the centre of the scale: SBR
Graduation Inlaid ivory scale from -2° to 99° every 20′, measuring to 90°. Ivory vernier (cracked) to 1′, zero at the right.
Dimensions Radius 13¾ inches (349 mm)
Description Ebony frame and limb with a brass index arm and fittings; a brass stop for the index arm; inlaid ivory plates on the crossbar and on the back of the frame. There is no tangent screw; the clamping screw is on the back of the index arm. Two socket shades (red and orange; the foot of the frame is broken). Index-glass adjustment by a screw (but the mount is incomplete); adjustment of both horizon glasses by levers, wing nuts, and milled clamping screws (one of which is a replacement). The sight vane has two pinholes and a swivelling shutter; the back sight vane has one pinhole. A pencil or screwdriver is missing from the crossbar. In a stepped oak box (repaired).
Provenance Presented by Sir James Caird in 1938; purchased as part of the Landau collection though Maggs Brothers.
Literature Clifton, *Directory*, 261.

074 NAV1263
Octant

Origin England
Maker Unknown
Date About 1800
Inscriptions None
Graduation Inlaid ivory scale from -2° to 100° every 20′, measuring to 92°. Ivory vernier to 1′, zero at the right.
Dimensions Radius 11¾ inches (298 mm)
Description Ebony frame and limb with a brass index arm and fittings; a brass stop for the index arm; inlaid ivory plates on the crossbar and on the back of the frame. There is no tangent screw; the clamping screw is on the back of the index arm. Three index shades (two red and one green). Index-glass adjustment by a screw; adjustment of horizon glass by a lever, wing nut, and milled clamping screw. The sight vane has two pinholes and a swivelling shutter. A brass-capped screwdriver is fitted in the crossbar. In a stepped mahogany box (repaired), painted white on the inside.
Provenance Adams collection; lent in 1936, bequeathed in 1974.

075 NAV1352
Octant

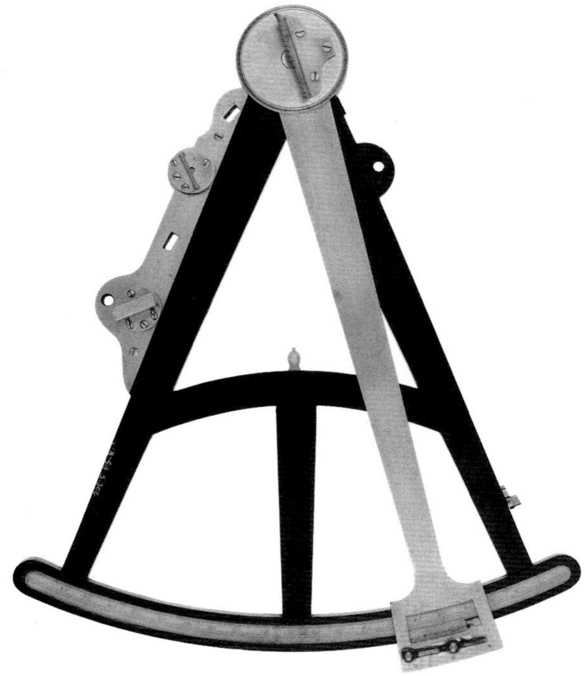

Origin London, England
Maker Unsigned, but with a scale mark for Edward Pritchard
Date About 1800
Inscriptions Marked (stamped) at the centre of the scale (at 47°): E [foul anchor] P
Graduation Inlaid ivory scale from -2° to 98° every 20′, measuring to 91°. Ivory vernier (broken) to 1′, zero at the right.
Dimensions Radius 13¾ inches (349 mm)
Description Ebony frame and limb with a brass index arm and fittings; a brass stop for the index arm; inlaid ivory plates on the crossbar and on the back of the frame (both missing). Tangent screw (knob missing) on the front of the index arm; the clamping screw is on the back. The socket shades and both glasses of the index and horizon glasses are missing. The sight vane and back sight vanes are missing. Index-glass adjustment by a screw; adjustment of both horizon glasses by levers, wing nuts, and clamping screws (both missing). An ivory-capped pencil is fitted in the crossbar. No box.
Provenance Presented by Rear Admiral Sir Edmund George Irving (1910–90), formerly Hydrographer of the Navy, and a Trustee of the National Maritime Museum from 1972 to 1985, in 1978.
Notes Edward Pritchard was an apprentice of Ramsden.
Literature McConnell, *Jesse Ramsden*, 46, 64, 247–8; Clifton, *Directory*, 224; Littlewood and Butler, *Of Ships and Stars*, Appendix 1.

076 NAV1272
Octant

Origin England
Maker Unknown
Date About 1800
Inscriptions Marked (stamped) at the centre of the scale (at 48°): [foul anchor]
Graduation Inlaid ivory scale from -3° to 98° every 20′, measuring to 90°. Ivory vernier to 1′, zero at the right.
Dimensions Radius 11¾ inches (298 mm)
Description Ebony frame and limb with a brass index arm and fittings; a brass stop for the index arm; inlaid ivory plates on the crossbar and on the back of the frame. There is no tangent screw; the clamping screw (missing) was on the back of the index arm. The socket shades are missing (the frame is damaged at the socket). Index-glass adjustment by a screw; adjustment of horizon glass by a lever, wing nut, and clamping screw (missing). The red copper sight vane is a replacement. A pencil or screwdriver is missing from the crossbar. No box.
Provenance Unknown; the typed record indicates that it was in the collection before November 1963, when it was cleaned.

077 NAV 1332
Octant

Origin London, England
Maker Unsigned, but with a scale mark for Spencer Browning and Rust
Date 1806
Inscriptions Marked (stamped) at the centre of the scale (at 47°): SBR
Marked on the ivory plate on the crossbar: John Sugden. 1806.
Graduation Inlaid ivory scale from -2° to 99° every 20′, measuring to 91°. Ivory vernier to 1′, zero at the right.
Dimensions Radius 13.4 inches (340 mm)
Description Ebony frame and limb with a brass index arm and fittings; stop for the index arm (missing); inlaid ivory plates on the crossbar and on the back of the frame (latter missing). Tangent screw (knob missing) on top of the index arm; the clamping screw is on the back. Three socket shades (all red). Index-glass adjustment by a screw; adjustment of horizon glass (glass missing) by a lever, worm gear, and milled clamping screw; adjustment of back horizon glass by a lever, wing nut, and milled clamping screw. The sight vane has two pinholes and a swivelling shutter; the back sight vane has one pinhole. A pencil or screwdriver is missing from the crossbar. In a shaped wooden box, containing a trade label for J.M. and H. Cronmire, 10 Bromehead Street and Commercial Road East, London (1865–69).
Provenance Presented by Lieutenant-Commander H.F.P. Albrecht RNR, Lymington, Hampshire, in 1960.
Literature Clifton, *Directory*, 71–2, 261.

078 NAV 1288
Octant

Origin London, England
Maker (William) Cook
Date About 1810
Inscriptions Signed on the ivory plate on the crossbar: . Cook . London .
Marked (stamped) at the centre of the scale (at 50°): [anchor]
Graduation Inlaid ivory scale from -3° to 100° every 20′, measuring to 90°. Ivory vernier to 1′, zero at the right.
Dimensions Radius 11.6 inch (295 mm)
Description Ebony frame and limb with a brass index arm and fittings; inlaid ivory plates on the crossbar and on the back of the frame. There are a tangent screw and clamping screw on the back of the index arm. Three socket shades (one green and two red). Index-glass adjustment by a screw; adjustment of horizon glass by a lever, milled screw, and worm gear; adjustment of back horizon glass by a lever, milled screw, and nut screw. The sight vane has two pinholes and a swivelling shutter; the back sight vane has one pinhole. An ivory-capped pencil is fitted in the crossbar. In a stepped oak box (damaged), tarred on the outside, painted blue-green on the inside.
Provenance Presented by Sir James Caird in 1938; purchased as part of the Landau collection through Maggs Brothers.
Literature Clifton, *Directory*, 64.

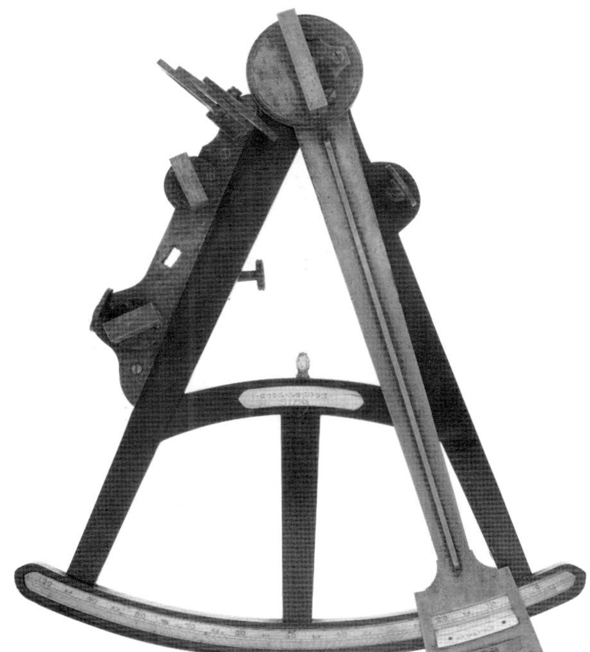

079 NAV1301
Octant

Origin London, England
Maker William Harris
Date About 1810
Inscriptions Signed on the ivory plate on the crossbar: Wm Harris 47 Holborn London [and partly erased] Harris London.
Marked (stamped) at the centre of the scale (at 48°): [foul anchor]
Graduation Inlaid ivory scale from -3° to 98° every 2′, measuring to 92°. Ivory vernier to 1′, zero at the right.
Dimensions Radius 13¾ inches (349 mm)
Description Ebony frame and limb with a brass index arm and fittings; a brass stop for the index arm; inlaid ivory plates on the crossbar and on the back of the frame. There is no tangent screw; the clamping screw is on the back of the index arm. Three index shades (green, orange, and red). Index-glass adjustment by a screw; adjustment of horizon glass by a lever, wing nut, and milled clamping screw. The sight vane has two pinholes and a swivelling shutter. A pencil or screwdriver is missing from the crossbar. In a keystone oak box with a stepped lid.
Provenance Presented by Mr Frederick Smith, London, in 1949.
Literature Clifton, *Directory*, 125–6.

080 NAV1260
Octant

Origin London, England
Maker Spencer Browning and Rust
Date About 1810
Inscriptions Signed on the ivory plate on the crossbar: Spencer Browning & Rust. London
Marked on the ivory plate on the back of the frame: W
Graduation Inlaid ivory scale from -4° to 101° every 20′, measuring to 90½°. Ivory vernier to 1′, zero at the right.
Dimensions Radius 11.85 inches (301 mm)
Description Ebony frame and limb with a brass index arm and fittings; inlaid ivory plates on the crossbar and on the back of the frame. There are a tangent screw and clamping screw on the back of the index arm. Three socket shades (two red and one green). Index-glass adjustment by a screw; adjustment of both horizon glasses by levers, wing nuts, and milled clamping screws. The sight vane (broken into two pieces) has two pinholes and a swivelling shutter; the back sight vane has one pinhole. A pencil or screwdriver is missing from the crossbar. In a stepped oak box, containing in the lid an oval label in MS '[illegible] . . . country Southampton 1908'.
Provenance Adams collection; lent in 1936, bequeathed in 1974. The label may indicate that Adams purchased the instrument in Southampton in 1908.
Literature Clifton, *Directory*, 261.

081 NAV1262
Octant

Origin London, England
Maker Spencer Browning and Rust
Date About 1810
Inscriptions Signed on the ivory plate on the crossbar: Spencer Browning & Rust
Marked (stamped) at the centre of the scale (at 47°): SBR
Graduation Inlaid ivory scale from -2° to 99° every 20′, measuring to 91°. Ivory vernier to 1′, zero at the right.
Dimensions Radius 11¾ inches (298 mm)
Description Ebony frame and limb with a brass index arm and fittings, a brass stop for the index arm; inlaid ivory plates on the crossbar and on the back of the frame. Tangent screw on the top of the index arm; the clamping screw (renewed by F.N. Adams) is on the back. Three socket shades (one green and two red). Index-glass adjustment by a screw; adjustment of horizon glass by a lever, worm gear, and milled clamping screw; adjustment of back horizon glass by a lever, wing nut, and milled clamping screw. The sight vane has two pinholes and a swivelling shutter; the back sight vane has one pinhole. A brass-capped screwdriver is fitted in the crossbar. In a stepped oak box.
Provenance Adams collection; lent in 1936, bequeathed in 1974.
Literature Clifton, *Directory*, 261.

082 NAV1286
Octant

Origin England
Maker Unknown
Date About 1810
Inscriptions None
Graduation Inlaid ivory scale from -2° to 100° every 20′, measuring to 91°. Ivory vernier to 1′, zero at the right.
Dimensions Radius 9¾ inches (248 mm)
Description Ebony frame and limb with a brass index arm and fittings; inlaid ivory plates on the crossbar and on the back of the frame. Tangent screw on the front of the index arm; the clamping screw is on the back. Three index shades (red, orange, and green, possibly replacements). Index-glass adjustment by a screw (the mount is incomplete, and the screw is a replacement); adjustment of horizon glass by a lever, milled screw, and worm gear. The sight vane has two pinholes and a swivelling shade (red). A pencil or screwdriver is missing from the crossbar. In a stepped oak box marked inside the lid in ink 'G. C. 1838' and illegible pencilled digits. In paint on the top of the lid: 'g. c.'
Provenance Presented by Sir James Caird in 1938; purchased as part of the Landau collection through Maggs Brothers.

083 NAV1333
Octant

Origin England
Maker Unknown
Date About 1810
Inscriptions Marked (stamped) at the centre of the scale (at 47°): [anchor]
Graduation Inlaid ivory scale from -1° to 99° every 20′, measuring to 90°. Ivory vernier to 1′, zero at the right.
Dimensions Radius 13¾ inches (349 mm)
Description Ebony frame and limb with a brass index arm and fittings; a brass stop for the index arm; inlaid ivory plates on the crossbar (missing) and on the back of the frame. There is no tangent screw; the clamping screw is on the back of the index arm. The socket shades are missing. Index-glass adjustment by a screw; adjustment of both horizon glasses (one glass missing) by levers, wing nuts, and milled clamping screws (one is missing). The sight vane has two pinholes and a swivelling shutter; the back sight vane is missing. One foot is missing from the frame; the right strut is cracked. No box.
Provenance Presented by Mr A.J. Woods (of Kelvin Hughes), Barkingside, Essex, in 1961.

084 NAV1338
Octant

Origin London, England
Maker Gilbert and Sons
Date About 1815
Inscriptions Signed at the bottom of the index arm: Gilbert & Sons | London
Graduation Inlaid ivory scale from -2° to 98° every 20′, measuring to 91°. Ivory vernier to 1′, zero at the right.
Dimensions Radius 11¾ inches (298 mm)
Description Ebony frame and limb with a brass index arm and fittings; inlaid plates on the crossbar and on the back of the frame (both missing). Tangent screw on top of the index arm; the clamping screw is on the back. Three shades (two red and one green); two horizon shades (red and green). Index-glass adjustment by a screw; adjustment of horizon glass (glass missing) by a lever, wing nut, and milled clamping screw. Threaded telescope bracket without adjustment. Telescope missing; a pencil or screwdriver is missing from the crossbar. No box.
Provenance Bequeathed by R.P. Lewthwaite, Norwich, Norfolk, in 1963.
Literature Clifton, *Directory*, 113.

085 NAV1314
Octant

Origin London, England
Maker Unsigned, but with a scale mark for Spencer Browning and Rust
Date About 1820
Inscriptions Marked (stamped) at the centre of the scale: SBR
Marked on the ivory plate on the crossbar: Alexander Yarmouth.
Graduation Inlaid ivory scale from -2° to 99° every 20′, measuring to 89°. Ivory vernier to 1′, zero at the right.
Dimensions Radius 8¾ inches (222 mm)
Description Ebony frame and limb with a brass index arm and fittings; a brass stop for the index arm; inlaid ivory plates on the crossbar and on the back of the frame. Tangent screw on top of the index arm; the clamping screw is on the back. Three index shades (two red and one green); no horizon shades; the socket hole has been filled in. Index-glass adjustment by a screw; adjustment of horizon glass by a lever, wing nut, and milled clamping screw. The sight vane has two pinholes and a swivelling shutter. A pencil or screwdriver is missing from the crossbar. No box.
Provenance Adams collection; lent in 1950, bequeathed in 1974.
Notes No other instrument by Alexander of Yarmouth is recorded; he may have been an owner or a retailer.
Literature Clifton, *Directory*, 5, 261; Stimson, 'Some Board of Longitude Instruments', 113; Mörzer Bruyns, 'The Navigating Instruments of the National Maritime Museum Greenwich', 26.

086 NAV1289
Octant

Origin London, England
Maker Joseph Hughes
Date About 1820
Inscriptions Signed on the ivory plate on the crossbar: Hughes ⋆ London
Graduation Inlaid ivory scale from -2° to 100° every 20′, measuring to 100°. Ivory vernier to 1′, zero at the right.
Dimensions Radius 9¾ inches (248 mm)
Description Ebony frame and limb with a brass index arm and fittings; inlaid ivory plates on the crossbar and on the back of the frame. There are a tangent screw and clamping screw on the back of the index arm. Three index shades (red, orange, and green). Index-glass adjustment by a screw; adjustment of horizon glass by a lever, wing nut, and milled clamping screw. The sight vane (the holding screw is missing) has two pinholes and a swivelling shade (red). An ivory pencil head is fitted in the crossbar (pencil missing). In a keystone oak box with stepped lid, containing in the lid a trade label for J. Hughes, 38 Queen Street, Ratcliffe Cross, London.
Provenance Presented by Sir James Caird in 1938; purchased from the Landau collection through Maggs Brothers.
Literature Clifton, *Directory*, 144.

087 NAV1310
Octant

Origin London, England
Maker (Isaac) Bradford
Date About 1820
Inscriptions Signed on the ivory plate on the crossbar: Bradford London
Graduation Inlaid ivory scale from -2° to 106° every 20′, measuring to 91°. Ivory vernier to 30″, zero at the right.
Dimensions Radius 9¾ inches (248 mm)
Description Ebony frame and limb with a brass index arm and fittings; wooden handle; an inlaid ivory plate on the crossbar. There are a tangent screw and clamping screw on the back of the index arm. Four shades (three red and one green); three horizon shades (two red and one green). Index-glass adjustment by a screw; adjustment of horizon glass by a lever, milled screw, and worm gear. Magnifier on an 83-mm swivelling arm. Threaded telescope bracket without adjustment. Telescope (88 mm), erect image (perhaps from a different instrument). Sight-tube (94 mm) and a shaded eyepiece (red); the adjusting pin and a second eyepiece are missing. In a mahogany keystone box, containing in the lid a trade label for John Lilley and Son, Jamaica Terrace, Commercial Road near the West India Docks, London (1846–65). A blank mother-of-pearl plate (incomplete) is inlaid on the top of the lid.
Provenance Presented by Mrs Stead, London, in 1952.
Literature Clifton, *Directory*, 36, 168.

088 NAV1337
Octant

Origin London, England
Maker William Cary
Date About 1825
Inscriptions Signed on the ivory plate on the crossbar: ★Cary . London★
Marked (scratched) on the back of the frame: Percival
Graduation Inlaid ivory scale from -2° to 100° every 20′, measuring to 92°. Ivory vernier to 1′, zero at the right.
Dimensions Radius 9½ inches (241 mm)
Description Ebony frame and limb with a brass index arm and fittings; inlaid ivory plates on the crossbar and on the back of the frame. There are a tangent screw and clamping screw on the back of the index arm. Three shades (two red and one green); three horizon shades (two red and one green). Index-glass adjustment by a screw; adjustment of horizon glass by a lever, milled screw, and worm gear. Threaded telescope bracket without adjustment. Telescope (182 mm), inverted image, four cross-wires. A pencil or screwdriver is missing from the crossbar. In a stepped keystone oak box, marked on the inside in ink 'S. Hayman 3-' [and] 'M. Sulivan', and on the top of the lid 'Hayman 3' [and] 'T Ager 10' and various illegible names.
Provenance Presented by Air Marshal Sir Edward Chilton, KBE, CB, RAF Northwood, Middlesex, in 1962.
Literature Clifton, *Directory*, 51.

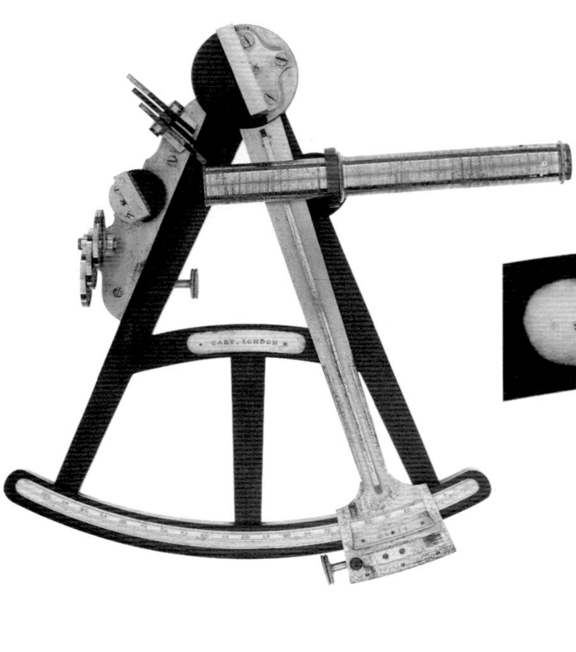

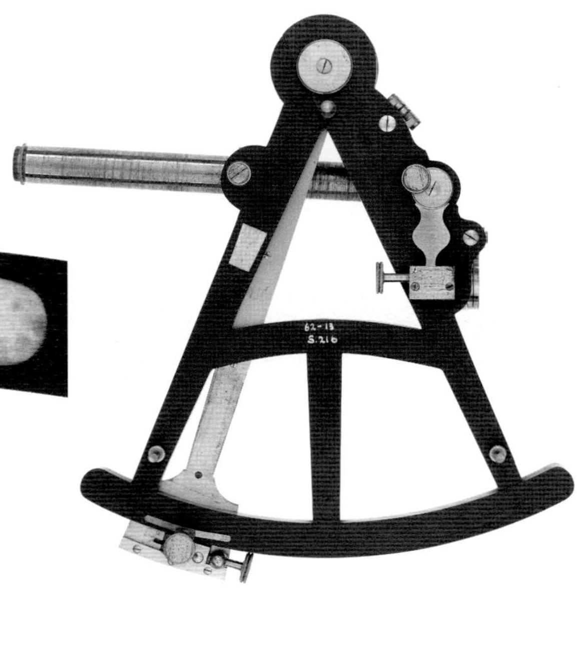

089 NAV1267
Octant

Origin London, England
Maker (William) Cary
Date About 1825
Inscriptions Signed on the ivory plate on the crossbar: Cary. London Marked (stamped) at the centre of the scale (at 47°): E[foul anchor]?
Graduation Inlaid ivory scale from -2° to 98° every 20′, measuring to 90°. Ivory vernier to 1′, zero at the right.
Dimensions Radius 8¾ inches (222 mm)
Description Ebony frame and limb with a brass index arm and fittings; an inlaid ivory plate on the crossbar. There are a tangent screw and clamping screw on the back of the index arm. Three shades (two red and one green); two horizon shades (one red; the other is missing). Index-glass adjustment by a screw; adjustment of horizon glass by a lever, wing nut, and milled clamping screw. The sight vane has two pinholes and a swivelling shutter. In a stepped oak box.
Provenance Adams collection; lent in 1950, bequeathed in 1974.
Notes The scale mark is probably for Edward Pritchard, who, like Cary, was an apprentice of Ramsden.
Literature Clifton, *Directory*, 51, 224; McConnell, *Jesse Ramsden*, 46, 64, 247–8; Stimson, 'Some Board of Longitude Instruments', 113.

090 NAV1264
Octant

Origin England
Maker Unknown
Date About 1825
Inscriptions Marked on the ivory plate on the crossbar: Goy & C^O^. 36. Leadenhall S^t^. London.
Graduation Inlaid ivory scale from -3° to 107° every 20′, measuring to 92°. Ivory vernier to 1′, zero at the right.
Dimensions Radius 9¾ inches (248 mm)
Description Ebony frame and limb with a brass index arm and fittings; inlaid ivory plates on the crossbar and on the back of the frame. There are a tangent screw and clamping screw on the back of the index arm. Three shades (one green and two red); two horizon shades (green and red). Index-glass adjustment by a screw; adjustment of horizon glass by a square-headed screw and detached key, a lever, a worm gear, and a milled clamping screw. Threaded telescope bracket without adjustment, with a swivelling sight vane with one pinhole. Telescope (86 mm), erect image, with a screw-on eyepiece. Milled adjusting key. An ivory-capped pencil is fitted in the crossbar. In a mahogany keystone box, containing in the lid a trade label for Goy and Co., Outfitters, Leadenhall Street, London, and a certificate for the Halstead Science and Art Club Prize for an oil painting, awarded to Edgar Tarry Adams at the Members' Work Competition, 23 January 1907.
Provenance Adams collection; lent in 1936, bequeathed in 1974.
Notes As outfitters, Goy and Co. were only retailers, not the makers of this instrument.
Literature Clifton, *Directory*, 117.

091 NAV1311
Octant

Origin England, possibly London
Maker Possibly Jones and Co.
Date About 1825
Inscriptions Marked on the ivory plate on the crossbar: Jones . & . C° London
Graduation Inlaid ivory scale from -3° to 108° every 20′, measuring to 89°. Ivory vernier to 1′, zero at the right.
Dimensions Radius 9½ inches (241 mm)
Description Ebony frame and limb with a brass index arm and fittings; inlaid ivory plate on the crossbar. There is no tangent screw; the clamping screw is on the back of the index arm. Three index shades (two red and one green). Index-glass adjustment by a screw; adjustment of horizon glass by a lever, wing nut, and milled clamping screw. The sight vane has two pinholes and a swivelling shutter. An ivory cap for a pencil (missing) is fitted in the crossbar. In a keystone mahogany box with a modern plywood lid.
Provenance Presented by Mr J.G.P. Jones, London, in 1952.
Notes Jones and Co. is not found in Clifton, *Directory*; they may have been retailers.

092 NAV1329
Octant

Origin London, England
Maker John William Norie and Co.
Date About 1825
Inscriptions Signed on the ivory plate on the crossbar: Norie & C° London
Graduation Inlaid ivory scale from -2° to 100° every 20′, measuring to 91°. Ivory vernier to 1′, zero at the right.
Dimensions Radius 12 inches (305 mm)
Description Ebony frame with a brass index arm and fittings; a brass stop for the index arm; inlaid ivory plates on the crossbar and on the back of the frame. There are a tangent screw and clamping screw on the back of the frame. Three index shades (two red and one green). Index-glass adjustment by a screw; adjustment of horizon glass by a detached key, lever, worm gear, and milled clamping screw. The sight vane has two pinholes and a swivelling shade (red). Adjusting key (missing). An ivory cap for a pencil (missing) is fitted in the crossbar. No box.
Provenance Presented by Mrs Doyle, Beckenham, Kent, in 1959.
Literature Clifton, *Directory*, 201; Fisher, *The Makers of the Blueback Charts*, 83–100.

093 NAV1290
Octant

Origin London, England
Maker Spencer and Co
Date About 1825
Inscriptions Signed on the ivory plate on the crossbar: Spencer & C° London
Graduation Inlaid ivory scale from -2° to 107° every 20′, measuring to 93°. Ivory vernier (damaged) to 1′, zero at the right.
Dimensions Radius 10.6 inches (269 mm)
Description Ebony frame and limb with a brass index arm and fittings; inlaid ivory plates on the crossbar and on the back of the frame. There are a tangent screw and clamping screw on the back of the index arm. Three index shades (one green, and two missing). Index-glass adjustment by a screw; adjustment of horizon glass by a lever and worm gear, and a square-headed screw and a detached key. The sight vane has two pinholes and a swivelling shutter. An ivory cap for a pencil (jammed) is fitted in the crossbar. Adjusting key (missing). One of the feet of the frame is an old replacement. In an oak keystone box, containing in the lid a trade label for Duren and Costigan, 20 Burling Slip, New York (established in 1855) stuck over an unidentified trade label. Illegible pencil notes in the lid.
Provenance Presented by Sir James Caird in 1938; purchased as part of the Landau collection through Maggs Brothers.
Literature Clifton, *Directory*, 261; Brewington, *The Peabody Museum Collection*, 125.

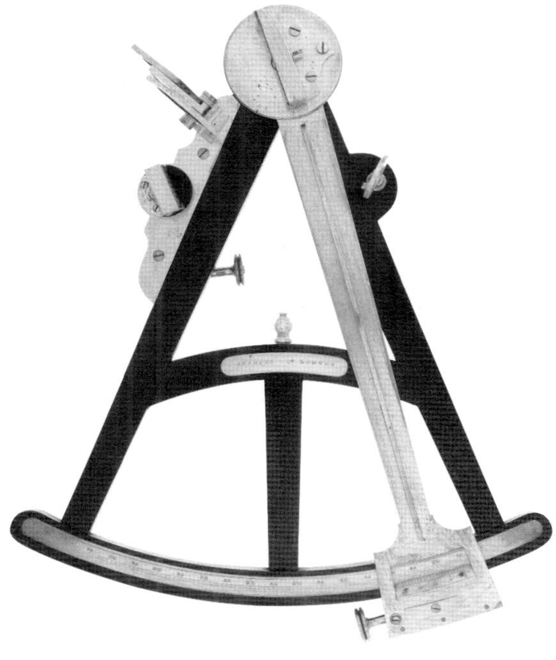

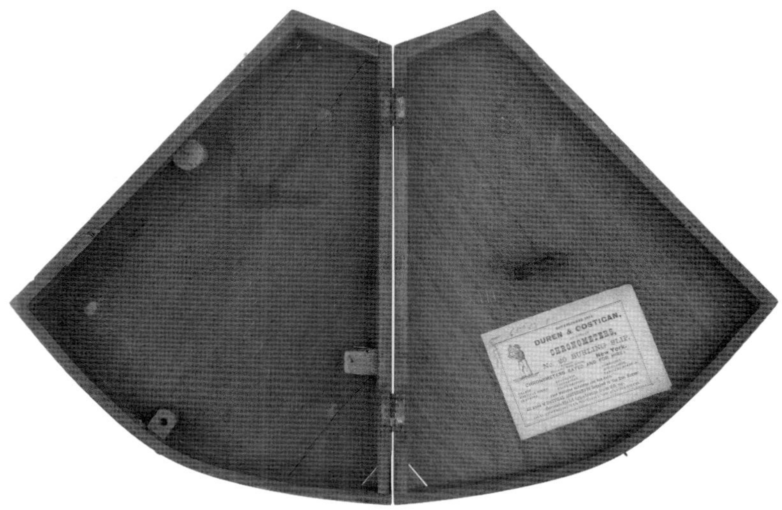

094 NAV1322
Octant

Origin England
Maker Unknown
Date About 1825
Inscriptions None
Graduation Inlaid ivory scale from -8° to 106° every 20′, measuring to 93°. Ivory vernier to 1′, zero at the right.
Dimensions Radius 9.6 inches (244 mm)
Description Ebony frame and limb with a black-lacquered brass index arm and fittings; inlaid ivory plates on the crossbar and on the back of the frame. There are a tangent screw and clamping screw (a replacement) on the back of the index arm. Three index shades (two red and one green). Index-glass adjustment by a screw; adjustment of horizon glass by a lever, worm gear, and milled clamping screw. The sight vane has one pinhole. An ivory cap for a pencil (missing) is fitted in the crossbar. No box.
Provenance Adams collection; lent in 1950, bequeathed in 1974.

095 NAV1321
Octant

Origin England
Maker Unknown
Date About 1825
Inscriptions None
Graduation Inlaid ivory scale from -2° to 100° every 20′, measuring to 96°. Ivory vernier to 1′, zero at the right.
Dimensions Radius 9.6 inches (244 mm)
Description Ebony frame and limb with a brass index arm and fittings; a brass stop for the index arm; inlaid plates on the crossbar and the back of the frame (both missing). There are a tangent screw and clamping screw on the back of the index arm. Three index shades (two red and one green). Index-glass adjustment by a screw; adjustment of horizon glass by a lever, wing nut, and milled clamping screw. The sight vane has two pinholes and a swivelling shutter. An ivory cap for a pencil (missing) is fitted in the crossbar. The left strut is cracked. No box.
Provenance Adams collection; lent in 1950, bequeathed in 1974.

096 NAV1305
Octant

Origin England
Maker Unknown
Date About 1825
Inscriptions None
Graduation Inlaid ivory scale from -2° to 100° every 20′, measuring to 92°. Ivory vernier to 1′, zero at the right.
Dimensions Radius 10.6 inches (269 mm)
Description Ebony frame and limb with a brass index arm and fittings; a brass stop for the index arm; inlaid ivory plates on the crossbar and the back of the frame. There is no tangent screw; the clamping screw is on the back of the index arm. Two sets of three socket shades (two red and one green; red, orange, and green); one set is probably from another instrument. Index-glass adjustment by a screw; adjustment of both horizon glasses by levers, wing nuts, and milled clamping screws. The sight vane has two pinholes and a swivelling shutter; the back sight vane has one pinhole. A pencil or screwdriver is missing from the crossbar. In a stepped keystone oak box, marked on the lid 'Alfred Terry', probably an owner's name.
Provenance Presented by Captain J.T. Shenton RN, Lydd, Kent, in 1944.

097 NAV1304
Octant

Origin London, England
Maker Joseph Fairey
Date About 1830
Inscriptions Signed on the ivory plate on the crossbar: Fairey. Maker. 150. Tooley Street London
Marked on the scale (at 48°): SBR
Graduation Inlaid ivory scale from -2° to 99° every 20′, measuring to 92°. Ivory vernier to 1′, zero at the right.
Dimension Radius 12 inches (305 mm)
Description Ebony frame and limb with a brass index arm and fittings, a brass stop for the index arm; inlaid ivory plates in the crossbar and on the back of the frame. Tangent screw (slightly bent) on the front of the index arm; the clamping screw is on the back. Three socket shades (two red and one green). Index-glass adjustment by a screw; adjustment of horizon glass by a lever, worm gear, and milled clamping screw; adjustment of back horizon glass by a lever, wing nut, and milled clamping screw. The sight vane has two pinholes and a swivelling shutter; the back sight vane has one pinhole. A pencil or screwdriver is missing from the crossbar. In an oak keystone stepped box, containing in the lid a trade label for Newmarch Lee, 166 High Street, Hull, ship chandler and late foreman to Mrs Fletcher.
Provenance Presented by Mr R.A. Bell, London, in 1937.
Notes Fairey had the scale of this octant divided by Spencer Browning and Rust.
Literature Clifton, *Directory*, 98, 261; Hill and Paget-Tomlinson, *Instruments of Navigation*, 20.

098 NAV1297
Octant

Origin London, England
Maker George Bradford
Date About 1830
Inscriptions Signed on the ivory plate on the crossbar: G. Bradford. 99 Minories London
Graduation Inlaid ivory scale from -3° to 109° every 30′, measuring to 90°. Ivory vernier to 1′, zero at the right.
Dimensions Radius 10 inches (254 mm)
Description Ebony frame and limb with a brass index arm and fittings; an inlaid ivory plate on the crossbar. There are a tangent screw and clamping screw on the back of the index arm. Three shades (two red and one green); three horizon shades (two red and one green). Index-glass adjustment by a screw; adjustment of horizon glass by a square-headed screw and a detached key, and by a lever, worm gear, and milled clamping screw. Threaded telescope bracket without adjustment, and with a swivelling sight vane with one pinhole. Telescope (83 mm), erect image. Sight-tube (79 mm), a shaded eyepiece (red), an adjusting key, and a magnifying glass (missing). In a wooden keystone box.
Provenance Adams collection; lent in 1946, bequeathed in 1974.
Literature Clifton, *Directory*, 36.

099 NAV 1302
Octant

Origin England
Maker Unknown
Date About 1830
Inscriptions None
Graduation Inlaid ivory scale from -4° to 103° every 20′, measuring to 91°. Ivory vernier to 1′, zero at the right.
Dimensions Radius 9¾ inches (248 mm)
Description Ebony frame with a brass index arm and fittings; inlaid ivory plates on the crossbar and the back of the frame. There are a tangent screw and clamping screw on the back of the index arm. Three index shades (two red and one green). Index-glass adjustment by a screw; adjustment of horizon glass by a square-headed screw and a detached key. The sight vane has two pinholes and a swivelling shade (red). Adjusting key (missing). An ivory-capped pencil is fitted in the crossbar. No box.
Provenance Unknown; the typed record indicates that it was found in the Museum's storeroom in December 1949.

100 NAV 1312
Octant

Origin England
Maker Unknown
Date About 1834
Inscriptions None
Graduation Inlaid ivory scale from -2° to 100° every 20′, measuring to 92°. Ivory vernier to 1′, zero at the right.
Dimensions Radius 9.6 inches (244 mm)
Description Ebony frame and limb with a black-lacquered brass index arm and fittings; inlaid ivory plates on the crossbar and on the back of the frame. There is no tangent screw; the clamping screw is on the back of the index arm. Three index shades (two red and one green). Index-glass adjustment by a screw; adjustment of horizon glass by a lever, wing nut, and milled clamping screw. The sight vane has two pinholes and a swivelling shade (glass missing). A pencil or screwdriver is missing from the crossbar. In a shaped, polished mahogany box, containing in the lid a trade label for Alexander Alexander, Optician (established in 1828), 6 High Street, Exeter (his address in 1834).
Provenance Presented by Mr G.E.L. Carter, Budleigh Salterton, South Devon, in 1953.
Literature Clifton, *Directory*, 5.

101 NAV1339
Octant

Origin Liverpool, England
Maker (James) Bassnett
Date About 1840
Inscriptions Signed on the ivory plate on the crossbar: Bassnett Liverpool
Graduation Inlaid ivory scale from -3° to 104° every 20′, measuring to 90°. Ivory vernier to 1′, zero at the right.
Dimensions Radius 7½ inches (191 mm)
Description Ebony frame and limb with a brass index arm and fittings; inlaid ivory plates on the crossbar and on the back of the frame. There are a tangent screw and clamping screw on the back of the index arm. Three index shades (two red and one green, cracked). Index-glass adjustment by a screw; adjustment of horizon glass by a square-headed screw, a lever, a worm gear, and a milled clamping screw. The sight vane has two pinholes and a swivelling shutter. A pencil or screwdriver is missing from the crossbar. No box.
Provenance Presented by Mr T.F. Peppitt, Jersey, Channel Islands, in 1963.
Literature Clifton, *Directory*, 21; Morrison-Low, *Making Scientific Instruments*, 324.

102 NAV1299
Octant

Origin Newcastle upon Tyne, England
Maker Septimus Anthony Cail
Date About 1840
Inscriptions Signed on the inlaid plate on the crossbar: S.A. Cail | Newcastle on Tyne
Graduation Inlaid ivory scale from -4° to 107° every by 20′, measuring to 90°. Ivory vernier to 1′, zero at the right.
Dimensions Radius 9.6 inches (244 mm)
Description Ebony frame and limb with a brass index arm and fittings; an inlaid ivory plate on the crossbar. There are a tangent screw and clamping screw on the back of the index arm. Three index shades (two red and one green). Index-glass adjustment by a screw; adjustment of horizon glass by a lever, worm gear, and milled clamping screw. The sight vane has two pinholes and a swivelling shutter. An ivory-capped pencil head is fitted in the crossbar (pencil missing). No box.
Provenance Adams collection; lent in 1946, bequeathed in 1974.
Literature Clifton, *Directory*, 47.

103 NAV 1340
Octant

Origin London, England
Maker George Whitbread
Date About 1840
Inscriptions Signed on the ivory plate on the crossbar: G. Whitbread. London.
Graduation Inlaid ivory scale from -2° to 107° every 20′, measuring to 93°. Ivory vernier to 1′, zero at the right.
Dimensions Radius 9½ inches (241 mm)
Description Ebony frame and limb with a brass index arm and fittings, wooden handle; an inlaid ivory plate on the crossbar. There are a tangent screw and clamping screw on the back of the index arm. Four shades (three red and one green); three horizon shades (two red and one green). Index-glass adjustment by a screw; adjustment of horizon glass by a lever, worm gear, and milled clamping screw. Threaded telescope bracket without adjustment. Telescope (89 mm), erect image. Sight-tube (81 mm), a shaded eyepiece (red), and a magnifying glass (missing). In a wooden keystone box.
Provenance Belonged to the Revd George Fisher (1794–1873) FRS, an FRGS, astronomer, who sailed to the Arctic and was later Chaplain and Principal of the Greenwich Hospital School. Presented by Miss Darnell in 1958 as part of the Fisher collection.
Literature Clifton, *Directory*, 295; Anita McConnell, 'Fisher, George (1794–1873) *astronomer*', in *Oxford Dictionary of National Biography* Oxford, 2004, <http://www.oxforddnb.com/.>

104 NAV 1259
Octant

Origin London, England
Maker Charles Wastell Dixey
Date About 1842
Inscriptions Signed at the bottom of the index arm: C.W. Dixey | Bond Street, London
Marked on the inlaid plate on the crossbar: J.P.G.
Graduation Inlaid ivory scale from -2° (off the arc) to 99° every 20′, measuring to 91°. Ivory vernier to 1′, zero at the right.
Dimensions Radius 11.6 inches (295 mm)
Description Ebony frame and limb with a brass index arm and fittings, a brass stop for the index arm; an inlaid ivory plate on the back of the frame. There is no tangent screw; the clamping screw is on the back of the index arm. Three socket shades (two red and one green). Index-glass adjustment by a screw; adjustment of both horizon glasses by screws, levers, wing nuts, and milled clamping screws. The sight vane has two pinholes and a swivelling shutter; the back sight vane has one pinhole. A brass-capped screwdriver is fitted in the crossbar. In a stepped oak box, containing two trade labels for C.W. Dixey, 3 New Bond Street, London (1839–62).
Provenance Adams collection; lent in 1936, bequeathed in 1974. Acquired by Adams from old, unused stock of C.W. Dixey in 1906.
Literature Clifton, *Directory*, 85.

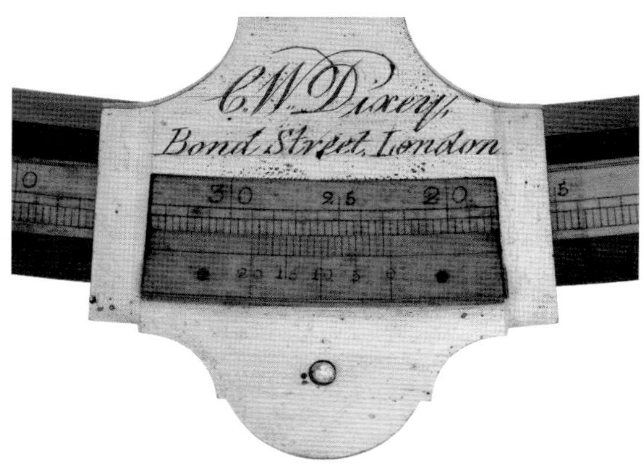

105 NAV1268
Octant

Origin England, possibly Leeds
Maker Possibly Gabriel Davis
Date About 1845
Inscriptions Marked on the inlaid plate on the crossbar: G: Davis Optician Leeds
Graduation Inlaid ivory scale from -2° to 101° every 20′, measuring to 95°. Ivory vernier to 1′, zero at the right.
Dimensions Radius 7½ inches (191 mm)
Description Ebony frame and limb with a brass index arm and fittings; inlaid ivory plates on the crossbar and on the back of the frame. Tangent screw on the front of the index arm; the clamping screw is on the back. Three socket shades (two red and one green). Index-glass adjustment by a screw; adjustment of horizon glass by a lever and worm gear; adjustment of back horizon glass by a lever, wing nut, and milled clamping screw. The sight vane has two pinholes and a swivelling shutter (missing); the back sight vane has one pinhole. An ivory-capped pencil is fitted in the crossbar. In an oak keystone box, containing in the lid a label in MS 'Given to me by Mr. Charles Wilson 1914'.
Provenance Adams collection; lent in 1936, bequeathed in 1974. Charles Wilson (1853–1921), who gave this instrument to Adams in 1914, led the firm of Imray, Laurie, Norie and Wilson Ltd, suppliers of nautical instruments and charts.
Notes Davis, an optician, may have been the retailer rather than the maker of this instrument.
Literature Clifton, *Directory*, 78; Fisher, *The Makers of the Blueback Charts*, 109–15.

106 NAV1292
Octant

Origin London, England
Maker George Whitbread
Date About 1845
Inscriptions Signed on the ivory plate on the crossbar: G. Whitbread . London
Graduation Inlaid ivory scale from -2° to 100° every 20′, measuring to 90°. Ivory vernier to 1′, zero at the right.
Dimensions Radius 9.4 inches (239 mm)
Description Ebony frame and limb with a brass black-lacquered index arm and fittings; inlaid ivory plates on the crossbar and on the back of the frame (both loose). There are a tangent screw and clamping screw are on the back of the index arm. Three index shades (two red and one green). Index-glass adjustment by a fixed milled key; adjustment of both horizon glasses by levers, worm gears, and milled clamping screws. The sight vane has two pinholes and a swivelling shade (red); the back sight vane has one pinhole. An ivory-capped pencil head is fitted in the crossbar. In a shaped box with a stepped lid, containing trade labels for George Whitbread, 2 Grenada Terrace, London (1842–74), and for the Magazzino di Carlo Ruspini, Strada Banchi and Strada Orefici, Genova.
Provenance Presented by Sir James Caird in 1938; purchased as part of the Landau collection through Maggs Brothers.
Literature Clifton, *Directory*, 295.

107 NAV1265
Octant

Origin London, England
Maker Robert Montague Barrett
Date About 1850
Inscriptions Signed on the ivory plate on the crossbar: R.M. Barrett. London
Graduation Inlaid ivory scale from -3° to 108° every 20′, measuring to 92°. Ivory vernier to 1′, zero at the right.
Dimensions Radius 9.6 inches (244 mm)
Description Ebony frame and limb with a brass index arm and fittings; inlaid ivory plate on the crossbar. There are a tangent screw and clamping screw on the back of the index arm. Three index shades (two red and one green). Index-glass adjustment by a screw; adjustment of horizon glass by a square-headed screw and a detached key, a lever, a worm gear, and a milled clamping screw. The sight vane has two pinholes and a swivelling shutter. An ivory pencil head is fitted in the crossbar (pencil missing). Milled adjusting key. In a mahogany keystone box.
Provenance Unknown; the typed record indicates that it was acquired in 1936.
Literature Clifton, *Directory*, 19.

108 NAV1335
Octant

Origin Southampton, England
Maker Joseph Rankin Stebbing
Date About 1850
Inscriptions Signed on the ivory plate on the crossbar: Stebbing Southampton
Graduation Inlaid ivory scale from 0° to 109° every 20′, measuring to 90°. Ivory vernier to 1′, zero at the right.
Dimensions Radius 9.7 inches (246 mm)
Description Ebony frame and limb with a brass index arm and fittings; wooden handle; an inlaid ivory plate on the crossbar. There are a tangent screw and clamping screw on the back of the index arm. Four shades (three red and one green); three horizon shades (two red and one green). Index-glass adjustment by a screw; adjustment of horizon glass by a square-headed screw and a detached key, a lever, a worm gear, and a milled clamping screw. Magnifier on an 85-mm swivelling arm. Threaded telescope bracket without adjustment. Telescope (82 mm), erect image. Sight-tube (82 mm), a shaded eyepiece (red), a milled adjusting key, and a magnifying glass (missing). In a polished mahogany box, containing a Henry Hughes and Son Ltd certificate of examination, dated 1937.
Provenance Belonged to Captain Edward John Molony (1850–97), master of *British Merchant* and *British Isles*. Presented by Surgeon Captain E. Bryan Kenny RN, London, in 1961.
Literature Clifton, *Directory*, 264.

PLATE 1 Mariner's astrolabe of possibly Spanish origin, dated about 1588 (cat. no. 0050); F5272-001, NAV0022.

PLATE 2 Ivory cross-staff by Thomas Tuttell of London, dated about 1700 (cat. no. 0070); F7088-012, NAV0505.

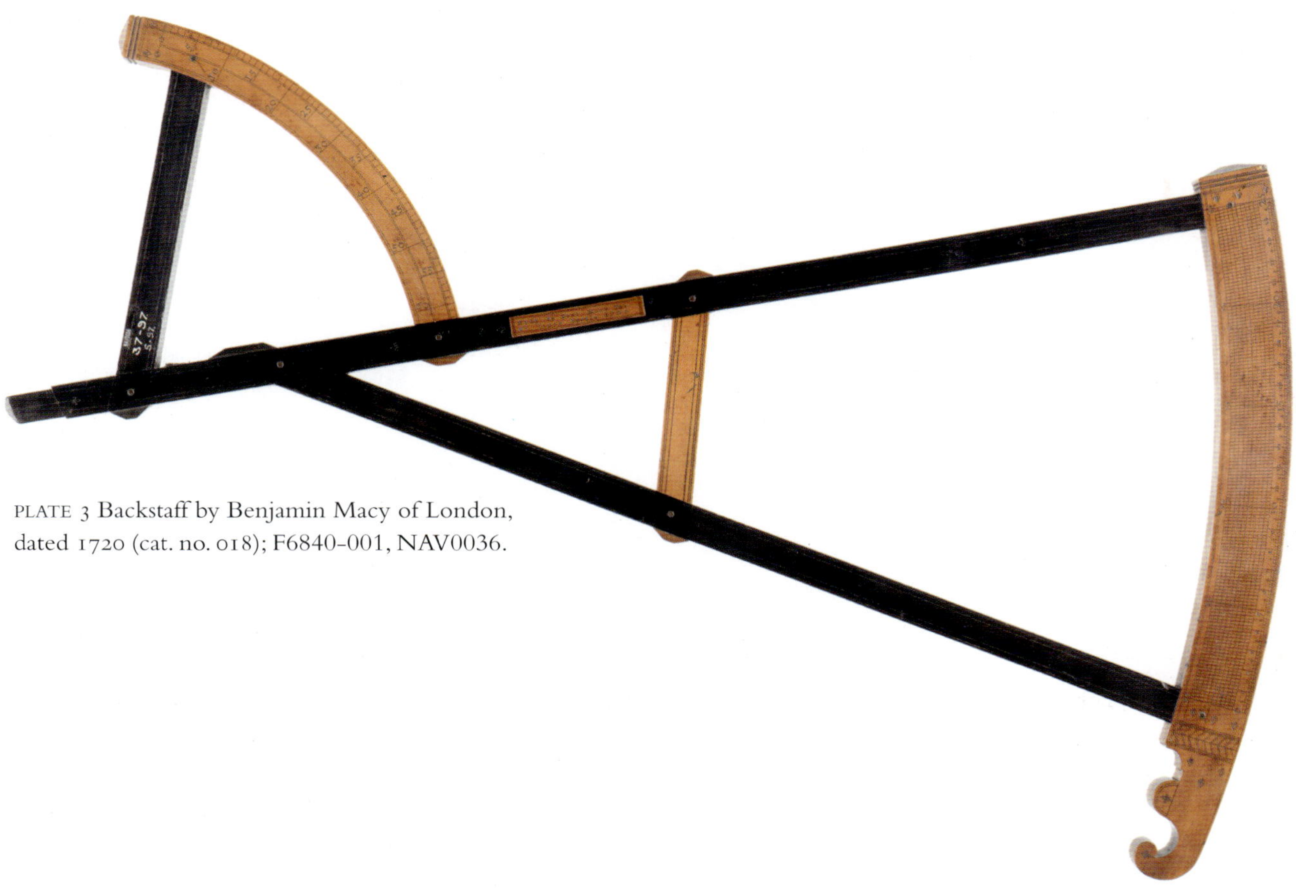

PLATE 3 Backstaff by Benjamin Macy of London, dated 1720 (cat. no. 018); F6840-001, NAV0036.

PLATE 4 Portrait of John Benbow, Sir Ralph Delavalle in the middle holding a backstaff, and Thomas Phillips. Oil painting on canvas by Thomas Murray, 1692–93. National Maritime Museum; BHC2547.

PLATE 5 A merchant captain holding a backstaff with the artificial horizon designed by John Elton. Oil painting on canvas by John Vanderbank, dated 1730s. National Maritime Museum; BHC3128.

PLATE 6 Portrait of John Hadley, presumably holding an octant. Oil painting on canvas attributed to Barholomew Dandridge, early 1730s. National Maritime Museum; BHC2731.

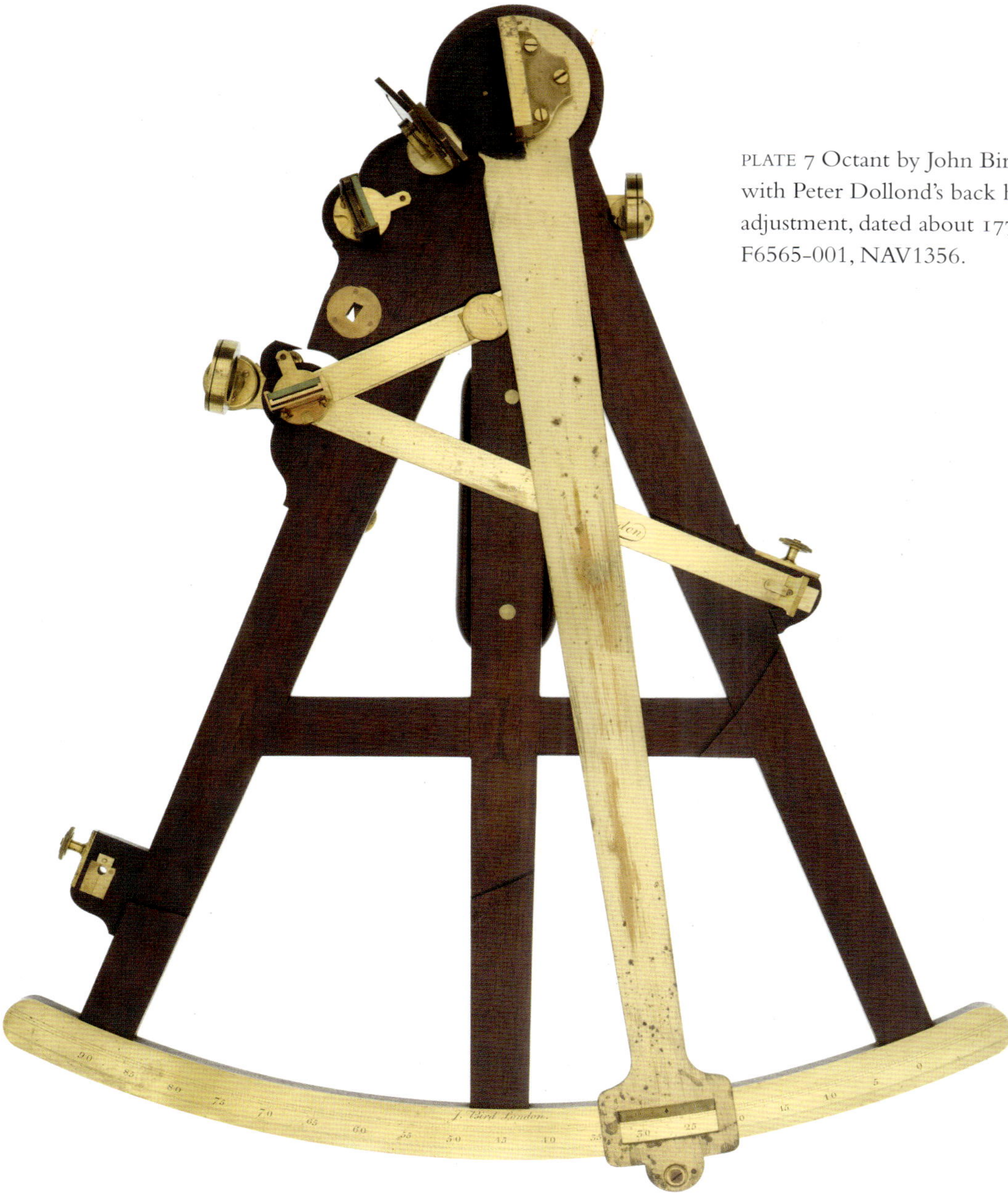

PLATE 7 Octant by John Bird of London, with Peter Dollond's back horizon-glass adjustment, dated about 1772 (cat. no. 051); F6565-001, NAV1356.

PLATE 8 Portrait of Nevil Maskelyne. Drawing in chalk attributed to John Russell, about 1776. National Maritime Museum; F5064, ZBA4305.

PLATE 9 Portrait of a Merchant Navy captain, possibly Robert Willoughby, holding a sea quadrant designed by Caleb Smith, about 1805. Anonymous oil painting on canvas. National Maritime Museum; BHC3130.

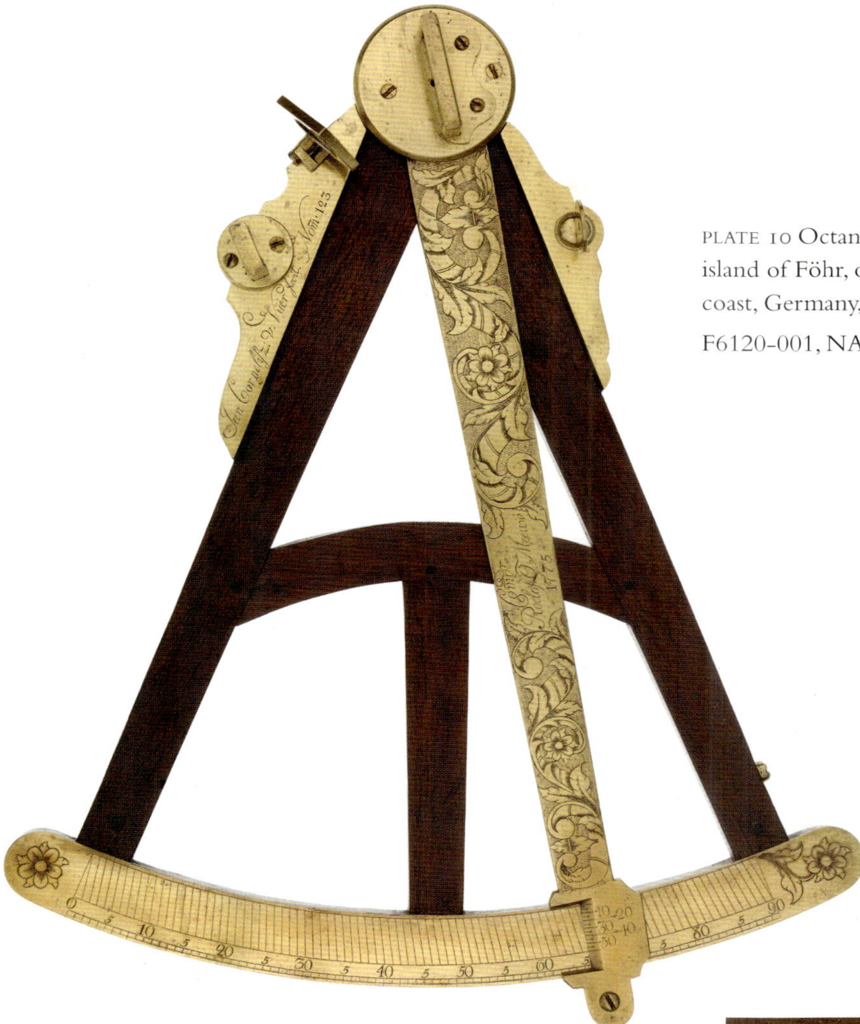

PLATE 10 Octant by Jan Corneliszoon of the island of Föhr, off the Schleswig-Holstein coast, Germany, dated 1775 (cat. no. 053); F6120-001, NAV1275.

PLATE 11 Portrait of a Hamburg sea captain with an octant by Spencer Browning and Rust, a pair of dividers, a telescope, and a chart of the river Elbe estuary up to Hamburg. Oil painting on canvas by Franz Xavier Riedmeyer, 1796. Courtesy of the Altonaer Museum in Hamburg, Norddeutsches Landesmuseum; 198/922.

PLATE 12 Octant by Spencer Browning and Rust of London, made for the French market and dated 1791 (cat. no. 068); F7447-001, NAV1276.

PLATE 13 Sextant by Jesse Ramsden of London, reputedly used on James Cook's third voyage to the Pacific, and dated about 1772 (cat. no. 135); F6420-001, NAV1236.

PLATE 14 Sextant with artificial horizon by Jesse Ramsden of London, dated about 1792/1834 or later. This artificial horizon was invented by A.B. Becher (cat. no. 150); F6918-005, NAV1242.

PLATE 15 Sextant box with figures of inlaid wood, dated about 1870. The shipping company house flags on either side of the sextant are for the Australian Steam Shipping Co Ltd, London (left), and Manchester Liners Ltd (cat. no. 200); F5452-005, NAV1158.

PLATE 16 Detail from 'The Usual Occupations of the Young Officers in the Steerage of a British Frigate at Sea'. An octant is stored against the deck beam, and another is lying on the deck. Oil painting on canvas by Augustus Earle, about 1820–37. National Maritime Museum; BHC1118.

PLATE 17 Angel holding a sextant, on the right-hand panel from an altar triptych titled 'God is Our Guide and Stay' that was commissioned for use on board USS aircraft carrier *Essex* built 1943. Oil painting on metal by Ethel Parsons Paullin, 1943. Courtesy of The Mariners' Museum, Newport News, Virginia; 1984.0045.000001.

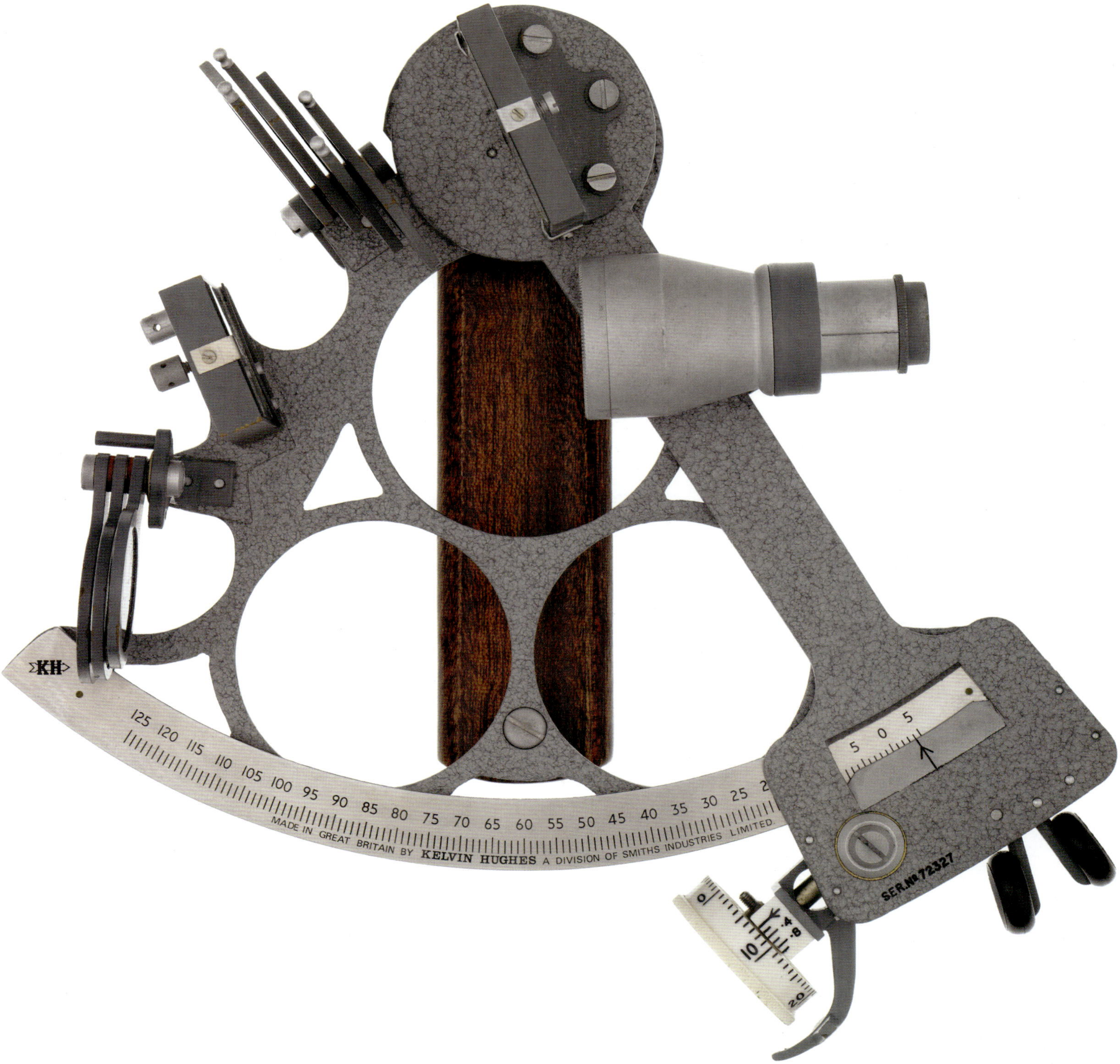

PLATE 18 Sextant by Kelvin Hughes of London, dated 1970 (cat. no. 272); F5997-001, NAV1209.

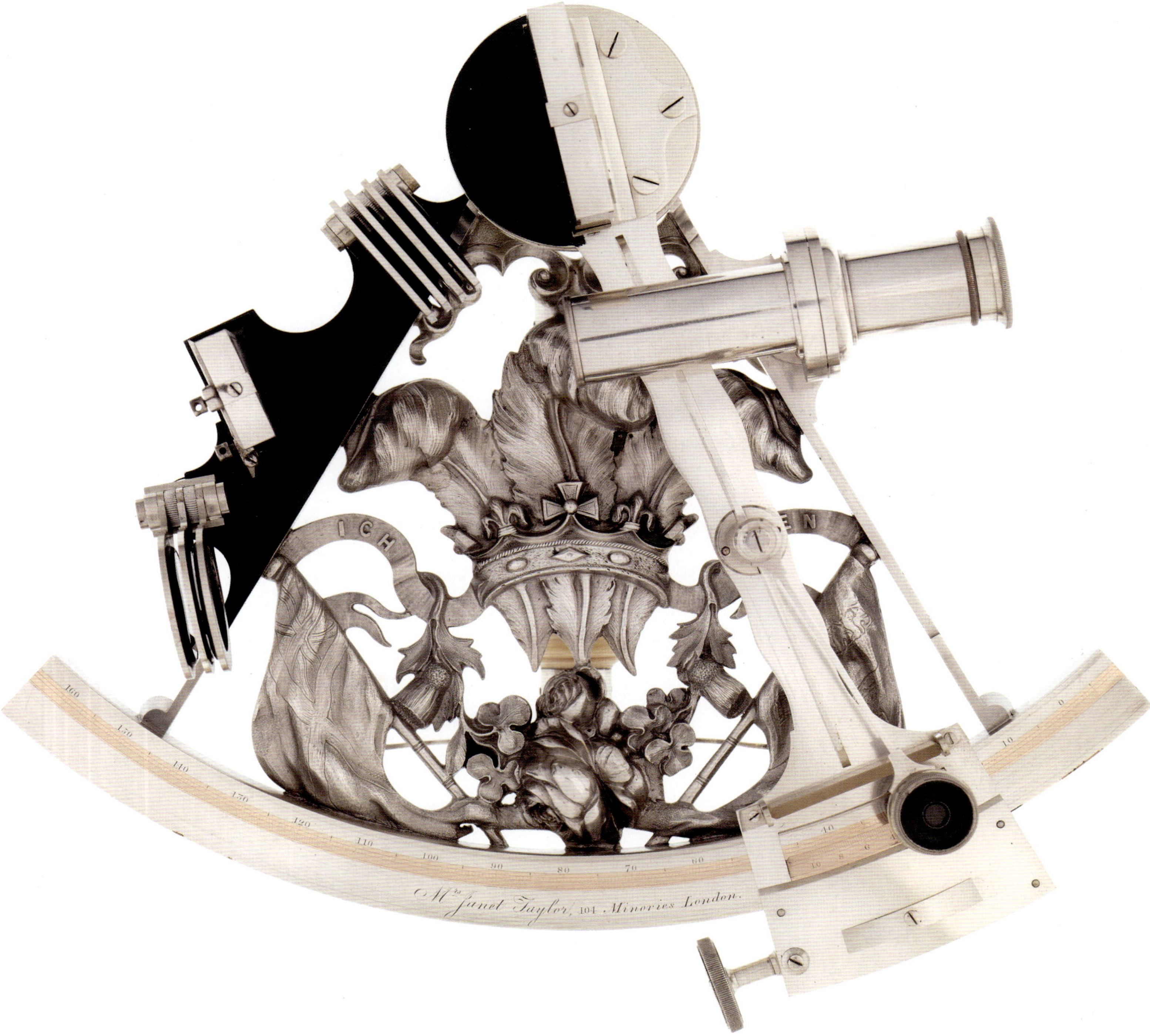

PLATE 19 Quintant by Mrs Janet Taylor of London, dated about 1850. This instrument belonged to the Prince of Wales, later King Edward VII (1841–1910), whose three-feather crest has replaced the inner framework (cat. no. 281); F6415-001, NAV1135.

PLATE 20 Sounding quintant by Henry Hughes & Son Ltd, dated 1952 (cat. no. 297); F5995-001, ACO1339.

PLATE 21 Repeating circle, probably by Jesse Ramsden of London, dated about 1787. This instrument was invented and commissioned by Jean Hyacinthe de Magellan (cat. no. 298); F6166-001, NAV0076.

PLATE 22 Artificial glass horizon by unknown English maker, dated about 1810 (cat. no. 317); F6827-001, NAV-0002.

109 NAV1327

Octant

Origin France
Maker Unknown
Date About 1850
Inscriptions None
Graduation Inlaid ivory scale from -5° to 112° every 20′, measuring to 104°. Ivory vernier to 1′, zero at the right.
Dimensions Radius 9¾ inches (248 mm)
Description Ebony frame and limb with a brass index arm and fittings. Tangent screw on the front of the index arm; the clamping screw is on the back. Three shades (one red and two green); three horizon shades (all red). Index-glass adjustment by a square-headed screw and a detached key; adjustment of horizon glass by a lever and worm gear, and a square-headed screw and a detached key. The sight vane has two pinholes and a swivelling shade with choice of two (red and green). Shaded eyepiece (missing) and a milled adjusting key. In a polished keystone box, containing in the lid a trade label for Isidore Bianchetti, 1 Rue Beauvau, Marseille, France.
Provenance Adams collection; purchased in 1958.
Notes The optician Isidore Bianchetti established his firm in 1827, and was later joined by two sons.
Literature Marcelin, *Dictionnaire des fabricants français*.

110 NAV1328

Octant

Origin London, England
Maker John Dring and William Fage
Date About 1850
Inscriptions Signed on the ivory plate on the crossbar: Dring & Fage | ⋆London⋆
Graduation The inlaid ivory scale is missing. Ivory vernier (broken) to 1′, zero at the right.
Dimensions Radius 9.6 inches (244 mm)
Description Ebony frame and limb with a black-lacquered brass index arm and fittings; inlaid ivory plates on the crossbar and on the back of the frame (latter missing). There are a tangent screw and clamping screw on the back of the index arm. Three shades (two red and one green); three horizon shades (red and green, one is missing). Index-glass adjustment by a screw (the mount is incomplete); adjustment of horizon glass (glass missing) by a milled screw, a lever, a worm gear, and a clamping screw (latter missing). The sight vane and two feet of the frame are missing; a pencil or screwdriver is missing from the crossbar. No box.
Provenance Presented by Mr Harold P. Wiggins, Henley-on-Thames, Oxfordshire, in 1959. See also cat. nos. 111 and 116.
Literature Clifton, *Directory*, 89.

111 NAV1330
Octant

Origin England
Maker Unknown
Date About 1850
Inscriptions None
Graduation The inlaid ivory scale is missing. Ivory vernier to 1′, zero at the right.
Dimensions Radius 9¾ inches (248 mm)
Description Ebony frame and limb with a black-lacquered brass index arm and fittings; inlaid ivory plates on the crossbar and on the back of the frame (both missing). The reinforcement bar is missing from the index arm. Remains of a tangent screw mounted on the back of the index arm; the clamp device and clamping screw are missing from the back. Index shades missing; three horizon shades (two red, cracked, and one green). Index-glass adjustment by a screw (glass and screw are missing); adjustment of horizon glass (glass cracked) by a milled knob, a lever, a worm gear, and a milled clamping screw. The sight vane has two pinholes and a swivelling shutter. In a keystone box (badly damaged), containing a trade label for Bernard Hennessy, 5 Wind Street Swansea.
Provenance Presented by Mr Harold P. Wiggins, Henley-on-Thames, Oxfordshire, in 1959. See also cat. nos. 110 and 116.
Literature Clifton, *Directory*, 132–3.

112 NAV1326
Octant

Origin England, possibly Ipswich
Maker Possibly Schulen and Boby
Date About 1850
Inscriptions Marked on the ivory plate on the crossbar: Schulen & Boby ⋆ Ipswich
Graduation Inlaid ivory scale from -3° to 108° every 20′, measuring to 92°. Ivory vernier to 1′, zero at the right.
Dimensions Radius 9.6 inches (244 mm)
Description Ebony frame and limb with a brass index arm and fittings; inlaid ivory plates on the crossbar and on the back of the frame. There are a tangent screw and clamping screw on the back of the index arm. Three shades (red, orange, and green), three horizon shades (two red and one green). Index-glass adjustment by a screw; adjustment of horizon glass by a lever, worm gear, and a milled clamping screw. The sight vane has two pinholes and a swivelling shade (red). No box.
Provenance Presented by Mr Nicholl Carnie, Merchant Navy, Liverpool, in 1956.
Notes Schulen and Boby may have been the retailers of this instrument; they are not recorded in Clifton, *Directory*.

113 NAV1273
Octant

Origin England
Maker Unknown
Date About 1850
Inscriptions None
Graduation Inlaid ivory scale from -2° to 100° every 20′, measuring to 92°. Ivory vernier to 1′, zero at the right.
Dimensions Radius 9¾ inches (248 mm)
Description Ebony frame and limb with a black-lacquered brass index arm and fittings; inlaid ivory plates on the crossbar and on the back of the frame. There are a tangent screw and clamping screw (broken, possibly a replacement) on the back of the index arm. Three index shades (two red and one green; one red glass is broken). Index-glass adjustment by a screw; adjustment of horizon glass by a milled screw, lever, and clamping screw. The sight vane and one foot of the frame are missing. In poor condition. No box.
Provenance Unknown; the typed record of the instrument pre-dates 1960.

114 NAV1271
Octant

Origin England
Maker Unknown
Date About 1850
Inscriptions None
Graduation Inlaid ivory scale from -2° to 105° every 20′, measuring to 88°. Ivory vernier to 30″, zero at the right.
Dimensions Radius 9¾ inches (248 mm)
Description Ebony frame and limb with a dark-brown-lacquered brass index arm and fittings, wooden handle; an inlaid ivory plate on the crossbar. There are a tangent screw and clamping screw on the back of the index arm. Three shades (two red and one green); three horizon shades. Index-glass adjustment by a screw (one side of the mount is missing); adjustment of horizon glass by a milled screw, and a lever, worm gear, and milled clamping screw. Threaded telescope bracket without adjustment. Telescope (78 mm), erect image. Sight-tube (83 mm). A shaded eyepiece and adjusting pin are missing. In a keystone polished mahogany box, containing in the lid a trade label for F. Smith, 18 Southside Street, Plymouth.
Provenance Unknown; the typed record indicates it was acquired in 1936, also mentioning that it came from the possession of the Griffin family, one of whose members sailed with James Cook in the *Resolution*.
Notes The reference to William Griffin (Griffith), who sailed as a cooper in the *Resolution* (1776–80), suggests that it came from him. However, because of the date of the instrument this is impossible. Possibly F. Smith was a retailer; his name is not recorded in Clifton, *Directory*.
Literature Beaglehole, *The Life of Captain James Cook*, 565, 642.

115 NAV1349
Surveyor's octant

Origin England, possibly Portsmouth
Maker Possibly E. and E. Emanuel
Date About 1850
Inscriptions Marked on the ivory plate on the crossbar: E & E Emanuel. 3 The Hard. Portsea
Graduation Inlaid ivory scale from -4° to 108° every 20′, measuring to 90°. Ivory vernier to 30″, zero at the right (cracked).
Dimensions Radius 9 inches (229 mm)
Description Ebony frame and limb with a brass index arm and fittings; wooden handle; inlaid ivory plates on the crossbar and on the front of the frame (the latter is probably a replacement). There are a tangent screw and clamping are on the back of the index arm. Two shades (red and green, both cracked); the horizon shades are missing. Index-glass adjustment by a screw; adjustment of horizon glass by a lever, worm gear, and milled clamping screw. The reflecting areas of the glasses are silvered paper (replacements). Threaded telescope bracket without adjustment; the telescope or swivelling sight is missing. No box.
Provenance Discovered in Basin No. 4, Portsmouth Dockyard, by Petty Officer Morley in 1968. Presented by Mr Mortimer, Receiver of Wrecks, HM Customs and Excise, Customs House, Plymouth, Devon, in 1974.
Notes Although the instrument is relatively heavy the large glass suggests this was a surveyor's octant. The Emanuels were goldsmiths and jewellers, and possibly only the retailers rather than the makers of this instrument.
Literature Clifton, *Directory*, 95.

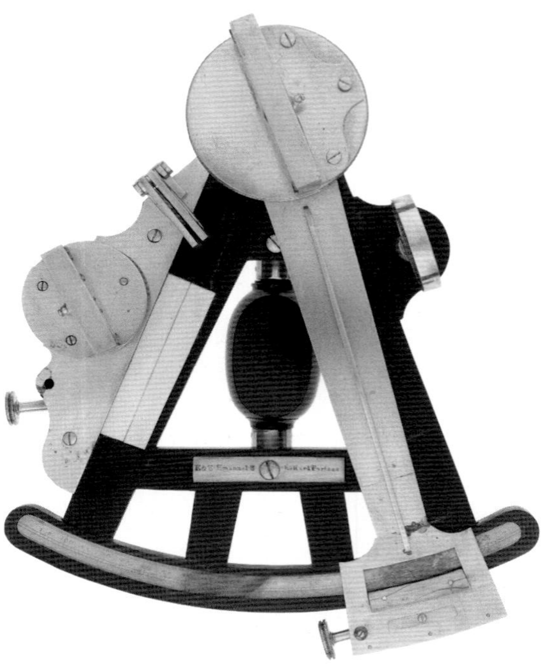

116 NAV1293
Octant

Origin London, England
Maker Janet Taylor
Date About 1850
Inscriptions Signed on the ivory plate on the crossbar: Mrs Janet. TAYLOR. IMP^{D} | 104.Minories. London | 815
Graduation Inlaid ivory scale from -3° to 109° every 20′, measuring to 99°. Ivory vernier to 1′, zero at the right.
Dimensions Radius 9¾ inches (248 mm)
Description Ebony frame and limb with a black-lacquered brass index arm and fittings; inlaid ivory plates on the crossbar and on the back of the frame. There are a tangent screw and clamping screw on the back of the index arm. Three shades (two red and one green); three horizon shades (two red and one green). Index-glass adjustment by a screw; adjustment of horizon glass by capstan screws. The sight vane has two pinholes and a swivelling shutter. An ivory pencil head is fitted in the crossbar (pencil missing). Adjusting pin. In a mahogany keystone box.
Provenance Presented by Mr Harold P. Wiggins (F. Wiggins and Sons, Minories, London), in 1938. His grandfather Frederick Wiggins was apprenticed to Mrs Janet Taylor as a nautical instrument maker. See also cat. nos. 110 and 111.
Literature Clifton, *Directory*, 275.

117 NAV0084
Octant with artificial horizon

Captain Beatson's patent altimeter
Origin London, England
Maker Probably Henry Hughes
Date 1850 or later
Inscriptions Marked on the cover of the valve: Captn. Beatson's | Patent Altimeter
Graduation Bronzed limb with inlaid silver scale from -2° to 100° and 92° to 10°, every 10′. Two silver verniers to 1′, zeros at left.
Dimensions Diameter 7.8 inches (198 mm)
Description Dark-brown-lacquered brass frame, wooden handle. Double-ended tangent screws (missing) on the front of the index arms; the clamping screws are on the back. Three shades (two red and one green); two horizon shades (red and green). Index-glass adjustment by a screw; adjustment of horizon glass by a screw and a capstan screw. Magnifier on a 63-mm swivelling arm serving both verniers. Threaded telescope bracket with perpendicular adjustment by rising-piece and a milled knob, the second telescope bracket is without adjustment. Telescopes and unidentified parts are missing. The instrument is in poor condition. The box or case is missing.
Artificial horizon This consists of a glass tube (missing) containing spirits or mercury, fitted in a groove on the inside of the limb, and ending in a finger-operated airtight valve. For an observation, the air valve was kept open, allowing the fluid to flow freely. At the instant the sight was taken, the valve was closed and the fluid was held in position. The altitude of the sight was taken as the average height of the two columns of fluid read against the two scales. The averaging eliminated any error due to expansion or contraction of the fluid caused by changes in temperature.
Provenance Presented by Kelvin Hughes, Barkingside, in 1961.
Notes Captain Donald Beatson patented this altimeter (no. 12 949 of 1850). The provenance of this instrument, and a signed, complete example (including its leather carrying case) preserved in the Nederlands Scheepvaartmuseum, Amsterdam, inv. no. B.0191(07), suggest that Hughes was the manufacturer.
Literature Ifland, *Taking the Stars*, 112–13; Clifton, *Directory*, 144.

118 NAV1274
Octant

Origin Glasgow and Sunderland, Great Britain
Maker Paul Cameron and John Blakeney
Date About 1855
Inscriptions Signed on the limb: Cameron & Blakeney, Glasgow & Sunderland.
Graduation Bronzed limb with inlaid silver scale from -5° to 125° every 15′, measuring to 101°. Silver vernier to 15″, zero at the right.
Dimensions Radius 7½ inches (191 mm)
Description Anodized brass plain-pattern frame, wooden handle. There are a tangent screw and clamping screw on the back of the index arm. Four shades (two green and two red); three horizon shades (one green and two red). Index-glass adjustment by a screw; adjustment of horizon glass by capstan screws. Magnifier on an 80-mm swivelling arm. Threaded telescope bracket without adjustment. Telescope (83 mm), erect image. Sight-tube (80 mm), and a shaded eyepiece (red). In a fitted wooden keystone box, containing in the lid trade labels for Francis M. Moore, The High Street, Belfast (1854–1906), and S. Thaxter and Son, Boston, USA (1822–1916).
Provenance Adams collection; lent in 1936, bequeathed in 1974.
Literature Clarke *et al.*, *Brass & Glass*, 225–7; Clifton, *Directory*, 48; Mollan, *Irish National Inventory*, 405; Brewington, *The Peabody Museum Collection*, 141; Warner, 'American Octants and Sextants', 111.

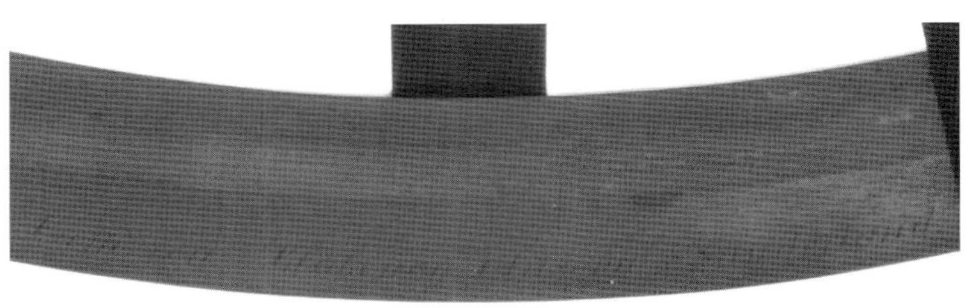

119 NAV1334
Octant

Origin London, England
Maker Robert Montague Barrett
Date About 1855
Inscriptions Signed on the ivory plate on the crossbar: R.M. Barrett. London | For. G. Sidstone. Dartmouth
Graduation Inlaid ivory scale from -2° to 108° every 20′, measuring to 93°. Ivory vernier to 1′, zero at the right.
Dimensions Radius 9¾ inches (248 mm)
Description Ebony frame and limb with a brass index arm and fittings; an inlaid ivory plate on the crossbar. Tangent screw on the front of the index arm; the clamping screw is on the back (both screws are broken). Three shades (two red and one green); one horizon shade (green), which is probably not original to this instrument. Index-glass adjustment by a screw; adjustment of horizon glass (glass missing) by a lever, worm gear and a milled clamping screw. Threaded telescope bracket without adjustment, with a swivelling sight vane with one pinhole. The telescope, adjusting pin, and shaded eyepiece are missing; the milled screws vary in type, and some must be replacements. In a shaped mahogany box, containing in the lid a trade label (badly damaged) for William Langford, Bristol (1828–73).
Provenance Presented by Sea Ranger Ship *Invicta*, Welling, Kent, in 1961.
Literature Clifton, *Directory*, 19, 162–3.

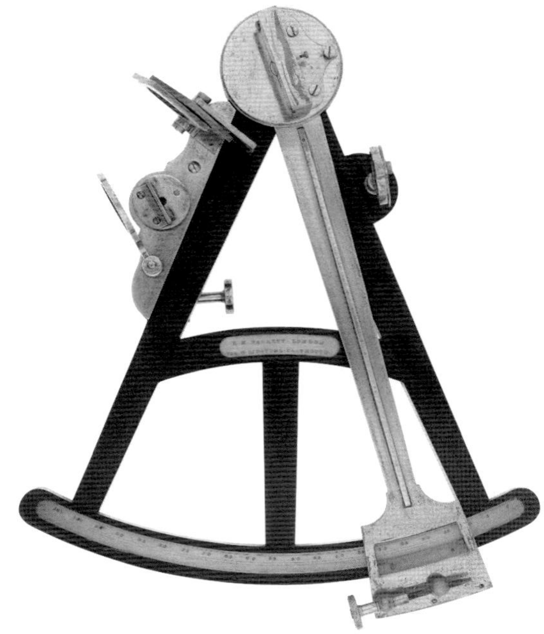

120 NAV1269
Octant

Origin New York, USA
Maker Henry Duren
Date About 1860
Inscriptions Signed on the ivory plate on the crossbar: H. Duren New-York
Graduation Inlaid ivory scale from -4° to 100° by 20′, measuring to 90°. Vernier to 1′, zero at the right.
Dimensions Radius 9.6 inches (244 mm)
Description Ebony frame and limb with a brass index arm and fittings; inlaid ivory plates on the crossbar and on the back of the frame. There are a tangent screw and clamping screw on the back of the index arm. Three index shades (two red and one green). Index-glass adjustment by a screw; adjustment of horizon glass by a lever, wing nut, and clamping screw. The sight vane has two pinholes and a swivelling shade (glass missing). An ivory pencil head is fitted in the crossbar (pencil missing). The left strut is cracked; the screw and washer of the sight vane are probably replacements. In an unusual triangular softwood box (stained), with a cracked lid.
Provenance Adams collection; lent in 1936, bequeathed in 1974.
Literature Brewington, *The Peabody Museum Collection*, 125.

121 NAV1307
Octant

Origin London, England
Maker (George) Dollond
Date About 1860
Inscriptions Signed on the ivory plate on the crossbar: Dollond London
Graduation Inlaid ivory scale from -4° to 107° every 20′, measuring to 92°. Ivory vernier to 30″, zero at the right.
Dimensions Radius 8.6 inches (218 mm)
Description Ebony frame and limb with a black-lacquered brass index arm and fittings; wooden handle; an inlaid ivory plate on the crossbar. There are a tangent screw and clamping screw on the back of the index arm. Three shades (two red and one green); three horizon shades (two red and one green). Index-glass adjustment by a screw; adjustment of horizon glass by a milled knob, and a lever, worm gear and a milled clamping screw. Threaded telescope bracket without adjustment, with a swivelling sight vane with one pinhole. Telescope (82 mm), erect image. One foot of the frame is a replacement. In a wooden keystone box, containing in the lid a trade label for John Bliss and Co., 110 Wall Street, near South Street, New York.
Provenance Purchased from Mr B.O. Woodcock, Harrow, Middlesex, in 1944.
Literature Barty-King, *Eyes Right*; Clifton, *Directory*, 86–7; Brewington, *The Peabody Museum Collection*, 119.

122 NAV1313
Octant

Origin London, England
Maker Henry Negretti and Joseph Warren Zambra
Date About 1860
Inscriptions Signed on the ivory plate on the crossbar: .Negretti & Zambra. London.
Graduation Inlaid ivory scale from -2° to 104° every 20′, measuring to 89°. Ivory vernier to 30″, zero at the right.
Dimensions Radius 17½ inches (445 mm)
Description Ebony frame and limb with a brass index arm and fittings, ebony handle; an inlaid ivory plate on the crossbar. There are a tangent screw and clamping screw on the back of the index arm. Three shades (two red and one green); three horizon shades (two red and one green). Index-glass adjustment by a square-headed screw and a detached key; adjustment of horizon glass by a square-headed key and a detached key, a lever, a worm gear, and a milled clamping screw. Threaded telescope bracket without adjustment. Telescope (86 mm), large objective lens; erect image. Sight-tube (96 mm), a shaded eyepiece (red), a magnifying glass, and an adjusting key. In a polished mahogany keystone box, containing in the lid a label in MS '1910. E. Company 5th Batt.^{n} Essex Reg.^{t} Honorary Member 2nd Prize won by E. T. Adams'.
Provenance Adams collection; lent in 1950, bequeathed in 1974.
Notes This is an unusually large octant for this date. It appears unused and may have been kept for display or demonstration purposes.
Literature Clifton, *Directory*, 198.

123 NAV1270
Octant

Origin London, England
Maker John Dennett Potter
Date About 1860
Inscriptions Signed on the ivory plate on the crossbar: J.D. Potter | Poultry . London
Graduation Inlaid ivory scale from -3° to 109° every 20′, measuring to 93°. Ivory vernier to 1′, zero at the right.
Dimensions Radius 9¾ inches (248 mm)
Description Ebony frame and limb with a dark-brown-lacquered brass index arm and fittings; an inlaid ivory plate on the crossbar. There are a tangent screw and clamping screw on the back of the index arm. Three index shades (two red and one green). Index-glass adjustment by a screw; adjustment of horizon glass by a lever, wing nut and a milled clamping screw. The sight vane has two pinholes and a swivelling shutter. A pencil or screwdriver is missing from the crossbar. In a keystone polished mahogany box, containing in the lid a trade label for J.D. Potter (successor to R.B. Bate), 31 Poultry, London, and a circular label in MS 'E. T. A. from Helen, Francis, Isabel, Gordon, & Pauline, Christmas 1909'.
Provenance Adams collection; lent in 1936, bequeathed in 1974. The label indicates that this instrument was given to Adams by his wife Helen and their children at Christmas in 1909.
Literature Clifton, *Directory*, 221; McConnell, *R B Bate of the Poulty*, 53–4.

124 NAV1323
Octant

Origin England
Maker Unknown
Date About 1860
Inscriptions None
Graduation Inlaid ivory scale from -4° to 109° every 20′, measuring to 91°. Ivory vernier to 1′, zero at the right.
Dimensions Radius 9¾ inches (248 mm)
Description Ebony frame and limb with a brass index arm and fittings, wooden handle; an inlaid ivory plate on the crossbar. There are a tangent screw and clamping screw on the back of the index arm. Three shades (two red and one green); three horizon shades (two red and one green). Index-glass adjustment by a screw; adjustment of horizon glass by a milled knob, and a lever, worm gear, and milled clamping screw. Threaded telescope bracket without adjustment, with a swivelling sight vane with one pinhole. Telescope (83 mm), erect image. No box.
Provenance Adams collection; lent in 1950, bequeathed in 1974.

125 NAV 1266
Octant

Origin London, England
Maker Norie and Wilson
Date About 1860
Inscriptions Signed on the ivory plate on the crossbar: Norie & Wilson, London
Marked (scratched) on the back of the frame: W.S.
Graduation Inlaid ivory scale from -4° to 108° every 20′, measuring to 88°. Ivory vernier to 1′, zero at the right.
Dimensions Radius 8¾ inches (222 mm)
Description Ebony frame and limb with a brass index arm and fittings; inlaid ivory plates on the crossbar and on the back of the frame. There are a tangent screw and clamping screw on the back of the index arm. Three index shades (one green and two red). Index-glass adjustment by a screw; adjustment of horizon glass by a square-headed screw and a detached key, a lever, a worm, gear and a milled clamping screw. The sight vane has two pinholes and a swivelling shutter. The silvering of the glasses is in poor condition; one foot of the frame is broken. Milled adjusting key. In a keystone mahogany box.
Provenance Adams collection; lent in 1936, bequeathed in 1974.
Notes The firm Norie and Wilson used the name on this instrument from 1883 to 1899, which is later than the instrument can be dated. Perhaps it was refurbished and fitted with a new nameplate.
Literature Clifton, *Directory*, 201; Fisher, *The Makers of the Blueback Charts*, 100–3.

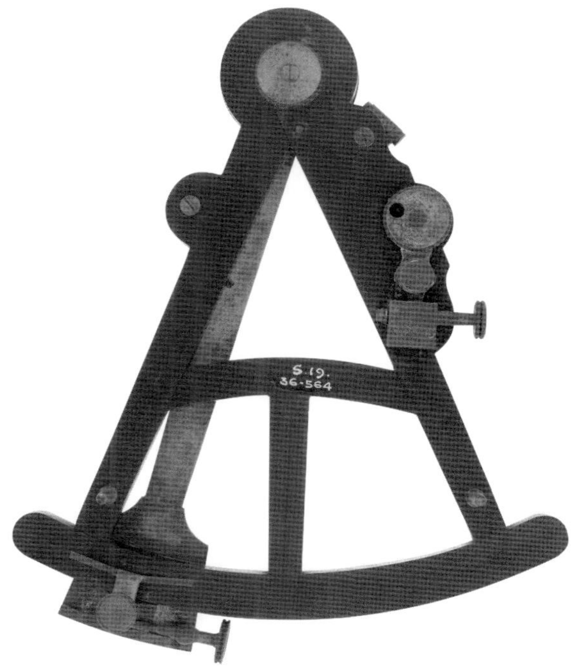

126 NAV1354
Octant with artificial horizon

Origin Leith, Scotland
Maker David Stalker
Date About 1870
Inscriptions Signed on the limb: D. Stalker. Leith.
Graduation Polished brass limb with inlaid silver scale from -5° to 120° every 15′, measuring to 105°. Silver vernier to 15″, zero at the right.
Dimensions Radius 7¼ inches (184 mm)
Description Black-lacquered brass plain-pattern frame, wooden handle. There are a tangent screw and clamping screw on the back of the index arm. Four shades (three red and one green); three horizon shades (two red and one green). Index-glass adjustment by a screw; adjustment of horizon glass by capstan screws. Magnifier on 89-mm swivelling arm. Threaded telescope bracket without adjustment. Telescope (82 mm), erect image. Sight-tube (82 mm), and a shaded eyepiece (red). In a polished wooden keystone box, containing in the lid a trade label for James White, 209 Sauchiehall Street, and 16, 18 Cambridge Street, Glasgow (1884–90).
Artificial horizon This consists of two spirit levels to allow the instrument to be held vertically. One level is fitted to the upper edge of the crossbar; a second spirit level is fitted at the centre of the lower handle-securing bracket (bubble missing).
Provenance Purchased from Relcy Antiques, Greenwich, London, in 1980.
Literature Clarke *et al.*, *Brass & Glass*, 108–10, 252–4.

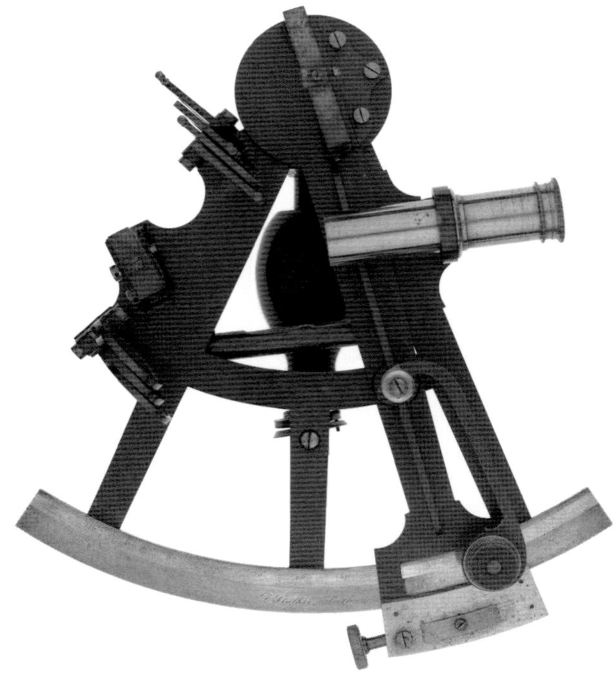

127 NAV1336
Octant

Origin London, England
Maker Janet Taylor and Co.
Date About 1875
Inscriptions Signed on the limb: Mrs. Janet Taylor & Co. 104 . Minories London.
Graduation Bronzed limb with inlaid ivory scale from -5° to 120° every 20′, measuring to 102°. Ivory vernier to 30″, zero at the right.
Dimensions Radius 9.4 inches (239 mm)
Description Anodized brass plain-pattern frame, wooden handle. There are a tangent screw and clamping screw on the back of the index arm. Four shades (three red and one green); three horizon shades (two red, one green). Index-glass adjustment by screw; adjustment of horizon glass by capstan screws. Magnifier on an 82-mm swivelling arm. Threaded telescope bracket without adjustment. Telescope (83 mm), erect image. Sight-tube (81 mm), a shaded eyepiece (red), and an adjusting pin. In a mahogany keystone box, containing in the lid a trade label for Henry Hughes and Son, 59 Fenchurch Street, London (after about 1875).
Provenance Caird Fund purchase from Mr J.T. Polak, Thorpe Bay, Essex, in 1961.
Literature Clifton, *Directory*, 144, 275.

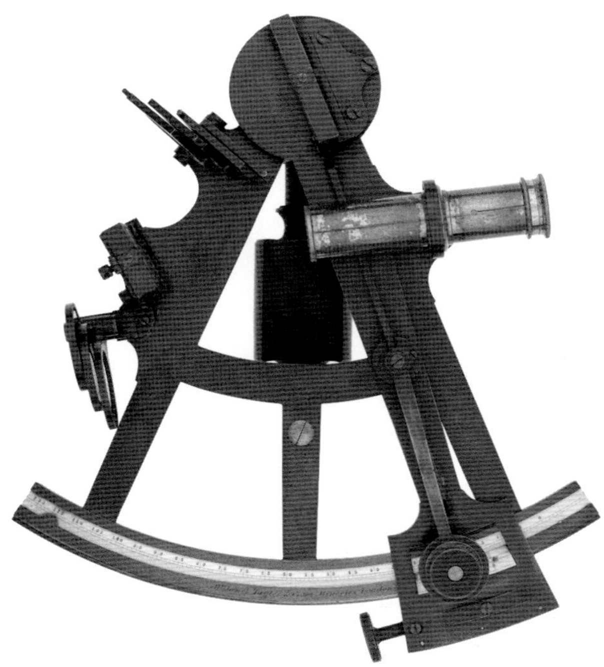

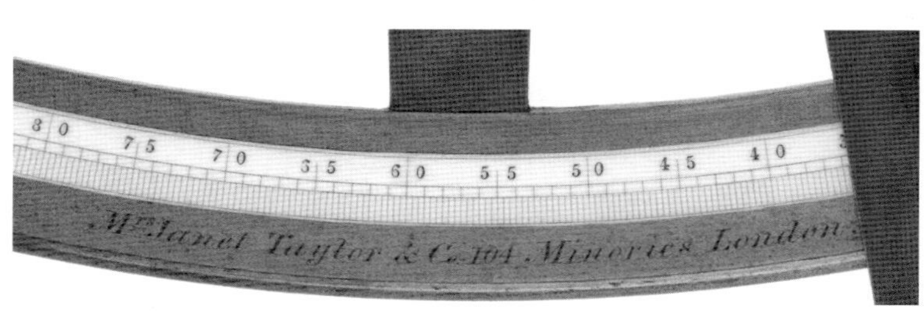

128 NAV1315
Octant

Origin Hamburg, Germany
Maker Thomas Downie
Date About 1880
Inscriptions Signed on the limb: Thomas Downie, Hamburg.
Graduation Bronzed limb with inlaid silver scale from -5° to 125° every 15′, measuring to 105°. Silver vernier to 15′, zero at the right.
Dimensions Radius 7½ inches (191 mm)
Description Brass plain-pattern frame, partly black-lacquered, wooden handle. There are a tangent screw and clamping screw on the back of the frame. Four shades (three red, one green); three horizon shades (two red and one green). Index-glass adjustment by a screw; adjustment of horizon glass by capstan screws. Magnifier on a 90-mm swivelling arm. Threaded telescope bracket without adjustment. Telescope (83 mm), erect image. Sight-tube (81 mm), a shaded eyepiece (red), and an adjusting pin (missing). In a square wooden box; inside the lid, in blue-pencil MS, 'J. S. Dare c/o S.S. NEDJED'.
Provenance Presented by Mr M.P. Dare, MA, Branshill Common, Eversley, Hampshire, in 1955.
Notes Nothing is known about Downie, who possibly came from Britain. Between 1875 and 1901 the Deutsche Seewarte in Hamburg tested 85 octants and sextants and one box sextant of his.
Literature Messerschmitt, 'Ergebnisse von Sextantenprüfungen', 37; Hill and Paget-Tomlinson, *Instruments of Navigation*, 36.

129 NAV1230
Octant

Origin London, England
Maker Probably Henry Hughes and Son Ltd
Date 1943
Inscriptions Marked on the index arm: AM | Ref No 6B / 177
Marked on the limb: 37782
Marked on the back of the frame: T | V | V 192
Marked on the back of the horizon shade mounting: 18 | V176
Marked on the index shade mounting: 27
Graduation Polished brass limb with inlaid silver scale from -5° to 120° every 1°, measuring to 100°. Micrometer to 1′.
Dimensions Radius 3½ inches (89 mm)
Description Polished brass straight-bar-pattern frame, black plastic handle. White plastic micrometer drum and pressure clamp at the bottom of the index arm. Four shades (all green); three horizon shades (all green). Index-glass adjustment by a screw; adjustment of horizon glass by capstan screws. Telescope bracket without adjustment. Fixed telescope (39 mm), with a rubber eyecup. Adjusting pin and a bottle for lubricating oil. In a fitted square Bakelite box, containing in the lid a Husun, Henry Hughes and Son Ltd certificate of examination, dated 27 November 1943.
Provenance Purchased from Relcy Antiques, Greenwich, London, in 1979.
Literature Clifton, *Directory*, 144; Hughes, *History of Air Navigation*.

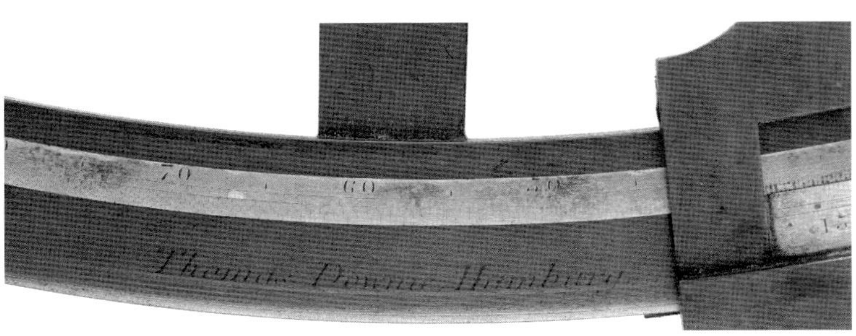

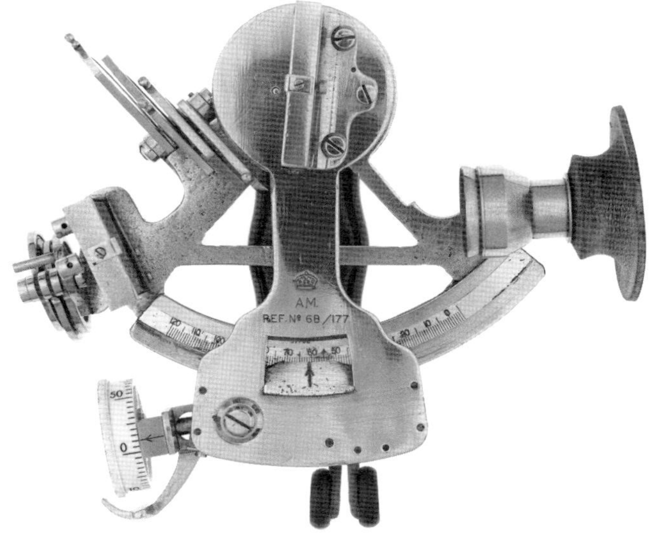

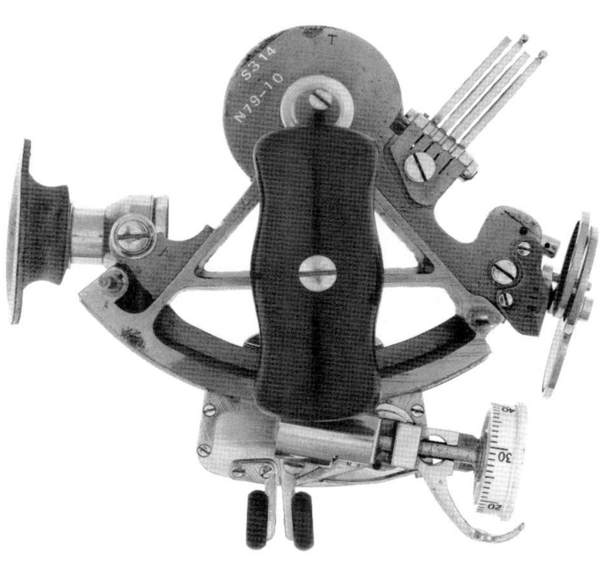

130 ACO1358
Octant

Origin London, England
Maker Probably Henry Hughes and Son Ltd
Date 1945
Inscriptions Marked on the index arm: [broad arrow] | Ref. No 6B/177
Marked on the limb: 45154
Graduation Polished brass limb with inlaid silver scale from -5° to 120° every 1°, measuring to 95°. Micrometer to 1′.
Dimensions Radius 3½ inches (89 mm)
Description Black-lacquered brass straight-bar-pattern frame, black plastic handle. White plastic micrometer drum and pressure clamp at the bottom of the index arm. Four shades (all green); three horizon shades (all green). Index-glass adjustment by a screw; adjustment of horizon glass by a screw and a capstan screw. Threaded telescope bracket without adjustment. Telescope (38 mm), erect image. Two unidentified parts are missing. In a fitted square Bakelite box, containing in the lid a Husun, H. Hughes and Son Ltd certificate of examination, dated 17 January 1945.
Provenance Transferred from the Admiralty Compass Observatory in 1969.
Literature Clifton, *Directory*, 144; Hughes, *History of Air Navigation*.

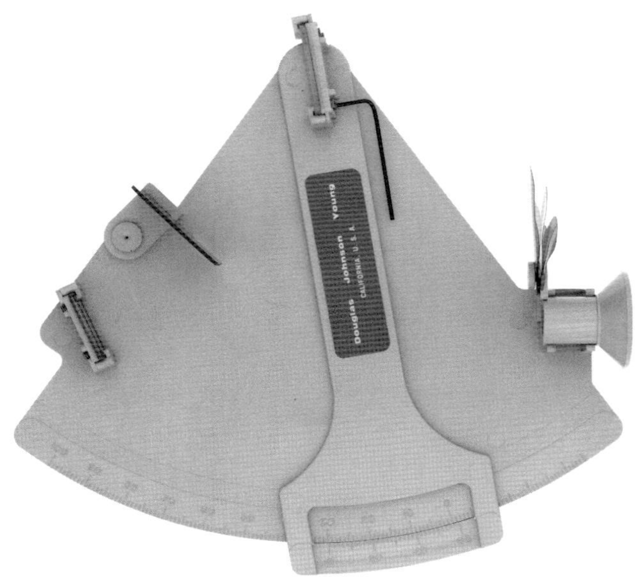

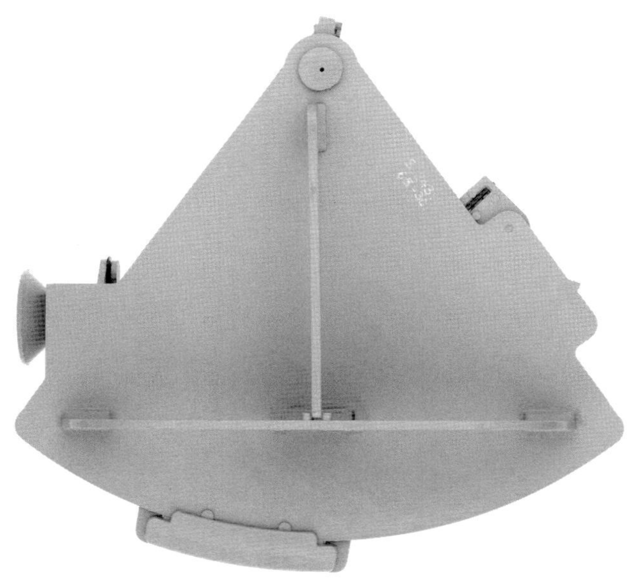

131 NAV1186
Octant

Davis Mark II Plastic Marine Sextant
Origin Oakland, California, USA
Maker Douglas Johnson Young
Date 1965
Inscriptions Signed on a stuck-on label on the index arm: Douglas Johnson Young | California, U.S.A.
Graduation Grey plastic scale from -35° to 100° every 1°, measuring from -14° to 100°. Plastic vernier to 2′, zero at the left.
Dimensions Radius 7½ inch (191 mm)
Description Grey plastic plate, a plastic handle and stiffening bar on the back. There is no tangent or clamping screw. One plastic index shade (blue). Index – and horizon-glass adjustment by screws with an Allen key. Plastic telescope bracket without adjustment. Fixed plastic sight-tube (32 mm), with three plastic shades (blue, yellow, and green). Allen key. No box.
Provenance Purchased from Douglas Johnson Young, Oakland, California, in 1965.
Notes William A. Davis of San Leandro, California, first manufactured Davis plastic sextants in 1963; in 2007, the Marks 15 and 25 were available.
Literature Weatherlake, *Davis Mark II Plastic Marine Sextant*; Leback, 'World War II: The Lifeboat Sextant'; Ifland, *Taking the Stars*, 41; Bartlett, 'Choosing & Using a Sextant'.

132 NAV 1187
Octant

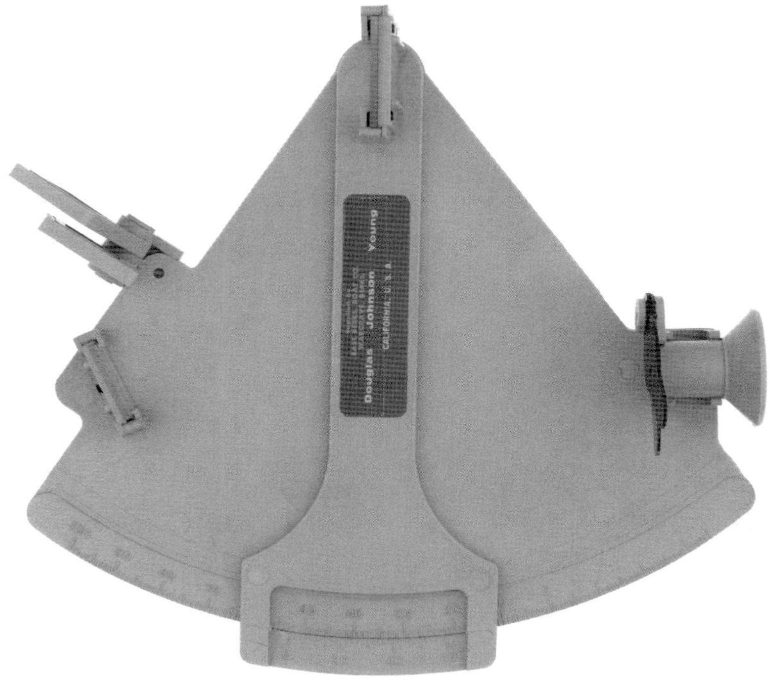

Davis Mark II Plastic Marine Sextant
Origin Oakland, California, USA
Maker Douglas Johnson Young
Date 1965
Inscriptions Signed on a stuck-on label on the index arm: Supplied by | East Berks. Boat Co. | Wargrave, Berks. | Douglas Johnson Young California, U.S.A.
Graduation Grey plastic scale from -35° to 100° every 1°, measuring from -14° to 100°. Plastic vernier to 2′, zero at the left.
Dimensions Radius 7½ inches (191 mm)
Description Grey plastic plate; a plastic handle and stiffening bar on the back. There is no tangent or clamping screw. Two index shades (both green). Index – and horizon-glass adjustment by screws with an Allen key. Plastic telescope bracket without adjustment. Fixed plastic sight-tube (32 mm), fitted with two plastic shades (green and red). Allen key (missing); an instruction booklet by Weatherlake.
Provenance Purchased from Douglas Johnson Young, Oakland, California, in 1965.
Notes William A. Davis of San Leandro, California, first manufactured Davis plastic sextants in 1963; in 2007, the Marks 15 and 25 were available.
Literature Weatherlake, *Davis Mark II Plastic Marine Sextant*; Leback, 'World War II: The Lifeboat Sextant'; Ifland, *Taking the Stars*, 41; Bartlett, 'Choosing & Using a Sextant'.

6. SEXTANTS

133 NAV1177
Sextant

Origin London, England
Maker John Bird
Date About 1758
Inscriptions Signed on the limb: J. Bird. London.
Graduation Polished brass scale from -13° (marked 77) to 128° every 20′, measuring to 122°. Brass vernier to 1′, zero at the centre.
Dimensions Radius 18¼ inches (464 mm)
Description Polished brass frame and limb, with a four-legged brass section on the back for a belt pole mounting; the detachable belt pole is of brass and wood. The tangent screw and clamping screw (modern replacements) are fitted to the limb on a separate shoe, to the left of the index arm. Three socket shades (all red). Index-glass adjustment by a screw; adjustment of horizon glass by a lever moved by a milled worm screw, with a milled clamping screw. Sight-tube (modern replacement). No box.
Provenance Presented by the Royal United Service Institution Museum in 1963, on its closure.
Notes The clamping screw, tangent screw, and sight-tube were missing and have been replaced by the Museum, modelled on the other surviving Bird sextants. According to Stimson, this was the first sextant to be made, which was ordered by Captain John Campbell RN. That was, however, a 16-inch instrument.
Literature Stimson, 'The First Marine Sextant?'; Warner, 'John Bird and the origin of the sextant'; Stimson, 'Some Board of Longitude Instruments', 95–100; Moskowitz, 'The World's First Sextants'; Wylly, *Official Catalogue*, 176, no. 2662; Hughes, *Official Catalogue*, 141, no. 2662; Clifton, *Directory*, 30.

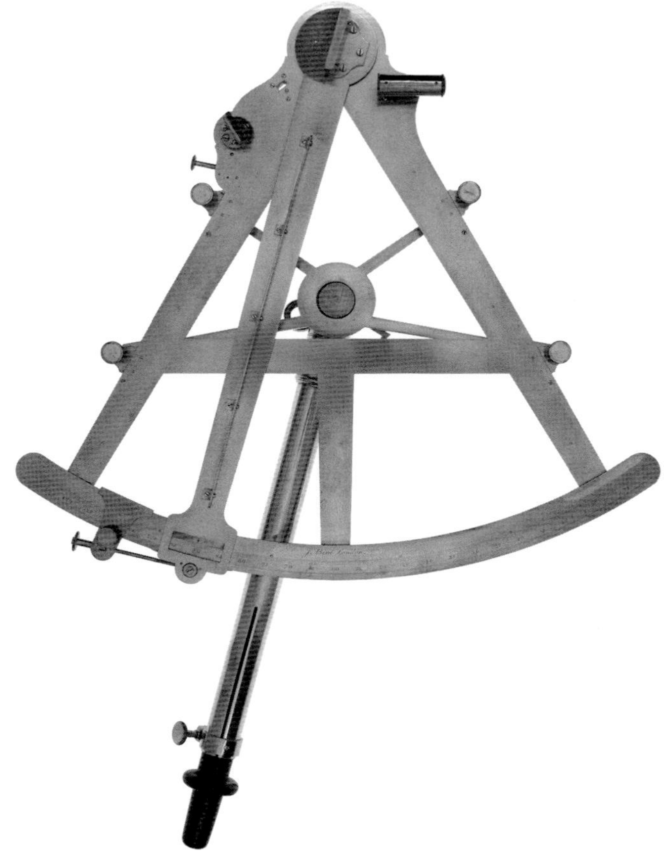

134 NAV1161
Sextant

Origin London, England
Maker Thomas Heath and Tycho Wing
Date About 1765
Inscriptions Signed on the index arm: Made by Heath & Wing London
Graduation Polished brass scale from -5° to 135° every 30′, measuring to 120°. Brass vernier to 1′, zero at the right. A second vernier to 1′, zero at the right, is fitted to the left of the index arm (the difference between the two zeros is 24° 30′).
Dimensions Radius 5.9 inches (150 mm)
Description Polished brass frame and limb, brass stop for the index arm. There is no tangent screw; the clamping screw is on the back of the index arm. Two index shades (both red); no horizon shades. Index-glass adjustment by a screw; adjustment of horizon glass (glass missing) by a lever wing nut and a milled clamping screw. The sight vane has two pinholes and a swivelling shade (red). In a mahogany keystone box, possibly of later date.
Provenance Presented by Commander Sir Peter Ogilvy Wedderburn Bt RN, in 1956.
Literature Clifton, *Directory*, 131.

135 NAV1236
Sextant

Origin London, England
Maker Jesse Ramsden
Date About 1772
Inscriptions Signed on the crossbar: Ramsden | London
Marked on the top end of the index arm: D.33
Graduation Polished brass scale from -2° to 134° every 20′, measuring to 120°. Brass vernier to 20″, zero at the right.
Dimensions Radius 15 inches (381 mm)
Description Polished brass frame and limb, wooden handle. The tangent screw (knob missing) is on the front of the index arm; the clamping screw is on the back. The lower part of the index glass is silvered and can be covered by a Maskelyne flap; the upper part is not silvered. Three shades (all red); one horizon shade (red). Index-glass adjustment by a screw; adjustment of horizon glass (glass missing) by screws and by Dollond's extended lever and clamp, operated from beneath the telescope bracket. Threaded telescope bracket in two parts, fitted for correcting collimation error; perpendicular adjustment by rising-piece and a milled knob. Telescope (127 mm), inverted image (possibly not original to this instrument). No box.
Provenance Reputed to have been used on Cook's third voyage to the Pacific (1776–80). Purchased from Trevor Philip and Sons, London, in 1982, with the assistance of the National Heritage Memorial Fund and an anonymous donor.
Notes Peter Dollond patented the horizon glass adjustment (no. 1017 of 1772). D.33 was the Hydrographic Office serial numbering adopted in February 1828, where D was the letter given to sextants.
Literature Maskelyne, 'Remarks on the Hadley's Quadrant', 57–8; Howse, 'The Principal Scientific Instruments Taken on Captain Cook's Voyages of Exploration', 120–3; Clifton, *Directory*, 227; Stimson, 'Some Board of Longitude Instruments', 114–15.
For image see colour section plate 13.

136 NAV1110
Sextant

Origin London, England
Maker Edward Nairne and Thomas Blunt
Date About 1780
Inscriptions Signed on the bottom of the index arm: Nairne & Blunt | London
Marked (stamped) at the centre of the scale (at 67°): I [foul anchor] R
Marked on the index arm: Once the Property of Admiral Sir John T. Duckworth, K.C.B., Bart. | of Wear, B.1748, D.1819.
Graduation Bronzed scale from -3° to 139° every 20′, measuring to 124°. Vernier to 30″, zero at the right.
Dimensions Radius 12 inches (305 mm)
Description Polished brass frame, wooden handle. The tangent screw (bent) is on the front of the index arm; the clamping screw is on the back. The lower part of the index glass is silvered, and can be covered by a Maskelyne flap; the upper part is not silvered. Three shades (two red and one green); one horizon shade (red). Index-glass adjustment by a screw; horizon-glass adjustment by screws and by Dollond's extended lever and clamp, operated from beneath the telescope bracket. Threaded telescope bracket in two parts, fitted for correcting collimation error; perpendicular adjustment by rising-piece and a milled knob. Telescope (129 mm), erect image, two parallel cross-wires. No box.
Provenance Admiral Sir John Thomas Duckworth (1748–1817), the original owner, defeated a French squadron off San Domingo in 1806 and commanded an expedition against Constantinople in 1807. On the index arm, his year of death is erroneously marked 1819. Presented by Mrs Picsse, Topsham, South Devon, in 1933.
Notes Peter Dollond patented the horizon glass adjustment (no. 1017 of 1772); the scale mark indicates that it was divided on Ramsden's engine.
Literature Maskelyne, 'Remarks on the Hadley's Quadrant', 57–8; Dollond and Dollond, *Directions for Using the New Improved Hadley's Quadrant*; Ifland, *Taking the Stars*, 84–5; Clifton, *Directory*, 196, 227; Stimson, 'Some Board of Longitude Instruments', 111–14; Stimson, 'The Influence of the Royal Observatory at Greenwich', 128; Hill and Paget-Tomlinson, *Instruments of Navigation*, 29; A.B. Sainsbury, 'Duckworth, Sir John Tomas, first baronet (1748–1817) *naval officer*', in *Oxford Dictionary of National Biography*, Oxford, 2004, <http://www.oxforddnb.com>.

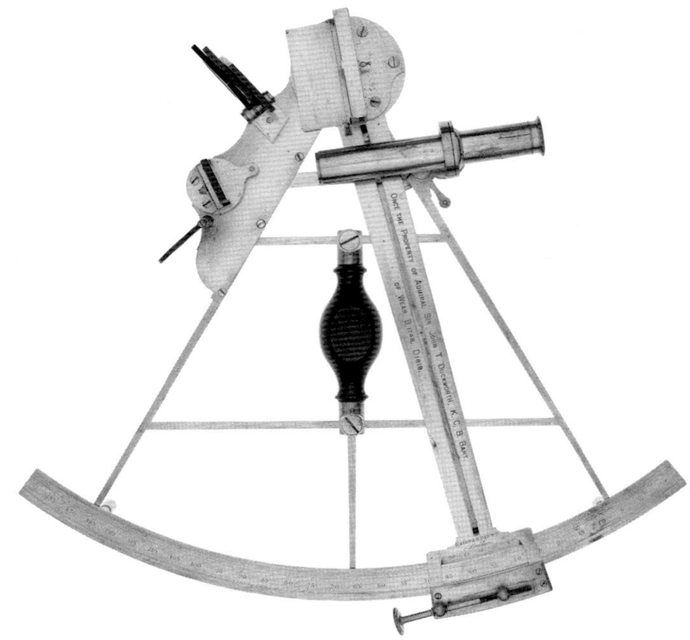

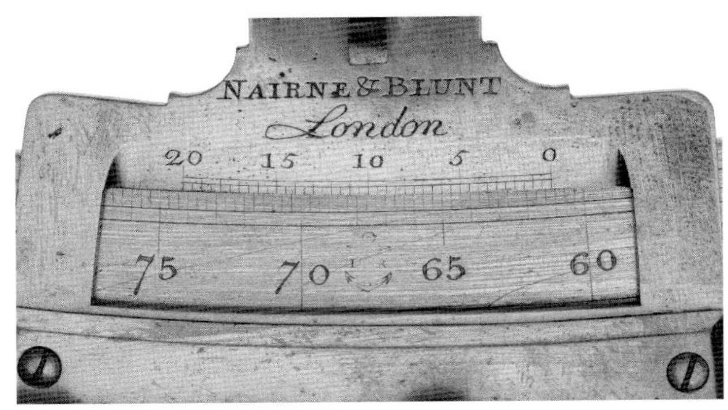

137 NAV1179
Sextant

Origin London, England
Maker Unsigned, but with a scale mark for Jesse Ramsden
Date About 1780
Inscriptions Marked (stamped) at the centre of the scale (at 65°): I [foul anchor] R
Graduation Inlaid ivory scale from -2° to 129° every 20′, measuring to 119°. Ivory vernier to 1′, zero at the right.
Dimensions Radius 15½ inches (394 mm)
Description Mahogany frame and limb with brass index arm and fittings, a brass stop for the index arm; an inlaid plate in the crossbar (missing). The double-ended tangent screw is on the front of the index arm; the clamping screw (a modern replacement) is on the back. Three index shades (one green and two red); no horizon shades. Index-glass adjustment by a screw; adjustment of horizon glass (glass broken) by a lever, wing nut, and clamping screw. Threaded telescope bracket with perpendicular adjustment by rising-piece and a milled knob. The telescope (74 mm), erect image, is in a wooden barrel. No box.

Provenance Purchased from Brigadier J.F. Walker, Thirsk, Yorkshire, in 1964.
Notes The scale mark indicates that the instrument was divided on Ramsden's engine.
Literature Clifton, *Directory*, 227; Stimson, 'Some Board of Longitude Instruments', 112–13.

138 NAV1296
Sextant with artificial horizon

Origin England
Maker Unknown
Date About 1780
Inscriptions None
Graduation Brass scale from -5° to 115° every 15′, measuring to 115°. Brass vernier to 15″, zero at the left.
Dimensions Radius 14½ inches (368 mm)
Description Anodized brass frame in two sections joined at the apex, with the limb at the top and an octagonal wooden handle at the bottom; the horizon glass, sight vane, spirit level, vernier, and shade are static. The limb and the index glass are on the movable part of the frame. One index shade (red); no horizon shades. Index- and horizon-glass adjustment by screws. The sight vane has one pinhole.
Artificial horizon This consists of a brass tube containing a spirit level. It is based on the principle of the double-reflected image of the celestial body being alongside the reflection of the level. The instrument is held limb uppermost by the handle (like a hand mirror).
Provenance Presented by Sir James Caird in 1937; purchased as part of the Gabb collection.
Literature Ifland, *Taking the Stars*, 100.

139 NAV1148
Box sextant

Pocket sextant
Origin London, England
Maker William Fraser
Date About 1780–90
Inscriptions Signed on the upper plate: Fraser, Mathema.l Ins.t Maker | to his MAJESTY. Bond 5 Str.t. LONDON. [read from the pivot] Marked on the lid: This Instrument | Used for many Years | By | Sir W.m Gell, | In | Asia Minor, Greece and Italy, | was at his Death in | Naples, | Given as a dear Memento | of Friendship to | John Auld Jo
Graduation Inlaid silver scale from -7° to 139° every 30′, measuring to 100° (the digits are read from the pivot). The silver vernier reads to 1′, zero at the left.
Dimensions Radius 1¾ inches (44 mm); diameter 67 mm
Description Circular polished brass plates with a lid that, when fitted on the back of the instrument, serves as a handle. The index arm is moved by a milled knob on the upper plate of the instrument. One index shade (red) moved by a lever; no horizon shades. Index and horizon glasses without adjustment. Magnifier (single lens) on a hinged 36-mm swivelling arm (bent). An eyepiece shutter with the choice of a plain pinhole or a shade (red).
Provenance Sir William Gell (1777–1836), the original owner, was a classical archaeologist and traveller. Presented by Sir James Caird in 1937; purchased as part of the Gabb collection.
Literature Clifton, *Directory*, 105; W.W. Wroth, 'Gell, Sir William (1777–1836) *classical archaeologist and traveller*' in *Oxford Dictionary of National Biography*, Oxford, 2004, <http://www.oxforddnb.com/>.

140 NAV 1103
Sextant

Origin London, England
Maker Jesse Ramsden
Date About 1783
Inscriptions Signed on the crossbar: Ramsden | London
Marked on the back of the frame, at the pivot: 724
Graduation Brass scale from -2° to 135° every 20′, measuring to 120°. Brass vernier to 30″, zero at the right, brass micrometer drum to 10″.
Dimensions Radius 12 inches (305 mm)
Description Polished brass frame and limb, wooden handle. The tangent screw, with a micrometer drum, is on the front of the index arm; the clamping screw is on the back. The lower part of the index glass is silvered and can be covered by a Maskelyne flap; the upper part is not silvered. Three shades (two red and one green); two horizon shades (red and green). Index-glass adjustment by a screw; adjustment of horizon glass by Dollond's extended lever and clamp, operated from beneath the telescope bracket. Threaded telescope bracket in two parts, fitted for correcting collimation error; perpendicular adjustment by rising-piece and a milled knob. Telescope (127 mm), inverted image. In a wooden keystone box.
Provenance Adams collection; lent in 1936, bequeathed in 1974.
Notes The earliest recorded use of a serial number by Ramsden. Peter Dollond patented the horizon glass adjustment (no. 1017 of 1772). The micrometer is defective.
Literature Maskelyne, 'Remarks on the Hadley's Quadrant', 57–8; Dollond and Dollond, *Directions for Using the New Improved Hadley's Quadrant*; Ifland, *Taking the Stars*, 84–5; Stimson, 'The Influence of the Royal Observatory at Greenwich', 128–30; Clifton, *Directory*, 227.

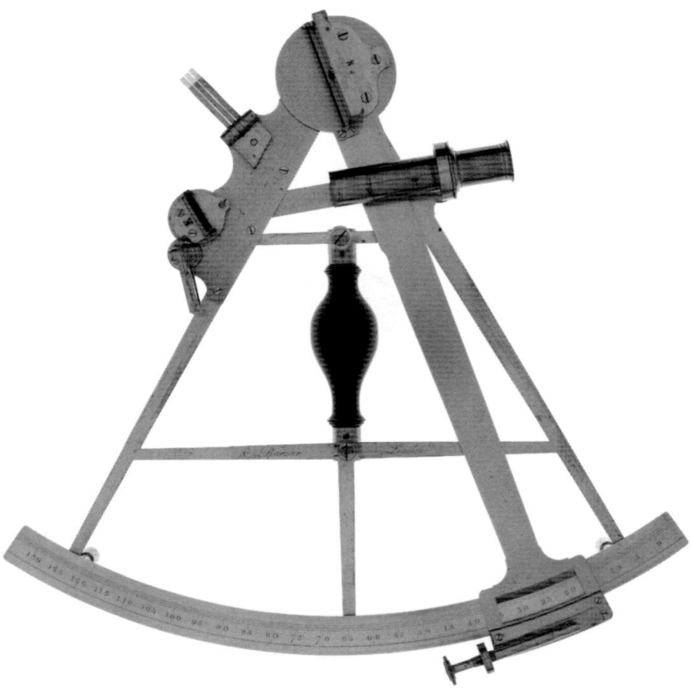

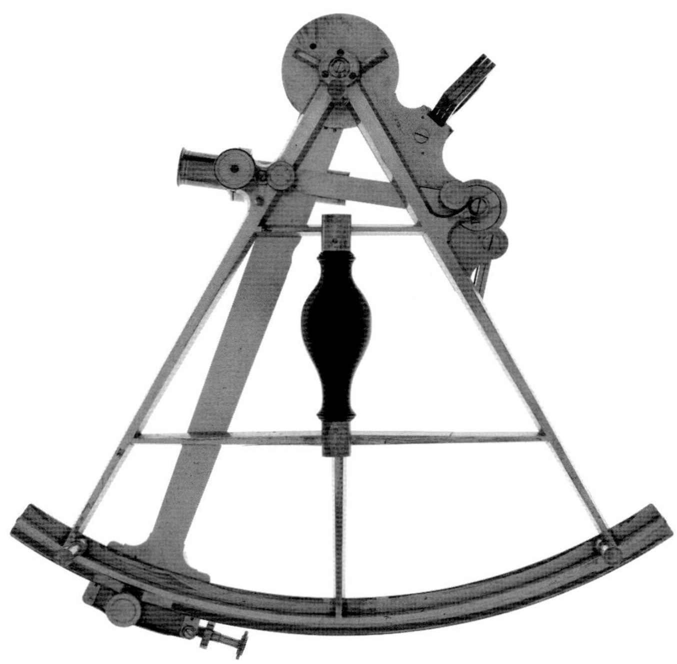

141 NAV1104
Sextant

Origin London, England
Maker Jesse Ramsden
Date About 1785
Inscriptions Signed on the limb: Ramsden London
Marked on the back of the frame at the pivot: 842
Graduation Brass scale from -2° to 137° every 20′, measuring to 121°. Brass vernier to 30″, zero at the right.
Dimensions Radius 14½ inches (368 mm)
Description Polished brass frame, wooden handle. The tangent screw and clamping screw are on the back of the index arm. The index glass is black rather than silvered, for use with Sun and Moon observations only. This must be a replacement as the instrument was originally fitted with shades, which are unnecessary with black glass, and there is no trace of a Maskelyne flap. Three shades (all red); two horizon shades (both red). Index-glass adjustment by a screw; adjustment horizon glass by a fixed milled key. Telescope bracket without adjustment, a sight-tube is mounted between the telescope bracket and the horizon glass mounting. Rotating variable shaded eyepiece with choice of two shades (both red). In a fitted wooden box.
Provenance Adams collection; lent in 1936, bequeathed in 1974.
Notes A telescope similarly mounted is found on a sextant by Ramsden preserved in the Whipple Museum of the History of Science, Cambridge, inv. no. Wh.2122.
Literature Stimson, 'The Influence of the Royal Observatory at Greenwich', 130; Bennett, *The Divided Circle*, 136; Clifton, *Directory*, 227.

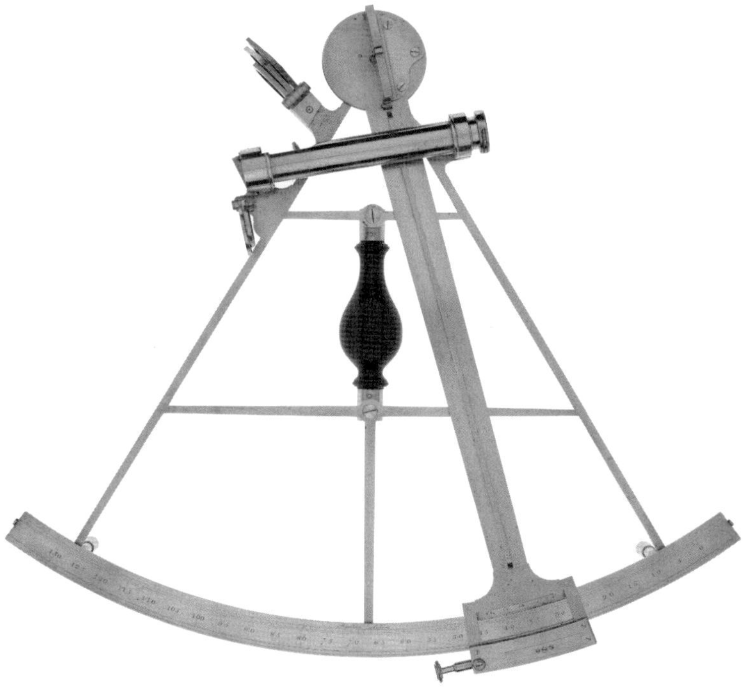

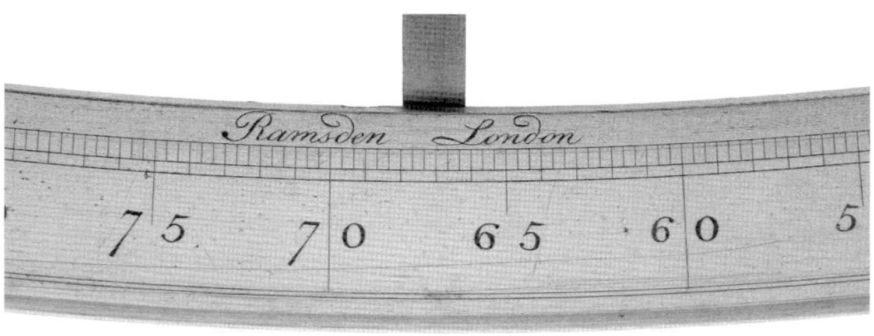

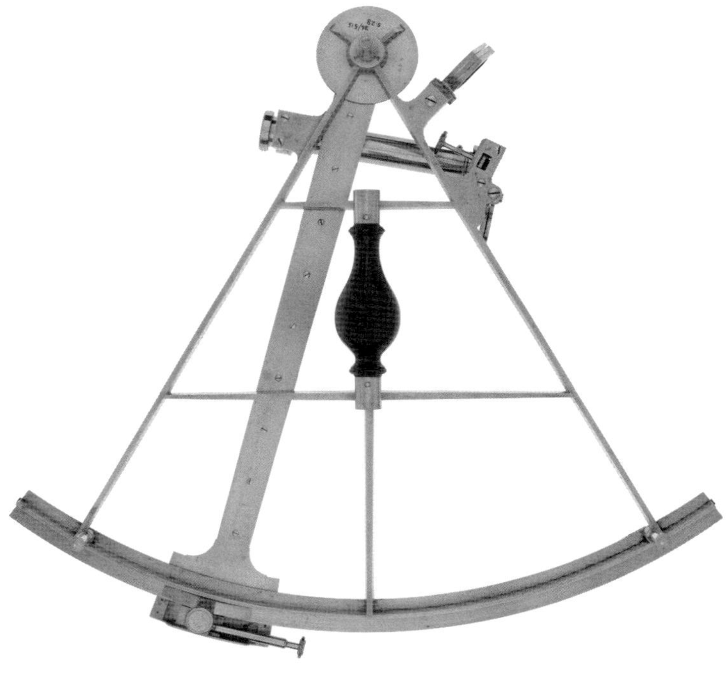

142 NAV1113
Sextant

Origin London, England
Maker (John or Peter) Dollond
Date About 1785
Inscriptions Signed at the bottom of the index arm: Dollond | London.
Graduation Bright brass scale from -4° to 136° every 20′, measuring to 120°. Vernier to 30″, zero at the right.
Dimensions Radius 12 inches (305 mm)
Description Black-lacquered brass diamond-pattern frame, wooden handle. The tangent screw and clamping screw are on the back of the index arm. Three shades (two red, one green); three horizon shades (two red and one green). Index-glass adjustment by a screw; adjustment of horizon glass by a fixed milled screw and two capstan screws. Magnifier on an 80-mm swivelling arm. Threaded telescope bracket with perpendicular adjustment by rising-piece and a milled knob. Telescope (129 mm), inverted image, two parallel cross-wires. Shaded eyepiece (missing) and an adjusting pin. In a polished mahogany keystone box, containing in the lid a trade label for Henry Hughes and Son, 59 Fenchurch Street, London.
Provenance Adams collection; lent in 1936, bequeathed in 1974. The presence of the Hughes label may suggest that Adams purchased this instrument from them.
Literature Clifton, *Directory*, 87.

143 NAV1105
Sextant

Origin London, England
Maker Jesse Ramsden
Date About 1790
Inscriptions Signed on the limb: Ramsden London
Marked on the back of the frame near the pivot: 974
Graduation Brass scale from -2° to 137° every 30′, measuring to 124°. Brass vernier to 30″, zero at the right.
Dimensions Radius 12 inches (305 mm)
Description Polished brass frame, wooden handle. The tangent screw and clamping screw are on the back of the index arm. Four shades (one red, two orange and one green); three horizon shades (two orange and one green). Index-glass adjustment by a screw; adjustment of horizon glass by a thumbscrew and worm gear. Threaded telescope bracket in two parts, fitted for correcting collimation error; perpendicular adjustment by rising-piece and a milled knob. Telescope (127 mm), inverted image, two parallel cross-wires; telescope (58 mm), erect image. Sight-tube (89 mm), and a rotating shaded

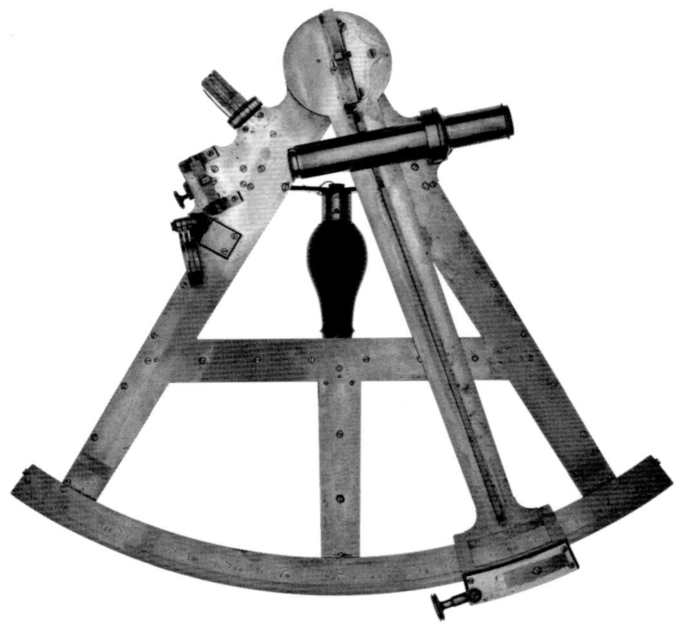

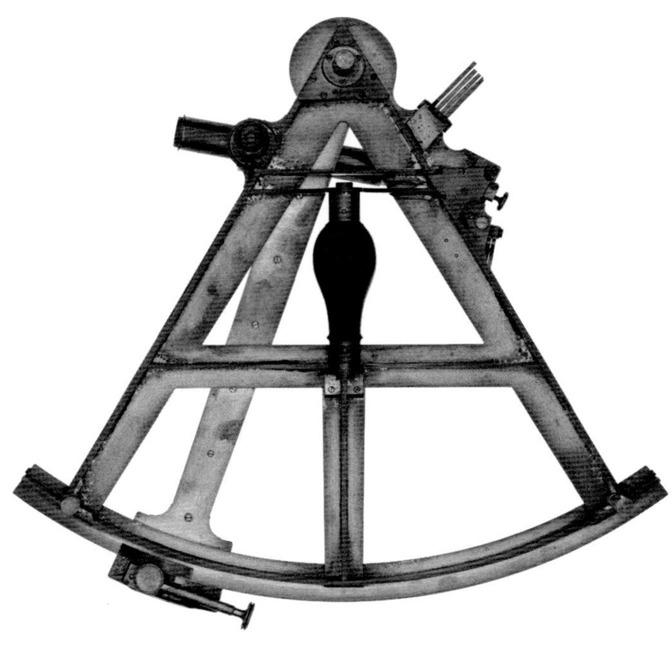

eyepiece with choice of three (all red). In a mahogany keystone box, containing in the lid a trade label for Henry Hughes and Son, 59 Fenchurch Street, London, and a circular label in MS 'ETA Helen, Francis, Isabel, Gordon & Pauline gave this instrument to Father on his birthday 8th December 1908'.

Provenance Adams collection; lent in 1936, bequeathed in 1974. The presence of the Hughes label may suggest that the instrument was purchased from them.

Notes The ink annotations in MS '5742 [and] Capt. Palmer' mentioned in the Museum's typed record were not found.

Literature Stimson, 'The Influence of the Royal Observatory at Greenwich', 130; Clifton, *Directory*, 227.

144 NAV1111
Sextant

Origin London, England
Maker Ritchingham
Date About 1790
Inscriptions Signed at the bottom of the index arm: Ritchingham | London
Graduation Brass scale from -1° to 130° every 20′, measuring to 120°. Brass vernier to 1′, zero at the right.
Dimensions Radius 11 inches (279 mm)
Description Polished brass frame and limb, wooden handle. The tangent screw is on the front of the index arm; the clamping screw is on the back. Three shades (two red and one green); two horizon shades (red and green). Index-glass adjustment by a screw; adjustment of horizon glass by a lever, knob, and milled clamping screw. Threaded telescope bracket with perpendicular adjustment by rising-piece and a milled knob. Telescope (108 mm), erect image. Sight-tube (107 mm). In a wooden keystone box, containing in the lid a trade label for Norie and Wilson, 156 Minories, London (1882–99).
Provenance Adams collection; lent in 1936, bequeathed in 1974.
Literature Hill and Paget-Tomlinson, *Instruments of Navigation*, 31; Clifton, *Directory*, 201, 233; Fisher, *The Makers of the Blueback Charts*, 100–3.

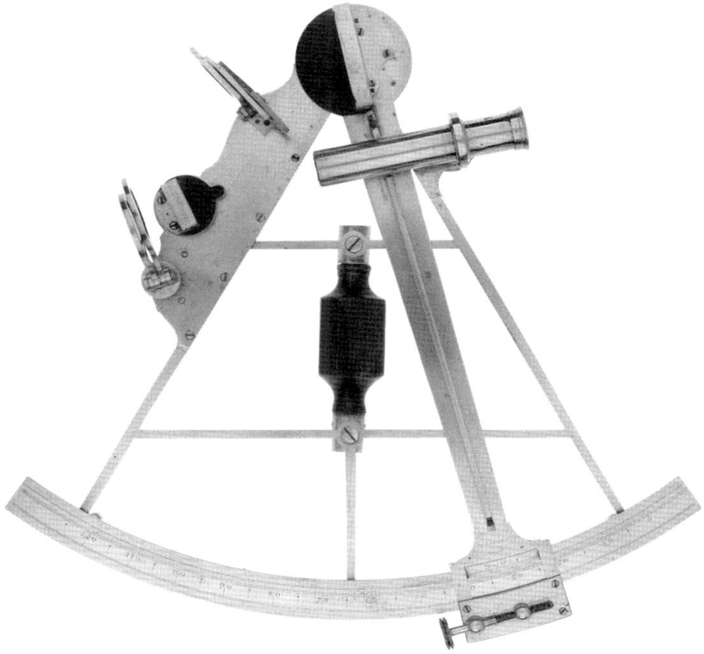

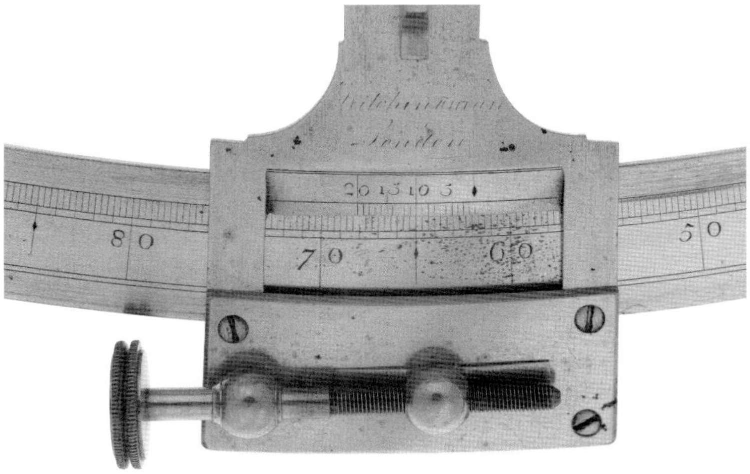

145 NAV1226
Sextant

Origin London, England
Maker (John and Edward) Troughton
Date About 1790
Inscriptions Signed on the limb: No 6 Invented & Made by Troughton London
Graduation Polished brass scale from -5° to 153° every 20′, measuring to 126°. Brass vernier to 30″, zero at the right.
Dimensions Radius 10 inches (254 mm)
Description Polished brass straight-bar-pattern pillar frame (24 pillars), wooden handle with a brass-lined threaded hole for a second handle or a stand. The tangent screw and clamping screw are on the back of the index arm. The lower part of the index glass is silvered, and can be covered by a Maskelyne flap; the upper part is blackened. Three shades (two red and one green); three horizon shades (two red and one green). Index-glass adjustment by a screw; adjustment of horizon glass by screws and a thumbscrew. Threaded telescope bracket in two parts, fitted for correcting collimation error; perpendicular adjustment by rising-piece and a milled knob. Telescope (184 mm), inverted image, two parallel cross-wires; a shaded eyepiece (red). Sight-tube (117 mm), and a shaded eyepiece (red). An Extra drawtube, a magnifying glass, and an unidentified part are missing. In a mahogany keystone box.
Provenance Purchased from David Weston Ltd, London, in 1979.
Notes Edward Troughton patented the pillar and plate frame (no. 1644 of 1788).
Literature Maskelyne, 'Remarks on the Hadley's Quadrant', 57–8; Stimson, 'The Influence of the Royal Observatory at Greenwich', 130; Clifton, *Directory*, 282; McConnell, *Instrument Makers to the World*.

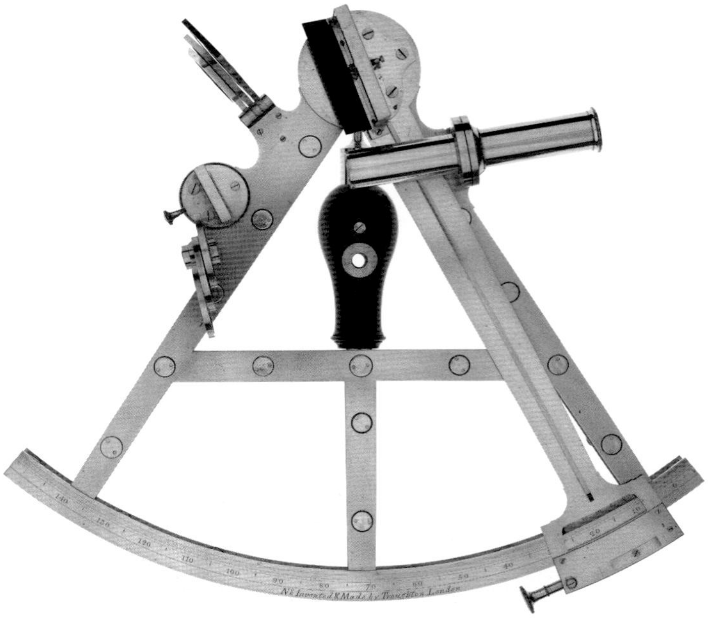

146 NAV1218
Sextant

Origin London, England
Maker (John and Edward) Troughton
Date About 1791
Inscriptions Signed on the limb: Invented & Made by Troughton London
Marked on the bottom of the index arm: No 69
Graduation Polished brass scale from -5° to 148° every 20′, measuring to 127°. Brass vernier to 20″, zero at the right.
Dimensions Radius 11½ inches (292 mm)
Description Polished brass straight-bar-pattern pillar frame (24 pillars), wooden handle. The tangent screw and clamping screw are on the back of the index arm. Three shades (red, orange, and green); three horizon shades (red, orange, and green). Index-glass adjustment by a screw; adjustment of horizon glass by screws and a thumbscrew. Telescope bracket (missing), with perpendicular adjustment by rising-piece and a milled knob. Telescopes and other parts are missing. No box.
Provenance Purchased at Sotheby's, London, March 1975, lot no. 108, through Maggs Brothers.
Notes Edward Troughton patented the pillar and plate frame (no. 1644 of 1788).
Literature Stimson, 'The influence of the Royal Observatory at Greenwich', 130; Clifton, *Directory*, 282; McConnell, *Instrument Makers to the World*.

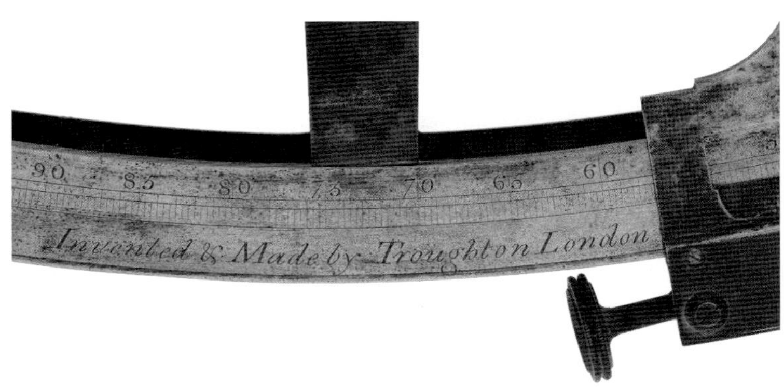

147 NAV1109
Sextant

Origin London, England
Maker Nicholas Meredith
Date About 1791
Inscriptions Signed on the limb: Meredith New Bond Street
Graduation Polished brass scale from -2° to 130° every 20′, measuring to 128°. Brass vernier to 1′, zero at the right.
Dimensions Radius 11½ inches (292 mm)
Description Polished brass frame with alloy fittings, wooden handle. The tangent screw is on the front, the clamping screw is on the back of the index arm. Three shades (two red, one green); three horizon shades (two red and one green). Index-glass adjustment by a screw; adjustment of horizon glass by a milled knob, and by an extended lever and clamp operated from beneath the telescope bracket. Threaded telescope bracket in two parts, fitted for correcting collimation error; perpendicular adjustment by rising-piece and a milled knob. Telescope (131 mm), inverted image, cross-wires (broken); telescope (122 mm), erect image. Star finder (85 mm), erect image (a modern addition). Sight-tube (89 mm). In a keystone polished mahogany box, containing trade labels for H.C. Blair (late Blair and Son), 45 Prince Street, Bristol and for William Gerrard, 35 South Castle Street, Liverpool (1862–90); and a label: '1908 Essex Regt. Edgar T. Adams Secundo Inter Omnes D.D.'
Provenance Adams collection; lent in 1936, bequeathed in 1974.
Notes Peter Dollond patented the horizon glass adjustment (no. 1017 of 1772).
Literature Dollond and Dollond, *Directions for Using the New Improved Hadley's Quadrant*; Ifland, *Taking the Stars*, 84–5; Clifton, *Directory*, 187.

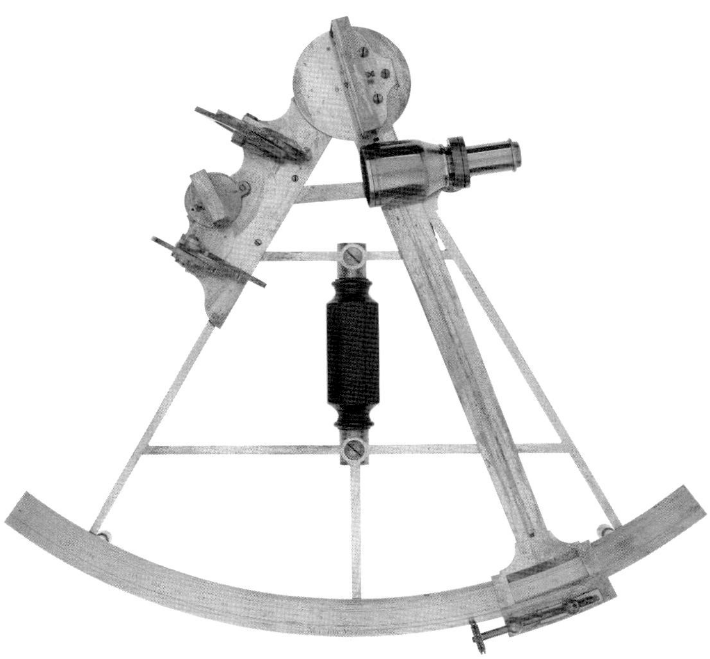

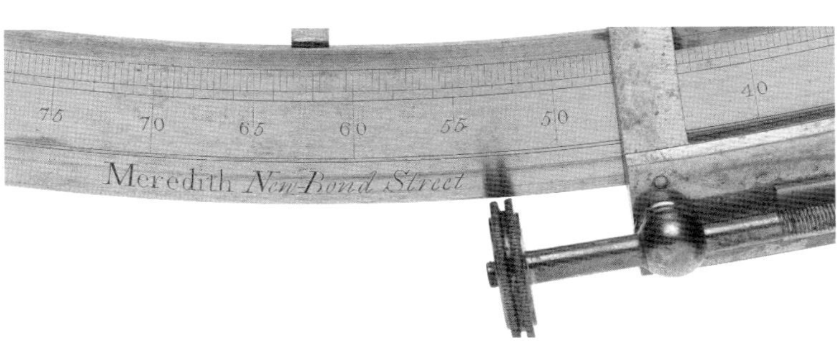

148 NAV1112
Sextant

Origin London, England
Maker William (?) Gilbert and Gabriel Wright
Date 1791
Inscriptions Signed on the limb: Kings Patent. Gilbert & Wright London
Graduation Inlaid silver scale from -4° 40′ to 141° 40′ every 20′, measuring to 122°. Silver vernier to 30″, zero at the right.
Dimensions Radius 11.6 inches (295 mm)
Description Polished brass straight-bar-pattern frame and limb, with a flat brass cross on the back for reinforcement, and a wooden handle with a brass-lined threaded hole for a second handle. The tangent screw and clamping screw are on the back of the index arm. Four round shades (three red and one green); two horizon shades (both red). Index-glass adjustment according to Gabriel Wright's patent; adjustment of horizon glass by a capped screw. Magnifier on a 90-mm swivelling arm. Threaded telescope bracket in two parts, fitted for correcting collimation error; perpendicular adjustment by rising-piece and a milled knob. Telescope (143 mm), inverted image and set focus, two parallel cross-wires; telescope (90 mm), erect image and set focus. In a keystone mahogany box.
Provenance Adams collection; lent in 1936, bequeathed in 1974.
Notes King's Patent' refers to Gabriel Wright's patent (no. 1229 of 1779) for an azimuth compass with an octant. The index glass is mounted on a platform on top of the index arm, and can be rotated round the pivot by screws for parallel adjustment. This patent adjustment is also found on Wright's instruments preserved in the Science Museum, London, inv. no. 1923.475, in The Mariners' Museum, Newport News, Virginia, inv. no. 1998.0039.00004, and in Mystic Seaport Museum, Mystic, Connecticut, inv. no. 1951.394.
Literature Ifland, *Taking the Stars*, 44; Clifton, *Directory*, 113.

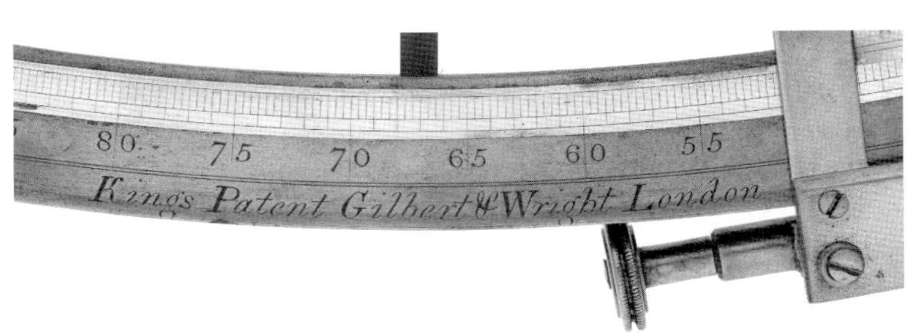

149 NAV 1140
Sextant

Origin London, England
Maker Jesse Ramsden
Date 1792
Inscriptions Signed on the limb: Ramsden London
Marked on the back of the frame at the pivot: 1123
Marked on the crossbar: This Sextant was presented to Admiral Sir John Ommaney [*sic*], | (then a Midshipman) by Ramsden himself in 1792; being graduated by his hand, | and afterwards given by Sir John to Capt. Blackwood, in 1851.
Graduation Polished brass scale from -2° to 137° every 10′, measuring to 125°. Brass vernier to 10″, zero at the right.
Dimensions Radius 10 inches (254 mm)
Description Polished brass straight-bar frame, wooden handle with brass-lined threaded hole for a second handle or a stand. The tangent screw and clamping screw are on the back of the index arm. Four shades (two red and two green); three horizon shades (two green and one red). Index-glass adjustment by a screw; adjustment of horizon glass by a capped milled knob. Magnifier, travelling on rack and pinion. Threaded telescope bracket in two parts, fitted for correcting collimation error; perpendicular adjustment by rising-piece and a milled knob. Telescope (126 mm), inverted image, four cross-wires. A second telescope, a sight-tube, a magnifying glass, and a shaded eyepiece are missing. In a polished keystone box, containing in the lid a Kew Observatory certificate of examination (date illegible), pasted over with a later note bearing MS, the inscription on the instrument as given above.
Provenance Admiral Sir John Acworth Ommanney, KCB (1773–1855), the original owner, was appointed to the *Lion* in 1792. Presented by Admiral Sir Herbert Edward Purey-Cust (1857–1938), in 1937.
Literature Stimson, 'The Influence of the Royal Observatory at Greenwich', 130; Hill and Paget-Tomlinson, *Instruments of Navigation*, 30; Clifton, *Directory*, 227; J.K. Laughton, 'Ommanney, Sir John Acworth (1773–1855), *naval officer*' in *Oxford Dictionary of National Biography*, Oxford, 2004, <http://www.oxforddnb.com/>.

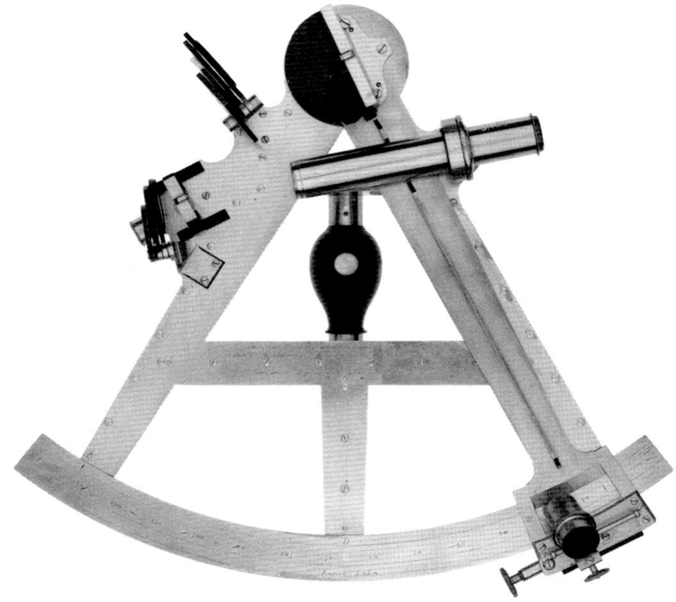

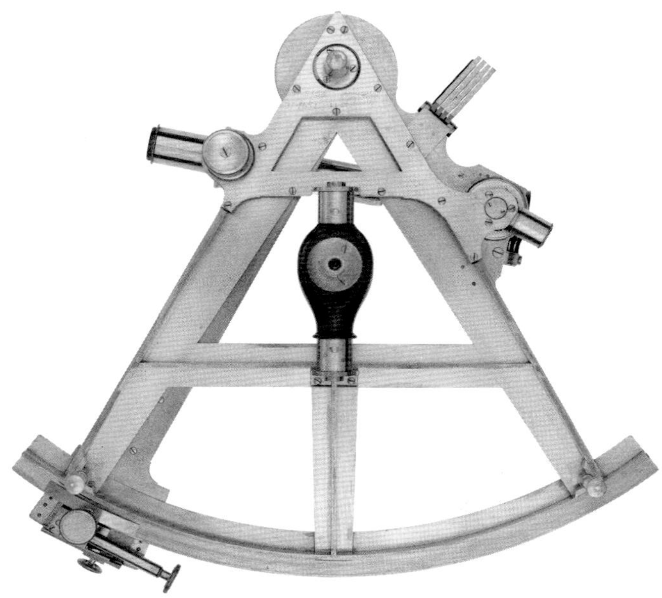

150 NAV1242
Sextant with artificial horizon

Origin London, England
Maker Jesse Ramsden
Date About 1792/1834 or later
Inscriptions Signed on the crossbar: Ramsden London
Marked on the back of the frame at the pivot: 1146
Marked on the body of the artificial horizon: Invented by A.B. Becher Lieu.t RN | 1834 No. [blank]
Graduation Polished brass limb with inlaid silver scale from -3° to 139° every 15′, measuring to 128°. Silver vernier to 1′, zero at the right.
Dimensions Radius 5¼ inches (133 mm)
Description Polished brass frame, detached wooden handle. The tangent screw is on the front of the index arm; the clamping screw is on the back. Three shades (one green and two red; the hinge of the middle shade is broken); three horizon shades (all red). Index-glass adjustment by a screw; adjustment of horizon glass (glass missing) by a square-headed screw and detached milled screw and by a fixed milled knob. Magnifier on a 76-mm swivelling arm. Detached threaded telescope bracket to be mounted on the back of the frame, in two parts, fitted for correcting collimation error; perpendicular adjustment by rising-piece and a milled knob. Telescope (108 mm), inverted image, four cross-wires; telescope (45 mm), erect image; shaded eyepiece (green) for use with an oil lamp. Sight-tube (73 mm), a rotating shaded eyepiece (the rotating mount with the shades is missing), and a milled adjusting key. Pendulum and two glass bottles (one containing oil for damping; the other bottle contains lamp oil). A label on the artificial horizon in MS reads '1° to add for correction of horizon'. In a fitted rectangular mahogany box which contains a second, square box for the oil bottles, and in the lid of that box a MS note '1° 49″ Add for correction of Horizon'.
Artificial horizon This consists of a 95-mm-long bracket for attaching the horizon to the frame. The pendulum is lowered into a black-lacquered brass conical attachment, filled with almond oil to damp the pendulum's movement. A separate oil lamp can be fitted to the bracket for illumination at night. Alexander Bridport Becher (1796–1876) invented his artificial horizon in 1834. This example is by an unknown maker and has been fitted on an older sextant.
Provenance Purchased from David Weston Ltd, London, in 1985.
Literature Becher, 'The Pendulum Marine Artificial Horizon'; Stimson, 'The Influence of the Royal Observatory at Greenwich', 130; Ifland, *Taking the Stars*, 104–05; Clifton, *Directory*, 227.
For image see colour section plate 14.

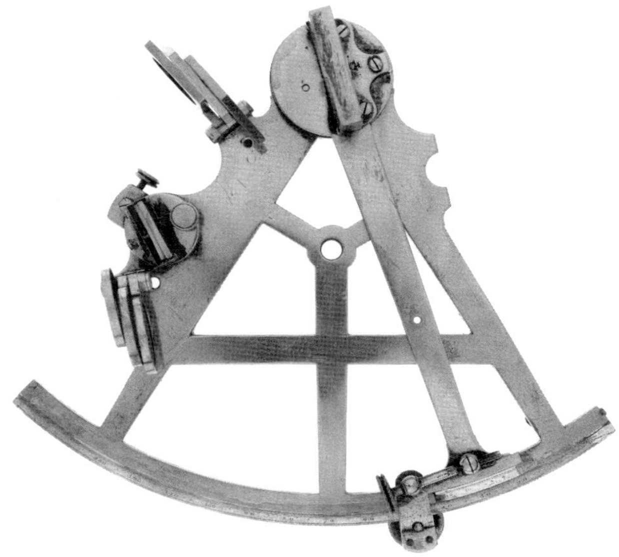

151 NAV 1106
Sextant

Double-frame bridge sextant
Origin London, England
Maker Jesse Ramsden
Date About 1795
Inscriptions Signed on the limb: Ramsden London
Marked on the right end of the limb: J. R. Napier
Marked on the back of the frame near the pivot: 1339
Graduation Bronzed limb with inlaid silver scale from -2° to 134° every 5′, measuring to 122°. Silver vernier to 5″, zero at the right.
Dimensions Radius 8¼ inches (210 mm)
Description Anodized brass straight-bar-pattern pillar frame (20 pillars) with a protective frame (bridge) over the fittings and mounts, wooden handle. The tangent screw and clamping screw are on the back of the index arm. Four shades (three red and one green); three horizon shades (two red and one green). Index-glass adjustment by screw; adjustment of horizon glass by capstan screws. Magnifier, travelling on a worm screw with a milled knob. Threaded telescope bracket with perpendicular adjustment by rising-piece and a milled knob. Telescope (129 mm), inverted image, two parallel cross-wires; telescope (126 mm), inverted image, two parallel cross-wires; telescope (59 mm), erect image. Sight-tube (89 mm), a rotating shaded eyepiece with choice of three shades (all blue), and an adjusting key. In a fitted polished mahogany keystone box with a brass plate on the lid marked 'R. Napier'. The lid contains a trade label for William Gerrard, 35 South Castle Street, Liverpool, and a circular label in MS 'Helen, Francis & Gordon gave this instrument to Father Christmas 1907'.
Provenance Belonged to Robert Napier (1791–1876), a marine engineer and shipbuilder on the river Clyde, and later to his son James Robert Napier, also a noted marine engineer. Adams collection; lent in 1936, bequeathed in 1974. The label indicates that the instrument was Adams's 1907 Christmas present from his wife Helen and their sons.
Notes Edward Troughton patented the pillar and plate frame (no. 1644 of 1788).
Literature Stimson, 'The Influence of the Royal Observatory at Greenwich', 130; Clifton, *Directory*, 227; Hill and Paget-Tomlinson, *Instruments of Navigation*, 30.

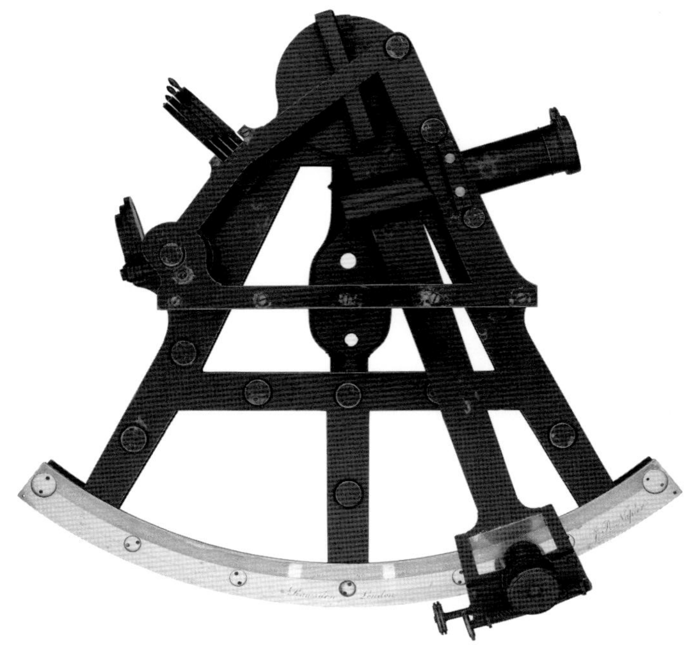

152 NAV1249
Box sextant

Pocket sextant
Origin London, England
Maker Jesse Ramsden
Date About 1796
Inscriptions Signed on the upper plate: Ramsden | London
Marked at the pivot of the magnifier: 1336
Graduation Brass scale from -2° to 106° every 30′, measuring to 89° (the digits are read from the pivot). Brass vernier to 10″, zero at the left.
Dimensions Radius 2¼ inches (57 mm)
Description Semicircular polished brass plates with on one side a sliding shutter, and on the other side a hinged plate. The index arm is moved along a tangent screw by a detached milled key fixed in the upper plate. Two index shades (red and orange) operated by brass studs moving in slots cut in the lower plate; no horizon shades. Index- and horizon-glass adjustment by screws. Magnifier (single lens) on a 51-mm swivelling arm. Telescope (26 mm), erect image, with a shaded eyepiece (red) fitted in the body. No box.
Provenance Purchased from Trevor Philip and Son Ltd, London, in 1986.
Literature Clifton, *Directory*, 227.

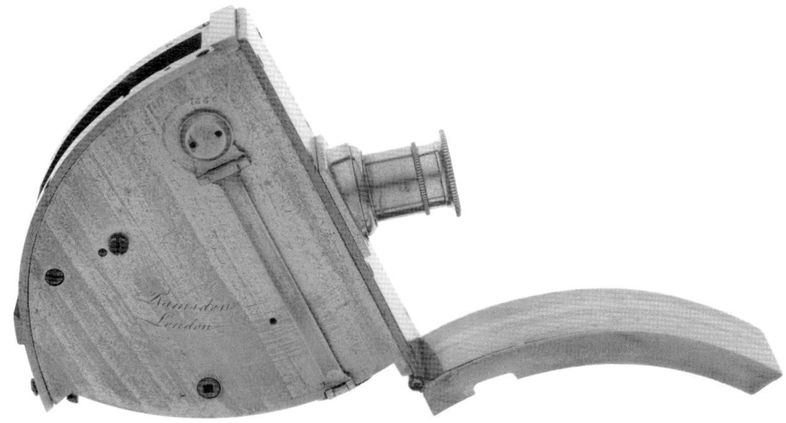

153 NAV1139
Sextant

Origin London, England
Maker (John and Edward) Troughton
Date About 1797
Inscriptions Signed on the limb: Troughton London
Marked on the central strut: 294
Marked on the bottom of the index arm: The Gift of Capt Jas Brisbane | To Mr T.H. Hoskins | Master of H.M. Ship Saturn | 1802
Graduation Polished brass scale from -5° to 145° every 15′, measuring to 125°. Brass vernier to 15″, zero at the right.
Dimensions Radius 11¾ inches (298 mm)
Description Anodized brass straight-bar-pattern pillar frame (24 pillars), wooden handle. The tangent screw and clamping screw are on the back of the index arm. Three shades (two red and one green); three horizon shades (two red and one green). Index-glass adjustment by a screw; adjustment of horizon glass by screw and a milled knob. Magnifier on a 111-mm swivelling arm. Threaded telescope bracket in two parts, fitted for correcting collimation error; perpendicular adjustment by rising-piece and a milled knob. Telescope (196 mm), inverted image, two parallel cross-wires. Sight-tube (131 mm). A magnifying glass and a shaded eyepiece are missing. In a polished mahogany keystone box, containing in the lid a printed notice by W. Heather pointing out that no parallel adjustment was made on the horizon glass and that correction must be allowed as directed by J. Norie. There are a number of illegible pencil notes inside the box.
Provenance Captain Sir James Brisbane RN (1774–1826), the original owner, entered the Navy in 1787 and was commissioned a lieutenant in 1794. Presented by Commander Douglas Hoskins RN, presumably a descendent of T.H. Hoskins, in 1937.
Notes William Heather (1793–1812) founded a navigational warehouse in Leadenhall Street, London, and John (William) Norie was a partner in the business that later became J.W. Norie and Co. Norie published the correction referred to by Heather in his navigational manuals. Edward Troughton patented the pillar and plate frame (no. 1644 of 1788).

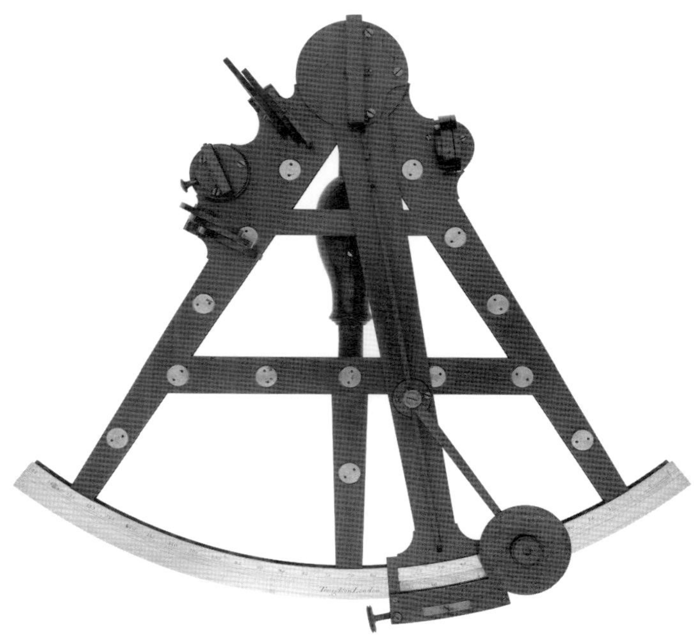

Literature Stimson, 'The Influence of the Royal Observatory at Greenwich', 130; Clifton, *Directory*, 282; McConnell, *Instrument Makers to the World*; Fisher, *The Makers of the Blueback Charts*, 74–83; J.K. Laughton, 'Brisbane, Sir James (1774–1826) *naval officer*'. In: *Oxford Dictionary of National Biography* Oxford, 2004 <http://www.oxforddnb.com/>.

154 NAV1216
Sextant

Bridge sextant
Origin London, England
Maker Jesse Ramsden
Date About 1797
Inscriptions Signed on the limb: Ramsden London
Marked on the back of the frame at the pivot: 1361
Graduation Polished brass limb with inlaid silver scale from -2° to 135° every 15′, measuring to 121°. Silver vernier to 20″, zero at the right.
Dimensions Radius 7 ½ inches (191 mm)
Description Polished brass diamond-pattern frame with a protective frame (bridge) over the fittings and mounts, wooden handle. The tangent screw and clamping screw are on the back of the index arm. Four shades (three red and one green); three horizon shades (two red and one green). Index-glass adjustment by a screw; adjustment of horizon glass by a clamp screw and a capstan screw. Magnifier, travelling on a worm screw with a milled knob (defective). Threaded telescope bracket in two parts, fitted for correcting collimation error; perpendicular adjustment by rising-piece and a milled knob. Telescope (119 mm), inverted image; extra drawtube (68 mm), inverted image, four cross-wires; extra drawtube (64 mm), erect image, with a rotating variable shaded eyepiece with choice of four shades (one clear and three red); this drawtube is probably not original to this instrument. Sight-tube (88 mm) and an adjusting pin. In a shaped, fitted wooden box, containing in the lid a trade label for J.W. Norie and Co., Leadenhall Street, London (1812–39) and illegible pencil notes dated June 1835.
Provenance Purchased from Mr D.R. Terry, Chellaston, Derby, in 1973.
Literature Stimson, 'The Influence of the Royal Observatory at Greenwich', 130; Clifton, *Directory*, 201, 227; Fisher, *The Makers of the Blueback Charts*, 83–96.

155 NAV 1107
Sextant

Double-frame bridge sextant
Origin London, England
Maker Jesse Ramsden
Date About 1798
Inscriptions Signed on the limb: Ramsden London
Marked on the back of the frame near the pivot: 1421
Graduation Polished brass limb with inlaid gold scale from -2° to 135° every 15′, measuring to 122°. Gold vernier to 20″, zero at the right.
Dimensions Radius 8 inches (203 mm)
Description Polished brass straight-bar-pattern pillar frame (18 pillars) with a protective frame (bridge) over the fittings and mounts, wooden handle. The tangent screw and clamping screw are on the back of the index arm. Four shades (three red and one green); three horizon shades (two red and one green). Index-glass adjustment by a screw; adjustment of horizon glass by a capstan screw. Threaded telescope bracket with perpendicular adjustment by rising-piece and a milled knob. Magnifier, travelling on a worm screw with a milled knob. Telescope (133 mm), inverted image. Rotating shaded eyepiece with choice of three shades (two red and one orange); telescope (175 mm), inverted image, four cross-wires. Sight-tube (89 mm) and an adjusting pin. In a shaped mahogany box.
Provenance Presented by Mr A.J. MacGregor to the Royal Naval Museum at the Royal Naval College before 1930, and on its closure in 1936 transferred to the NMM.
Notes Edward Troughton patented the pillar and plate frame (no. 1644 of 1788).
Literature Stimson, 'The Influence of the Royal Observatory at Greenwich', 130; Clifton, *Directory*, 227.

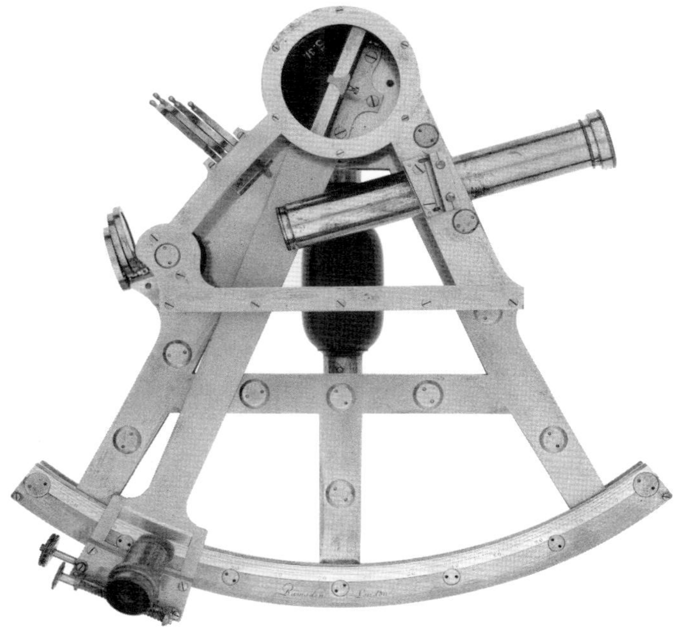

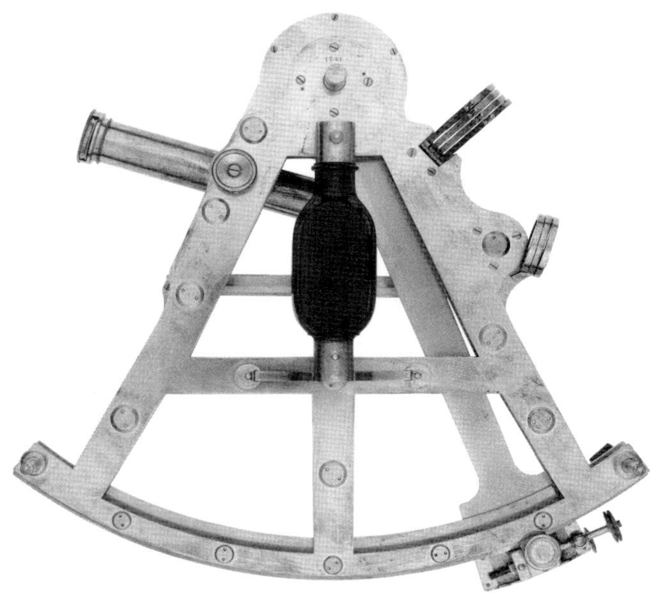

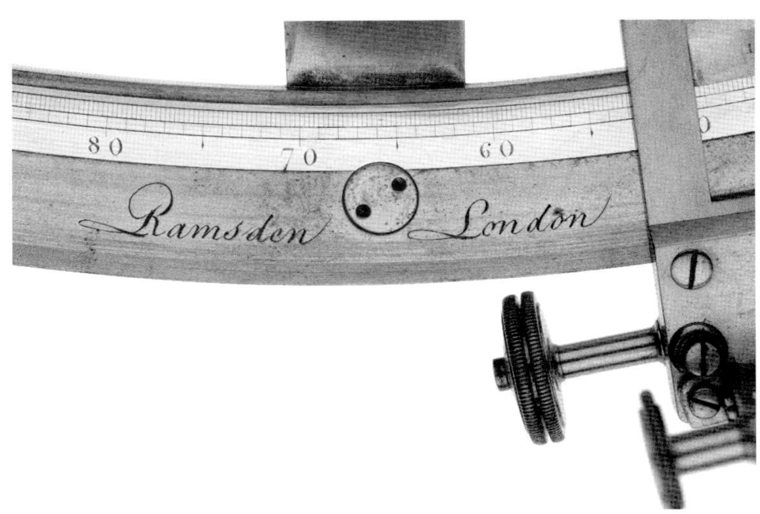

156 NAV1142
Surveyor's sextant

Origin Vienna, Austria
Maker Johann M. Ekling
Date About 1800
Inscriptions Signed on the index arm: Joh. M. Ekling in Wien
Graduation Polished brass limb with inlaid silver scale from 0° to 120° every 1°, measuring to 119°. Silver vernier to 3′, zero at the right.
Dimensions Radius 4½ inches (114 mm)
Description Polished brass frame, detached wooden handle. The clamping screw is mounted on a second silver arc on top of the frame; a milled knob at the end of that arc provides slow-motion adjustment of the index arm. No shades. Index glass without adjustment; adjustment of horizon glass by a screw. Threaded telescope bracket without adjustment. Telescope (63 mm), inverted image, shaded eyepiece (green). Sight-tube (44 mm). In a leather-covered case that is decorated with stamped gilt flower motifs and lined with red velvet.
Provenance Presented by Sir James Caird in 1938; purchased as part of the Landau collection through Maggs Brothers.

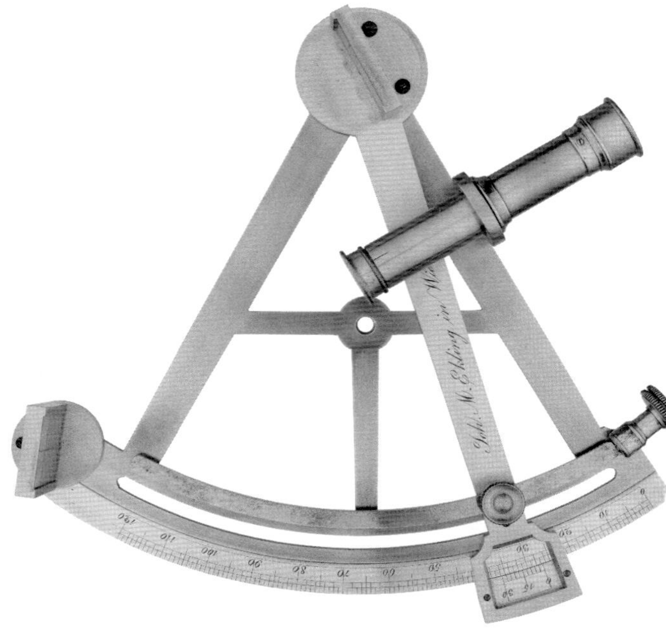

157 NAV1199
Sextant

Origin London, England
Maker Spencer Browning and Rust
Date About 1800
Inscriptions Signed on the inlaid plate on the crossbar: Spencer Browning & Rust London
Marked (stamped) on the scale: SBR
Graduation Inlaid ivory scale from -2° to 136° every 20′, measuring to 120°. Ivory vernier to 1′, zero at the right.
Dimensions Radius 11¾ inches (298 mm)
Description Ebony frame and limb with brass index arm and fittings; wooden handle, brass stop for the index arm; an inlaid ivory plate on the crossbar. The tangent screw is on the front of the index arm; the clamping screw is on the back. Three shades (two red and one green); three horizon shades (two red and one green). Index-glass adjustment by a screw; adjustment of horizon glass by a milled knob, and an extended brass lever and clamp operated from beneath the telescope bracket. Threaded telescope bracket with a swivelling sight vane and one pinhole; perpendicular adjustment by rising-piece and a milled knob. Telescopes are missing. In a shaped mahogany box (probably not original to this instrument) with a stepped lid.
Provenance Presented to HMS *Dryad* Museum by Captain P.J. Wyatt RN in 1958. Mr Karl Ekborg, a Swedish shipowner, had given it to Captain Wyatt; it had been used on Swedish ships owned by Mr Ekborg's father. Presented by HMS *Dryad* Museum, Southwick, Hampshire, in 1967.
Notes Peter Dollond patented the horizon glass adjustment (no. 1017 of 1772).
Literature Dollond and Dollond, *Directions for Using the New Improved Hadley's Quadrant*; Ifland, *Taking the Stars*, 84–5; Clifton, *Directory*, 261.

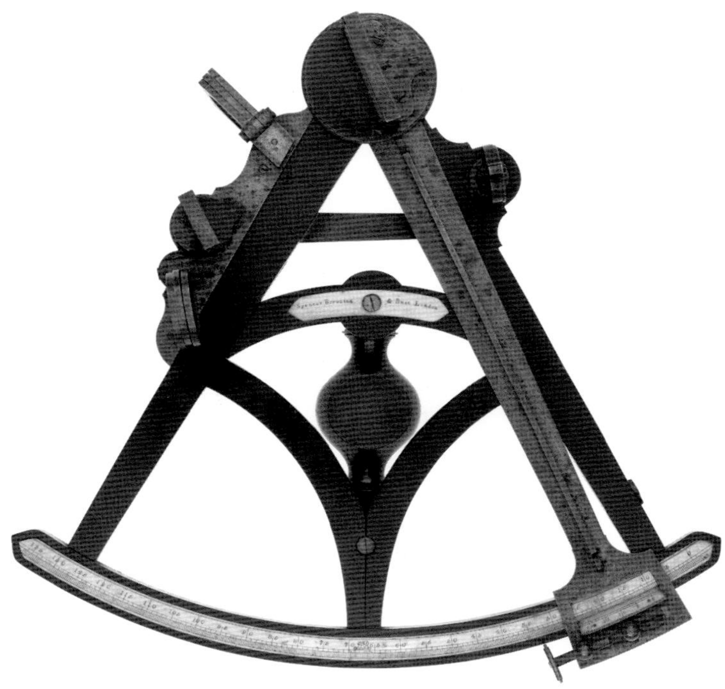

158 NAV1234
Box sextant

Pocket sextant
Origin London, England
Maker Dudley Adams
Date About 1800
Inscriptions Signed on the upper plate: Adams London [read from the pivot]
Graduation Polished brass scale from -3° to 149° every 30′, measuring to 135° (the digits are read from the pivot). Brass vernier to 1′, zero at the left.
Dimensions Radius 3½ inches (89 mm); diameter 101 mm
Description Circular polished brass plates with a lid that, when fitted on the back of the instrument, serves as a handle. Index arm moved by a milled knob, fixed in the upper plate. One index shade (red) moved by a lever; no horizon shades. Index glass without adjustment; adjustment of horizon glass by screws. Magnifier (single lens) on a 65-mm swivelling arm. Pinhole sight with sliding shaded shutter (red). No box or case.
Provenance Purchased from Mr A. Gray, London, in 1981.
Literature Clifton, *Directory*, 2; Millburn, *Adams of Fleet Street*, 273–306.

159 NAV1143
Sextant

Origin London, England
Maker Ripley and Son
Date About 1805
Inscriptions Signed on the limb: Ripley and Son, Hermitage, London
Graduation Polished brass limb with inlaid silver scale from -5° to 140° every 10′, measuring to 126°. Silver vernier to 10″, zero at the right.
Dimensions Radius 8¾ inches (222 mm)
Description Polished brass oval-pattern frame, wooden handle. The tangent screw and clamping screw are on the back of the index arm. Four shades (one green and three red); three horizon shades (one green and two red). Index-glass adjustment by a screw; adjustment of horizon glass by thumbscrew and capstan screws. Magnifier on an 80-mm swivelling arm. Threaded telescope bracket in two parts, fitted for correcting collimation error; perpendicular adjustment by rising-piece and a milled knob. Telescope (83 mm), erect image. Sight-tube (81 mm). A Shaded eyepiece and adjusting pin are missing. In a mahogany keystone box.
Provenance Adams collection; lent in 1946, bequeathed in 1974.
Literature Clifton, *Directory*, 233; Dunn, 'Touching and Cleaning'.

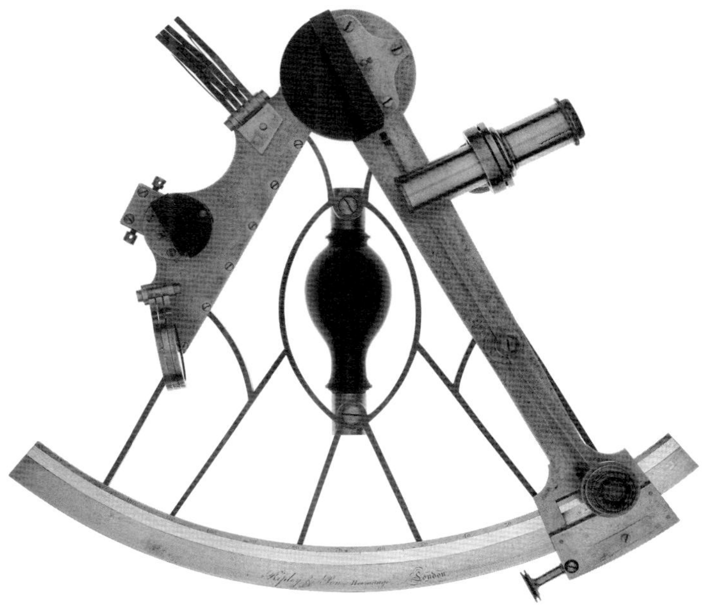

160 NAV1235
Sextant

Hoppe's improved sextant
Origin London, England
Maker John Hinde
Date About 1805
Inscriptions Signed on the upper index arm: Ino Hinde London
Graduation The lower inlaid silver scale is from -2° to 144° every 15′, measuring to 125°; the upper inlaid silver scale is from 143° to 0° to 137° every 20′, measuring to 102° and 116°. Lower vernier to 15′, zero at the right; upper vernier to 10′, zero at the centre.
Dimensions Radius of the lower arc 8¾ inches (222 mm); radius of the upper arc 3 inches (76 mm)
Description Polished brass diamond-pattern frame with two polished limbs and two index arms, wooden handle. The tangent screws and clamping screws are on the back of both index arms. Four shades (three red and one green); three horizon shades (two red and one green). Index-glass adjustment by a screw; adjustment of horizon glass by a worm screw operated by a capstan screw. The swivelling arm of the lower magnifier is broken, the magnifier is missing, and the upper magnifier is on a 45-mm swivelling arm. Threaded telescope bracket with perpendicular adjustment by rising-piece and a milled knob. Telescope (162 mm), inverted image; telescope (92 mm), erect image. Sight-tube (86 mm), a shaded eyepiece (red), an adjusting pin, and a screwdriver. In a wooden keystone box.
Provenance Purchased from Mr P.A. Monks, Eastleigh, Hampshire, in 1982.
Notes Hoppe's improvement consists of an index glass that can be rotated by a short index arm that moves over the second scale, mounted on top of the frame. By this means, the theoretical range of the instrument is increased to 265 degrees, but in practice probably not to over 160 degrees.
Literature NN, *An Explanation of E Hoppe's Improved Sextant*; Ifland, *Taking the Stars*, 29–31; Clifton, *Directory*, 137.

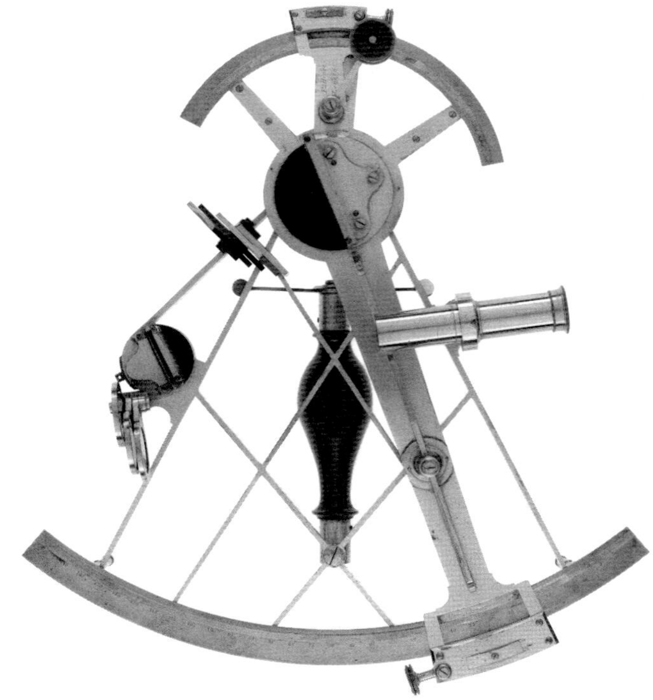

161 NAV1141
Sextant

Origin France
Maker Unknown
Date About 1810
Inscriptions None
Graduation Inlaid ivory scale from -7° to 133° every 20′, measuring to 117°. Ivory vernier to 1′, zero at the right.
Dimensions Radius 9 inches (229 mm)
Description Ebony frame and limb with brass index arm and fittings; wooden handle; an inlaid ivory plate on the crossbar. The tangent screw and clamping screw are on the back of the index arm. Three shades (two red and one green); two horizon shades (red and green). Index- and horizon-glass adjustment by square-headed screws and a detached key. Magnifier (single lens) on a 90-mm hinged swivelling arm. Threaded telescope bracket in two parts, fitted for correcting collimation error; perpendicular adjustment by rising-piece and a milled knob. Sight-tube (81 mm). The telescope, shaded eyepiece, and adjusting key are missing. In a wooden keystone box, containing in the lid a trade label for Hubé, Méchanicien, Rue Basse Riche no. 6 (the street and number have been added in MS), Brest.
Provenance Presented by Sir James Caird in 1938; purchased as part of the Landau collection through Maggs Brothers.
Notes Hubé Frères (rather than Hubé) are recorded as being established at this address in the nineteenth century.
Literature Marcelin, *Dictionnaire des fabricants français*.

162 NAV 1144
Sextant

Origin London, England
Maker Matthew Berge
Date About 1810
Inscriptions Signed on the limb: Berge London, late Ramsden
Marked on the back of the frame: 1695
Graduation Polished brass limb with inlaid silver scale from -2° to 148° every 20′, measuring to 135°. Silver vernier to 30″, zero at the right.
Dimensions Radius 5 inches (127 mm)
Description Polished solid ¼-inch-thick brass frame, detached wooden handle. The tangent screw is on the front of the index arm; the clamping screw is on the back. Three shades (two red and one green); three horizon shades (two red and one green). Index- and horizon-glass adjustment by screws. Magnifier (detached) on a 55-mm swivelling arm. Threaded telescope bracket with perpendicular adjustment by rising-piece and a milled knob. Telescope (133 mm), inverted image, four cross-wires; telescope (59 mm), erect image. Sight-tube (57 mm), and a silver reflector fitting into the index arm clamp to throw light on the scale. In a fitted mahogany box.
Provenance Lent by Major F.A. Yorke, London, in 1947 and presented by him in 1959.
Literature Stimson, 'The Influence of the Royal Observatory at Greenwich', 130; Clifton, *Directory*, 28.

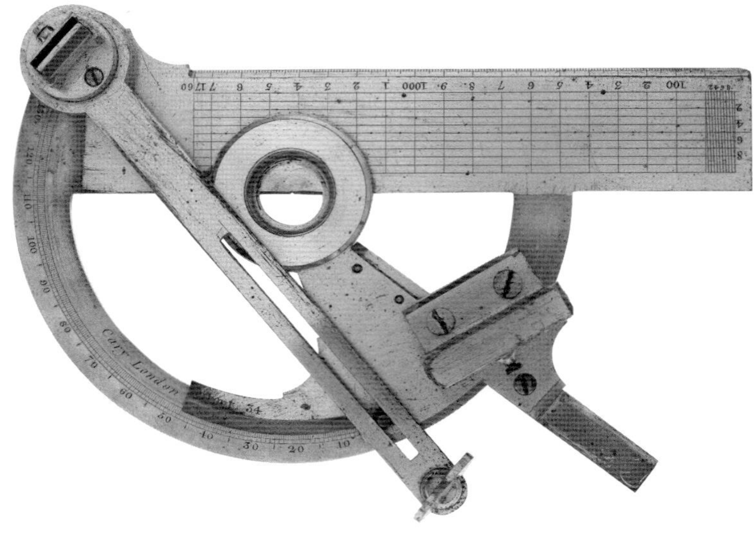

163 NAV0140
Sextant protractor

Douglas reflecting protractor
Origin London, England
Maker William Cary
Date About 1811
Inscriptions Signed on the limb of the protractor: Cary London Marked on the index glass mounting plate: Patent 34
Graduation Brass scale from 0° to 130° every 30′, measuring to 121°. Brass vernier to 1′, zero at the right.
Dimensions Overall length 148 mm
Description Semi-circular brass protractor and straight edge. There is no tangent or clamping screw. The base is a straight edge is divided from 0 to 1760 and carries a diagonal scale to 1 yard. No shades. Index- and horizon-glass adjustment by screws. The index glass is mounted on an arm that moves along the protractor and carries the vernier; the horizon glass and the sight vane are mounted at either end of a slotted index arm. The slot rides over a pin that is fixed to the arm of the index glass. In a fitted leather case marked in ink 'G Talbot'.
Provenance Purchased from Delehar Antiques, London, in 1985.
Notes Sir Howard Douglas (1776–1861) designed this instrument (patent no. 3461 of 1811) for land and marine surveying; with this instrument, the magnitude of an angle may be copied without reading it off. Sir Howard, son of the equally inventive Admiral Sir Charles Douglas, was an army officer who studied navigation shortly before he designed this instrument, for which he was elected an FRS in 1816.
Literature Ifland, *Taking the Stars*, 134; Clifton, *Directory*, 51; H.M. Chichester, 'Douglas, Sir Howard, third baronet (1776–1861) *army officer and colonial officer*' in *Oxford Dictionary of National Biography*, Oxford, 2004.,<http://www.oxforddnb.com/>.

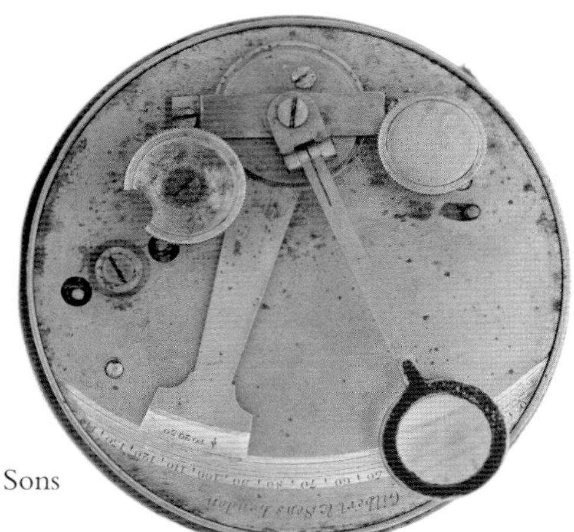

164 NAV1220
Box sextant

Pocket sextant
Origin London, England
Maker Gilbert and Sons
Date About 1815
Inscriptions Signed on the upper plate: Gilbert & Sons London
Graduation Polished brass limb with inlaid silver scale from -5° to 145° every 30′, measuring to 129° (the digits are read from the pivot). Silver vernier to 1′, zero at the left.
Dimensions Radius 1¾ inches (44 mm); diameter 67 mm
Description Circular polished brass plates with a lid that, when fitted on the back of the instrument, serves as a handle. Index arm moved by a milled knob on the upper plate. One index shade (red) moved by a lever; no horizon shades. Index glass without adjustment; adjustment of horizon glass by a milled knob fixed in the upper plate. Magnifier (single lens) on a 45-mm swivelling arm. Sliding shutter with choice of a pinhole or a shade (red). Two screws to hold the upper plate are missing. In a rectangular wooden box, also containing an artificial glass horizon (cat. no. 318) and a pocket compass (NAV0466).
Provenance Belonged to Edward Newell Harrison, clerk of HMS *Assistance*, who went to the Arctic in 1850–1, in search of Sir John Franklin, and died in the West Indies before 1855. Presented by Miss E.V. Trenerry, Perivale, Middlesex, a great-granddaughter of Harrison, in 1975. She also presented cat. no. 318.
Literature Clifton, *Directory*, 113; Savours, *The Search for the North West Passage*, 193–4.

165 NAV1155
Sextant

Bridge sextant
Origin London, England
Maker Matthew Berge
Date About 1819
Inscriptions Signed on the limb: M. Berge, London
Marked on the back of the frame near the pivot: 1966
Graduation Polished brass limb with inlaid silver scale from -2° to 136° every 5′, measuring to 121°. Silver vernier to 20″, zero at the right.
Dimensions Radius 7¾ inches (197 mm)
Description Anodized brass diamond-pattern frame with a protective frame (bridge) over the fittings and mounts; wooden handle. The tangent screw and clamping screw are on the back of the index arm. Four shades (three red and one green); three horizon shades (two red and one green). Index-glass adjustment by a screw; adjustment of horizon glass by capstan screws. Magnifier, travelling on a worm screw with a milled knob. Threaded telescope bracket with perpendicular adjustment by rising-piece and a milled knob. Telescope (135 mm), inverted image, four cross-wires; telescope (178 mm), inverted image, four cross-wires. Rotating shaded eyepiece with choice of three shades (two red and one green). Sight-tube (92 mm) and an adjusting pin. In a mahogany keystone box, containing in the lid a trade label for W.C. Cox, 87 Fore Street, Devonport (1851), stuck over a trade label for George Stebbing (d 1846), High Street, Portsmouth, in which he announces that his son George James Stebbing was continuing the business.
Provenance Formerly the property of A.H. Oliver and Admiral R.A. Oliver; on loan from one of their descendants since 1939.
Literature Stimson, 'The Influence of the Royal Observatory at Greenwich', 130; Clifton, *Directory*, 28, 69, 263–4.

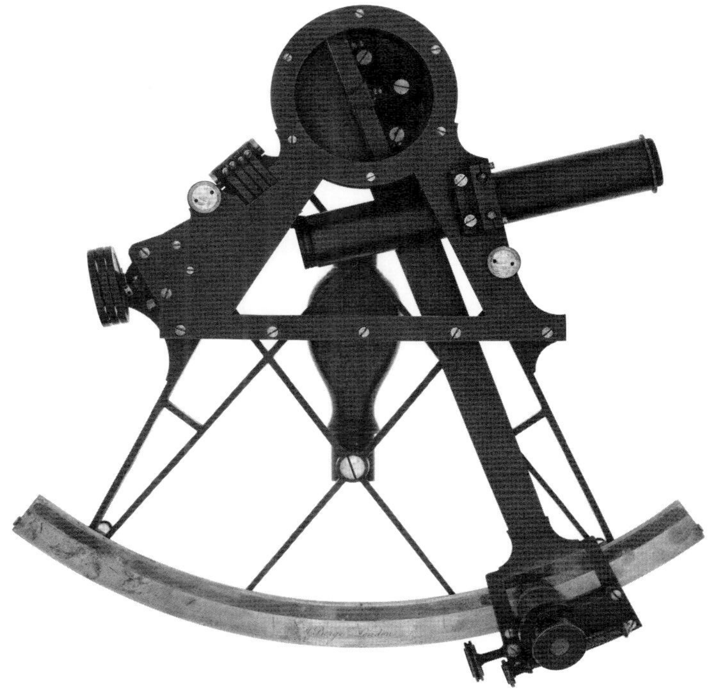

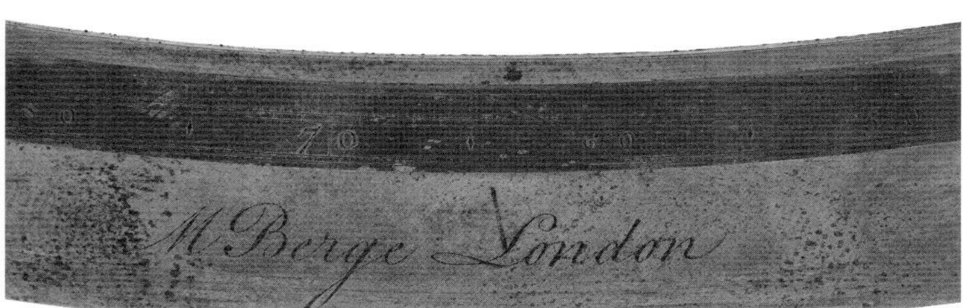

166 NAV1156
Sextant

Origin London, England
Maker Edward Troughton
Date About 1820
Inscriptions Signed on the limb: Troughton London
Marked on the limb: Platina
Marked on the central strut: 1003
Graduation Polished brass limb with inlaid platinum scale from -5° to 150° every 10′, measuring to 125°. Platinum vernier to 10″, zero at the right.
Dimensions Radius 8¼ inches (210 mm)
Description Polished brass straight-bar-pattern pillar frame (20 pillars), wooden handle. The tangent screw and clamping screw are on the back of the index arm. Four shades (two red, one orange and one green); three horizon shades (red, orange, and green). Index-glass adjustment by a screw; adjustment of horizon glass (oval glass in square mount) by a square-headed screw and a detached key. Magnifier on a 70-mm swivelling arm. Threaded telescope bracket in two parts, fitted for correcting collimation error; perpendicular adjustment by rising-piece and a milled knob. Telescope (74 mm), erect image; telescope (182 mm), inverted image, two parallel cross-wires; extra drawtube (90 mm), erect image, four cross-wires. Sight-tube (126 mm), two shaded eyepieces (both orange), and a milled adjusting key. In a polished mahogany keystone box, containing in the lid a trade label for W.C. Cox, 93 Fore Street, Devonport and 35 South Side Street, Plymouth, and a repair label for Troughton and Simms, 138 Fleet Street, London, dated September 1915.
Provenance Formerly the property of A.H. Oliver and on loan from a descendant since 1939.
Notes Edward Troughton patented the pillar and plate frame (no. 1644 of 1788).
Literature Stimson, 'The Influence of the Royal Observatory at Greenwich', 130; Clifton, *Directory*, 281; McConnell, *Instrument Makers to the World*; Chaldecott, 'Platinum and Palladium in Astronomy and Navigation'.

167 NAV 1198
Sextant

Origin London, England
Maker James Gilkerson and Co.
Date About 1820
Inscriptions Signed on the index arm: Gilkerson & C°. | Tower Hill London.
Graduation Inlaid ivory scale from -3° to 136° every 20′, measuring to 118°. Ivory vernier to 30′, zero at the right.
Dimensions Radius 7.85 inches (199 mm)
Description Ebony frame and limb with a black-lacquered brass index arm and fittings; wooden handle; an inlaid ivory plate on the crossbar. The tangent screw and clamping screw are on the back of the index arm. Four shades (three red and one green); three horizon shades (two red and one green). Index-glass adjustment according to Gabriel Wright's patent, by which the mount can be rotated around its pivot by screws for parallel adjustment; horizon-glass adjustment by a screw and a thumbscrew. Threaded telescope bracket with perpendicular adjustment by rising-piece and a milled knob. Telescope (171 mm), inverted image, four cross-wires; extra drawtube (47 mm), erect image, with a swivelling shaded eyepiece (red). Sight-tube (67 mm), a rotating shaded eyepiece with choice of four shades (clear, two red, one green). One foot of the frame is broken and several unidentified parts are missing. In a polished mahogany keystone box, containing in the lid a trade label for Wilson and Gillie, New Quay, North Shields.
Provenance Presented to HMS *Dryad* Museum by Mr G. R. Holloway (Extra Master) in 1961: it had belonged to his grandfather. Presented by HMS *Dryad* Museum, Southwick, Hampshire, in 1967.
Notes Gabriel Wright patented the index-glass adjustment (no. 1229 of 1779).
Literature Clifton, *Directory*, 113.

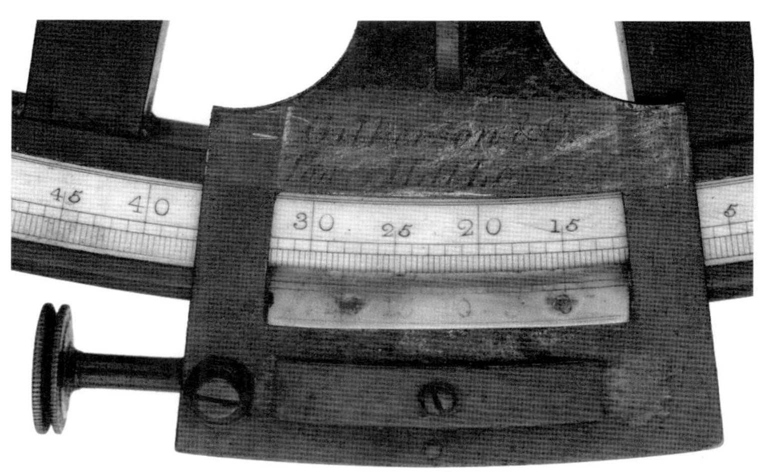

168 NAV1203
Box sextant

Pocket sextant
Origin London, England
Maker Charles Augustus Schmalcalder
Date About 1820
Inscriptions Signed on the upper plate: Schmalcalder, 82, Strand, London. [read from the pivot]
Marked on the back of the instrument: J. Lort Stokes
Graduation Inlaid silver scale from -5° to 152° every $30'$, measuring to 139° (the digits are read from the pivot). Silver vernier to $1'$, zero at the left.
Dimensions Radius 1¾ inches, (44 mm); diameter 79 mm
Description Circular polished brass plates with a lid that, when screwed on the back of the instrument, serves as a handle. Index arm moved by a milled knob on the upper plate. Two index shades (red and green) moved by levers; no horizon shades. Index-glass adjustment by a screw; adjustment of horizon glass by a square-headed screw, using a milled key fixed in the upper plate. Magnifier (single lens) on a 75-mm swivelling arm attached at the pivot. Sight-tube (19 mm) fitted in the body of the instrument. In a leather case marked 'D.146', with a sling strap.
Provenance John Lort Stokes (1812–85), the original owner, was an officer in the Royal Navy who spent most of his career as ahydrographic surveyor. The instrument remained in his family until purchased as part of the Lort Stokes collection, through Evans Roach and Co., Solicitors, Haverfordwest, in 1960.
Notes D.146 was the Hydrographic Office serial numbering adopted in February 1828, where D was the letter given to sextants.
Literature Clifton, *Directory*, 245–6; Stimson, 'Some Board of Longitude instruments', 114–15; J.K. Laughton, 'Stokes, John Lort (1812–85) *naval officer*', in *Oxford Dictionary of National Biography*, Oxford, 2004, <http://www.oxforddnb.com/>.

169 NAV1355
Sextant with artificial horizon

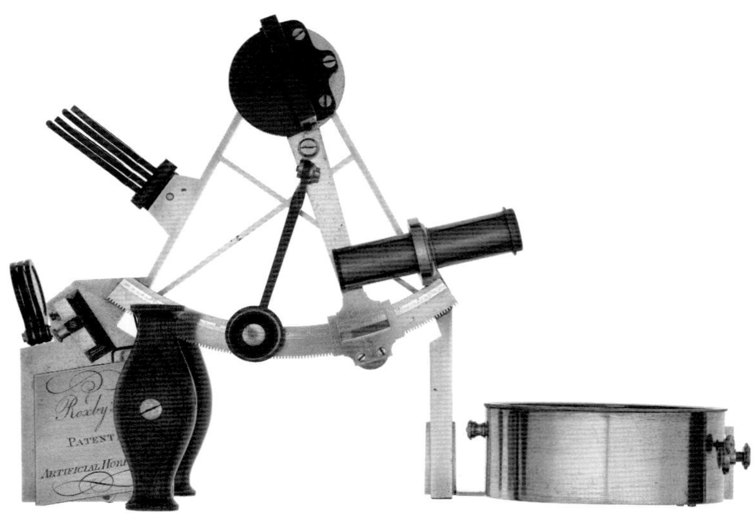

Roxby's Pendulum Quadrant
Origin England
Maker Unknown
Date 1822 or later
Inscriptions Marked on the cover of the artificial horizon: Roxby's | Patent, | Artificial Horizon.
Graduation Polished limb with inlaid silver scale from -5° to 135° every $10'$, measuring to 116°. Silver vernier to $10''$, zero at the right.
Dimensions Radius 5 inches (127 mm)
Description Polished brass single cross-pattern frame and fittings; two wooden handles on either side of the frame. The limb has rack teeth cut in its underside; the pinion wheel (engaging in the rack on the limb) is on the back of the index arm. Four shades (three red and one green); three horizon shades (two red and one green). Index-glass adjustment by a screw; adjustment of horizon glass by a capstan screw and a capped capstan screw. Magnifier on an 80-mm swivelling arm. Threaded telescope bracket without adjustment. Telescope (160 mm), inverted image, with four cross-wires; telescope (81 mm), erect image. Sight-tube (81 mm), shaded eyepiece (red), an eyepiece without glass, and a spare indicator for the artificial horizon. In a fitted mahogany box with a brass carrying handle.
Artificial horizon This consists of a pendulum attached to the lower part of the instrument frame, with an indicator that extends above the pendulum weight and is visible in front of the horizon glass. The compass bowl (diameter 106 mm) is fitted with a mounting bracket to attach it to the instrument; it has a card stop mechanism and two lubber lines. Paper card (diameter 89 mm) covered with mica, and with a jewelled pivot cap at the centre. The card is graduated into eight points, and in degrees in quadrants on a silver ring. The needle (106 mm) is a flat-bar type attached below the card, floating on a steel pin in a brass mounting.
Provenance Purchased from Mr. T. Mackay, London, in 1983.
Notes Robert Benton Roxby patented his artificial horizon (no. 4695 of 1822).
Literature Clifton, *Directory*, 239.

170 NAV1116
Sextant

Origin Liverpool, England
Maker Thomas Jones
Date About 1825
Inscriptions Signed on the limb: Thomas Jones Liverpool.
Graduation Polished brass limb with inlaid silver scale from -3° to 155° every 15′, measuring to 130°. Silver vernier to 15″, zero at the right.
Dimensions Radius 5 inches (127 mm)
Description Anodized brass diamond-pattern frame, wooden handle. The tangent screw and clamping screw are on the back of the index arm. Four shades (three red and one green); three horizon shades (two red and one green). Index-glass adjustment by a screw; adjustment of horizon glass by a square-headed screw and a detached key, and two capstan screws. Magnifier on an 80 mm swivelling arm. Threaded telescope bracket in two parts, fitted for correcting collimation error; perpendicular adjustment by rising-piece and a milled knob. Telescope (83 mm), erect image; telescope (162 mm), inverted image, two parallel cross-wires. Sight-tube (75 mm), a shaded eyepiece (red), a second shaded eyepiece (grey) from another instrument, a screwdriver, an adjusting pin, and a milled adjusting key; two unidentified parts are missing. In a wooden keystone box, containing in the lid a trade label for John Lilley and Son, London. Inside the lid in MS 'E.T. Adams 1906' and a small label in MS '40 TAL'.
Provenance Adams collection; lent in 1936, bequeathed in 1974. The MS label may suggest that Adams acquired the instrument in 1906.
Literature Clifton, *Directory*, 154; Morrison-Low, *Making Scientific Instruments*, 330.

171 NAV1146
Box sextant

Pocket sextant
Origin Edinburgh, Scotland
Maker Alexander Adie
Date About 1830
Inscriptions Signed on the upper plate: ADIE, Edinburgh
Graduation Inlaid silver scale from -5° to 165° every 30′, measuring to 137° (the digits are read from the pivot). Silver vernier to 1′, zero at the left.
Dimensions Radius 1.85 inches (47 mm); diameter 76 mm
Description Circular polished brass plates with a lid that, when screwed on the back of the instrument, serves as the handle. The tangent screw and clamping screw are on the front of the index arm, which is operated on a brass quadrant mounted at the pivot point. Two index shades (green and red) moved by levers; no horizon shades. Index glass without adjustment; adjustment of horizon glass by a screw and by a milled key fixed in the upper plate. Magnifier (single lens) on a 50-mm swivelling arm. Telescope (51 mm), erect image, fitted within the body of the instrument. The pinhole sight is operated by a lever contained in the instrument for use when the telescope is removed.
Provenance Presented to the Royal Naval Museum by Captain R.C. Davenport RN, and transferred from the Royal Naval College, Greenwich, in 1936 on its closure.
Literature Clarke *et al.*, *Brass & Glass*, 25–84; Clifton, *Directory*, 3–4.

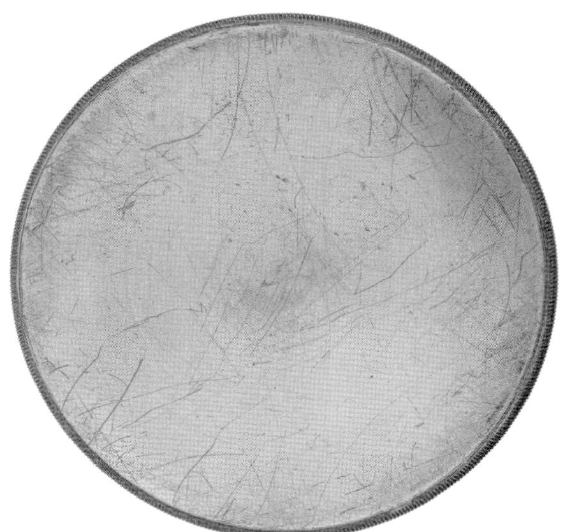

172 NAV1201
Sextant

Origin London, England
Maker Thomas Jones
Date About 1830
Inscriptions Signed on the limb: Thomas Jones 62 Charing Cross London
Graduation Polished brass limb with inlaid silver scale from -5° to 145° every 10′, measuring to 130°. Silver vernier to 10″, zero at the right.
Dimensions Radius 8 inches (203 mm)
Description Anodized brass straight-bar-pattern pillar frame (20 pillars), wooden handle. The tangent screw and clamping screw (latter incomplete) are on the back of the index arm. Four shades (two red, one orange, and one green); three horizon shades (green, orange, and red). Index- and horizon-glass adjustment by screws. Magnifier (incomplete) on a 95 mm swivelling arm (not original to this instrument). Threaded telescope bracket in two parts, fitted for correcting collimation error; perpendicular adjustment by rising-piece and a milled knob. Telescope (175 mm, incomplete); telescope (84 mm, incomplete). Sight-tube (76 mm). Two further telescopes and shaded eyepieces are missing. In a wooden keystone box with a brass plate on the lid marked 'Lord John Hay R.N.' The lid contains a trade label for J.C. Dennis, 122 Bishopsgate Street, London (1850–62).
Provenance Two Lords John Hay could have owned this instrument. The elder (1793–1851) rejoined the Royal Navy in 1832 and his nephew (1827–1916) joined in 1839. Presented by HMS *Dryad* Museum, Southwick, Hampshire, in 1967.
Notes Edward Troughton patented the pillar and plate frame (no. 1644 of 1788).
Literature Clifton, *Directory*, 154; J.K. Laughton, 'Hay, Lord John (1793–1851) *naval officer and politician*', and Andrew Lambert, 'Hay, Lord John (1827–1916) *naval officer and politician*', in *Oxford Dictionary of National Biography*, Oxford, 2004, <http://www.oxforddnb.com/>.

173 NAV1233
Box sextant

Pocket sextant
Origin London, England
Maker Edward Troughton and William Simms
Date About 1830
Inscriptions Signed on the upper plate: Troughton & Simms, London [read from the pivot]
Graduation Inlaid silver scale from -5° to 160° every 30′, measuring to 123° (the digits are read from the pivot). Silver vernier to 30″, zero at the left.
Dimensions Radius 3 inches (76 mm); diameter 108 mm
Description Circular polished brass plates with a lid that, when screwed on the back of the instrument, serves as a handle. Index arm moved by a brass stud at the end of the arm, and fitted with a clamping screw and tangent screw, which operate in a slot cut in the upper plate. No shades. Index glass without adjustment; adjustment of horizon glass by square-headed screws with a milled screw fixed in the upper plate. Magnifier (single lens) on an 89-mm swivelling arm attached at the pivot. A pinhole sight.
Provenance Purchased from Delehar Antiques, London, in 1980.
Literature Clifton, *Directory*, 282; McConnell, *Instrument Makers to the World*.

174 NAV1174
Sextant

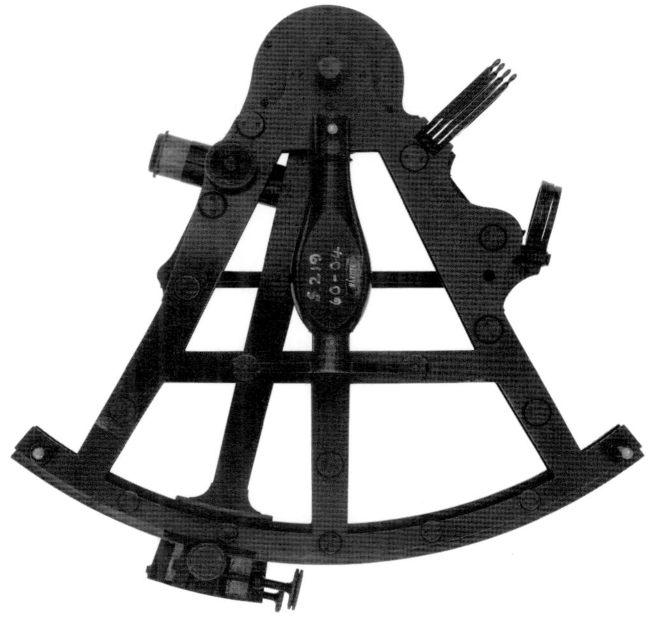

Double-frame bridge sextant
Origin London, England
Maker Nathaniel Worthington and James Allan
Date About 1831
Inscriptions Signed on the limb: Worthington & Allan. London
Marked on the back of the frame near the pivot: 2029
Marked on the bridge: Presented to J. Lort Stokes by Capt. R. Fitzroy. 1834
Graduation Polished brass limb with inlaid silver scale from -2° to 139° every 15′, measuring to 123°. Silver vernier to 15″, zero at the right.
Dimensions Radius 8 inches (216 mm)
Description Black-lacquered brass straight-bar-pattern pillar frame (20 pillars) with a protective frame (bridge) over the fittings and mounts; wooden handle. The tangent screw and clamping screw are on the back of the index arm. Four shades (two red, one orange, and one green); three horizon shades (red, orange, and green). Index-glass adjustment by a screw; adjustment of horizon glass by a capstan screw and a capped screw. Magnifier, travelling on worm screw with a milled knob. Threaded telescope bracket with perpendicular adjustment by rising-piece and a milled knob. Telescope (173 mm), inverted image, four cross-wires; telescope (145 mm), inverted image, four cross-wires; telescope (65 mm), erect image. Rotating shaded eyepiece with choice of three shades (two red and one green), an adjusting pin, and a piece of chamois leather. In a mahogany keystone box, containing in the lid a trade label for J.D. Potter, 31 Poultry, successor to R. Bate (1851–80). Illegible pencilled calculations are found in the lid. The box is fitted in a leather carrying case with a leather strap.
Provenance Captain (later Admiral) Robert Fitzroy, the original owner, commanded the survey vessel *Beagle* on two major voyages, from 1828 to 1830 and from 1831 to 1836. John Lort Stokes (1812–85) served under Fitzroy on both voyages, and himself became a notable surveyor. Perhaps Fitzroy presented this instrument because 1834 marked the tenth anniversary of Stokes's entry into the Navy.
It remained in his family until purchased as part of the Lort Stokes collection, through Evans Roach and Co., Solicitors, Haverfordwest, in 1960.
Notes The serial number 2029 indicates the year of manufacture as being around 1831, which suggests that Fitzroy bought the instrument for the second *Beagle* voyage, during which he gave it to Stokes. Edward Troughton patented the pillar and plate frame (no. 1644 of 1788).
Literature Stimson, 'The Influence of the Royal Observatory at Greenwich', 130; Clifton, *Directory*, 305; McConnell, *R B Bate of the Poultry*, 53–4; J.K. Laughton, 'Stokes, John Lort (1812–85) *naval officer*, in *Oxford Dictionary of National Biography*, Oxford, 2004, <http://www.oxforddnb.com/>.

175 NAV1151
Sextant

Origin London, England
Maker William Parnell
Date About 1835
Inscriptions Signed on the inlaid plate in the crossbar: Parnell. London.
Marked (stamped) at the centre of the scale (at 67°): [foul anchor]
Graduation Inlaid ivory scale from -2° to 134° every 20′, measuring to 120°. Ivory vernier to 30″, zero at the right.
Dimensions Radius 8¾ inches (222 mm)
Description Ebony frame and limb with brass index arm and fittings; wooden handle; an inlaid ivory plate in the crossbar. The tangent screw is on the front of the index arm; the clamping screw is on the back. Four shades (three red and one green); three horizon shades (two red and one green). Index-glass adjustment by a screw; adjustment of horizon glass by a screw, lever, worm gear, and milled screw. Threaded telescope bracket with perpendicular adjustment by rising-piece and a milled screw. Telescope (125 mm), inverted image, two parallel cross-wires. Sight-tube (79 mm). In a wooden keystone box, containing a trade label for William Parnell, 94 near the Bridge, Lower East Smithfield, London (1816–39).
Provenance Presented by Commander Sinclair Thompson RN, Exmouth, in 1945. Correspondence in the box suggested he might have purchased it from B.O. Woodcock in 1944.
Notes According to Hill and Paget-Tomlinson, this instrument came from an old Dundee whaler.
Literature Clifton, *Directory*, 209–10; Hill and Paget-Tomlinson, *Instruments of Navigation*, 32.

176 NAV1154
Box sextant

Pocket sextant
Origin London, England
Maker Thomas Jones
Date About 1840
Inscriptions Signed on the upper plate: Thomas Jones, 62 Charing Cross, London [read from the pivot]
Graduation Inlaid silver scale from -2° to 134° every 30′, measuring to 125° (the digits are read from the pivot). Silver vernier to 1′, zero at the left.
Dimensions Diameter 2½ inches (64 mm); radius 36 mm
Description Circular dark-lacquered brass plates with a lid that, when screwed on the back of the instrument, serves as a handle. The index arm is moved by a central screw. Two index shades (red and green) moved by levers; no horizon shades. Index glass without adjustment; adjustment of horizon glass by a milled key fixed in the upper plate. Magnifier (single lens) on a 50 mm swivelling arm. Telescope (51 mm), erect image, fixed in the instrument; fitted pinhole, with sighting shield with sliding retraction. Shaded eyepiece (red). In a round leather carrying case (strap missing) marked on the lid 'R. A. Oliver'. A paper label stuck on the side of the case is marked 'A. H. Oliver R.N'.
Provenance Formerly the property of Admiral R.A. Oliver and A.H. Oliver; on loan from one of their descendants since 1939.
Literature Clifton, *Directory*, 154.

177 NAV1157
Sextant

Origin London, England
Maker Edward Troughton and William Simms
Date About 1840
Inscriptions Signed on the limb: Troughton & Simms, London.
Marked on the limb: Silver
Marked on the central strut: 2739
Graduation Polished brass limb with inlaid silver scale from -5° to 157° every 10′, measuring to 131°. Silver vernier to 10″, zero at the right.
Dimensions Radius 7.9 inches (201 mm)
Description Anodized brass straight-bar-pattern pillar frame (20 pillars), wooden handle. The tangent screw and clamping screw are on the back of the index arm (both defective and incomplete). Four shades (one green and three red); three horizon shades (two red and one green). Index-glass adjustment by a screw; adjustment of horizon glass by a square-headed screw and a detached key. Magnifier on an 89 mm swivelling arm; a frosted glass shade. Threaded telescope bracket in two parts, fitted for correcting collimation error; perpendicular adjustment by rising-piece and a milled knob. Sight-tube (130 mm). Adjusting key (missing). One foot is missing from the frame. No box.
Provenance Presented by Mr G.E.L. Carter, Budleigh Salterton, South Devon, in 1953.
Notes Edward Troughton patented the pillar and plate frame (no. 1644 of 1788).
Literature Stimson, 'The Influence of the Royal Observatory at Greenwich', 130; Clifton, *Directory*, 282; McConnell, *Instrument Makers to the World*.

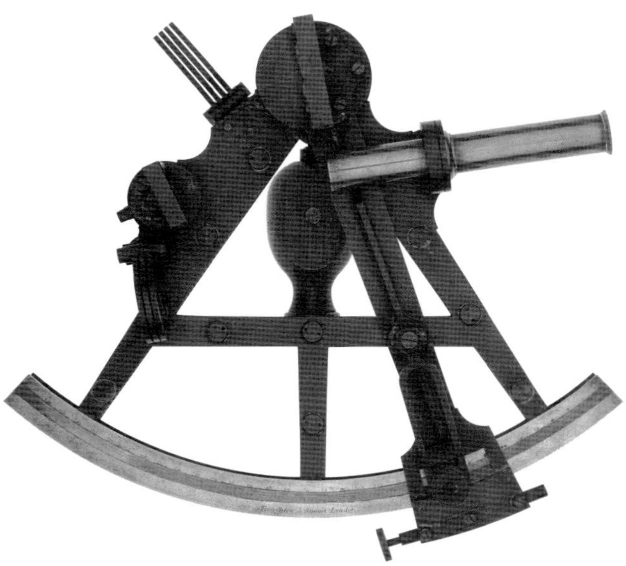

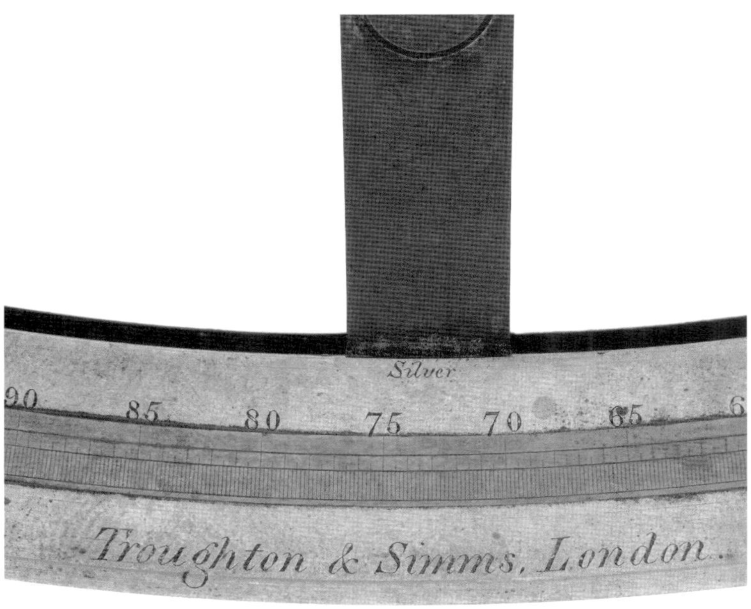

178 AAA2230
Sextant

Origin England, possibly Portsmouth
Maker Possibly E. and E. Emanuel
Date About 1840
Inscriptions Signed on the limb: E & E Emanuel, Portsmouth
Marked on the frame: Frederick Hornby, R.N.
Graduation Inlaid silver scale up to 150° every 10′. Silver vernier to 10″.
Dimensions Radius 6 inches (152 mm)
Description Polished brass straight-bar-pattern pillar frame (20 pillars), wooden handle. The tangent screw and clamping screw are on the back of the index arm. All the shades are missing. Magnifier on a 76 mm swivelling arm. Threaded telescope bracket in two parts, fitted for correcting collimation error; perpendicular adjustment by rising-piece and a milled knob. The tangent screw knob, the clamp, and the knob for the rising-piece are covered with chamois leather for protection of the skin in polar regions. No box.
Provenance Lieutenant Frederick Hornby RN, the original owner, served under Commander Francis R.M. Crozier in HMS *Terror* on Franklin's ill-fated 1845 expedition in search of the North-West Passage. There were no survivors, and this instrument was found by Captain (Sir) Leopold McClintock at Ross Cairn, Point Victory, King William Island, on 2 June 1859, during the *Fox* search expedition (find no. 2230). It was subsequently presented to the Greenwich Hospital collection by Vice Admiral W.W. Hornby, and transferred as part of that collection in 1936, on indefinite loan. It was stolen from the National Maritime Museum in 1982 and remains missing.
Notes Edward Troughton patented the pillar and plate frame (no. 1644 of 1788). The Emanuels were goldsmiths and jewellers, and possibly only the retailers rather than the makers of this instrument.
Literature Savours, *The Search for the North West Passage*, 335; Clifton, *Directory*, 95; Hill and Paget-Tomlinson, *Instruments of Navigation*, 23.

179 NAV1246
Sextant

Origin London, England
Maker George Whitbread
Date About 1840
Inscriptions Signed on the limb: G. Whitbread, London
Marked on the limb: 2165
Graduation Polished brass limb with inlaid silver scale from -5° to 155° every 10′, measuring to 137°. Silver vernier to 1′, zero at the right.
Dimensions Radius 6 inches (152 mm)
Description Anodized brass curved-bar-pattern frame, wooden handle. The tangent screw and clamping screw are on the back of the index arm. Four shades (three red and one green); three horizon shades (two red and one green). Index-glass adjustment by screw; adjustment of horizon glass by capstan screws. Magnifier on an 83 mm swivelling arm. Threaded telescope bracket in two parts, fitted for correcting collimation error; perpendicular adjustment by rising-piece and a milled knob. Telescope (73 mm), erect image; telescope (142 mm), inverted image, four cross-wires. Sight-tube (68 mm) and a shaded eyepiece (red); a magnifying glass and adjusting pin are missing. In a square fitted wooden box lined with textile; leather strap. A brass plate on the lid is marked 'Lord Charles Beresford H.M.S. Condor'. The box contains a trade label for George Lee and Son, Portsea (about 1847).
Provenance Captain (later Admiral) Lord Charles Beresford (1846–1919), the original owner, was commander of HMS *Condor* in the battle of Alexandria in 1882. In 1913 Beresford presented the instrument to Eric Ebbage, the donor, then a Sea Scout in the Second Kingston-on-Thames Troupe, as an award for the best ship model at a Birmingham exhibition. Presented by Mr E.L. Ebbage, London, in 1986.
Literature Clifton, *Directory*, 295; V.W. Baddeley, 'Beresford, Charles William de la Poer, Baron Beresford (1846–1919) *naval officer*', in *Oxford Dictionary of National Biography*, Oxford, 2004, <http://www.oxforddnb.com/>; *Daily Mirror*, 29 July 1913.

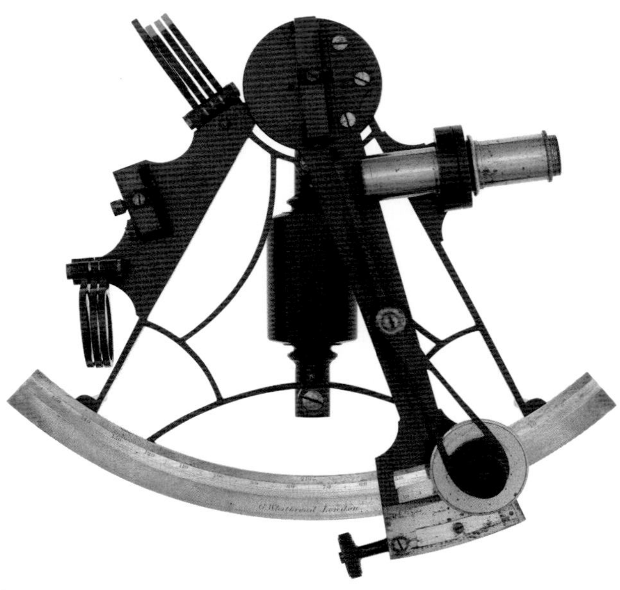

180 NAV1241
Sextant with artificial horizon

Dennis's Lunar Sextant
Origin London, England
Maker John Charles Dennis
Date 1842
Inscriptions Signed on the limb: Dennis's Lunar Sextant. | London.
Marked on the frame: Register | N° 1287 | June 6, 1842
Graduation Polished brass limb with inlaid silver scale from -5° to 150° every 10′, measuring to 128°. Silver vernier to 10″, zero at the right.
Dimensions Radius 7¼ inches (184 mm)
Description Anodized brass three-circle-pattern frame, wooden handle. The tangent screw and clamping screw are on the back of the index arm. Four shades (three grey and one orange); three horizon shades (all grey). Index-glass adjustment by a screw; adjustment of horizon glass by a square-headed screw and a detached key. Magnifier on an 89 mm swivelling arm; a frosted glass shade. Threaded telescope bracket in two parts, fitted for correcting collimation error; perpendicular adjustment by rising-piece and a milled knob. Telescope (181 mm), inverted image, two parallel cross-wires, focused manually or by a milled screw and worm gear; extra drawtube (74 mm), two parallel cross-wires; telescope (83 mm), erect image. Sight-tube (80 mm), a shaded eyepiece (red), a magnifying glass, an

adjusting key, and a piece of chamois leather (the last not found). In a square fitted wooden box lined with green textile, and with an ivory plate on the lid inscribed 'J.C. Dennis[s] Patent Sextant London'. The lid contains a trade label for J.C. Dennis, 122 Bishopsgate Street, London (1850–62).

Artificial horizon This consists of a rectangular cast iron tray and a brass roof-shaped cover, both black-lacquered. Two sides of the cover have clear glass panels, at an angle of 90°. A boxwood jar for storing mercury (filled) when not in use.

Provenance Purchased from David Weston Ltd, London, in 1985.

Notes Dennis's patent was not found in NN, *Patents for Inventions 1636–1866*.

Literature Clifton, *Directory*, 82.

181 NAV 1195

Sextant

Origin London, England

Maker Spencer Browning and Rust

Date 1844

Inscriptions Signed on the limb: Spencer Browning & Rust, London Marked on the limb: Presented by Greenwich Hospital to Edmund Swain August 1844

Graduation Polished brass limb with inlaid silver scale from -5° to 145° every 10′, measuring to 133°. Silver vernier to 10″, zero at the right.

Dimensions Radius 8.15 inches (207 mm)

Description Black-lacquered brass oval-pattern frame, wooden handle. The tangent screw and clamping screw are on the back of the index arm. Four shades (three red and one green); three horizon shades (two red and one green). Index-glass adjustment by a screw; adjustment of horizon glass by a capstan screw. Magnifier on a 101 mm swivelling arm. Threaded telescope bracket in two parts, fitted for correcting collimation error; perpendicular adjustment by rising-piece and a milled knob. Telescope (171 mm, ocular section missing), probably with inverted image; telescope (171 mm, ocular section missing), probably with inverted image. Sight-tube (92 mm), sight-tube (76 mm, incomplete), and an adjusting pin. A shaded eyepiece, a magnifying glass, and an unidentified part are missing. In a wooden keystone box, containing in the lid a trade label for Henry Hughes and Son.

Provenance Edmund Swain, the original owner, is recorded in the *Navy List* from 1847 to 1871 as rising from Assistant Master to Staff Commander. He died on active service on the China Station in HMS *Ocean*. Presented by the Greenwich Sea Cadets, Royal Naval College, Greenwich, London, in 1966.

Notes For the presentation of instruments to Greenwich Hospital boys, see the note with cat. no. 072.

Literature Clifton, *Directory*, 261.

182 NAV1118
Double box sextant

Pocket sextant
Origin London, England
Maker William Cary
Date About 1845
Inscriptions Signed on the upper plate: Cary, London.
Marked on the upper plate: Capt. George's R.N. | C.G. | Registered | Double Sextant | No. 61
Marked (stamped) on the lower plate: WD[broad arrow]
Graduation Two inlaid silver scales, both from -5° to 180° every 30′, measuring to 156° (the digits are read from the pivot). Two gold verniers to 1′, zeros at the left.
Dimensions Radius 4.85 inches (123 mm); diameter 126 mm
Description Circular black-lacquered brass plates with a hinged wooden handle. The two index arms are on the outside of the upper and lower plates, and separate milled knobs move them. No shades. The glasses are contained in the instrument; index-glass adjustment is with a single screw; adjustment of horizon glasses by a milled key fixed in the upper plate. Two magnifiers (both single-lens), each on a 70-mm swivelling arm. Telescope (104 mm), erect image; eyepiece with three shades (all green). The lower plate can be removed for cleaning. In a fitted rectangular mahogany box, containing in the lid a certificate for the Halstead Science and Art Club Prize for an oil painting, awarded to E.T. Adams, 15 March 1911.
Provenance Adams collection; lent in 1936, bequeathed in 1974.
Notes George's double sextant could be used to measure two altitudes of the same celestial body or of two different celestial bodies in quick succession. This was useful when a body was visible only briefly, for example between moving clouds. According to Raper, it could also measure the angular distance between opposite points of the horizon, thus serving as a dip sector; but for with this example, the range is limited to 156°. For this sextant and for his artificial mercurial horizon (cat. no. 330), George (1809–87) received a medal at the International Geographical Congress in Paris, 1875. The broad arrow indicates that the instrument belonged to the Royal Navy.
Literature George, 'Description of the Improved Pocket Sextant'; Raper, *The Practice of Navigation*, 166; NN, 'Captain C. George, R.N.'; Clifton, *Directory*, 51.

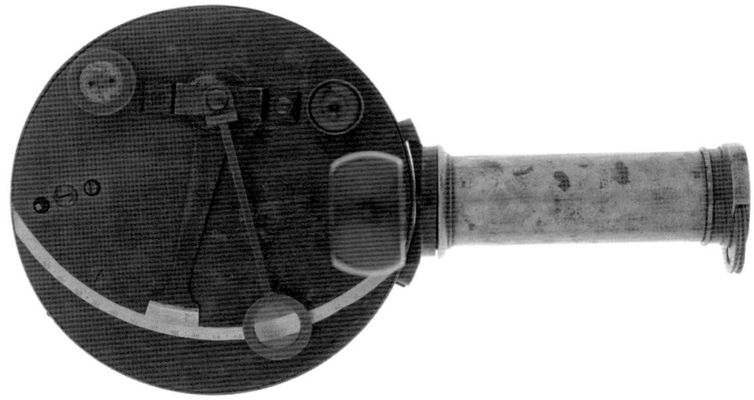

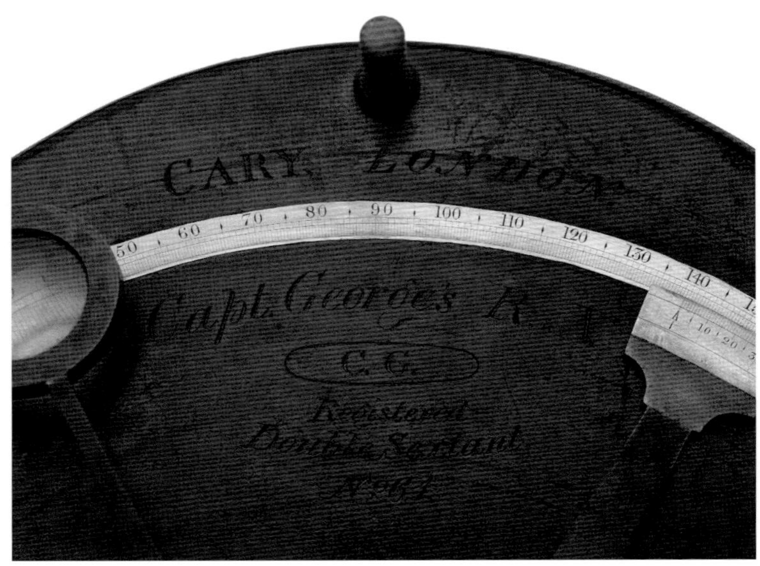

183 AAA2141.1 and AAA2141.2
Sextant eyepieces

Origin Probably England
Maker Unknown
Date 1845 or earlier
Inscriptions None
Dimensions Diameter of both 23 mm (diameter of clear glass 7 mm; diameter of yellow glass 4 mm)
Description Two eyepiece lens caps for sextant tubes. One has a clear lens; the other is a yellow shade. The clear-glass eyepiece has a screw thread on the outside of the brass fitting; the other has a thread on the inside.
Provenance Relics of Sir John Franklin's ill-fated expedition in search of the North-West Passage in 1845–*c.*1848. There were no survivors, and Lieutenant W.R. Hobson found these eyepieces at Northern Cairn Campsite, Cape Felix, King William Island, 25 May 1859, during the *Fox* search expedition under Captain (Sir) Francis Leopold McClintock (find no. 2141). They were subsequently presented to the Greenwich Hospital collection and transferred in 1936 as part of it, on indefinite loan.
Literature Savours, *The Search for the North West Passage*, 333.

184 NAV1223
Sextant

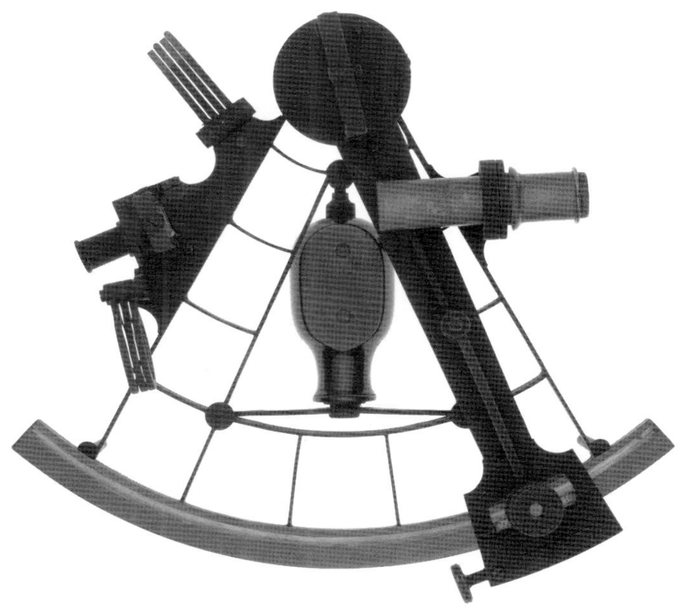

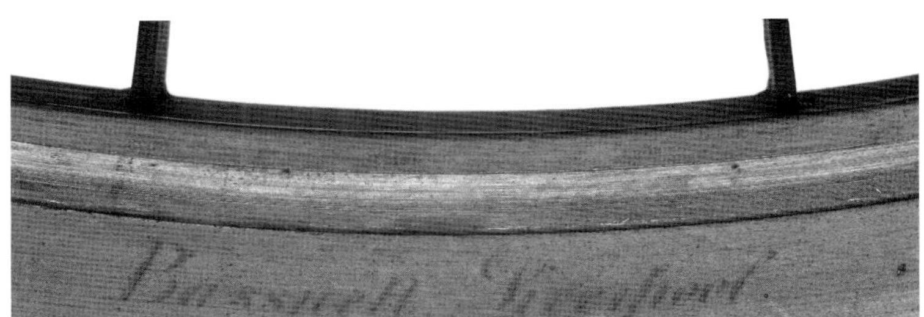

Origin Liverpool, England
Maker (James) Bassnett
Date About 1845
Inscriptions Signed on the limb: Bassnett, Liverpool.
Graduation Polished brass limb; the inlaid scale and the vernier are missing.
Dimensions Radius 8 inches (203 mm)
Description Blackened brass triangle-pattern frame, wooden handle. The tangent screw and clamping screw are on the back of the index arm. Four shades (two red, one orange, and one green); three horizon shades (two red and one green). Index-glass adjustment by a screw; adjustment of horizon glass by a capped milled knob and worm gear. Magnifier on an 85 mm swivelling arm. Threaded telescope bracket in two parts, fitted for correcting collimation error; perpendicular adjustment by rising-piece and a milled screw. Telescope (185 mm), inverted image; extra drawtube (78 mm), inverted image, two parallel cross-wires; telescope (83 mm), erect image. Sight-tube (83 mm); shaded eyepiece (green). In a square fitted wooden box (the top of the lid has been replaced by plywood), containing a trade label for Hayes Brothers, James Street, Cardiff.
Provenance Presented by Mr J.H. Allen, Lymington, Hampshire, in 1976.
Literature Clifton, *Directory*, 21; Morrison-Low, *Making Scientific Instruments*, 324.

185 NAV1120
Double sounding sextant

Origin London, England
Maker Nathaniel Worthington
Date About 1847
Inscriptions Signed on the limb: Worthington London
Marked on the limb: D.176
Graduation Polished brass limb with two parallel inlaid silver scales; the lower scale runs from the left from -5° to 160° every 20′, measuring to 132°, and the upper scale runs from the right from -10° to 160° every 20′, measuring to 126°. Two silver verniers to 20″, zeros at the left (lower) and right (upper).
Dimensions Radius 8 inches (203 mm)
Description Anodized brass diamond-pattern frame, wooden handle. The upper tangent screw and clamping screw are on the front of the upper index arm; the lower tangent screw and clamping screw are on the back of the lower index arm. Two index glasses and one large horizon glass. Four shades (three red and one green); three horizon shades (two red and one green). Index-glass adjustment by a thumbscrew and a capstan screw; adjustment of horizon glass by a capstan screw. Upper magnifier (single lens) on an 83 mm swivelling arm, lower magnifier (single lens) travelling on a bar. Threaded telescope bracket in two parts, fitted for correcting collimation error; perpendicular adjustment by rising-piece and a milled knob. Telescope (102 mm), erect image; telescope (67 mm), erect image. Rotating shaded eyepiece with choice of three shades (two green and one red). In a square fitted wooden box, containing in the lid a trade label for William Gerrard, Liverpool, and a circular label in MS 'A/e 1907'.
Provenance Adams collection; lent in 1936, bequeathed in 1974. The MS label may suggest that Adams acquired this instrument in 1907.
Notes Invented by Rear Admiral Frederick William Beechey (1796–1856) in about 1847 for use in surveying, where angles between three objects at once may be required (compare cat. no. 279). D.176 was the Hydrographic Office serial numbering adopted in February 1828, where D was the letter given to sextants.
Literature Clifton, *Directory*, 305; Stimson, 'Some Board of Longitude Instruments', 114–15; J.K. Laughton, 'Beechey, Frederick William (1796–1856) *naval officer and hydrographer*', in *Oxford Dictionary of National Biography*, Oxford, 2004, <http://www.oxforddnb.com/>.

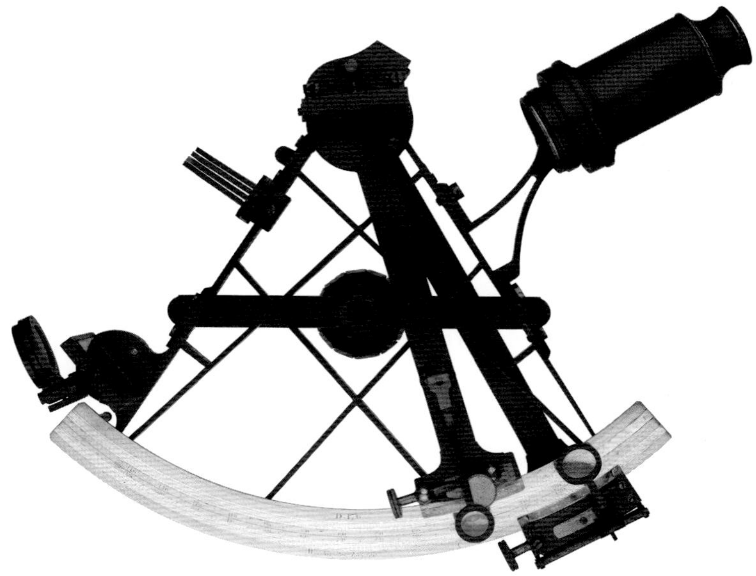

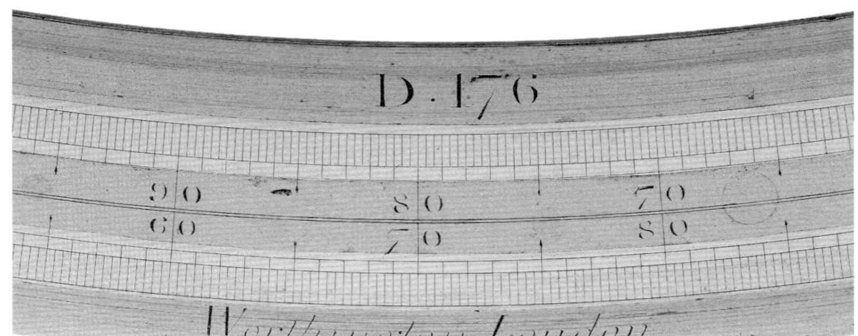

186 NAV1121
Sextant

Origin Munich, Germany
Maker Ertel and Sohn
Date About 1850
Inscriptions Signed on the limb: Ertel & Sohn in München
Graduation Polished limb with inlaid silver scale from -12° to 150° every 20′, measuring to 132°. Silver vernier to 20″, zero at the right.
Dimensions Radius 8 inches (203 mm)
Description Bronze lacquered frame, wooden handle. The tangent screw, working against a spring load, and the clamping screw are on the back of the index arm. Three shades (one orange and two green); four horizon shades (one orange and three green). Index-glass adjustment by a screw; adjustment of horizon glass by a square-headed screw and a detached key and a screw. Magnifier on a 190 mm swivelling arm; a frosted glass shade. Threaded telescope bracket in two parts, fitted for correcting collimation error; perpendicular adjustment by rising-piece and a milled knob. Telescope (193 mm), inverted image, three cross-wires, swivelling shaded eyepiece (green); dust cap; extra drawtube (80 mm), inverted image, three cross-wires.

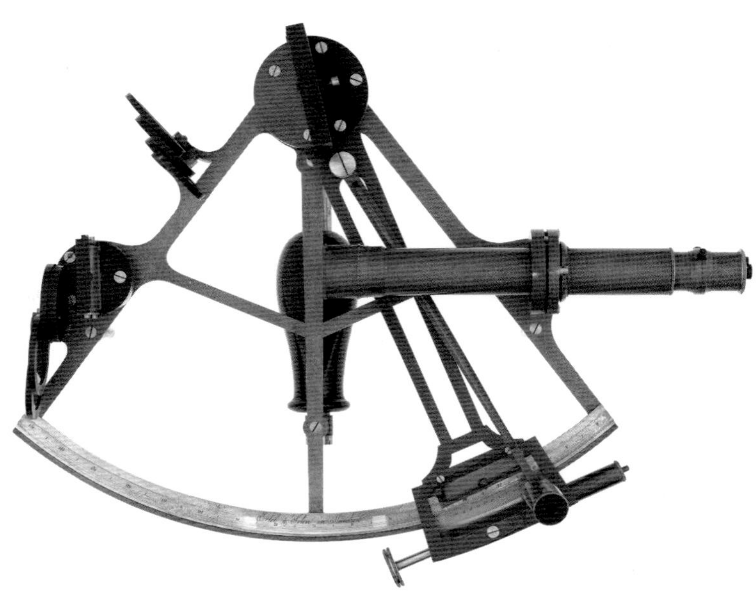

Sight-tube (131 mm), a screwdriver, and two adjusting keys, two rectangular sight vanes, one with a cross-wire and the other with a pinhole. In a mahogany keystone box.
Provenance Adams collection; lent in 1936, bequeathed in 1974.
Literature Brachner, 'German Nineteenth-Century Scientific Instrument Makers', 129, 139.

187 NAV1150
Box sextant

Pocket sextant
Origin England
Maker Unknown
Date About 1850
Inscriptions None
Graduation Inlaid silver scale from -5° to 170° every 30′, measuring to 139° (the digits are read from the pivot). Silver vernier, reading to 1′, zero at the left.
Dimensions Radius 1.6 inches (41 mm); diameter 70 mm
Description Circular polished brass plates with a lid that, when screwed on the back of the instrument, serves as a handle. The index arm is moved by a milled screw on the upper plate. Two index shades (red and green) moved by levers; no horizon shades. Horizon-glass adjustment by a milled key fixed in the upper plate. Magnifier (single lens) on a 45 mm swivelling arm. Telescope (65 mm), erect image, with an angled bracket for attaching it to the instrument by a milled screw. Sliding pinhole shutter. In a fitted leather case, lined with velvet.
Provenance Presented by Commander D. Edwards RN, Wincanton, Somerset, in 1949.

188 NAV1160
Sextant

Origin England
Maker Unknown
Date About 1850
Inscriptions Marked on the scale (at 65°): F W
Graduation Inlaid ivory scale from -2° to 137° every 20′, measuring to 118°. Ivory vernier to 30″, zero at the right.
Dimensions Radius 7.9 inches (201 mm)
Description Ebony frame and limb with brass index arm and fittings, wooden handle; an inlaid ivory plate in the crossbar. The tangent screw and clamping screw are on the back of the index arm. All the shades are missing. Index-glass adjustment by a screw; adjustment of horizon glass by a square-headed screw and a detached key, a lever, a worm gear, and a milled clamping screw. Magnifier on a 90-mm swivelling arm. Threaded telescope bracket in two parts, fitted for correcting collimation error; perpendicular adjustment by rising-piece and a milled knob. The telescope is missing. Sight-tube (82 mm), a shaded eyepiece (missing), a milled adjusting key, and a magnifying glass (missing). In a keystone polished wooden box (badly damaged), containing in the lid a trade label (stuck over two others) for W.F. Cannon, 175 High Street, Shadwell, London (*c.* 1885–1901).
Provenance Presented by Commander C.E.N. Frankcom, OBE, RD, RNR, Marine Superintendent of the Meteorological Office, Middlesex, in 1955.
Literature Stimson, 'Some Board of Longitude Instruments', 114.

189 NAV1145
Surveyor's pocket sextant

Origin London, England
Maker William Cary
Date 1850
Inscriptions Signed on the crossbar: Cary London [read from the pivot]
Graduation Polished brass scale from -4° to 155° every 1°, measuring to 127° (the digits are read from the pivot). Brass vernier to 1°, zero at the left.
Dimensions Radius 2 inches (51 mm)
Description Lacquered brass straight-bar-pattern frame, detached wooden handle. There is no tangent or clamping screw. No shades. Index-glass adjustment by a screw; adjustment of horizon glass by a milled knob. The sight vane has one pinhole. In a shaped leather case (damaged) with, on the bottom in MS, 'J. Smith [illegible] 1850'.
Provenance Sir Allen Young (1827–1915), the original owner, went in 1857–9 with Sir Francis Leopold McClintock as a navigating officer in the yacht *Fox*, in search of Sir John Franklin. Young probably used the sextant during sledging operations. On loan from a descendant of Sir Allen Young since 1948.
Literature Clifton, *Directory*, 51; Hill and Paget-Tomlinson, *Instruments of Navigation*, 34; Savours, *The Search for the North West Passage*, 282–99; R.N. Rudmose Brown, 'Young, Sir Allen (1827–1915) *merchant navy officer and polar explorer*', in *Oxford Dictionary of National Biography*, Oxford, 2004, <http://www.oxforddnb.com/>.

190 NAV1164
Box sextant

Pocket sextant
Origin London, England
Maker Elliott Brothers
Date 1854–56
Inscriptions Signed on the upper plate: Elliott Brothers, 56 Strand London. [read from the pivot]
Marked on the lid: J.A. Papillon
Marked on the side: H.S. & A.C. | E.T. Adams | 1st Pr: Ins. Ph: | 1909
Graduation Inlaid silver scale from -5° to 165° every 30′, measuring to 136° (the digits are read from the pivot). Silver vernier to 1′, zero at the left.
Dimensions Radius 1.9 inches (48 mm); diameter 77 mm
Description Circular polished brass plates with a lid that, when screwed on the back of the instrument, serves as a handle. The index arm is moved by a milled knob on the upper plate. Two index shades (red and green) moved by levers; no horizon shades. Index-glass adjustment by a screw; adjustment of horizon glass by a square-headed screw and a milled key fixed in the upper plate. Magnifier (single lens) on a 50 mm swivelling arm. Telescope (63 mm), erect image. Sliding pinhole shutter. Shaded eyepiece (green) and an angled bracket for attaching it to the instrument. The case contains an adjusting key (probably from another instrument). In a fitted velvet-lined leather carrying case (strap missing) marked in MS 'E.T. Adams R. Terrence'.
Provenance Adams collection; purchased in 1958.
Notes The abbreviation in the inscription on the side stands for 'Halstead Science and Art Club', of which Adams was the founder.
Literature Clifton, 'An Introduction to the History of Elliott Brothers'; Clifton, *Directory*, 95.

191 NAV1163
Box sextant

Pocket sextant
Origin London, England
Maker Elliott Brothers
Date 1854–6
Inscriptions Signed on the upper plate: Elliott Brothers, 56 Strand London [read from the pivot]
Marked on top of the lid: Honorius S. Sitwell
Graduation Inlaid silver scale from -5° to 155° every 30′, measuring to 131° (the digits are read from the pivot). Silver vernier to 1′, zero at the left.
Dimensions Radius 1.9 inches (48 mm); diameter 76 mm
Description Circular polished brass plates with a lid that, when screwed on the back of the instrument, serves as a handle. The index arm is moved by a milled knob on the upper plate. Two index shades (red and green) moved by levers; no horizon shades. Index glass without adjustment; adjustment of horizon glass by two screws and a milled key fixed in the upper plate. Magnifier (single lens) on a 50 mm swivelling arm. The telescope and bracket are missing. Sliding pinhole shutter. No box or case.
Provenance Belonged to Colonel Honorius Sisson Sitwell RE. Presented by Mr D. McMichael, Leatherhead, Surrey, in 1957.
Literature Clifton, 'An Introduction to the History of Elliott Brothers'; Clifton, *Directory*, 95; Hill and Paget-Tomlinson, *Instruments of Navigation*, 90.

192 NAV1153
Box sextant

Pocket sextant
Origin London, England
Maker Elliott Brothers
Date About 1855
Inscriptions Signed on the upper plate: Elliott Brothers, London [read from the pivot]
Marked on the lower plate: W.A. Urquhart. | 62
Graduation Inlaid silver scale from -5° to 155° every 30′, measuring to 120° (the digits are read from the pivot). Silver vernier to 1′, zero at the left.
Dimensions Radius 1¼ inches (32 mm); diameter 70 mm
Description Circular anodized brass plates with a lid with a bayonet fitting that, when attached to the back of the instrument, serves as a handle. The index arm is moved by a milled knob on the upper plate. No shades. Index glass without adjustment; adjustment of horizon glass by a milled key fixed in the upper plate. The lid has a small brass handle on the back, and contains a paper table of natural tangents to a base of 100. Magnifier (single lens) on a hinged 39-mm swivelling arm. No telescope. Sliding pinhole sight with choice of a small plain glass or a larger pinhole. In a fitted leather case lined with red velvet, and stamped on the side of the case 'W.A.U.'
Provenance Presented by The Marine Society, London, in 1950.
Literature Clifton, 'An Introduction to the History of Elliott Brothers'; Clifton, *Directory*, 95.

193 NAV 1114
Sextant

Origin England, possibly Sunderland
Maker Possibly A. Baharie
Date About 1855
Inscriptions Signed on the limb: A. Baharie, Sunderland
Graduation Polished brass limb with inlaid silver scale from -5° to 135° every 15′, measuring to 114°. Silver vernier to 15″, zero at the right.
Dimensions Radius 7½ inches (191 mm)
Description Polished brass tulip-pattern frame, wooden handle. The tangent screw and clamping screw are on the back of the index arm. Four shades (three red and one green); three horizon shades (two red and one green). Index-glass adjustment by a screw; adjustment of horizon glass by capstan screws. Magnifier on a 77-mm swivelling arm. Threaded telescope bracket without adjustment. Telescope (83 mm), erect image. Sight-tube (82 mm), two shaded eyepieces (both red), an adjusting pin, and a screwdriver by A. Ridge and Sons. In a mahogany keystone box, containing in the lid a trade label for John Nixon, Commercial Road, London (1850–69).
Provenance Adams collection; lent in 1936, bequeathed in 1974.
Notes Baharie may have been a retailer rather than a manufacturer.
Literature Mörzer Bruyns, 'The Navigating Instruments of the National Maritime Museum Greenwich', 27; Clifton, *Directory*, 201.

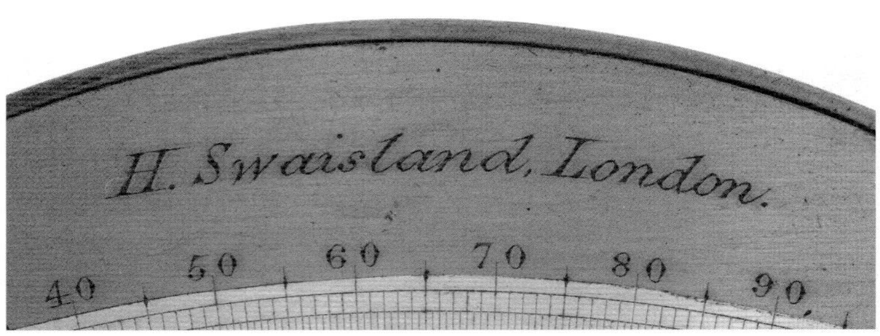

194 NAV 1245
Box sextant

Pocket sextant
Origin London, England
Maker Henry Swaisland
Date About 1859
Inscriptions Signed on the upper plate: H. Swaisland, London. [read from the pivot]
Graduation Inlaid silver scale from -5° to 140° every 30′, measuring to 120° (the digits are read from the pivot). Silver vernier to 1′, zero at the left.
Dimensions Radius 2 inches (51 mm); diameter 76 mm
Description Circular polished brass plates with a lid that, when screwed on the back of the instrument, serves as a handle. The index arm is moved by a milled knob on the upper plate. Two index shades (green and red) moved by levers; no horizon shades. Index glass without adjustment; adjustment of horizon glass by square-headed screws with a milled key fixed in the upper plate. Magnifier (single lens) on a hinged 51 mm swivelling arm. Telescope (74 mm), erect image, with a shaded eyepiece (red) fitted in the body. A pinhole sight for use when the telescope is removed. A compass (diameter 32 mm) is fitted at the centre of the upper plate, graduated into points, and per quadrant in degrees. The compass is fitted with a stop mechanism and a card filter, and two small sight vanes attached to the upper plate. In a fitted leather carrying case.
Provenance Purchased from Delehar Antiques, London, in 1986.
Notes This combination of sextant and compass enables the observer to take a bearing and an angle of a line simultaneously. Henry Swaisland provisionally patented this pocket sextant (no. 452 of 1859).

195 NAV1137
Sextant

Origin Newcastle upon Tyne, England
Maker John Cail
Date About 1860
Inscriptions Signed on the limb: John Cail, Grey St. Newcastle on Tyne.
Graduation Polished brass limb with inlaid silver scale from -5° to 160° every 15′, measuring to 134°. Silver vernier to 15″, zero at the right.
Dimensions Radius 7½ inches (191 mm)
Description Anodized brass oval-pattern frame, wooden handle. The tangent screw and clamping screw are on the back of the index arm. Four shades (three red and one green); three horizon shades (one red and two grey). Index-glass adjustment by a screw; adjustment of horizon glass by capstan screws. Magnifier on a 93-mm swivelling arm. Threaded telescope bracket in two parts, fitted for correcting collimation error; perpendicular adjustment by rising-piece and a milled knob. Telescope (183 mm), inverted image, four cross-wires; telescope (83 mm), erect image. Sight-tube (81 mm), a shaded eyepiece (red), a spare clamping screw for the index arm (later addition), and an adjusting pin (missing). In a keystone box, containing in the lid a trade label for John Cail, Newcastle-on-Tyne; a rubber stamp; and pencil notes (both of the latter illegible).
Provenance Presented by Sir James Caird in 1936; purchased as part of the Landau collection through Maggs Brothers.
Literature Callender, *The Caird Collection: Supplement* 96, no. 19; Clifton, *Directory*, 47; Morrison-Low, *Making Scientific Instruments*, 122, 124.

196 NAV1117
Sextant

Origin England, possibly Liverpool
Maker Possibly Owen Owens
Date About 1860
Inscriptions Signed on the inlaid plate on the crossbar: Owen.Owens | 4.South.Castle.St. | Liverpool
Graduation Inlaid ivory scale from -5° to 137° every 10′, measuring to 118°. Ivory vernier to 10″, zero at the right.
Dimensions Radius 7½ inches (191 mm)
Description Ebony frame and limb with brass index arm and fittings, wooden handle; an inlaid ivory plate on the crossbar. The tangent screw and clamping screw are on the back of the index arm. Four shades (three red and one green); three horizon shades (two red and one green). Index-glass adjustment by a screw; adjustment of horizon glass by a square-headed screw and a detached key, a lever, a worm gear, and a milled knob. Magnifier on a 93-mm swivelling arm. Threaded telescope bracket in two parts, fitted for correcting collimation error; perpendicular adjustment by rising-piece and a milled knob. Telescope (83 mm), erect image; telescope (173 mm), inverted image, four cross-wires. Sight-tube (82 mm), a shaded eyepiece (green), and a milled adjusting key. In a mahogany keystone box, containing in the lid a trade label for John Parkes and Sons, Liverpool, and a label in MS '1913 E. Company 5th Batt. Essex Reg. Hon. Members 1st Prize Awarded to E.T. Adams'.
Provenance Adams collection; lent in 1936, bequeathed in 1974.
Notes Owens was a clockmaker and watchmaker and perhaps the retailer rather than the manufacturer of this instrument.
Literature Banfield, *Barometer Makers and Retailers*, 161.

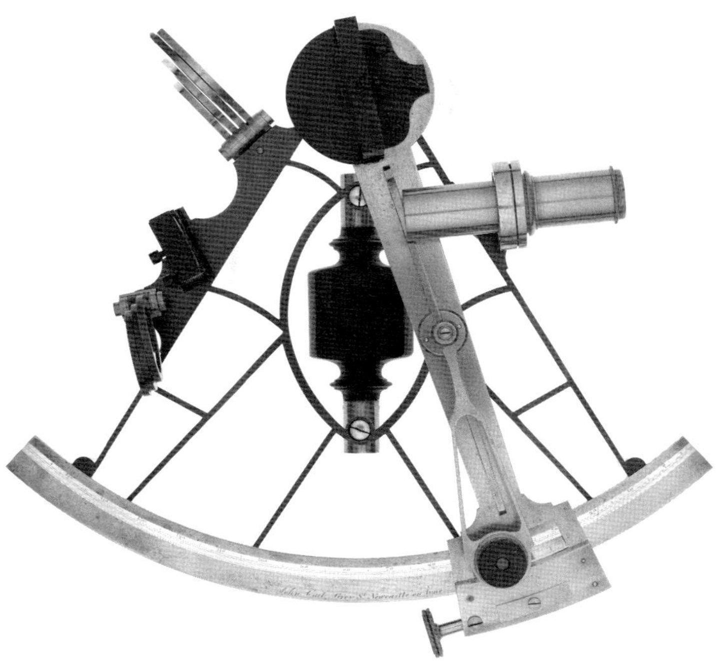

197 NAV 1134
Sextant

Origin London, England
Maker (George or William) Dollond
Date About 1865
Inscriptions Signed on the limb: Dollond, London
Graduation Polished brass limb with inlaid silver scale from -5° to 149° every 10′, measuring to 126°. Silver vernier to 10″, zero at the right.
Dimensions Radius 8 inches (203 mm)
Description Grey/green-lacquered brass curved-pattern frame, wooden handle. The tangent screw, working against a spring load, and the clamping screw are on the back of the index arm. Four shades (one green, two orange, and one red); three horizon shades (red, orange, and green). Index-glass adjustment by a screw; adjustment of horizon glass by capstan screws. Magnifier, travelling on a worm wheel; a frosted glass shade. Threaded telescope bracket with perpendicular adjustment by rising-piece and a milled knob. Telescope (159 mm), inverted image, four cross-wires; extra drawtube (46 mm), erect image, four cross-wires; telescope (92 mm), erect image. Sight-tube (83 mm), a shaded eyepiece (green), and an adjusting pin. In a mahogany keystone box.
Provenance Presented by HM Queen Mary in 1936.
Literature Barty-King, *Eyes Right*; Clifton, *Directory*, 86–7.

198 NAV 1277
Night sextant

Night octant
Origin Nantes, France
Maker Leon Dubas
Date About 1868
Inscriptions Signed on the limb: Dubas, Nantes, Breveté. s.g.d.g.
Graduation Polished brass limb with inlaid ivory scale from -5° to 125° every 20′, measuring to 106°. Ivory vernier to 30″, zero at the right.
Dimensions Radius 8½ inches (216 mm)
Description Anodized brass frame, wooden handle. The tangent screw is on top of the index arm; the clamping screw is on the back. No shades, but a star glass (*stellar lenticular*). Index- and horizon-glass adjustment by a square-headed screw and a detached key. Magnifier (single lens) on a 108 mm swivelling arm. Threaded telescope bracket without adjustment. Telescope (160 mm), erect image, with a wide objective lens (star finder). Milled adjusting key, screwdriver. In a square fitted wooden box, containing a note on the index error in English, dated 1888.
Provenance Belonged to Prince George, later King George V (1865–1936), who was sent to train as a Royal Naval officer in HMS *Britannia* in 1877. His wife, HM Queen Mary, presented it in 1936.
Notes Captain Pierre Michel Albert Laurent, of the Compagnie Générale Trans-Atlantique, Le Havre, patented the star glass for elongating the image of a star into a horizontal line (provisional patent no. 2426 of 1859 and no. 2665 of 1864, and, in the USA, no. 47 778 of 1865). 'S.g.d.g.' (*sans garantie de gouvernement*) means that although the instrument was patented, there was no government guarantee, and only the inventor was responsible for the object's quality.
Literature Laurent, *Explanation, Adjustment and Use of Laurent's Patent Night Octant*; Mörzer Bruyns, 'The Introduction of the Elongating Glass on Sextants'; Ifland, *Taking the Stars*, 80, 195; Marcelin, *Dictionnaire des fabricants français*; H.C.G. Matthew, 'George V (1865–1936), *king of Great Britain, Ireland, and the British dominions beyond the seas, and emperor of India*', in *Oxford Dictionary of National Biography*, Oxford, 2004, <http://www.oxforddnb.com/>.

199 NAV1126
Sextant

Origin London, England
Maker Reynolds and Wiggins
Date About 1870
Inscriptions Signed on the limb: Reynolds and Wiggins, 10 King St. Tower Hill, London
Graduation Polished brass limb with inlaid silver scale from -5° to 160° every 10′, measuring to 134°. Silver vernier to 10″, zero at the right.
Dimensions Radius 7¼ inches (184 mm)
Description Blackened brass oval-pattern frame, wooden handle. The tangent screw and clamping screw are on the back of the index arm. Four shades (three red and one green); three horizon shades (two red and one green). Index-glass adjustment by a screw; adjustment of horizon glass by capstan screws. Magnifier on an 82 mm swivelling arm; a frosted glass shade. Threaded telescope bracket in two parts, fitted for correcting collimation error; perpendicular adjustment by rising-piece and a milled knob. Telescope (185 mm), inverted image, four cross-wires; telescope (84 mm), erect image. Sight-tube (82 mm), a shaded eyepiece (red), and an adjusting pin (missing). In a wooden keystone box lined with green textile. A brass disc on the lid is marked 'T.M. Hardy. R.N. 1793'.
Provenance The inscription on the brass disc suggests that this instrument belonged to Vice Admiral Sir Thomas Masterman Hardy (1769–1839) who was on *Amphitrite* in 1793, was later Nelson's flag captain, including at Trafalgar, and was Governor of Greenwich Hospital from 1834 until his death. The sextant is, however, of much later date. Presented to the Greenwich Hospital collection by Mr Ernest A. Howard and in 1936 transferred as part of it, on indefinite loan.
Literature J.K. Laughton, 'Hardy, Sir Thomas Masterman, baronet (1769–1839) *naval officer*', in *Oxford Dictionary of National Biography*, Oxford, 2004, <http://www.oxforddnb.com/>.

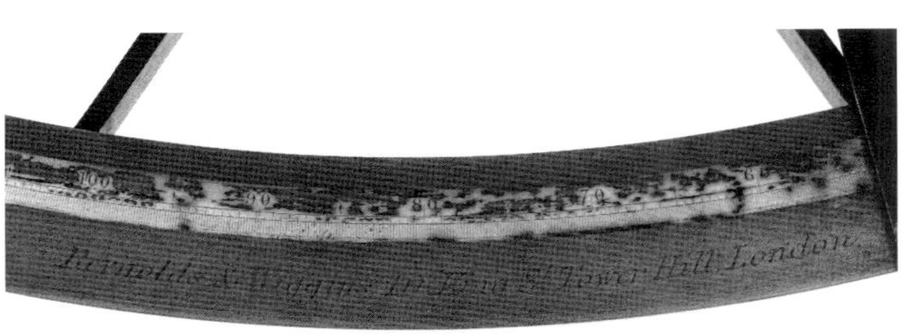

200 NAV1158
Sextant

Origin London, England
Maker William Parkinson and William James Frodsham
Date About 1870
Inscriptions Signed on the limb: Parkinson and Frodsham . . . [illegible] London
Graduation Polished brass limb with inlaid silver scale from -5° to 150° every 10′, measuring to 129°. Silver vernier to 10″, zero at the right.
Dimensions Radius 8.15 inches (207 mm)
Description Polished brass straight-bar-pattern pillar frame (20 pillars), wooden handle. The tangent screw and clamping screw are on the back of the index arm. Four shades (two red and two grey); three horizon shades (one red and two grey). Index-glass adjustment by a screw; adjustment of horizon glass (large) by capstan screws. Magnifier on an 80-mm swivelling arm; a frosted glass shade. Threaded telescope bracket in two parts, fitted for correcting collimation error; perpendicular adjustment by rising-piece and a milled knob. Telescope (184 mm), inverted image, no cross-wires; telescope (81 mm), inverted image, four-cross wires; extra drawtube (49 mm), erect image; telescope (68 mm), erect image (star finder), that does not fit in the telescope bracket. Sight-tube (82 mm), six shaded eyepieces (three red, one orange, one green, and one grey), one part missing (probably a magnifying glass). In a square fitted wooden box, probably not original to this instrument.
Provenance Presented by Mr J.K. Bryon, Manchester, in 1955.
Notes A figure of a sextant, made of inlaid wood, is found in the lid of the box, with shipping company house flags on either side, also of inlaid wood. These are for the Australian Steam Shipping Co. Ltd, London, on the left, and for Manchester Liners Ltd, Manchester, on the right. Edward Troughton patented the pillar and plate frame (no. 1644 of 1788).
Literature Clifton, *Directory*, 209.
For image see colour section plate 15.

201 NAV1162
Sextant

Origin Glasgow, Scotland
Maker James Whyte and Co.
Date About 1870
Inscriptions Signed on the limb: J. Whyte & C°. Glasgow.
Graduation Polished limb with inlaid silver scale from -5° to 150° every 10′, measuring to 125°. Silver vernier to 10″, zero at the right.
Dimensions Radius 7 inches (178 mm)
Description Anodized brass diamond-and-curve-pattern frame, wooden handle. The tangent screw and clamping screw are on the back of the index arm. Four shades (three red and one green); three horizon shades (two red and one green). Index-glass adjustment by a screw; adjustment of horizon glass by capstan screws. Magnifier on a 78 mm swivelling arm; a frosted glass shade. Telescope bracket (missing) with perpendicular adjustment by rising-piece and a milled knob. Telescope(s) and unidentified parts are missing. Two feet are missing from the frame. No box.
Provenance Presented by Mrs E.J. Nightingale, London, in 1956.
Literature Clarke *et al.*, *Brass & Glass*, 281–7.

202 NAV1128
Sextant with artificial horizon

Colby's horizon instrument
Origin Unknown
Maker Unknown
Date 1870 or later
Inscriptions Marked on the sighting crosspiece: Colby's Horizon Inst. Patented Feb. 22 1870.
Graduation Brass scale to 105° on either side of 0° every 30′, measuring to 92°. Two verniers; one for altitude, to 1′, zero at the centre, and the other for zenith distance, to 1′, zero at the right.
Dimensions Radius 3½ inches (89 mm)
Description Black-lacquered circular brass plate; the upper segment is cut away; wooden handle. There is a spring-loaded pin for locking when the observation is made. A fixed bracket with a 53 mm sight-tube at both ends, and a spirit level centred between the two. The object end of the sight-tubes has cross-wires and two shades (red and green). In a square fitted wooden box.
Artificial horizon The horizontal line of reference is established by an arc-shaped pendulum, pivoted at the centre of the instrument, which hangs freely during a sighting. The pendulum is then locked by a trigger mechanism before the sextant is read.
Provenance Adams collection; lent in 1936, bequeathed in 1974.
Notes Hall Colby of New York patented 'Colby's altimeter or self-adjusting quadrant or sextant' in the UK (no. 1200 of 1854 and no. 1736 of 1855) and in the USA (no. 100, 119 of 1870).
Literature Ifland, *Taking the Stars*, 107–8, 193.

203 NAV 1132
Sextant

Origin London, England
Maker Joseph Hughes
Date About 1875
Inscriptions Signed on the limb: J. Hughes, London.
Graduation Polished brass limb with inlaid silver scale from -5° to 150° every 15′, measuring to 131°. Silver vernier to 15″, zero at the right.
Dimensions Radius 8 inches (203 mm)
Description Anodized brass curved-pattern frame, wooden handle. The tangent screw and clamping screw are on the back of the index arm. Four shades (two red, one orange, and one green); three horizon shades (two red and one green). Index-glass adjustment by a screw; adjustment of horizon glass by capstan screws. Magnifier on an 89 mm swivelling arm. Threaded telescope bracket in two parts, fitted for correcting collimation error; perpendicular adjustment by rising-piece and a milled knob. Telescope (184 mm), inverted image, four cross-wires; telescope (83 mm), erect image. Sight-tube (81mm), a shaded eyepiece (red), an adjusting pin, and a magnifying glass. In a mahogany keystone box.
Provenance Presented by HM Queen Mary in 1936.
Literature Clifton, *Directory*, 144.

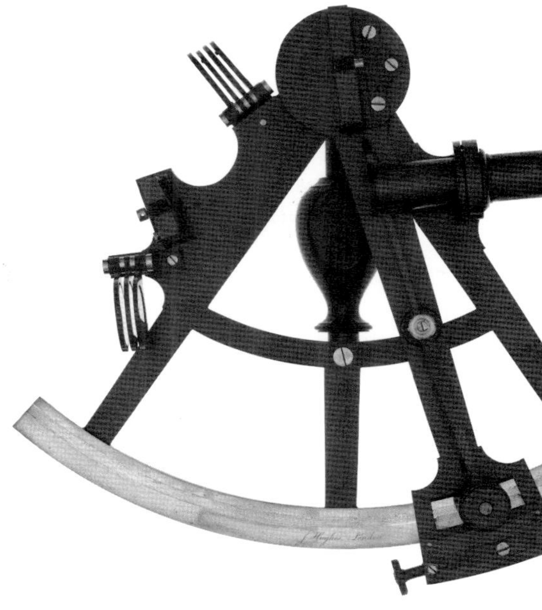

204 NAV 1200
Sextant

Origin London, England
Maker Edward Smith
Date After 1875
Inscriptions Signed on the limb: E. Smith, from Whitbreads, 69 Fenchurch St. London
Marked (stamped) on the limb: 2076
Graduation Polished brass limb with inlaid silver scale from -5° to 160° every 10′, measuring to 125°. Silver vernier to 10″, zero at the right.
Dimensions Radius 7¾ inches (197 mm)
Description Varnished brass straight-bar-pattern pillar frame (18 pillars), wooden handle with a brass-lined threaded hole for a second handle or a stand. The tangent screw and clamping screw are on the back of the index arm. Four shades (all green); three horizon shades (all green). Index-glass adjustment by a screw; adjustment of horizon glass by capstan screws. Magnifier on an 80 mm swivelling arm. Threaded telescope bracket in two parts, fitted for correcting collimation error; perpendicular adjustment by rising-piece and a milled knob. Telescope (193 mm), inverted image, four cross-wires; extra drawtube (78 mm), inverted image, two parallel cross-wires; telescope (83 mm), erect image (star finder). Rotating shaded eyepiece with choice of three shades (all green), an adjusting pin, and a piece of chamois leather. In a wooden keystone box (an inlaid plate is missing from the lid) lined with green textile, containing a trade label for George Lee and Son, Portsea (1850–1900), and in the lid a National Physical Laboratory certificate of examination, dated June 1905.
Provenance Presented by HMS *Dryad* Museum, Southwick, Hampshire, in 1967.
Notes Edward Smith and George Lee worked for George Whitbread. Edward Troughton patented the pillar and plate frame (no. 1644 of 1788).
Literature Clifton, *Directory*, 165, 255.

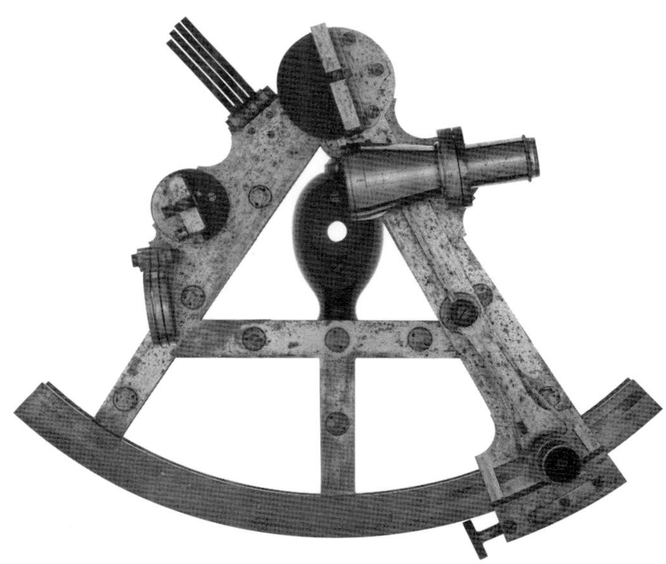

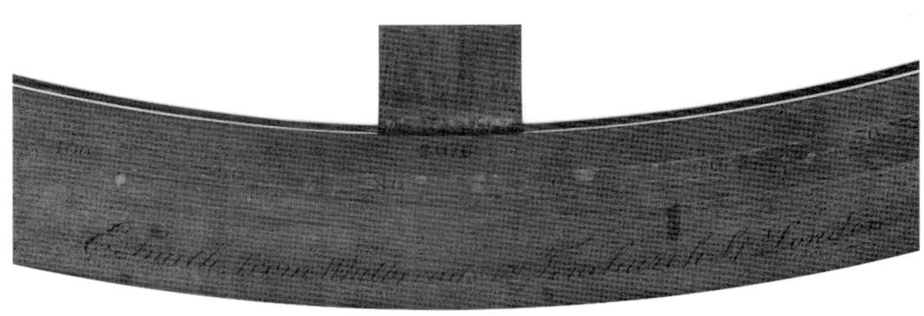

205 NAV1188
Sextant

Origin London, England
Maker Henry Hughes and Son
Date 1877
Inscriptions Signed on the limb: Henry Hughes & Son, 59 Fenchurch St. London.
Marked on the limb: 2451
Graduation Polished brass limb with inlaid silver scale from -5° to 155° every 10′, measuring to 132°. Silver vernier to 10″, zero at the right.
Dimensions Radius 6½ inches (165 mm)
Description Anodized brass three-circle-pattern frame, wooden handle. The tangent screw and clamping screw are on the back of the index arm. Four shades (two grey, one green, and one red); three horizon shades (green, grey, and red). Index-glass adjustment by a screw; adjustment of horizon glass by capstan screws. Magnifier on a 76 mm swivelling arm; a frosted glass shade. Threaded telescope bracket in two parts, fitted for correcting collimation error; perpendicular adjustment by rising-piece and a milled knob. Telescope (172 mm), inverted image, four cross-wires; telescope (83 mm), erect image. Sight-tube (81 mm), a shaded eyepiece (red), and an adjusting pin (missing). In a fitted wooden box, containing in the lid a trade label for Dobbie, Son and Hutton, London E.C., Glasgow House, Alexander Dobbie and Son (after 1894); also, on the lid, a silver plate inscribed 'Thames Nautical Training College HMS Worcester, Port of London. Prize for proficiency in the use of Sextant, Midsummer 1877'.
Provenance Belonged to Arthur Redmond Pilkington (born 1860), who won this instrument; he was later master of the barque *Catalina*, the clipper *Mikado*, the *Nurembidgee*, the *Wooloomooloo*, and the *Fifeshire*. HMS *Worcester* was the training ship of the Thames Nautical Training College, founded in 1862. Presented by Mrs B. Bennett, Gloucester, in 1966.
Literature Clifton, *Directory*, 144; Clarke *et al.*, *Brass & Glass*, 229.

206 NAV1125
Night sextant

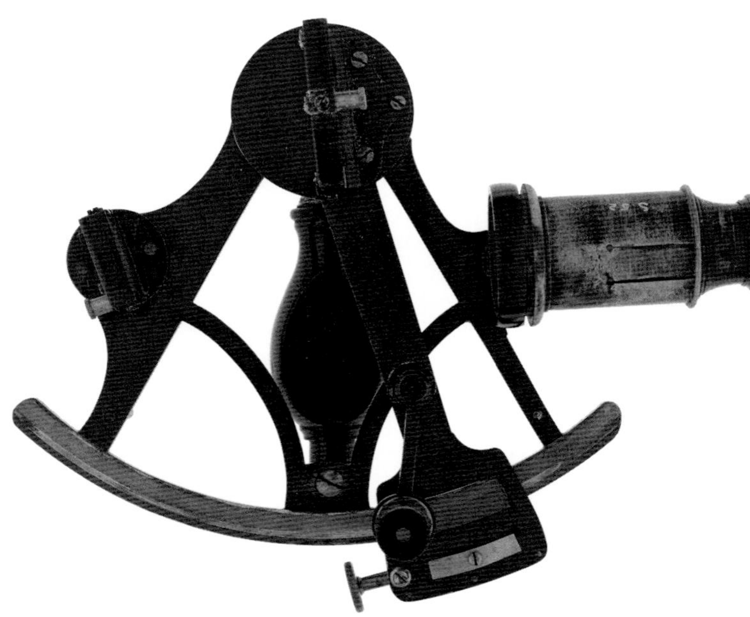

Origin England
Maker Unknown
Date About 1880
Inscriptions Marked on the left curve: H.O. 10
Marked on the left strut: 10
Graduation Polished brass limb with inlaid silver scale from -5° to 155° every 30′, measuring to 132°. Silver vernier to 1′, zero at the right.
Dimensions Radius 5 inches (127 mm)
Description Black-lacquered brass curved-pattern frame, wooden handle. The tangent screw and clamping screw are on the back of the index arm. No shades. Index-glass adjustment by a capped screw; adjustment of horizon glass by a square-headed screw, a detached key, and a capped screw. Magnifier on a 45-mm swivelling arm. Threaded telescope bracket without adjustment. Telescope (77 mm), erect image, with wide objective lens (star finder). Detached sight vane with one pinhole. Milled adjusting key. In a square fitted wooden box marked 'H.O. [broad arrow] 10', and containing in the lid a trade label for Cary, Porter, Ltd, 7 Pall Mall, London.
Provenance On indefinite loan from the Hydrographer of the Navy, since 1935.
Notes 'H.O.' stands for 'Hydrographic Office'.
Literature Clifton, *Directory*, 51.

207 NAV1159
Sextant

Origin England
Maker Unknown
Date About 1880
Inscriptions None
Graduation Polished brass limb with inlaid silver scale from -5° to 135° every 10′, measuring to 130°. Silver vernier to 10″, zero at the right.
Dimensions Radius 7¼ inches (184 mm)
Description Anodized brass triangle-pattern frame, wooden handle. The tangent screw and clamping screw are on the back of the index arm. Four shades (three red and one green); three horizon shades (two red and one green). Index-glass adjustment by a screw; adjustment of horizon glass by a capstan screw and a milled knob that is swivel-mounted. Magnifier on a 95 mm swivelling arm; a frosted glass shade. Threaded telescope bracket in two parts, fitted for correcting collimation error; perpendicular adjustment by rising-piece and a milled knob. Telescope (184), inverted image, four cross-wires; extra drawtube (81 mm), inverted image, two parallel cross-wires; telescope (83 mm), erect image. Sight-tube (80 mm), a shaded eyepiece (green), and an adjusting pin; an unidentified part is missing. In a keystone box, lined with green textile and with a blank brass plate on the lid. The lid contains a certificate for the index error, issued by 'Observatory', Devonport, dated 10 November 1898, with the name J. Coombes, Admiralty Agent, and signed 'Capt. J. I. Graham RN'.
Provenance Presented by Commander Ian Mudie RN, Stockbridge, Hampshire, in 1953.
Notes According to an advertisement in the *Post Office Directory of Plymouth, Devonport and Stonehouse* of 1906–1907, Messrs Coombes were optical and scientific instrument makers at 'The Observatory', Fore Street, Devonport (established 1806). They rated chronometers by transit observations and apparently also examined sextants. A similar certificate for a sextant, signed by Coombes and dated 1897, is preserved in the Peabody Essex Museum, Salem, Massachusetts, inv. no. M.11046.
Literature Brewington, *The Peabody Museum Collection*, 48.

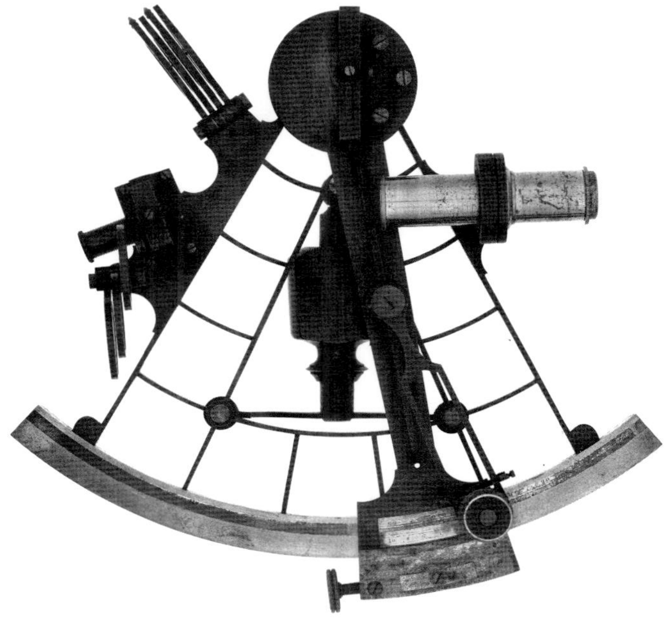

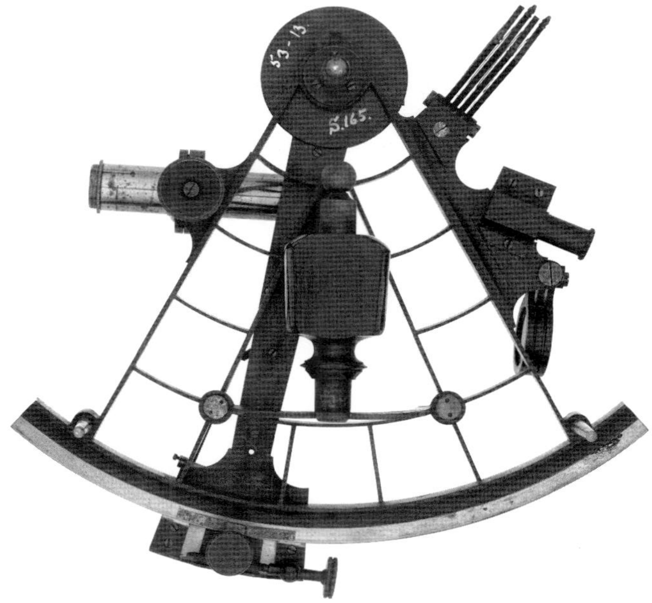

208 NAV1152
Sextant

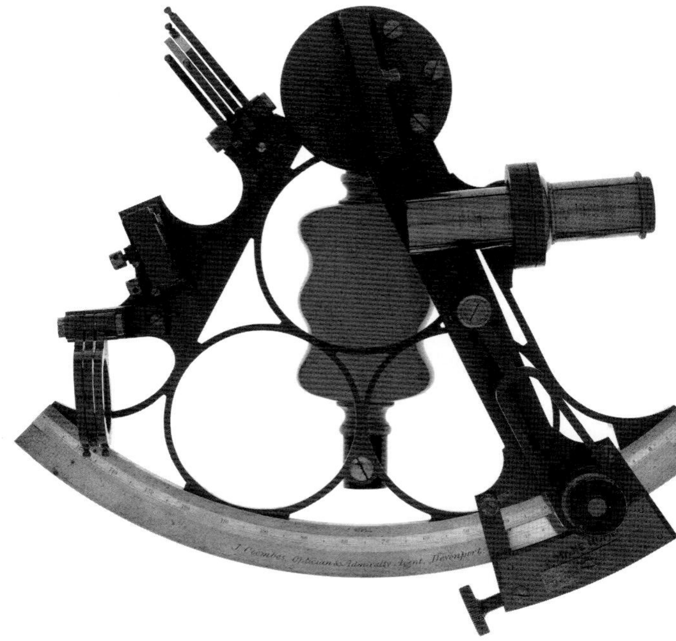

Origin Devonport, England
Maker J. Coombes
Date About 1891
Inscriptions Signed on the limb: J. Coombes, Optician & Admiralty Agent, Devonport.
Marked on the limb: 4525
Marked at the bottom of the index arm: A. Milne Home
Graduation Polished brass limb with inlaid silver scale from -5° to 160° every 10′, measuring to 135°. Silver vernier to 10″, zero at the right.
Dimensions Radius 6½ inches (165 mm)
Description Dark-grey-lacquered brass three-circle-pattern frame, wooden handle. The tangent screw and clamping screw are on the back of the index arm. Four shades (all grey); three horizon shades (all grey). Index-glass adjustment by a screw; adjustment of horizon glass by capstan screws. Magnifier on an 80 mm swivelling arm; a frosted glass shade (broken). Threaded telescope bracket in two parts, fitted for correcting collimation error; perpendicular adjustment by rising-piece and a milled screw. Telescope (84 mm), erect image; telescope (182 mm), inverted image, four cross-wires; extra drawtube (81 mm), two parallel cross-wires. Sight-tube (81 mm), a shaded eyepiece (red), an adjusting pin, and a magnifying glass (missing). In a square fitted mahogany box, containing a small ivory plate for Coombes Optican and Admiralty Agent Devonport. The lid contains a Kew Observatory certificate of examination, dated February 1891. An inlaid brass plate in the lid is marked 'A. Milne Home'.

Provenance A. Milne Home, the original owner, may have been the son of David Milne (1805–90), who assumed the name Milne Home in 1852. Presented by Commander Milne Home RN, Petersfield, Hampshire, in 1949.
Notes According to an advertisement in the *Post Office Directory of Plymouth, Devonport and Stonehouse* of 1906–07, Messrs Coombes were optical and scientific instrument makers at 'The Observatory', Fore Street, Devonport (established 1806), and manufactured sextants for the Royal Navy and the Britannia Royal Naval College.
Literature Marjory G. Roy, 'Home, David Milne [*formerly* David Milne] (1805–90), *advocate and scientist*', in *Oxford Dictionary of National Biography*, Oxford, 2004, <http://www.oxforddnb.com/>.

209 NAV1127
Sextant

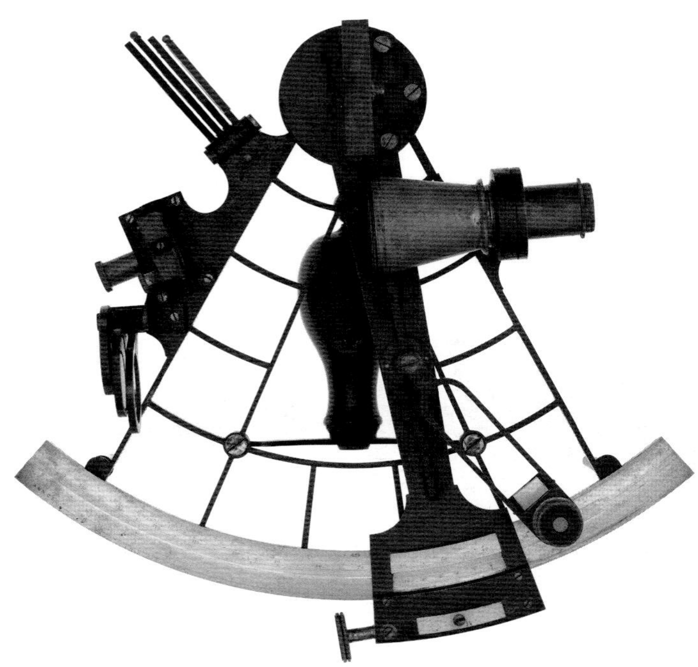

Origin England, possibly London
Maker Unknown
Date About 1892
Inscriptions Marked on the left end of the limb: A. & N.C.S.L
Marked (stamped) at the centre of the limb: 459
Graduation Polished brass limb with inlaid silver scale from -5° to 150° every 10′, measuring to 125°. Silver vernier to 10″, zero at the right.
Dimensions Radius 7¼ inches (184 mm)
Description Anodized brass triangle-pattern frame, wooden handle. The tangent screw and clamping screw are on the back of the index arm. Four shades (all green); three horizon shades (all green). Index-glass adjustment by a screw; adjustment of horizon glass by worm gear and a capped milled knob. Magnifier on an 83 mm swivelling arm; a frosted glass shade. Threaded telescope bracket in two parts, fitted for correcting collimation error; perpendicular adjustment by rising-piece and a milled knob. Telescope (186 mm), inverted image, two parallel

cross-wires (broken); telescope (84 mm), erect image (star finder). Shaded eyepiece (green); an adjusting pin. Several unidentified parts are missing. In a square fitted wooden box with a blank brass plate on the lid, lined with green textile, and containing a trade label for George Lee and Son, Portsmouth (before 1912), and also, in the lid, a Class A Kew Observatory certificate of examination, dated 1892.

Provenance This instrument was captured by the donor, then a midshipman on HMS *Centurion*, from the naval school at the Pei Yang Arsenal, near Tientsin, China, during the Boxer Rebellion in 1900. The Army and Navy Co-operative Society Ltd, London, was the retailer. It is likely that the retailer obtained its sextants from a local manufacturer such as Elliott Brothers, who, like the Society, were situated in the Strand, or from Henry Hughes and Son or Heath and Co. Presented by Commander W.B.C. Ross RN, London, in 1935.

Literature Clifton, 'An Introduction to the History of Elliott Brothers'. Clifton, *Directory*, 165; Hill and Paget-Tomlinson, *Instruments of Navigation*, 37.

210 NAV 1224
Box sextant

Pocket sextant
Origin London, England
Maker J.H. Steward
Date 1893
Inscriptions Signed on the upper plate: J. H. Steward. Strand, London. | Provly Patented [read from the pivot]
Graduation Inlaid silver scale from -5° to 205° every 30′, measuring to 186° (the digits are read from the pivot). Silver vernier 0′-30′, zero at the left.
Dimensions Radius 2¼ inches (57 mm); diameter 102 mm
Description Circular anodized brass plates with a detached wooden handle for attachment to the back of the instrument by bayonet fitting, and a plumb and line. The index arm is moved by a milled knob on the lower plate. Two index shades (both green); no horizon shades. Two index glasses; one rigidly fixed, the other moved by the index arm. Index-glass adjustment by a key fixed in the wooden handle; the horizon glass is completely silvered, with adjustment by a milled knob on the back of the instrument. Magnifier (single lens) on a 58 mm swivelling arm. Threaded telescope bracket without adjustment. Telescope (62 mm), erect image. In a fitted brown leather case with a carrying strap, lined with brown velvet, and containing in the lid a stamp for J.H. Steward, 406 Strand, London, and marked on the outside 'AJW'.
Provenance Purchased from Delehar Antiques Ltd, London, in 1977.
Notes The angle between the two index glasses may be increased to up to 180°. This instrument can be used for reading vertical angles, and was primarily meant for use on land. According to the patent it could, for example, be used when setting out railway curves. J.H. Blakesley and J.H. Steward patented this pocket sextant (no. 9440 of 1893).

211 NAV1115
Sextant

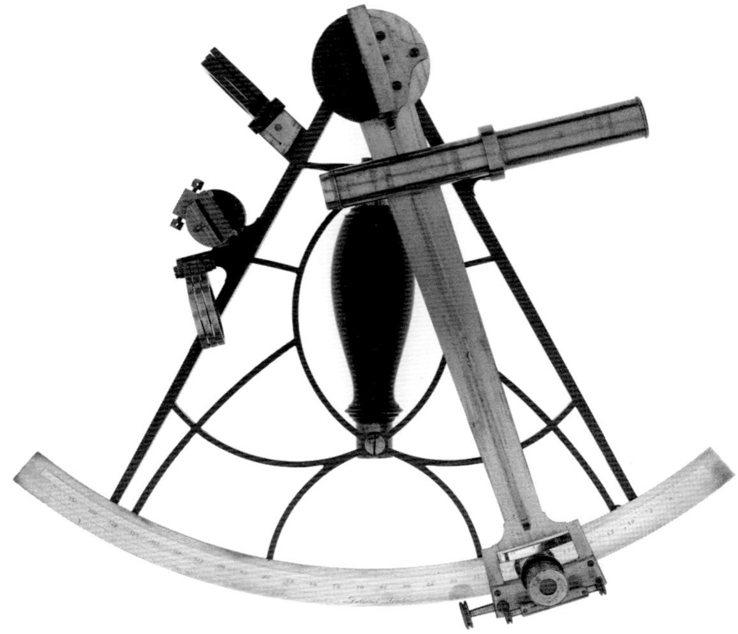

Origin London, England
Maker Dollond and Co. (Tyson Crawford)
Date About 1895
Inscriptions Signed on the limb: Dollond London
Graduation Polished brass limb with inlaid gold scale from -2° to 139° every 10′, measuring to 122°. Gold vernier to 10″, zero at the right.
Dimensions Radius 9½ inches (241 mm)
Description Anodized brass oval-curve-pattern frame, wooden handle. The tangent screw and clamping screw are on the back of the index arm. Three shades (one green and two red); three horizon shades (two red and one green). Index-glass adjustment by a screw; adjustment of horizon glass by capstan screws. Magnifier moved on a brass-covered screw by milled knobs at either end. Threaded telescope bracket with perpendicular adjustment by rising-piece and a milled knob. Telescope (150 mm), inverted image, four cross-wires; telescope (89 mm), erect image; extra drawtube (40 mm), inverted image, four cross-wires. Sight-tube (82 mm) and an adjusting pin. In a mahogany keystone box, containing in the lid trade labels for Dollond and Co., London, and for William Gerrard, Liverpool (1862–90), and an oval label in MS 'A. Bulmermill 1907'.
Provenance Adams collection; lent in 1936, bequeathed in 1974. The MS label may suggest the acquisition of this instrument by Adams in 1907.
Literature Barty-King, *Eyes Right*.

212 NAV1168
Sextant

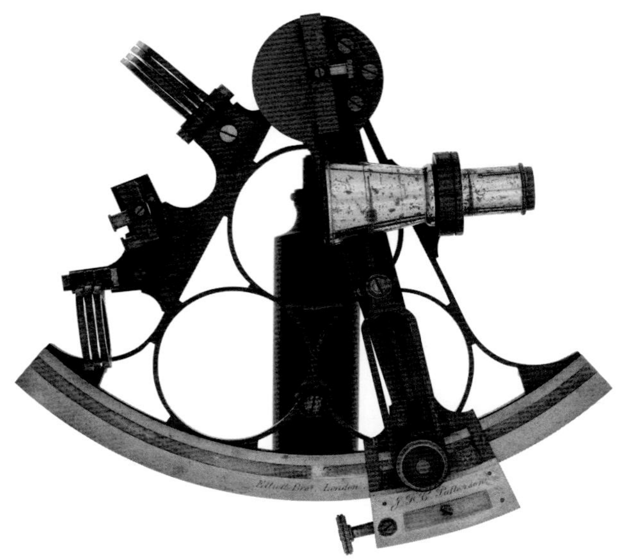

Origin London, England
Maker Elliott Brothers
Date About 1898
Inscriptions Signed on the limb: Elliott Bro[s], London.
Marked on the limb: 1012
Marked at the bottom of the index arm: J. F. C. Patterson.
Graduation Polished brass limb with inlaid silver scale from -5° to 155° every 10′, measuring to 123°. Silver vernier to 10″, zero at the right.
Dimensions Radius 6½ inches (165 mm)
Description Anodized brass three-circle-pattern frame, wooden handle with an ivorine plate on the back. The tangent screw and clamping screw are on the back of the index arm. Four shades (all green, one cracked); three horizon shades (all green). Index- and horizon-glass adjustment by capped capstan screws. Magnifier on a 75 mm swivelling arm; a frosted glass shade. Threaded telescope bracket in two parts, fitted for correcting collimation error; perpendicular adjustment by rising-piece and a milled knob. Telescope (186 mm), inverted image, two parallel cross-wires; extra drawtube (78 mm), inverted image, four cross-wires; telescope (82 mm), erect image (star finder). Sight-tube (82 mm), two shaded eyepieces (both green), and an adjusting pin; one unidentified part is missing. In a square fitted wooden box, containing in the lid a Kew Observatory certificate of examination, dated December 1898. An inlaid plate is missing from the top of the lid.
Provenance Purchased from Mrs Greville Clarke, Hurdle, Hampshire, in 1959.
Literature Clifton, 'An Introduction to the History of Elliott Brothers'.

213 NAV1136
Surveyor's double sextant and station pointer

Spiegel-Goniograph
Origin Trieste (Austro-Hungarian empire)
Maker H. and F. Müller
Date About 1900
Inscriptions Signed on the upper index glass mount: H. & F. Müller | Trieste | No 136
Graduation Steel scale from 143° to 0° to 143° every 30′, measuring to 130° on either side. Steel verniers to 1′, zeros at the right and left.
Dimensions Radius 3 inches (76 mm)
Description Green/black-lacquered brass frame, detached wooden handle. Two index glasses (one above the other), adjustment by screws; two horizon glasses (one above the other), adjustment by a square-headed screw and a detached key. No shades. The glasses are linked to the station pointer arms by rods, each of which has a clamping screw on top and a tangent screw at the bottom. The two horizon glasses have a reflecting portion only. Threaded telescope bracket without adjustment. Sight-tube (78 mm), a threaded pinhole eyepiece, three adjusting pins, and an adjusting key. In a rectangular fitted mahogany box with a brass carrying handle.
Provenance Presented by HM Queen Mary in 1936.
Notes With this instrument, the two angles that are measured may immediately be copied onto paper, without reading off. Konstantin Edler von Pott of the Imperial Austro-Hungarian Navy probably designed it around 1875.
Literature Pott, *Der doppelte Spiegel-Goniograph*.

214 NAV1173
Sextant

Origin London, England
Maker Cary
Date About 1900
Inscriptions Signed on the limb: Cary, London.
Marked on the limb: Gold & Platina | 2493
Graduation Polished bronze limb with inlaid gold scale from -5° to 154° every 10′, measuring to 129°. Platinum vernier to 10″, zero at the right.
Dimensions Radius 8 inches (203 mm)
Description Black-lacquered bronze straight-bar-pattern frame; wooden handle with a brass-lined hole for placing it on a stand. The tangent screw and clamping screw are on the back of the index arm. Four shades (all green); three horizon shades (all green). Index-glass adjustment by a screw; hinge-mounted horizon glass, adjustment by a capped milled screw. Magnifier on a 75 mm swivelling arm with a brass tube for holding a light; a frosted glass shade. Threaded telescope bracket in two parts, fitted for correcting collimation error; perpendicular adjustment by rising-piece and a milled knob. Telescope (232 mm), inverted image, focusing by milled knob, no cross-wires; extra drawtube (49 mm), inverted image, two parallel cross wires; telescope (84 mm), erect image (star finder). Two shaded eyepieces (both green), an adjusting pin, and a light for the scale (latter missing). In a square fitted box (an inlaid plate is missing from the lid), containing in the lid a Class A National Physical Laboratory certificate of examination, dated May 1914.
Provenance Presented by Commander E.L. Pawsey RN, Tonbridge, Kent, in 1962.
Literature Clifton, *Directory*, 51; Chaldecott, 'Platinum and Palladium in Astronomy and Navigation'.

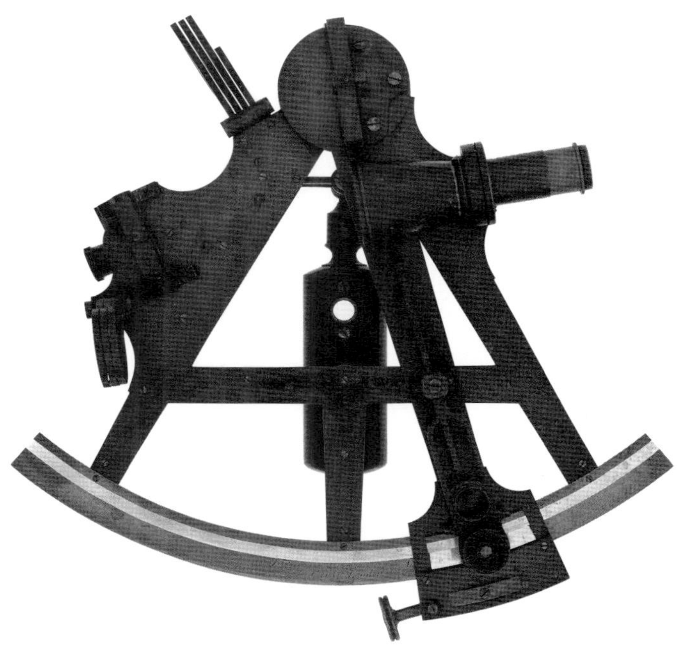

215 NAV1205
Box sextant

Pocket sextant
Origin London, England
Maker J.H. Steward Ltd
Date About 1900
Inscriptions Signed on the upper plate: J.H. Steward. Ltd. London. [read from the pivot]
Marked on the upper plate: 3366 [broad arrow].
Graduation Inlaid silver scale from -5° to 135° every 30′, measuring to 117° (the digits are read from the pivot). Silver vernier to 1′, zero at the left.
Dimensions Radius $1\frac{3}{4}$ inches (44 mm); diameter 79 mm
Descriptions Circular anodized brass plates with a lid that when screwed on the back of the instrument, serves as a handle. The lid contains a paper table of natural tangents to a base of 100. Index arm moved by a milled knob on the upper plate. Two index shades (red and green) moved by levers; no horizon shades. Index-glass adjustment by a screw; adjustment of horizon glass by square-headed screws, with a key fixed in the upper plate. Magnifier (single lens) on a 57 mm swivelling arm attached at the pivot. Telescope (50 mm), erect image, fitted in the body of the instrument. Shaded eyepiece (red); a pinhole sight shutter is fitted for use when the telescope is removed. No box or case.
Provenance Presented by Mr A.H. Hammond, Harrow, Middlesex, in 1969.
Notes The broad arrow indicates that the instrument belonged to the Royal Navy.

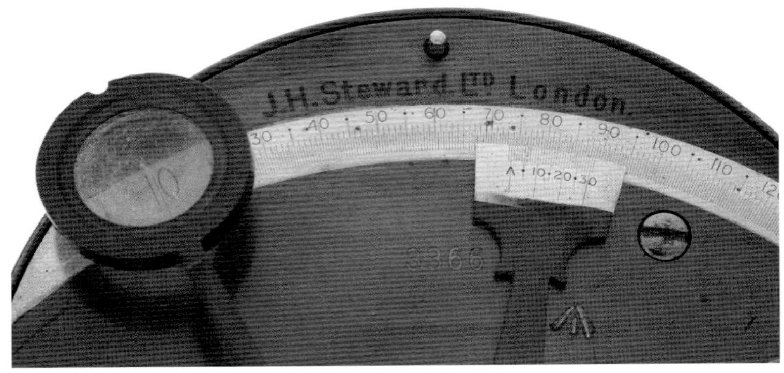

216 ZBA4568
Sextant

Origin Liverpool, England
Maker Frodsham and Keen
Date About 1904
Inscriptions Signed on the limb: Frodsham & Keen, Liverpool
Graduation Polished brass limb with inlaid silver scale from -5° to 150° by 10′, measuring to 133°. Silver vernier to 5″, zero at the right.
Dimensions Radius 6.2 inch (155 mm)
Description Anodized brass three-circle pattern frame, wooden handle. The tangent screw and clamp screw are on the back of the index arm. Four shades (blue, orange, red, grey); three horizon shades (red, orange, blue). Index glass adjustment by a screw; adjustment of horizon glass by capstan screws. Magnifier on an 80 mm swivelling arm; a frosted glass shade. Threaded telescope bracket in two parts for correcting collimation error; perpendicular adjustment by rising piece and a milled knob. Telescope (83 mm), erect image; telescope (186 mm) inverted image, four cross wires. Sight-tube (80 mm), a shaded eyepiece (red) and an adjusting pin. In a fitted square wooden box with felt lining, one part missing (probably an eyepiece with variable shades). Brass plaque on top of the case lid inscribed: H.M.S. "Conway" | The | Brocklebank Prize | Presented to | J.D. Campbell | July 1904

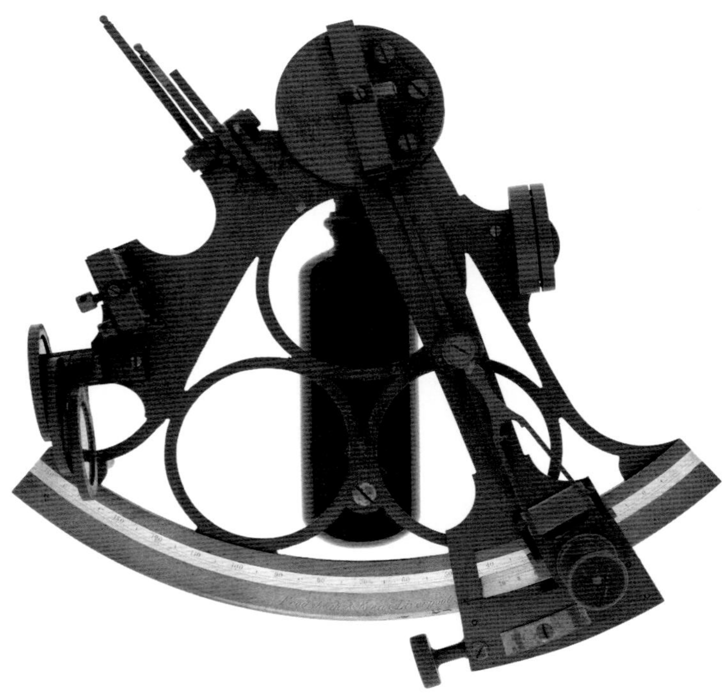

Provenance John Duncan Campbell (born 1886), the original owner, was admitted to HMS Conway in 1902 and left in 1904. It was founded in 1859 by the Mercantile Marine Service Associations to train boys for service in the British merchant navy. The training ship was moored on the river Mersey until 1941. The sextant was one of a number of prizes won by Campbell while he was at the school. Purchased via Gillian Gould, Antique Scientific Instruments and Collectables, London, in 2008.
Literature Clifton, *Directory*, 106.

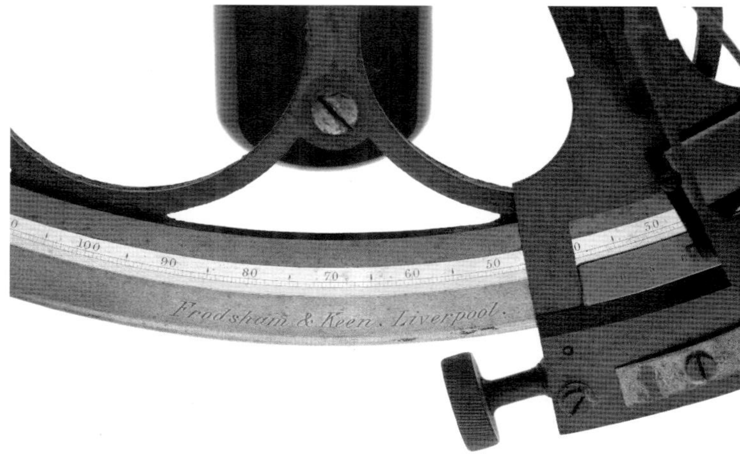

217 ZBA4538
Surveying sextant

Origin London, England
Maker The Soho Scientific Instrument Company
Date 1909 or later
Inscriptions Signed on the side of the left strut: Prov$^{l.}$ Patent 9082–09 | THE SOHO | SCIENTIFIC INSTRUMENT C$^{o.}$ | LONDON | No 0972
Graduation Inlaid silver scale from -5° to 140° every 1°, measuring to 120°. Silver vernier reading to 15′, zero at the left.
Dimensions Radius 5 inches (127 mm)
Description Partly black-lacquered brass folding frame, the frame serves as the handle. The two struts are pivoted and act through a collar sliding on a third pivoted arm. No tangent screw; the clamping screw is on the front of the frame. No shades. Index- and horizon-glass adjustment by capstan screws. The lower half of the index glass serves to view to the observer's rear, along a bubble level. Fixed cylindrical telescope tube. Sight-tube (43 mm). A hinged compass (diameter 50 mm) is fitted to the left strut; it is graduated into points, and per quadrant in degrees. A bubble level adjustable by capstan screws is fitted on the left strut. No box.
Provenance Purchased from Tesseract, Hastings-on-Hudson, New York, USA, in October 2007.
Notes Thomas James Rickford patented this surveying instrument (no. 9082 of 1909); the patent was granted in 1910. The instrument can be used to measure horizontal and vertical angles, and the measured angle may be protracted in actual reading off.
Literature Tesseract Early Scientific Instruments, *Catalogue Eighty-Four*, Summer 2007, no. 28.

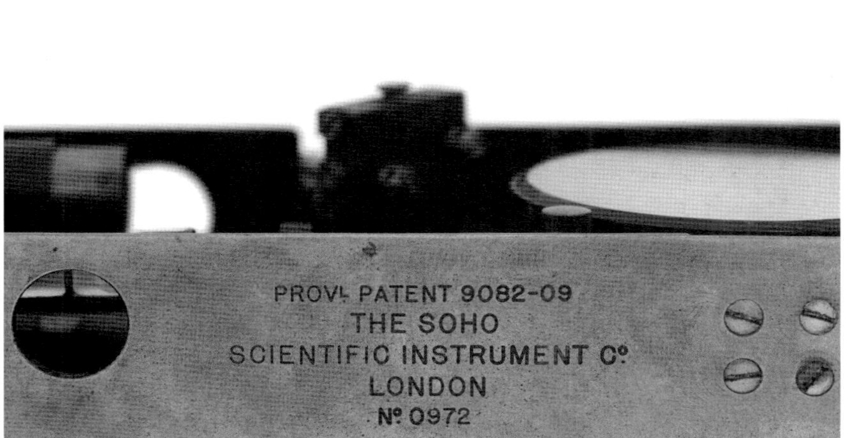

218 NAV1196
Sextant

Origin Neukölln, Germany
Maker H. Haecke
Date About 1910
Inscriptions Signed on the upper part of the index arm: H Haecke Neukölln 7477
Marked on a brass plate on the handle: [crown] | M721
Graduation Polished brass inlaid silver scale, of which the division survives, every 10′, but the digits are missing. Silver vernier to 10″, zero at the right.
Dimensions Radius 7 inches (178 mm)
Description Anodized brass diamond-pattern frame, wooden handle. The tangent screw and clamping screw (both bent) are at the bottom of the index arm. Four round shades (all grey); three horizon shades (all grey). The index and horizon glass and their mounts are missing. The magnifier is missing. Threaded telescope bracket in two parts, fitted for correcting collimation error; perpendicular adjustment by rising-piece and a milled knob. The telescopes are missing. In poor condition. No box.
Provenance Recovered from the wreck of a German warship scuttled at Scapa Flow at the end of World War I, and given to Vice Admiral Sir Henry Francis Oliver (1865–1965), Commander-in-Chief of the Home Fleet, in 1919. Presented by his widow, Dame Beryl Oliver, GBE, RRC, London, in 1966.
Notes The crowned M refers to the Imperial German Navy.
Literature Brachner, 'German Nineteenth-Century Scientific Instrument Makers', 141.

219 NAV1222
Sextant

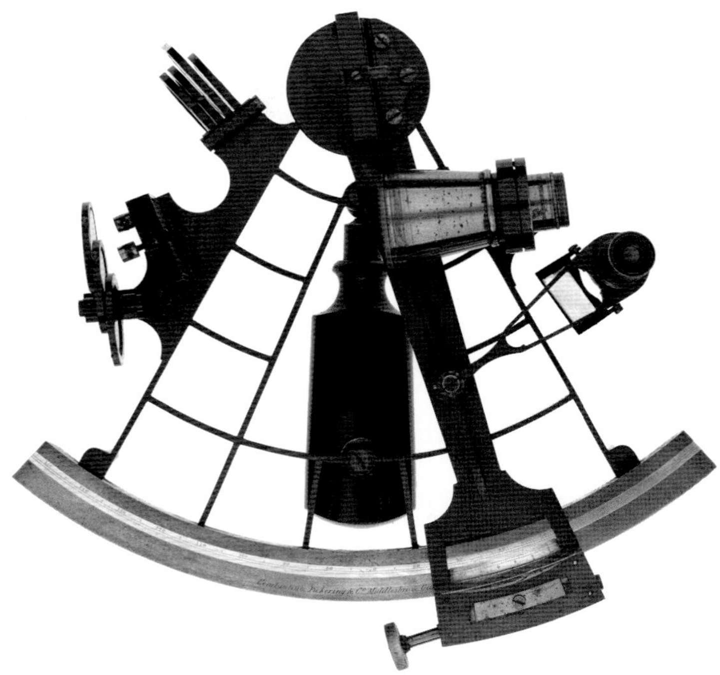

Origin England
Maker Unknown
Date About 1915
Inscriptions Marked on the limb: Constantine Pickering & C°. Middlesbro & Cardiff
Graduation Polished brass limb with inlaid silver scale from -5° to 150° every 10′, measuring to 129°. Silver vernier reading to 10″, zero at the right.
Dimensions Radius 7.7 inches (196 mm)
Description Anodized brass triangle-pattern frame, wooden handle. The tangent screw (the thread has a protective brass cover) and clamping screw are on the back of the index arm. Four shades (three red and one green); three horizon shades (two red and one green). Index-glass adjustment by a screw; adjustment of horizon glass by capstan screws. Magnifier on an 80 mm swivelling arm; a frosted glass shade. Threaded telescope bracket in two parts, fitted for correcting collimation error; perpendicular adjustment by rising-piece and a milled knob. Telescope (180 mm), inverted image, four cross-wires; extra drawtube (77 mm), inverted image, two parallel cross-telescope (89 mm), erect image; telescope (68 mm), erect image (star finder). Sight-tube (82 mm), a shaded eyepiece (red), an adjusting pin, and a magnifying glass. Marked on a brass plate (detached) 'Special prize presented to J.B. Hewson for executive ability & all-round efficiency'. In a square fitted wooden box.
Provenance Bequeathed by Sir Joseph Bushby Hewson, Radlett, Hertfordshire, in 1976.

Notes The Constantine and Pickering Steamship Company probably awarded this prize. Commander Sir Joseph Bushby Hewson, FRGS, RD, RNR, Extra Master (1902–76), was the author of *A History of the Practice of Navigation* (Glasgow, 1951).

220 NAV1213
Surveyor's sextant

Paget angle sextant
Origin London and York, England
Maker T. Cooke and Sons Ltd
Date About 1915
Inscriptions Signed on the upper plate: T. Cooke & Sons Ltd. | London & York.
Marked on the upper plate: The Paget | Angle Sextant | No. 687
Graduation Ivorine scale along the rim of the upper plate, from 0° to 90° to 0° every 30′. The scale is read by use of a black indicating line on an ivorine index mark at the end of the index arm.
Dimensions Diameter 4½ inches (114 mm)
Description Black-lacquered aluminium frame consisting of two circular plates, the upper having an ivorine overlay; a wooden handle, to be screwed into the lower plate. There is no tangent or clamping screw; the index arm is moved by hand. No shades. Index- and horizon-glass adjustment by capstan screws with an adjusting pin fixed inside the handle. The sight vane has one pinhole mounted within the frame. In a fitted wooden box containing a Hydrographic Department packing note dated 3 June 1918.
Provenance Presented by the Navigation and Direction Division of the Admiralty in 1961.
Notes The Paget sextant was designed around 1892, and brought onto the market by H. Hughes and Son.
Literature NN, 'A New Angle Sextant'; McConnell, *Instrument Makers to the World*; Ifland, *Taking the Stars*, 131–2.

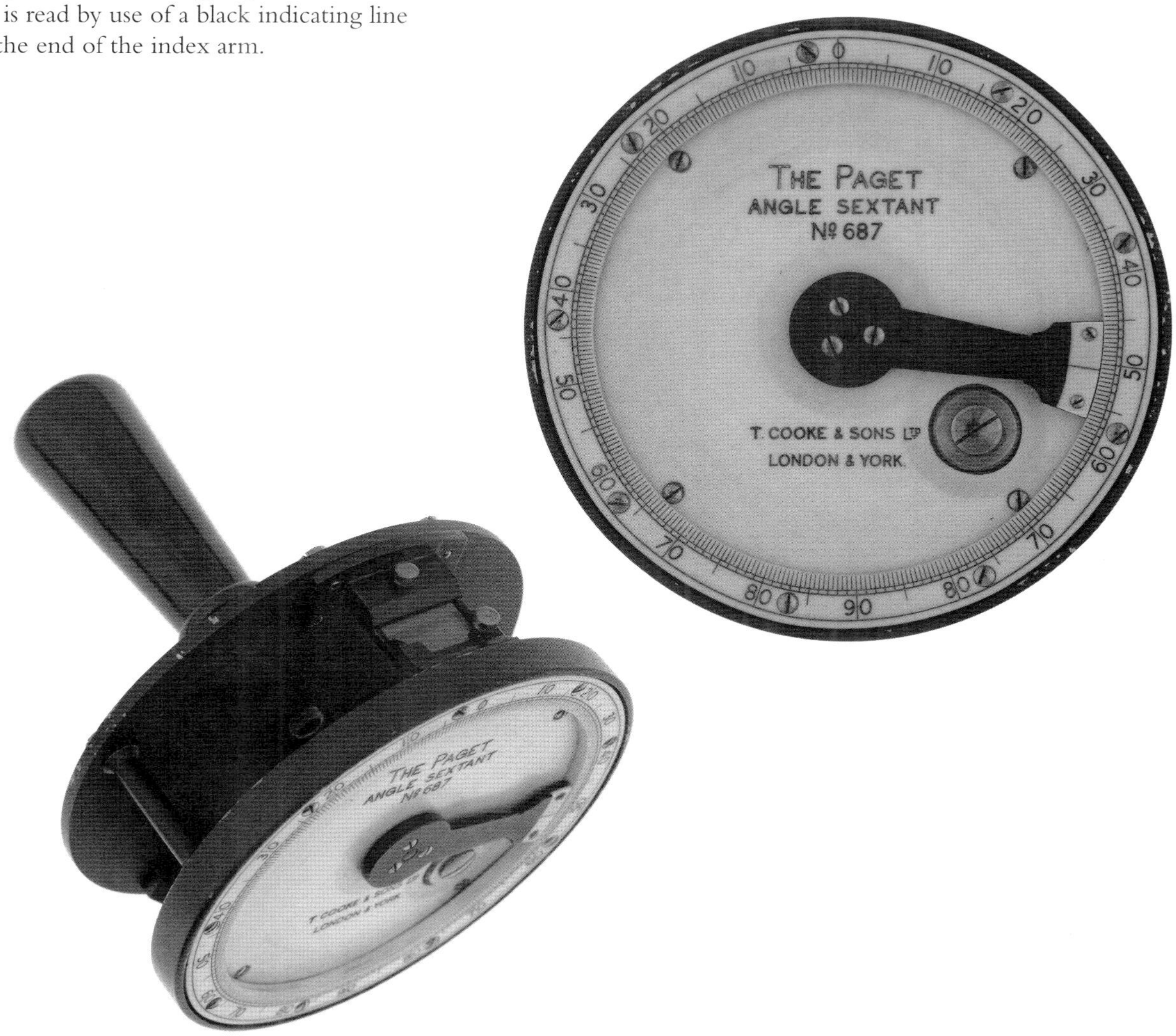

221 NAV1239
Sextant with artificial horizon

Origin Paris, France
Maker Hurlimann, Ponthus and Therrode
Date About 1915
Inscriptions Signed on the limb: Hurlimann, Ponthus & Therrode, succ.rs à Paris (2975)
Signed on the artificial horizon: PONTHUS & THERRODE | Paris
Graduation Polished brass limb with inlaid silver scale from -5° to 130° every 20′, measuring to 108°. Silver vernier to 10″, zero at the right.
Dimensions Radius 5½ inches (140 mm)
Description Black-lacquered brass frame, wooden handle, and a tube on the back of the frame for placing the instrument on a stand. The double-ended tangent screw is on the front of the index arm; the clamping screw is on the back. Three index shades (red, green, and orange); two horizon shades (both orange) on the outside of the artificial horizon. Index glass (large) and horizon glass (large with circular opening in the centre of the silvering); adjustment of both by square-headed screws and a detached key. Magnifier (single lens) on a 70-mm swivelling arm; a frosted glass shade. Threaded telescope bracket without adjustment. Telescope (10 mm), erect image, with a wide objective lens (star finder), perhaps not original to this instrument. Adjusting key (missing). Box missing.
Artificial horizon This consists of an airtight housing containing a gyroscope that can be locked with a lever mechanism. The gyroscope has a glass piece through the body, with a series of horizontal lines, and is propelled by an air pump (missing). The air gauge is calibrated from 0 to 76 cm of mercury.
Provenance Purchased at Sotheby's, London, June 1984, lot no. 500, through Harriet Wynter Ltd.
Notes Admiral Georges Ernest Fleuriais (1840–95) first patented this artificial horizon in 1886. An earlier version in the collection is cat. no. 331.
Literature Ifland, *Taking the Stars*, 115–17; Marcelin, *Dictionnaire des fabricants français*; Mars, *Zeevaartkundige hoekmeetinstrumenten*, 12–20; Hughes, *The Book of the Sextant*.

222 NAV1166
Sextant

Bell pattern Mark III sextant
Origin New Eltham, London, England
Maker Heath and Co. Ltd
Date About 1917
Inscriptions Signed on the limb: Heath & Co. Ltd. New Eltham. London.
Marked on the limb: P.350
Marked on the frame: Rd.517155
Marked on bottom of the index arm: J. R. Grant
Graduation Polished brass limb with inlaid silver scale from -5° to 155° every 10′, measuring to 131°. Silver vernier to 10″, zero at the right.
Dimensions Radius 7 inches (178 mm)
Description Anodized brass bell-pattern frame, grey-green finish, wooden handle with a brass-lined hole for the locking mechanism. The tangent screw and clamping screw are on the back of the index arm. Four shades (all green); three horizon shades (all green). Index- and horizon-glass adjustment by a square-headed screw and a detached key, and capped square-headed screws. Magnifier on a 95 mm swivelling arm; a frosted glass shade. Threaded telescope bracket in two parts, fitted for correcting collimation error; perpendicular adjustment by rising-piece and a milled knob. Telescope (176 mm),

inverted image, four cross-wires; extra drawtube (95 mm), inverted image, two parallel cross-wires; telescope (72 mm), erect image (star finder). Sight-tube (82 mm), and two shaded eyepieces (both green), a patent greatest-angle clamp screw, a Heath's patent definition shade, and a milled adjusting key. One part is missing (possibly a screw-driver). In a square fitted wooden box with a patent locking mechanism for securing the instrument when not in use. A white ivorine label on the side of the lid is marked 'Bell pattern Mark III sextant'. An inlaid brass plate on top of the lid is marked 'J. R. Grant'. The lid contains a National Physical Laboratory certificate of examination, dated 8 September 1917.
Provenance Presented by Mrs Doyle, Beckenham, in 1959.
Notes Heath's patent definition shade was to prevent the eye being dazzled during an observation in bright sunlight.
Literature Watts, *The Sextant Simplified*, 79.

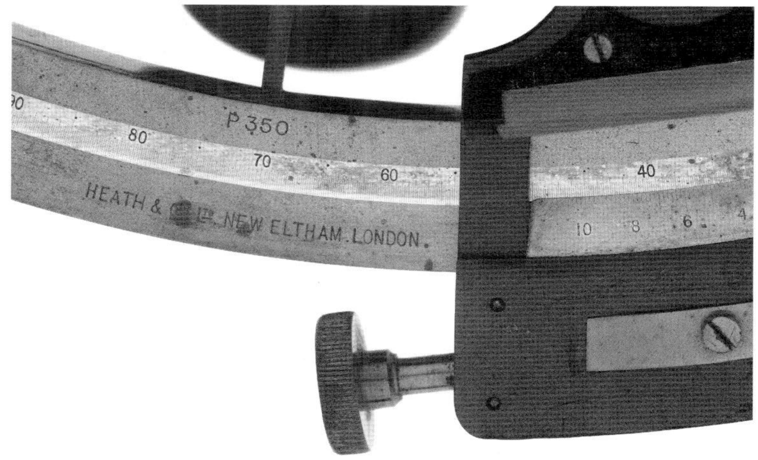

223 NAV1221
Box sextant

Pocket sextant
Origin London, England
Maker W.F. Stanley and Co. Ltd
Date 1918
Inscriptions Signed on upper plate: Stanley, London. | N°. 377 | 1918 [read from the pivot]
Signed on the lid: Stanley, London | No. 3776. | 1918
Marked on the upper plate: [broad arrow]
Marked on the lower plate: 88
Graduation Inlaid silver scale from -5° to 130° every 30′, measuring to 115° (the digits are read from the pivot). Silver vernier to 1′, zero at the left.
Dimensions Radius 1¾ inches (44 mm); diameter 68 mm
Description Circular brown-lacquered brass plates with a lid with a bayonet fitting that when attached to the back of the instrument, serves as a handle. The lid has a brass handle, and contains a paper table of natural tangents to a base of 100. The index arm is moved by a milled knob on the upper plate. Two index shades (red and green) moved by levers; no horizon shades. Index glass without adjustment; adjustment of horizon glass by a milled key fixed in the upper plate. Magnifier (single lens) on a 45-mm swivelling arm. Sliding sight vane with one pinhole and a shade (blue). In a fitted leather carrying case with a strap, and lined with red velvet. The case is marked 'D. Mason and Sons, Ltd. Birmingham 1918'.
Provenance Bequeathed by Sir Joseph Bushby Hewson, Radlett, Hertfordshire, in 1976.
Notes The broad arrow indicates that the instrument belonged to the Royal Navy.
Literature NN, *Stanley: Great Turnstile, Holborn, London*, 54.

224 ACO1257
Surveyor's sextant

Paget angle sextant
Origin London, England
Maker Henry Hughes and Son Ltd
Date About 1920
Inscriptions Signed on the upper plate: H. Hughes & Son | 59 Fenchurch St | London
Marked on the upper plate: The Paget | Angle Sextant | N° 104 | [in MS] Z57
Graduation Ivorine scale in black from 0° to 90° to 0° every 30′. The scale is read by use of a black indicating line on an ivorine index mark at the end of the index arm.
Dimensions Diameter 4½ inches (114 mm)
Description Black-lacquered alloy frame consisting of two circular plates, the upper having an ivorine overlay; a wooden handle, to be screwed in to the lower plate. There is no tangent or clamping screw; the index arm is moved by hand. No shades. Index- and horizon-glass adjustment by screws. The sight vane has one pinhole mounted within the frame. In a fitted mahogany box lined with green textile.
Provenance Transferred from the Admiralty Compass Observatory in 1969.
Notes The Paget sextant was designed around 1892 and brought onto the market by H. Hughes and Son.
Literature NN, 'A New Angle Sextant'; Ifland, *Taking the Stars*, 131–2.

225 ACO1357
Surveyor's sextant

Paget angle sextant
Origin London and York, England
Maker T. Cooke and Sons Ltd
Date About 1920
Inscriptions Signed on the upper plate: T. Cooke & Sons, Ltd | London & York. | No. 124
Graduation Ivorine scale on the upper plate with black marking from 0° to 90° to 0° every 30′. The scale is read by use of a black indicating line on an ivorine index mark at the end of the index arm.
Dimensions Diameter 4½ inches (114 mm)
Description Black-lacquered alloy frame consisting of two circular plates, the upper having an ivorine overlay; a wooden handle, to be screwed into the lower plate. There is no tangent or clamping screw; the index arm is moved by hand. No shades. Index- and horizon-glass adjustment by screws. The sight vane has one pinhole mounted within the frame. In a fitted wooden box, the lid is lined with green textile.
Provenance Transferred from the Admiralty Compass Observatory in 1993.
Notes The Paget sextant was designed around 1892 and brought onto the market by H. Hughes and Son.
Literature NN, 'A New Angle Sextant', McConnell, *Instrument Makers to the World*; Ifland, *Taking the Stars*, 131–2.

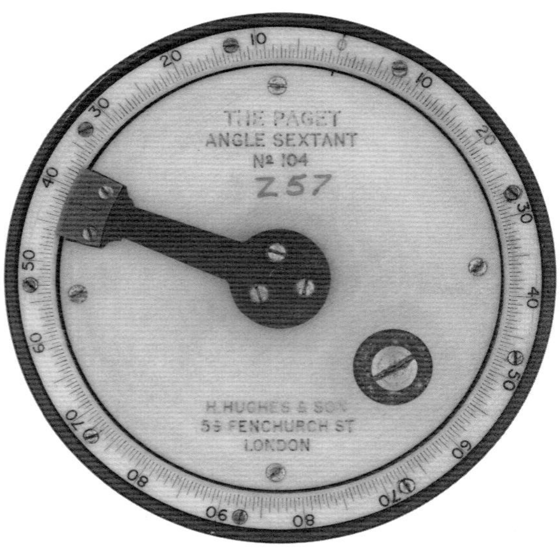

226 NAV 1169
Sextant

Origin Cardiff, Newport, and Swansea, Wales
Maker Wilson Fletcher Bruce and Sons Ltd
Date About 1920
Inscriptions Signed on the limb: Wilson.Fletcher.Bruce & Sons Ltd. Cardiff. Barry Dock. Newport & Swansea.
Marked on the limb above the silver scale: S 31
Graduation Polished brass limb with inlaid silver scale from -5° to 156° every 10′, measuring to 129°. Silver vernier to 1′, zero at the right.
Dimensions Radius 6½ inches (165 mm)
Description Anodized brass three-circle-pattern frame, wooden handle. The tangent screw and clamping screw are on the back of the index arm; a protective brass cylinder covers the thread. Four shades (two red, one green, and one blue); three horizon shades (blue, green, and red). Index-glass adjustment by a screw; adjustment of horizon glass by capstan screws. Magnifier on an 85 mm swivelling arm; a frosted glass shade. Threaded telescope bracket in two parts, fitted for correcting collimation error; perpendicular adjustment by rising-piece and a milled knob. Telescope (181 mm), inverted image, four cross-wires; telescope (68 mm), erect image (star finder). Sight-tube (85 mm) with a fitted shade (green), an adjusting pin, and a shaded eyepiece (missing). In a square fitted wooden box, containing in the lid a Hezzanith Observatory Works certificate of examination, dated January 1920.
Provenance Purchased from Mr H.T. Pockett, Greenwich, London, in 1960.
Notes Pockett's was a long-standing family-owned Greenwich pawnbroker, which in the late nineteenth and early twentieth century lent money against instruments to officers at the Royal Naval College. Pockett's lasted as a souvenir shop into the 1980s.

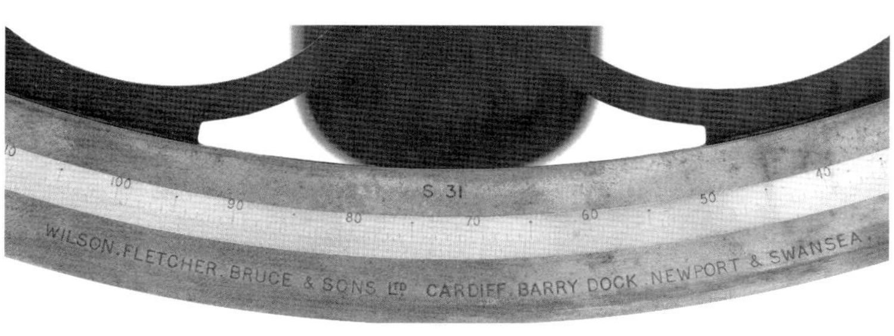

227 NAV 1228
Surveyor's sextant

Paget angle sextant
Origin London, England
Maker Henry Hughes and Son Ltd
Date About 1920
Inscriptions Signed on the upper plate: H. Hughes & Son, Ltd. | London
Marked on the upper plate: The Paget | Angle Sextant | No 896
Graduation Ivorine scale from 0° to 90° to 0° every 30′. The scale is read by use of a black indicating line on an ivorine index mark at the end of the index arm.
Dimensions Diameter 4½ inches (114 mm)
Description Black-lacquered alloy frame consisting of two circular plates, the upper having an ivorine overlay; a wooden handle, to be screwed into the lower plate. There is no tangent or clamping screw; the index arm is moved by hand. No shades. Index- and horizon-glass adjustment by capstan screws, with an adjusting pin fixed in the handle. The sight has one pinhole mounted within the frame. In a square fitted wooden box, containing in the lid a trade label for H. Hughes and Son. Scratched on the lid 'Paget Angle Sextant Used with Station Pointer'.
Provenance Presented by the Directorate of Naval Warfare, London, in 1979.
Notes The Paget sextant was designed around 1892 and brought onto the market by H. Hughes and Son.
Literature NN, 'A New Angle Sextant', Ifland, *Taking the Stars*, 131–2.

228 NAV1248
Sextant

Origin New Eltham, London, England
Maker Heath and Co. Ltd
Date 1921
Inscriptions Signed on the limb: Heath & Co Ltd. New Eltham. London.
Marked on the limb: U556
Marked on a plate at the end of the index arm: "Hezzanith" | Endless Tangent Screw [and] Automatic Clamp | Pat. Nº 17840. [and] Semper | Paratus
Marked on the back of the horizon glass mounting: Nº 517155
Graduation Polished brass limb with inlaid silver scale from -5° to 156° every 10′, measuring to 131°. Silver vernier to 1′, zero at the right.
Dimensions Radius 7 inches (178 mm)
Description Anodized; wooden handle with a brass-lined hole for the locking mechanism. The endless tangent screw and clamping screw are on the back of the index arm. Four shades (all green); a Wollaston prism; three horizon shades (all green). Index- and horizon-glass adjustment by capped capstan screws. Magnifier on an 80-mm swivelling arm; a frosted glass shade. Threaded telescope bracket in two parts, fitted for correcting collimation error; perpendicular adjustment by a clamping screw. Telescope (219 mm), inverted image, large objective lens; telescope (179 mm), inverted image; telescope (73 mm), erect image (star finder); telescope (180 mm), inverted image, four cross-wires; extra drawtube (70 mm, incomplete). Sight-tube (82 mm), two shaded eyepieces (green, and mauve), two L-shaped brackets, a patent greatest angle clamp, a screwdriver, and an adjusting spanner. In a square fitted wooden box with a patent locking mechanism for securing the instrument when not in use, and a blank inlaid brass plate on the lid. Contained in the lid is a National Physical Laboratory certificate of examination, dated January 1921.
Provenance Bequeathed by Captain E.L. Davies RN, Brighton, in 1986, through Donne, Mileham and Haddock, Solicitors.
Notes The Wollaston prism was used for star observations at night, when the horizon was not well defined. It produces two images of a star, one above and the other below the original position of the star, with the horizon exactly in between them.
Literature Watts, *The Sextant Simplified*, 82.

229 NAV1250
Sextant

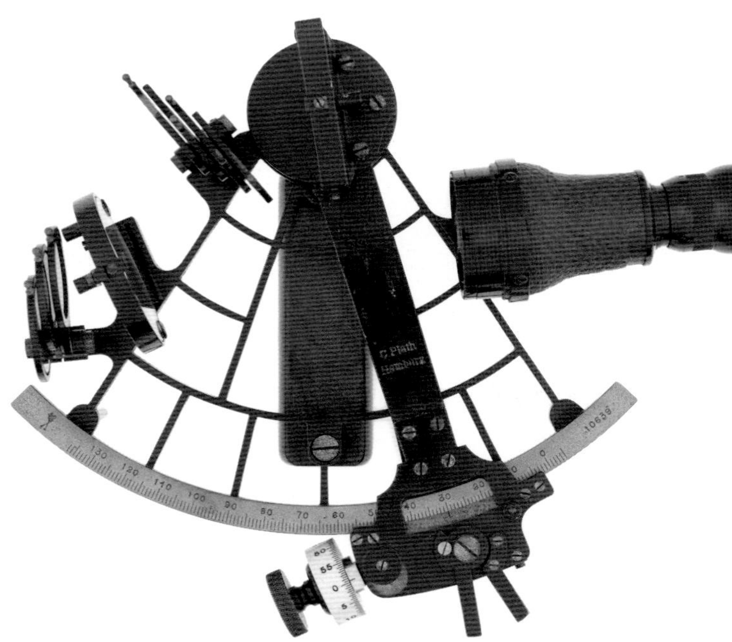

Origin Hamburg, Germany
Maker Carl Plath
Date About 1925
Inscriptions Signed on the index arm: C. Plath | Hamburg
Marked on the limb [Plath logo] | 10639
Graduation Polished brass scale from -5° to 137° every ½′, measuring to 132°. Micrometer to ½′
Dimensions Radius 6¼ inches (159 mm)
Description Anodized brass frame, wooden handle. White ivorine micrometer drum, pressure clamp and worm screw at the bottom of the index arm. Four shades (all green); three horizon shades (all green). Index-glass adjustment by a capped square-headed screw and a separate spanner; large round horizon glass, adjustment by a capped square-headed screw and a detached spanner. Threaded telescope bracket in two parts, fitted for correcting collimation error; perpendicular adjustment by rising-piece and a milled knob. Telescope

(103 mm), erect image; telescope (76 mm), erect image. Sight-tube (83 mm), a spanner, a shaded eyepiece (green) and a brush. In a square fitted wooden box, containing in the lid a metal trade label for C. Plath, Hamburg, Stubbenhuk 25. The box contains a Kelvin Hughes certificate of examination, dated 18 April 1955, and a trade label for Lilley and Reynolds, London.
Provenance Presented by Captain W.A. Howgego RN, Royal Alfred House, Eastbourne, East Sussex, in 1988.
Literature Schaafhausen *et al.*, *C Plath 1862–1962*, 96–8.

230 ACO1267
Bubble sextant

Type 2
Origin Japan
Maker Shimadzu Corporation
Date After 1926
Inscriptions Signed in Japanese characters on a plate on the averager: Shimadzu Corporation
Marked in Japanese characters on the plate on the averager: Air Bubble Sextant, Type 2 | No. 3148 | Date of manufacture, in Japanese characters: Showa [the Showa era began in 1926]
Marked in Japanese characters on the circular handle: Spare bulb in[side]
Dimensions 254 × 159 × 108 mm (length × height × width)

Description Casing of black-lacquered alloy and artificial materials, housing optical equipment, clockwork averaging mechanism, electrical equipment, and a rubber eyecup. Two index shades (both green); one detached shade (red). A white plastic plate fitted to the side of the instrument is marked in Japanese characters 'Instrumental error chart.' Adjusting pin. In a fitted wooden box, the lid is missing. The instrument is incomplete.
Provenance Transferred from the Admiralty Compass Observatory in 1969.

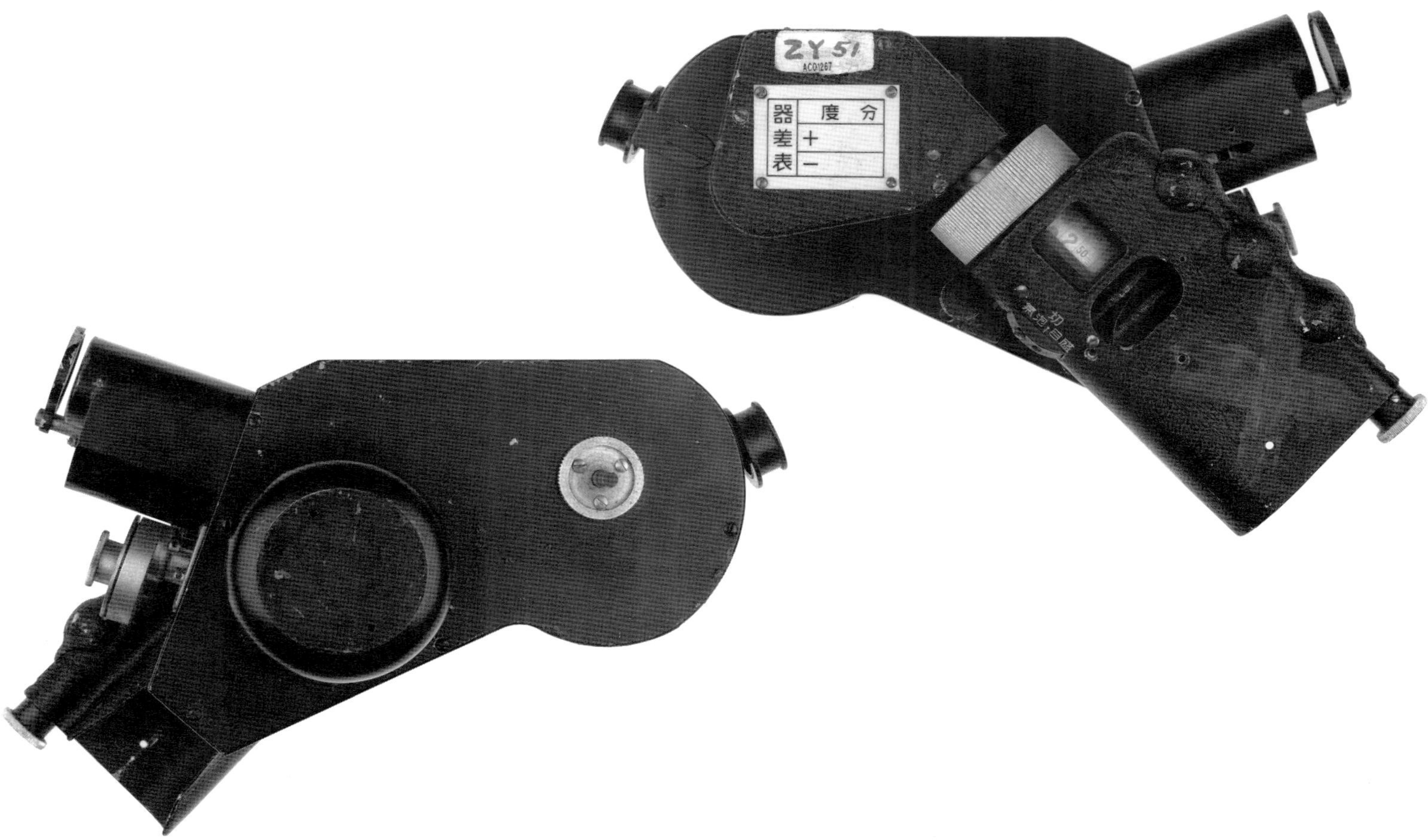

231 NAV1219
Sextant

Origin New Eltham, London, England
Maker Heath and Co. Ltd
Date 1934
Inscriptions Signed on the limb: Heath & Co New Eltham London S.E.9.
Marked on the limb: D 242 | Made in England | Bell Trade Mark Regd' | D S Tibbits R.N.
Marked on a plate on the index arm: "Hezzanith" | Endless Tangent Screw [and] Automatic Clamp. | Patent [and] Semper | Paratus
Marked on the back of the frame near the horizon glass: Rgd 517155
Graduation Polished brass limb with inlaid silver scale from -5° to 130° every 1°, measuring to 124°. Micrometer to 12″.
Dimensions Radius 7 inches (178 mm)
Description Black-lacquered brass bell-pattern frame, wooden handle with a brass-lined hole for the locking mechanism. An automatic clamp and an endless tangent screw with a white plastic micrometer drum ('Hezzanith' Rapid Reader patent) are at the bottom of the index arm. Four shades (all green); a Wollaston prism; three horizon shades (all green). Index-glass adjustment by a capped square-headed screw and detached spanner; adjustment of horizon glass by capstan screws. Detached threaded telescope bracket with perpendicular adjustment by rising-piece and a milled knob. Telescope (120 mm), erect image, with wide objective lens (star finder); telescope (187 mm), inverted image, two parallel cross-wires; extra drawtube (96 mm), inverted image, four cross-wires. Sight-tube (82 mm). Two shaded eyepieces (both green), a brush, an adjusting pin, a screw-driver, and a spanner. One part is missing (perhaps a battery or lamp). In a square fitted mahogany box with a patent locking mechanism for securing the instrument when not in use, and contained in the lid a Hezzanith Instrument Works certificate of examination (not dated), and a repair label from Heath and Co. for 1940.
Provenance Captain Sir David Stanley Tibbits RN, the original owner, used it at sea continually between 1934 and 1961, and again in 1977 on board *Ayesha* on a trip from Bermuda to the US and back. For the latter voyage, it was apparently lent by the Museum. Presented by Tibbits, a Trustee of the National Maritime Museum from 1974 to 1976, in 1975.
Notes The Wollaston prism was used for star observations at night, when the horizon was not well defined. It produces two images of a star, one above and the other below the original position of the star, with the horizon exactly in between them.
Literature Watts, *The Sextant Simplified* 82; Littlewood and Butler, *Of Ships and Stars*, Appendix 1.

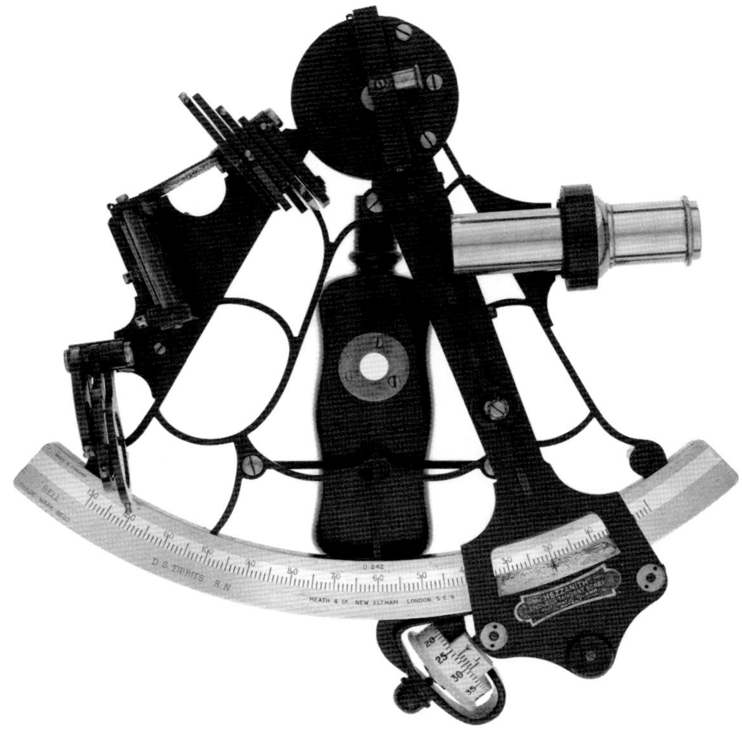

232 ACO1341
Sextant

Origin England
Maker Unknown
Date About 1938
Inscriptions Marked on the index arm: Jac. P. Vrolijk | Scheveningen Marked on the limb: XX 238
Graduation Bronzed limb with inlaid silver scale from -5° to 125° every 20′, measuring to 108°. Silver vernier to 30″, zero at the right.
Dimensions Diameter 6.9 inches (175 mm)
Description Black-lacquered brass Dollond-pattern frame, wooden handle. The tangent screw and clamping screw are on the back of the index arm. Two shades (both green); one horizon shade (green). Index- and horizon-glass adjustment by screws and capstan screws. Magnifier on a 100 mm swivelling arm. Threaded telescope bracket without adjustment. Telescope (87 mm), erect image; second telescope missing. Sight-tube (88 mm), a shaded eyepiece (green), a brush, and an adjusting pin. A second magnifier is from another instrument. In a square fitted wooden box.
Provenance Transferred from the Admiralty Compass Observatory in 1969.
Notes The firm of Jacob P. Vrolijk was a ship chandlery founded in 1872, catering mainly for the sea fishery industry in the Dutch North Sea port of Scheveningen.

233 ACO1340
Sextant

Origin Plymouth, England
Maker G.E. Hicks
Date 1938
Inscriptions Signed on the limb: G. E. Hicks. Plymouth
Graduation Bronzed limb with inlaid silver scale from -5° to 150° every 10′, measuring to 125°. Silver vernier to 2″, zero at the right.
Dimensions Radius 7½ inches (191 mm)
Description Anodized brass triangle-pattern frame, wooden handle. The tangent screw and clamping screw are on the back of the index arm. Four shades (two red, one green, and one orange); three horizon shades (one green and two red). Index-glass adjustment by a capped screw; adjustment of horizon glass by capstan screws. Magnifier (missing) on an 80 mm swivelling arm; a frosted glass shade. Threaded telescope bracket in two parts, fitted for correcting collimation error; perpendicular adjustment by rising-piece and a milled knob. Telescope (190 mm), inverted image, with cross-wires; telescope (76 mm), erect image, with wide objective lens (star finder); telescope (84 mm), erect image. Sight-tube (82 mm); a shaded eyepiece (green); a second eyepiece and an adjusting pin are missing. In a square fitted wooden box, containing in the lid a certificate of examination of the Deutsche Seewarte Hamburg, dated 18 February 1938.
Provenance Transferred from the Admiralty Compass Observatory in 1969.

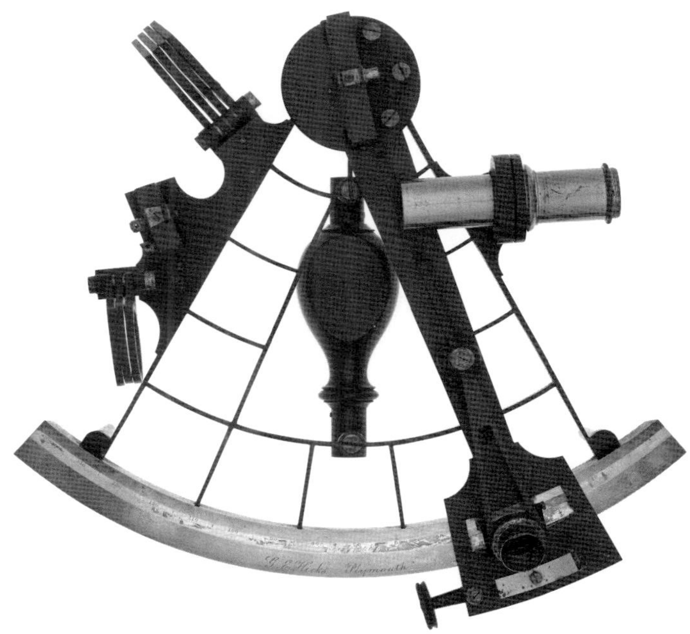

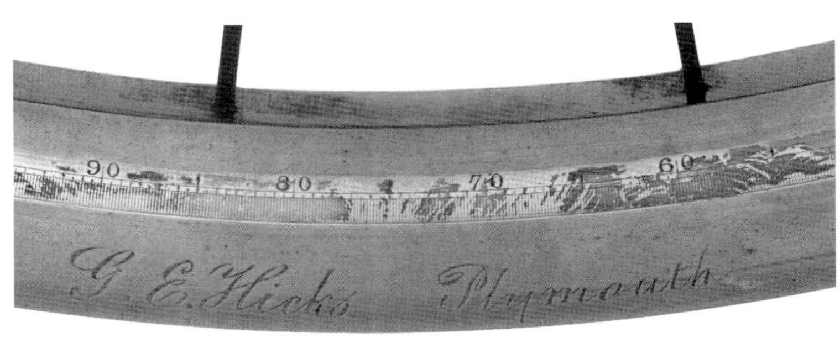

234 ACO1256
Sextant

Origin New Eltham, London, England
Maker Heath and Co. Ltd
Date About 1939
Inscriptions Signed on the limb: Heath & Co New Eltham, London S.E.9.
Marked on the limb: Made in England | E 56
Marked on a plate at the bottom of the index arm: Hezzanith | Rapid-Reader | Patent [and] "Hezzanith" | Endless Tangent Screw | Automatic Clamp. | Patent [and] Semper | Paratus
Graduation Bronzed limb with inlaid silver scale from -5° to 130° every 1°, measuring to 129°. Micrometer to 0.2′.
Dimensions Diameter 6½ inches (165 mm)
Description Anodized brass three-circle-pattern frame, wooden handle. White plastic micrometer drum and clamp at the bottom of the index arm. Four shades (all green); three horizon shades (all green). Index- and horizon-glass adjustment by capped capstan screws. Threaded telescope bracket in two parts, fitted for correcting collimation error; perpendicular adjustment by rising-piece and a milled knob. Telescope (172 mm), inverted image, four cross-wires; telescope (73 mm), erect image, wide objective lens (star finder); extra drawtube (96 mm), two parallel cross-wires. Sight-tube (82 mm), two shaded eyepieces (both green), an adjusting pin, and a spanner. In a square fitted wooden box, containing in the lid a National Physical Laboratory Teddington certificate of examination, dated 30 December 1939; with an inlaid blank brass plate on the top of the lid.
Provenance Transferred from the Admiralty Compass Observatory in 1969.

235 NAV1229
Sextant

Admiralty sextant
Origin London, England
Maker Henry Hughes and Son Ltd
Date About 1939
Inscriptions Signed on the limb: H. Hughes & Son Ltd. 59 Fenchurch St. London. E.C.
Marked on the index arm: Husun Trade mark
Graduation Polished brass scale from -5° to 135° every 1°, measuring to 135°. Micrometer to 1′.
Dimensions Radius 6½ inches (165 mm)
Description Brass Admiralty-pattern frame, wooden handle that can contain a battery for scale illumination. Plastic micrometer drum; the pressure clamp and worm wheel (damaged) are on the back of the index arm. Four shades (all green); two horizon shades (both green). Round index glass without adjustment; round horizon glass, adjustment by capped screws and by a square-headed screw and a separate key. Threaded telescope bracket with perpendicular movement by rising-piece and a milled knob. Telescope (82 mm), star finder (lens missing), erect image; telescope (146 mm), inverted image; extra drawtube (70 mm). Two shaded eyepieces (both green), oil bottle (for lubrication). Adjusting key (missing). In a square fitted wooden box with an inlaid plate in the lid (missing); inside the lid, a metal frame for holding a certificate of examination (certificate missing).
Provenance Lost in HMS *Royalo* (FY 714), a Fowey trawler built in 1916 and taken over by the Navy in 1940, which was sunk by a mine off Penzance, Cornwall, on 1 September 1940. Divers recovered the instrument in 1962 at a depth of 60 feet, and apparently gave it to Kelvin Hughes, who presented it to the Directorate of Naval Warfare, London in 1963. Presented by the Directorate in 1979.

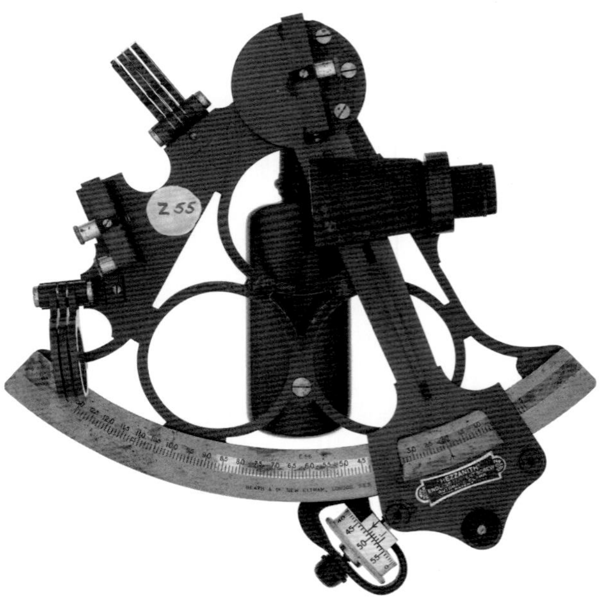

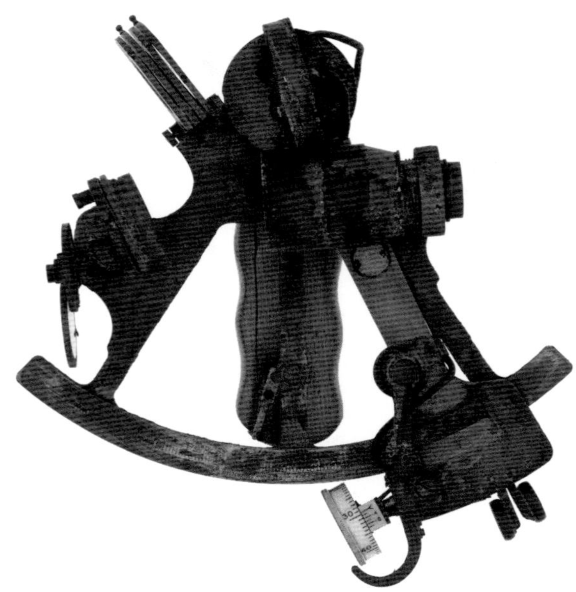

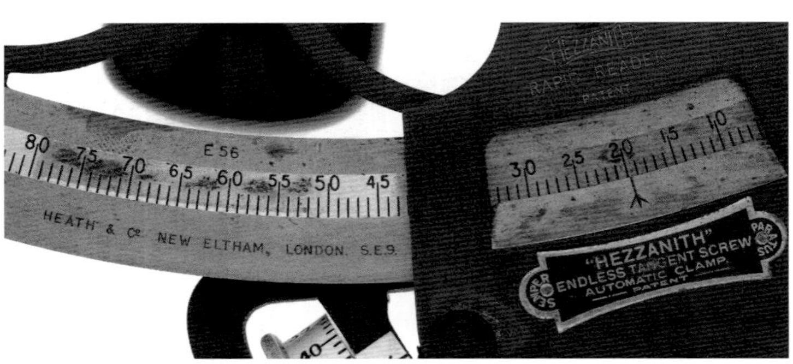

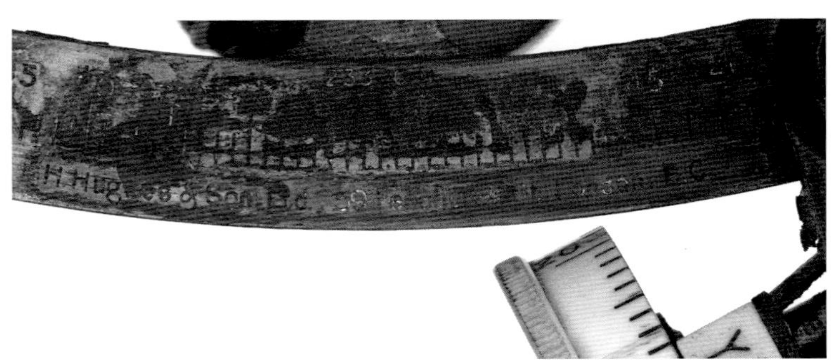

236 ACO1268
Bubble sextant

Libellen-Oktant
Origin Hamburg, Germany
Maker Carl Plath
Date About 1940
Inscriptions Signed on the left of the casing: Libellen-Oktant | mit Mittlungseinrichtung | Bauart: C. Plath-De Te We | Gerät-Nr. 127–134-B-2 | Hersteller: C. Plath
Marked on the right side of the casing: J.B.
Graduation Scale from 0° to 80° every 30′
Dimensions 330 × 260 × 140 mm (length × height × width)
Description Casing of grey-lacquered aluminium and artificial materials, with black plastic fittings, housing optical equipment, a clockwork averaging mechanism, electrical equipment, and a rubber eyecup. Two index shades (green and red). In a rectangular grey-lacquered aluminium case containing a reflecting flat (mirror missing), a light bulb, a plastic suspension strap, instructions for use, and a contents list in German. A telescope and various parts are missing.
Provenance Transferred from the Admiralty Compass Observatory in 1969.
Literature Schaafhausen *et al.*, *C Plath 1862–1962,* 112–16.

237 ACO1269
Bubble sextant

Libellen-Oktant
Origin Hamburg, Germany
Maker Probably Carl Plath
Date About 1940
Inscriptions Marked on the left side of the casing: Libellen-Oktant | mit Mittlungseinrichting | Gerät-Nr 127–134 B-2 | Hersteller [name erased]
Marked on the right side of the casing: J.B. | O
Graduation Scale from 0° to 80° every 30′
Dimensions 330 × 260 × 140 mm (length × height × width)
Description Casing of grey-lacquered aluminium and artificial materials, with black plastic fittings, housing optical equipment, a clockwork averaging mechanism, electrical equipment, and arubber eyecup. Two index shades (red and green). In a grey-lacquered wooden carrying case containing a reflecting flat, four light bulbs, assorted pieces of electrical equipment, instructions for use, and a contents list in German. The case is marked on the outside 'Gerät Nr: 127–134B'.
Provenance Transferred from the Admiralty Compass Observatory in 1969.
Literature Schaafhausen *et al.*, *C Plath 1862–1962,* 112–16.

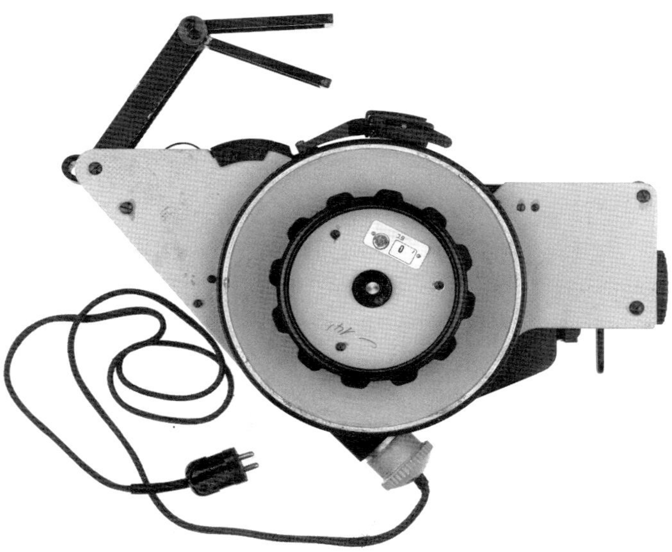

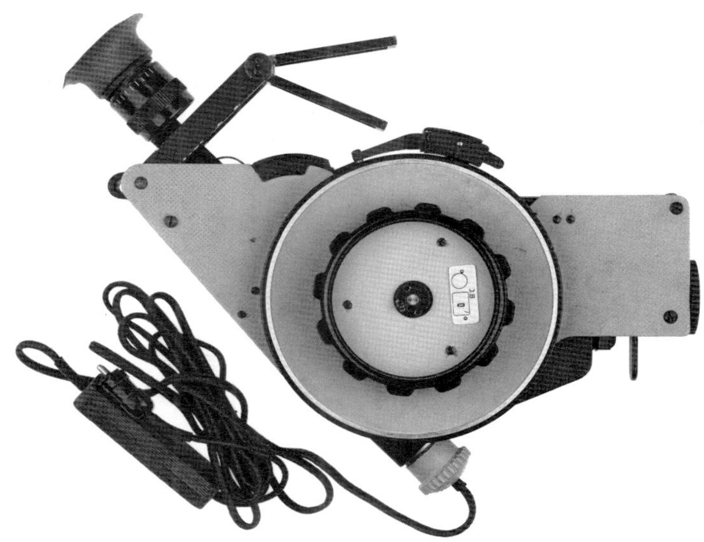

238 ACO1273
Bubble sextant

Sold-Libellensextant
Origin Hamburg, Germany
Maker Probably Carl Plath
Date About 1940
Inscriptions Marked on the left side of the casing: Sold-Libellensextant | Gerät Nr. 127–134 | Hersteller [name erased] | M [eagle & swastika]
Marked on the right side of the casing: J.B. | Geprüft | Deutsche Seewarte
Graduation Scale from 0° to 80° every 30′.
Dimensions 330 × 260 × 140 mm (length × height × width)
Description Casing of grey-lacquered aluminium and artificial materials, with black plastic fittings, housing optical equipment, a clockwork averaging mechanism, electrical equipment, and a rubber eyecup. Two index shades (green and red). In a black-lacquered metal carrying case containing a reflecting flat, a battery charger, a battery, two light bulbs (four others are missing) and a contents list in German. The case is marked on a plate on the side 'Sold-Libellensextant M [eagle & swastika]'.
Provenance Transferred from the Admiralty Compass Observatory in 1969.
Literature Schaafhausen *et al.*, *C Plath 1862–1962*, 112–16.

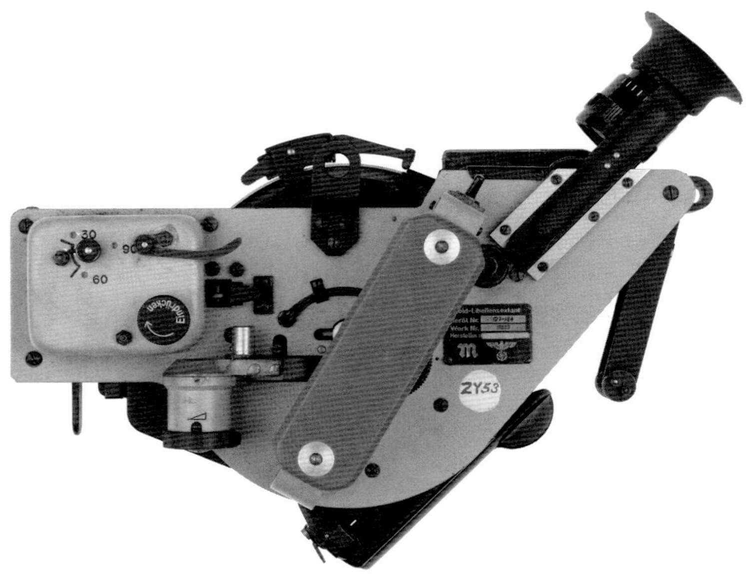

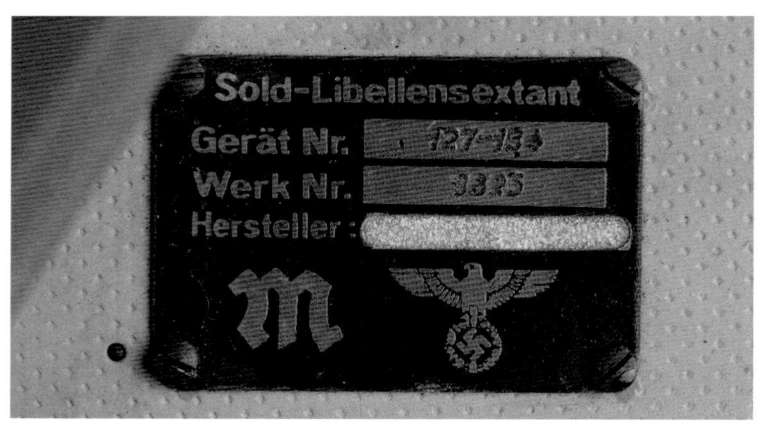

239 ACO1274
Bubble sextant

Sold-Libellensextant
Origin Hamburg, Germany
Maker Probably Carl Plath
Date About 1940
Inscriptions Marked on the left side of the casing: Sold-Libellensextant | Gerät Nr. 127–134 | Hersteller [name erased] | M [eagle & swastika]
Marked on the right side of the casing J.B. | Geprüft | Deutsche Seewarte
Graduation Scale from 0° to 80° every 30′
Dimensions 330 × 260 × 140 mm (length × height × width)
Description Casing of grey-lacquered aluminium and artificial materials, with black plastic fittings, housing optical equipment, a clockwork averaging mechanism, electrical equipment, and a rubber eyecup. Two index shades (red and green). In a rectangular black-lacquered metal carrying case containing a reflecting flat, a battery charger, a battery, two light bulbs (four others are missing), and a contents list in German. The telescope and suspension strap are missing. The carrying case is marked on the side 'Sold-Libellensextant M [eagle & swastika]'.
Provenance Transferred from the Admiralty Compass Observatory in 1969.
Literature Schaafhausen *et al.*, *C Plath 1862–1962,* 112–16.

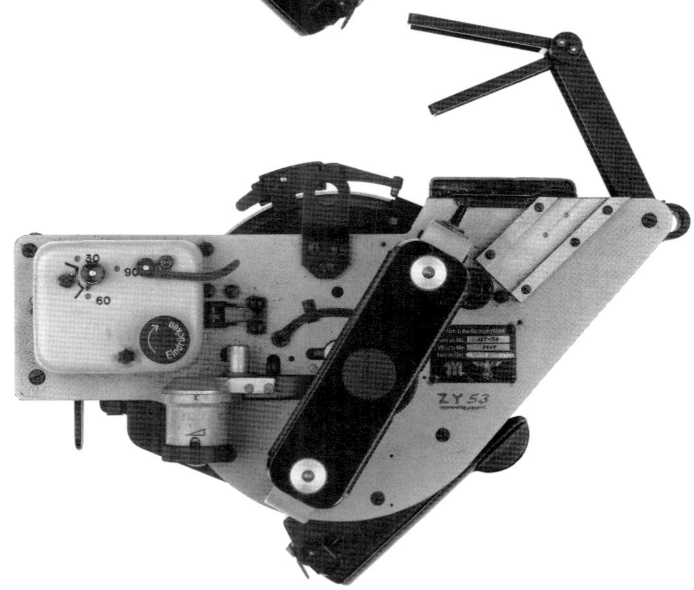

240 ACO1349
Surveyor's sextant

Paget angle sextant
Origin London, England
Maker Henry Hughes and Son Ltd
Date About 1940
Inscriptions Signed on the upper plate: H. Hughes & Son, Ltd | London
Marked on the upper plate: The Paget | Angle Sextant | No. 763
Graduation Ivorine scale with black markings, from 0° to 90° to 0° every 30′. The scale is read by use of a black indicating line on an ivorine index mark at the end of the index arm.
Dimensions Diameter 4½ inches (114 mm)
Description Black-lacquered alloy frame consisting of two circular plates, the upper having an ivorine overlay; a wooden handle (missing) to be screwed into the lower plate. There is no tangent or clamping screw; the index arm is moved by hand. No shades. Index- and horizon-glass adjustment by capstan screws. The sight vane has one pinhole mounted within the frame. In a fitted wooden box partly lined with green textile, and containing in the lid instructions for use by H. Hughes and Son Ltd, 59 Fenchurch Street, London.
Provenance Transferred from the Admiralty Compass Observatory in 1969.
Notes The Paget sextant was designed around 1892 and brought onto the market by H. Hughes and Son.
Literature NN, 'A New Angle Sextant', 444, 448; Ifland, *Taking the Stars*, 131–2.

241 NAV1202
Bubble sextant

Sold-Libellensextant
Origin Hamburg, Germany
Maker Carl Plath
Date About 1940
Inscriptions Signed on the handle: C. Plath Hamburg [Plath logo]
Signed on the left side of the casing: Sold-Libellensextant | Gerät Nr. | Hersteller: C. Plath, Hamburg | M [eagle & swastika]
Marked on a plate at the centre of the knob on the right side of the casing: J.B. Geprüft Deutsche Seewarte
Graduation Scale from 0° to 80° every 10′
Dimensions 330 × 260 × 140 mm (length × height × width)
Description Casing of grey-lacquered aluminium and artificial materials, housing optical equipment, a clockwork averaging mechanism, electrical equipment, and a rubber eyecup. Two index shades (green and red). Telescope with a rubber eyecup. In a fitted rectangular grey-lacquered aluminium carrying case containing a reflecting flat, an adaptor, six light bulbs, and, inside the lid, a contents list in German.
Provenance Presented by HMS *Dryad* Museum, Southwick, Hampshire, in 1967.
Literature Schaafhausen *et al.*, *C Plath 1862–1962*, 112–16.

242 NAV1210
Bubble sextant

Sold-Libellensextant KM 2
Origin Hamburg, Germany
Maker Carl Plath
Date About 1940
Inscriptions Signed on the handle: C. Plath Hamburg [Plath logo]
Signed on the left side of the casing: Sold-Libellensextant | Gerät Nr. | Hersteller C. Plath, Hamburg | M [eagle & swastika]
Marked on a plate at the centre of the knob on the right side of the casing: J.B. Geprüft Deutsche Seewarte
Graduation Scale from 0° to 80° every 30′
Dimensions 330 × 260 × 140 mm (length × height × width)
Description Casing of grey-lacquered aluminium and artificial materials, housing optical equipment, clockwork averaging mechanism, electrical equipment, and a rubber eyecup. Two index shades (green and red). In a rectangular fitted grey-lacquered aluminium carrying case containing a reflecting flat, a battery charger, six light bulbs (one is broken), and a contents list in German; also an instruction manual, *Beschreibung und Bedienungsvorschrift für den Sold-Libellensextanten KM 2*, published by the Oberkommando der Kriegsmarine Deutsche Seewarte in 1944.
Provenance Presented by the Navigation and Direction Division of the Admiralty in 1961.
Literature NN, *Beschreibung und Bedienunsvorschrift für den Sold-Libellensextanten KM 2;* Schaafhausen *et al., C Plath 1862–1962,* 112–16.

243 NAV1212
Bubble sextant

Sold-Libellensextant
Origin Hamburg, Germany
Maker Carl Plath
Date About 1940
Inscriptions Signed on the handle: C. Plath [Plath logo]
Signed on the left side of the casing: Sold-Libellensextant | Gerät Nr | Hersteller: C. Plath, Hamburg | M [eagle & Swastika]
Marked on a plate at the centre of the knob on the right side of the casing: J.B. Geprüft Deutsche Seewarte
Graduation Scale from 0° to 80° every 10′.
Dimensions 330 × 235 × 140 mm (length × height × width)
Description Casing of grey-lacquered aluminium and artificial materials, housing optical equipment, a clockwork averaging mechanism, electrical equipment, and a rubber eyecup. Two index shades (red and green). In a rectangular fitted grey-lacquered aluminium carrying case containing a reflecting flat, a battery charger, one battery, six light bulbs, two plastic suspension straps, and a contents list in German.
Provenance Presented by the Navigation and Direction Division of the Admiralty in 1961.
Literature Schaafhausen *et al., C Plath 1862–1962,* 112–16.

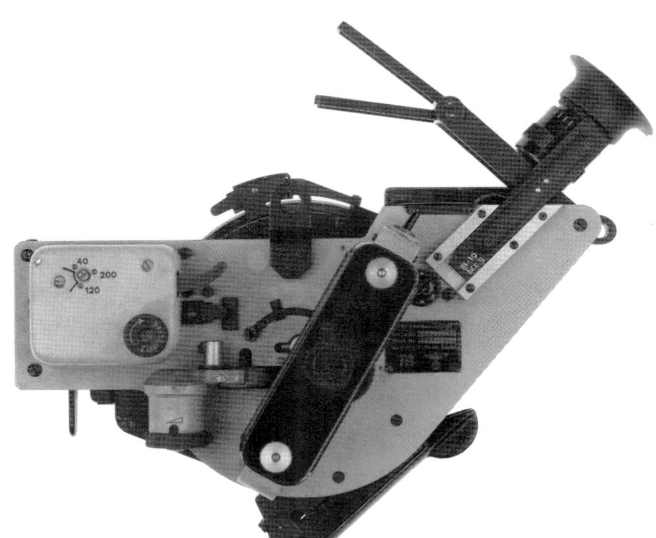

244 ZBA0151
Sextant

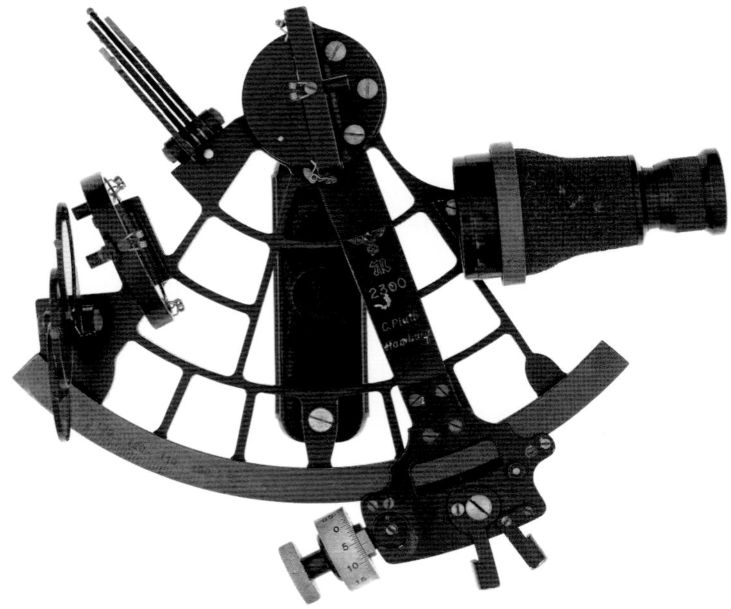

Origin Hamburg, Germany
Maker Carl Plath
Date About 1940
Inscriptions Signed on the index arm: C. Plath | Hamburg
Signed on the handle: C. Plath [Plath logo] Hamburg
Marked on the index arm: [eagle & Swastika] | M | 2300
Marked on the limb: D.S. [Plath logo] | 18170
Graduation Brass scale from -5° to 137° every 1°, measuring to 130°. Micrometer to ½′.
Dimensions Radius 6¼ inches (159 mm)
Description Black-lacquered alloy ladder-pattern frame, black Bakelite or similar handle. White plastic micrometer drum and clamp with worm wheel on the bottom of the index arm. Four shades (three grey and one green); three horizon shades (two grey and one green). Large index glass, adjustment by a capped square-headed screw; large round horizon glass, adjustment by capped screws, a detached pin and a lever. Telescope fixed to its bracket without adjustment, with large objective (98 mm), erect image. Adjusting pin, an adjusting lever, a shaded eyepiece (grey), two bottles of oil (for lubrication) and spare screws. In a square fitted pine box, containing a certificate of examination from the Deutsche Seewarte, dated 21 November 1940, and a brass trade label for Plath.
Provenance Belonged to Lieutenant-Commander W. Ashton RN, who, according to documentation that came with the donation, used it during World War II, when serving in the Hydrographic Branch of the Royal Navy. Considering the date of the German certificate of examination, the instrument must have been taken during or even after the war. Purchased from Mrs. B.E. Ashton, Devizes, Wiltshire, in 1995.
Notes 'D.S.' stands for 'Deutsche Seewarte'.
Literature Schaafhausen *et al.*, *C Plath 1862–1962*, 112–16.

245 NAV1895
Bubble sextant

Mark IX
Origin London, England
Maker Probably Henry Hughes and Son Ltd
Date About 1940
Inscriptions Marked on the left of the casing: A.M. | Bubble Sextant | Mark IX | 6B/218 | N° 19506/40 | Brit pats | 480112 490621 | and pending
Graduation Scale from 0° to 80° every 1°. Vernier to 1′.
Dimensions 200 × 240 × 180 mm (length × height × width)
Description Black crackle-lacquered alloy casing incorporating artificial materials, housing optical equipment, a clockwork averaging mechanism, electrical equipment, and a rubber eyecup. Incomplete and in poor condition. In a fitted Bakelite case.
Provenance Presented by Chevalier F.K. Gollcher, Valletta, Malta, in 1962.
Notes The Mark IX was probably developed around 1940. 'A.M.' stands for 'Air Ministry'.
Literature Hughes, *History of Air Navigation*, 119–29.

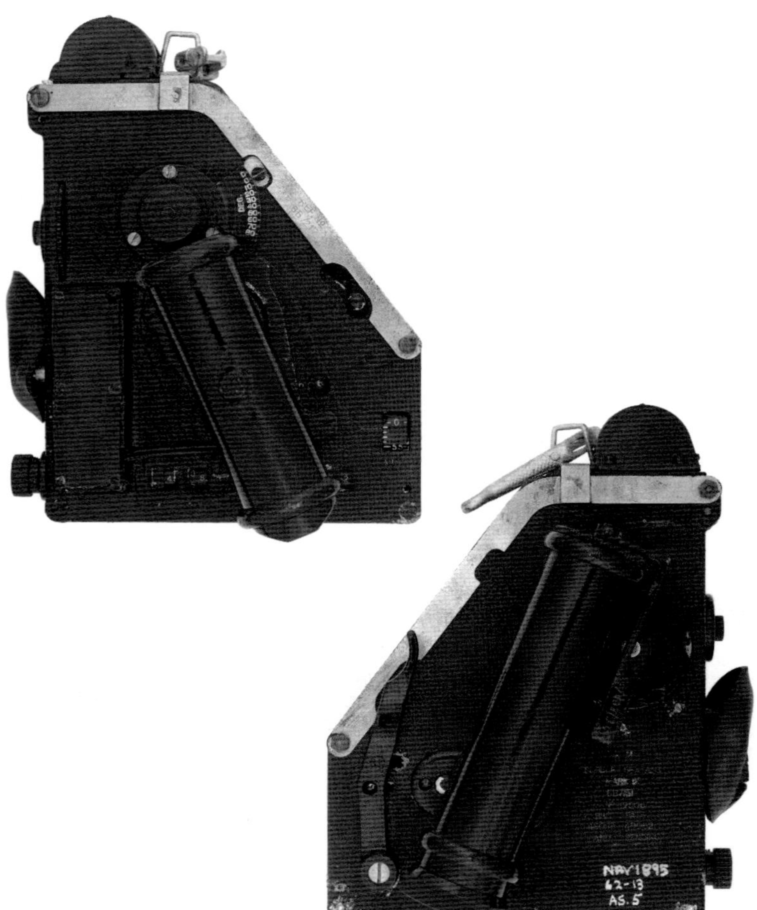

246 ACO1254
Sextant

Origin New Eltham, London, England
Maker Heath and Co. Ltd
Date About 1940
Inscriptions Signed on the limb: Heath & C° New Eltham. London. S.E.9.
Marked on the limb: Made in England | F130
Marked on a plate on the end of the index arm: "Hezzanith" | Endless Tangent Screw [and] Automatic Clamp | Patent [and] Semper | Paratus
Graduation Bronzed limb with inlaid silver scale from -5½° to 150° every 10′, measuring to 130°. Silver vernier to 2″ zero at the right.
Dimensions 6 ½ inches (165 mm)
Description Black-lacquered brass three-circle-pattern frame, wooden handle. The endless tangent screw and clamping screw are on the back of the index arm. Four shades (all green); three horizon shades (all green). Index-glass adjustment by a screw; adjustment of horizon glass by capstan screws. Magnifier on an 85 mm swivelling arm. Threaded telescope bracket with perpendicular adjustment by rising-piece and a milled knob. Telescope (71 mm), erect image, with wide objective lens (star finder). Sight-tube (82 mm), a shaded eyepiece (red), and an adjusting pin. In a square fitted wooden box, containing in the lid a Hezzanith Instrument Works certificate of examination, dated 23 December 1940.
Provenance Transferred from the Admiralty Compass Observatory in 1969.

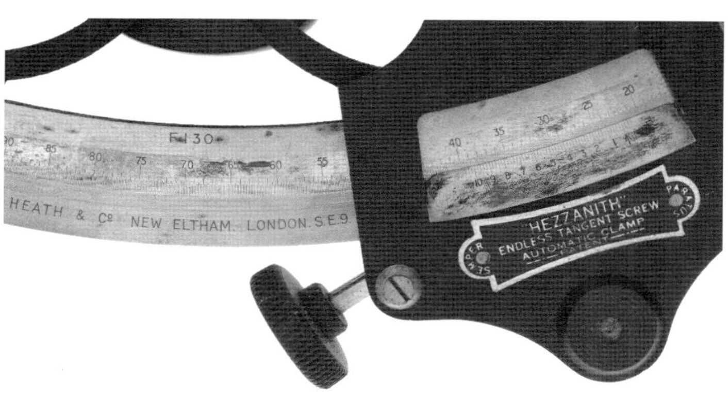

247 ZBA0086
Bubble sextant

Mark IXA
Origin London, England
Maker Probably Henry Hughes and Son Ltd
Date About 1941
Inscriptions Marked on the left side of the casing: H/P 3 | [broad arrow] | Bubble Sextant | Mark IX.A. | Ref No. 6B/218 | Brit Patts | 480112 490621 | and pending
Marked on the right side of the casing: Serial No 62/41
Graduation Scale from 0° to 80° every 1°. Vernier to 1′.
Dimensions 144 × 170 × 50 mm (length × height × width)
Description Black crackle-lacquered alloy casing incorporating artificial materials, housing optical equipment, a clockwork averaging mechanism, electrical equipment, and a rubber eyecup. Three shades (all red). The cover of the averaging mechanism is missing; no box or parts found. In poor condition.
Provenance Transferred from the Ministry of Defence, Chertsey, in 1996.
Notes The Mark IXA was probably developed around 1941.
Literature Hughes, *History of Air Navigation*, 119–29.

248 NAV1238
Bubble sextant

Mark IXA
Origin London, England
Maker Probably Henry Hughes and Son Ltd
Date About 1942
Inscriptions Marked on the averager: A.M. [crown] Serial No 5836/42
Marked: A.M. | Bubble Sextant | Mark | 6B | 218 | 9818 | BR1 | 420112 90 | A . . . ng
Graduation Scale from 0° to 80° every 1°. Vernier to 1′.
Dimensions 165 × 229 × 153 mm (length × height × width)
Description Alloy casing incorporating artificial materials, housing optical equipment, a clockwork averaging mechanism, and electrical equipment, a rubber eyecup. In a fitted shaped plastic case with a leather carrying strap. In poor condition.
Provenance Presented by the Fairmile Construction Co. Ltd, Cobham, Surrey, in 1984.
Notes The Mark IXA was probably developed around 1941. 'A.M'. stands for 'Air Ministry'.
Literature Hughes, *History of Air Navigation*, 119–29.

249 NAV1929
Bubble sextant

Mark IX
Origin London, England
Maker Probably Henry Hughes and Son Ltd
Date About 1943
Inscriptions Marked on the left side of the casing, illegible through corrosion, but very likely to read: A.M. | Mark IX | SERIAL NO. 12819/43
Marked on the averaging machine: SERIAL NO 179/44
Graduation Scale from 0° to 80° every 1°. Vernier to 1′.
Dimensions 190 × 254 × 154 mm (length × height × width)
Description Black crackle-lacquered alloy casing incorporating artificial materials, housing optical equipment, a clockwork averaging mechanism, electrical equipment, and a rubber eyecup. Badly corroded, various parts missing. In poor condition. No box or case.
Provenance Presented by Kelvin Hughes, Barkingside, Essex, in 1961.
Notes The Mark IX was probably developed around 1940. 'A.M'. stands for 'Air Ministry'.
Literature Hughes, *History of Air Navigation*, 119–29.

250 ACO1270
Gyroscopic sextant

Seekreiselsextant
Origin Hamburg, Germany
Maker Probably Carl Plath
Date About 1943
Inscriptions Marked on the left of the casing: Seekreiselsextant | Gerät. 616–42A | Hersteller gtl
Marked on the right side of the casing: J.B.O. | Geprüft | Deutsche Seewarte
Graduation Scale from 0° to 80° every 30′. Vernier to 1′.
Dimensions 356 × 279 × 152 mm (length × height × width)
Description Casing of grey-lacquered aluminium and artificial materials, with black plastic fittings, housing optical equipment, a clockwork averaging mechanism, electrical equipment, and a rubber eyecup. The shades are missing. In a fitted wooden carrying case containing an air pump connector, a battery connector, a battery charger (one battery is missing), six light bulbs, a plastic suspension strap, a detached handle, and instructions for use. Marked on the side of the carrying case '1952'.
Provenance Transferred from the Admiralty Compass Observatory in 1969.
Literature Schaafhausen *et al.*, *C Plath 1862–1962,* 112–16.

251 ACO1271
Gyroscopic sextant

Kreiselsextant
Origin Hamburg, Germany
Maker Probably Carl Plath
Date About 1943
Inscriptions Marked on the left of the casing: Kreiselsextant | Gerät. 616–42A | Hersteller gtl
Marked on the right of the casing: J.B.O. | Geprüft | Deutsche Seewarte
Graduation Scale from 0° to 80° every 30′. Vernier to 1′.
Dimensions 356 × 297 × 152 mm (length × height × width)
Description Casing of grey-lacquered aluminium and artificial materials, with black fittings, housing optical equipment, a clockwork averaging mechanism, electrical equipment, and a rubber eyecup. Four index shades (three grey and one red). In a fitted wooden carrying case containing the shades, a battery charger, three light bulbs (two other bulbs are missing, one is broken), a detached handle, and instructions for use in German. Marked on the side of the carrying case '3533'.
Provenance Transferred from the Admiralty Compass Observatory in 1969.
Literature Schaafhausen *et al.*, *C Plath 1862–1962,* 112–16.

252 ACO1272
Gyroscopic sextant

Kreiselsextant
Origin Hamburg, Germany
Maker Probably Carl Plath
Date About 1943
Inscriptions Marked on the left side of the casing: Kreiselsextant | Gerät. 616–42A | Hersteller gtl
Marked on the right side of the casing: J.B.O. | Geprüft | Deutsche Seewarte
Graduation Scale from 0° to 80° every 30′. Vernier to 1′.
Dimensions 356 × 279 × 152 mm (length × height × width)
Description Casing of grey-lacquered aluminium and artificial materials, with black plastic fittings, housing optical equipment, a clockwork averaging mechanism, electrical equipment, and a rubber eyecup. Four index shades (three grey and one red). In a fitted wooden carrying case containing the shades, a battery charger, two batteries, six light bulbs, a plastic suspension strap, a handle and instructions for use in German. Marked on the side of the case '3383'.
Provenance Transferred from the Admiralty Compass Observatory in 1969.
Literature Schaafhausen *et al.*, *C Plath 1862–1962,* 112–16

253 ACO1411
Gyroscopic sextant

Seekreiselsextant
Origin Hamburg, Germany
Maker Probably Carl Plath
Date About 1943
Inscriptions Marked on the left side of the casing: Seekreiselsextant | Gerät 616–42A | Hersteller gtl
Marked on the left side of the casing: J.B. | Geprüft | Deutsche Seewarte
Graduation Scale from 0° to 80° every 1°. Vernier to 1′.
Dimensions 356 × 279 × 152 mm (length × height × width)
Description Casing of aluminium and artificial materials, housing optical equipment, a clockwork averaging mechanism, electrical equipment, and a rubber eyecup. The shades and various unidentified parts are missing. No box or case.
Provenance Transferred from the Admiralty Compass Observatory in 1993.
Literature Schaafhausen *et al.*, *C Plath 1862–1962,* 112–16.

254 NAV1167
Gyroscopic sextant

Seekreiselsextant
Origin Hamburg, Germany
Maker Probably Carl Plath
Date About 1943
Inscriptions Marked on the left side of the casing: Seekreiselsextant | Gerät 616–42A | Hersteller gtl
Marked on a plate at the centre of the knob on the right side of the casing: J.B. Geprüft Deutsche Seewarte
Graduation Scale from 0° to 80° every 1°. Vernier to 1′.
Dimensions 356 × 279 × 152 mm (length × height × width)
Description Casing of aluminium and artificial materials, housing optical equipment, a clockwork averaging mechanism, electrical equipment, and a rubber eyecup. In a square fitted wooden carrying case containing eight shades (four green and four red), a battery charger, eight light bulbs, a plastic suspension strap, a detached handle, and instructions for use in German; also an issue of *Navigation: Journal of the Institute of Navigation* containing a report of tests performed on this kind of instrument.
Provenance Presented by the Ministry of Transport and Aviation in 1959.
Literature Schaafhausen *et al.*, *C Plath 1862–1962*, 112–16; Hunt, 'The German Gyro-Sextant'.

255 NAV1181
Gyroscopic sextant

Origin Hamburg, Germany
Maker Carl Plath
Date About 1943
Inscriptions Signed on the handle: C. Plath Hamburg [Plath logo]
Marked on the left side of the casing: Gerät-Nr 616–42 | Hersteller [blank]
Marked on a plate at the centre of the right side of the casing: J.B. Geprüft Deutsche Seewarte
Graduation Scale from 0° to 80° every 10′. Vernier to 1′.
Dimensions 356 × 279 × 152 mm (length × height × width)
Description Casing of grey-lacquered aluminium and artificial materials, housing optical equipment, a clockwork averaging mechanism, electrical equipment, and a rubber eyecup. Two index shades (green and red). In a square fitted wooden carrying box, containing one shade (grey), a battery charger, two batteries, two light bulbs, and instructions for use in German. Several unidentified parts are missing. The box is marked on the outside '4753'.
Provenance Presented by the Navigation and Direction Division of the Admiralty in 1961.
Literature Schaafhausen *et al.*, *C Plath 1862–1962*, 112–16.

256 NAV1183
Gyroscopic sextant

Seekreiselsextant 2D
Origin Hamburg, Germany
Maker Probably Carl Plath
Date About 1943
Inscriptions Marked on the left side of the casing: Seekreiselsextant | Gerät 616–42A | Hersteller gtl
Marked on a plate at the centre on the right side of the casing: J.B. Geprüft Deutsche Seewarte
Graduation Scale from 0° to 80° every 10′. Vernier to 1′.
Dimensions 356 × 279 × 152 mm (length × height × width)
Description Casing of aluminium and artificial materials, housing optical equipment, a clockwork averaging mechanism, electrical equipment, and a rubber eyecup. In a square fitted wooden carrying case containing four shades (one red, two grey, and one missing), a battery charger (two batteries are missing), six light bulbs, a plastic suspension strap, and instructions for use in German. The case is marked on the side: '5185'.
Provenance Presented by the Navigation and Direction Division of the Admiralty in 1961.
Literature Schaafhausen *et al.*, *C Plath 1862–1962*, 112–16.

257 NAV1184
Gyroscopic sextant

Seekreiselsextant 2D
Origin Hamburg, Germany
Maker Probably Carl Plath
Date About 1943
Inscriptions Marked on the left side of the casing: Seekreiselsextant | Gerät 616–42A | Hersteller gtl
Marked on a plate on the right side of the casing: J.B. Geprüft Deutsche Seewarte
Graduation Scale from 0° to 80° every 10′. Vernier to 1′.
Dimensions 356 × 279 × 152 mm (length × height × width)
Description Casing of aluminium and artificial materials, housing optical equipment, a clockwork averaging mechanism (winding knob missing), electrical equipment, and a rubber eyecup. The clamp is broken. In a square fitted wooden carrying case containing four shades (three grey and one red), a battery charger, two batteries, four light bulbs (two broken), a suspension strap (missing), and instructions for use in German. The case marked on the side: '3384'.
Provenance Presented by the Navigation and Direction Division of the Admiralty in 1961.
Literature Schaafhausen *et al.*, *C Plath 1862–1962*, 112–16.

258 NAV1185
Gyroscopic sextant

Seekreiselsextant
Origin Hamburg, Germany
Maker Probably Carl Plath
Date About 1943
Inscriptions Marked on the left side of the casing: Seekreiselsextant | Gerät 616–42A | Hersteller gtl
Marked on a plate on the right side of the casing: J.B. Geprüft Deutsche Seewarte
Graduation Scale from 0° to 80° every 30′. Vernier to 1′.
Dimensions 356 × 279 × 152 mm (length × height × width)
Description Casing of aluminium and artificial materials, housing optical equipment, a clockwork averaging mechanism, electrical equipment, and a rubber eyecup. In a rectangular fitted wooden carrying case containing four shades (three grey and one red), two batteries, five light bulbs, a plastic suspension strap, and instructions for use in German. The case is marked on the side '2006'.
Provenance Presented by the Navigation and Direction Division of the Admiralty in 1961.
Literature Schaafhausen *et al.*, *C Plath 1862–1962*, 112–16.

259 ACO1255
Sextant

Origin London, England
Maker Henry Hughes and Son Ltd
Date 1943
Inscriptions Signed on the limb: H. Hughes & Son Ltd London
Marked on the limb: Made in England | 34785
Marked on the index arm: Huson Trade Mark
Graduation Brass scale from -5° to 130° every 1°, measuring to 122°. Micrometer to 0.2° = 5 12′.
Dimensions Radius 6½ inches (165 mm)
Description Black-lacquered brass three-circle- pattern frame, black plastic handle. Plastic micrometer drum and clamp release knobs at the bottom of the index arm. Four shades (all green); three horizon shades (all green). Index-glass adjustment by a screw; adjustment of horizon glass by capstan screws. Threaded telescope bracket with perpendicular adjustment by rising-piece and a milled knob. Telescope (73 mm), inverted image, with wide objective lens (star finder). Sight-tube (83 mm), a shaded eyepiece (green), and an adjusting pin. An electric light at the end of the index arm; the batteries are stored in the handle. Several unidentified parts are missing. In a square fitted wooden box lined with green textile, and containing in the lid a Husun certificate of examination, dated 1943.
Provenance Transferred from the Admiralty Compass Observatory in 1969.

260 ZBA1701
Gyroscopic sextant

Seekreiselsextant
Origin Hamburg, Germany
Maker Probably Carl Plath
Date 1943
Inscriptions Marked on the left side of the casing: Seekreiselsextant | Gerät 616–42A | Hersteller gtl
Marked on the right of the casing: J.B. Geprüft Deutsche Seewarte
Graduation Scale from 0° to 80°, every 30′. Vernier to 1′.
Dimensions 356 × 260 × 152 mm (length × height × width)
Description Casing of aluminium and artificial materials, housing optical equipment, a clockwork averaging mechanism, electrical equipment, and a rubber eyecup. Four index shades (three grey and one red). In a square fitted wooden carrying box, containing a detached handle, an electric adaptor, six light bulbs, a battery (missing), electric wire, and instructions for use in German on the inside of the lid.
Provenance Acquired as part of the Royal Greenwich Observatory collection in 1998.
Literature Schaafhausen *et al.*, *C Plath 1862–1962,* 112–16.

261 NAV1171
Sextant with artificial horizon

Origin Long Island City, New York, USA
Maker Einson-Freeman Co.
Date About 1945
Inscriptions Signed on the central strut: Einson-Freeman Co., L.I. City, N.Y.
Graduation Cardboard scale from -3° to 123° every 30′, measuring to 120°. White cardboard vernier to 1′.
Dimensions Radius 9 inches (229 mm)
Description Black cardboard frame and index arm. No clamping screw or tangent screw. Two index shades (red and green). Glasses without adjustment. Wooden telescope bracket without adjustment. Fixed cardboard telescope (101 mm).
Artificial horizon This consists of a spirit level mounted on a wooden block attached to the frame below the horizon glass.
Provenance Presented by Kelvin Hughes, Barkingside, Essex, in 1961.
Notes Einson-Freeman started in 1909 as a lithographic company and after 1933 made jigsaw puzzles. The production of simple cardboard sextants was probably part of World War II efforts.

262 NAV 1172
Gyroscopic sextant

Seekreiselsextant 3D
Origin Hamburg, Germany
Maker Probably Carl Plath
Date 1945
Inscriptions Marked on the left side of the casing: Seekreiselsextant | Gerät 616–42A | Hersteller gtl
Marked on a plate at the centre of the right side of the casing: J.B. Geprüft Deutsche Seewarte
Graduation Scale from 0° to 80° every 10′. Vernier to 1′.
Dimensions 356 × 279 × 152 mm (length × height × width)
Description Casing of aluminium and artificial materials, housing optical equipment, a clockwork averaging mechanism, electrical equipment, and a rubber eyecup, and artificial materials. In a square fitted wooden carrying case containing four shades (two grey and two red), a battery charger, two batteries, two blue light bulbs, three small light bulbs, a suspension strap (missing), and instructions for use in German dated January 1945. It has a hand-operated air pump for spinning the gyroscope (length 570 mm).
Provenance Presented by Kelvin Hughes, Barkingside, Essex, in 1961.
Literature Schaafhausen *et al.*, *C Plath 1862–1962,* 112–16.

263 NAV1247
Sextant with artificial horizon

Hezzanith bubble horizon sextant
Origin New Eltham, London, England
Maker Heath and Co. Ltd
Date About 1946
Inscriptions Signed on the limb: Heath & Co New Eltham, London. S.E.9.
Marked on the limb: Hezzanith | B878
Marked on the index arm: Hezzanith | Rapid-Reader | Patent [and] "Hezzanith" | Bubble Horizon | Sextant
Graduation Alloy scale from -5° to 120° every 1°, measuring to 116°. Micrometer to 1′.
Dimensions Radius 4¼ inches (108 mm)
Description Grey-lacquered alloy frame, wooden handle. Alloy micrometer drum, pressure clamp and worm screw on the back of the index arm. Three index shades (all green); no horizon shades. Index- and horizon-glass adjustment by capped capstan screws. Prismatic telescope (147 mm), inverted image, four cross-wires, mounted on the frame without adjustment. A battery, a brush, a glass eyedropper, a spare light bulb, and two rubber washers. In a square fitted wooden box, containing in the lid a Heath and Co. Ltd certificate of examination, dated January 1946.
Artificial horizon The telescope incorporates an artificial bubble-type horizon and a detached lamp (with an on–off switch) with a battery holder that can be mounted on the telescope for use with the artificial horizon at night.
Provenance Purchased from Mr D. Essex, London, in 1986.

264 ZBA0747
Pocket bubble sextant

Bubble circle sextant
Origin Chicago, Illinois, USA
Maker G. Felsenthal and Sons
Date About 1946
Inscriptions Signed on the front of the frame: G. Felsenthal & Sons Chicago
Marked on the front of the frame: Bubble | Circle | Sextant | Bu Aero. U.S. Navy | Sextant-Bubble Circle | Stock No. R88-S-340 Contract No. NOa(s) 7505 | MFR'S. Part NO. FNA-29
Graduation Four quadrantal plastic scales from 0° to 90° by 1°. No vernier.
Dimensions Radius 2.4 inches (61 mm)
Description Circular plastic frame with a clear plastic sighting circle, carrying two white plastic folding sights and black sighting lines on the front. The back of the frame holds a bubble level and a folding mirror in a plastic housing. In a fitted plastic wallet containing a pencil; a transparent plastic 6 inch (157 mm) ruler; a pencil marked 'Made in U.S.A. Dixon'; a booklet titled *Pocket Bubble Sextant: Instructions for Use in Finding Latitude and Longitude*, published by the Nautical Almanac Office, U.S. Naval Observatory, 1 May 1946; and two paper circular protractors, made by Eastern Science Supply Co., Boston, Mass.
Provenance Acquired as part of the Royal Greenwich Observatory collection in 1998.
Notes These plastic sextants were made for the US Bureau of Aeronautics in the mid 1940s, probably for emergency use.
Literature NN, *Pocket Bubble Sextant: Instructions for Use in Finding Latitude and Longitude.*

265 ACO1258
Bubble sextant

Marine bubble sextant
Origin London, England
Maker Henry Hughes and Son Ltd
Date 1949
Inscriptions Signed on the right side of the casing: H. Hughes and Son Ltd. | London.
Marked on the right side of the casing: Marine Bubble Sextant | Ser. No 114. | Brit Pats 490621.522215. | 534913 and Pending.
Graduation Scale from 0° to 80° every 1′. Vernier to 1′.
Dimensions 165 × 215 × 172 mm (length × height × width)
Description Casing of black-lacquered alloy and artificial materials, with black plastic fittings, housing optical equipment, a clockwork averaging mechanism, electrical equipment, and a rubber eyecup. Three index shades (all green). In a square fitted wooden box with leather carrying strap, and containing a Henry Hughes and Son Ltd certificate of examination, dated 3 March 1949.
Provenance Transferred from the Admiralty Compass Observatory in 1969.

266 ZBA0089
Sextant

Sextant pattern 702
Origin London, England
Maker Henry Hughes and Son Ltd
Date About 1950
Inscriptions Signed on the limb: H. Hughes & Son Ltd. London. Marked on the limb: 40742 | Made in England
Graduation Bronzed limb with inlaid silver scale from -5° to 130° every 1°, measuring to 123°. Micrometer to 10′.
Dimensions Radius 6 inches (152 mm)
Description Black crackle-lacquered brass or bronze three-circle-pattern frame, black plastic handle containing battery. White plastic micrometer drum and quick-release thumb clamp at the bottom of the index arm, disengaging from a rack. Scale lit by battery- powered lamp, battery stored in the handle (with an on-off switch). Four shades (all green); three horizon shades (all green). Index-glass adjustment by a screw; adjustment of horizon glass by capstan screws. Detached threaded telescope bracket with perpendicular adjustment by rising-piece and a milled knob. Telescope (73 mm), erect image; telescope (188 mm), inverted image, four cross-wires. Sight-tube (82 mm), a brush, and a shaded eyepiece (green). A second telescope and an adjusting key are missing. In a square fitted mahogany box (probably not original to this sextant) lined with green textile, containing in the lid an Admiralty Compass Observatory certificate of serviceability for marine sextant no. 28485 (a different instrument) dated 13 May 1968.
Provenance Transferred from the Ministry of Defence Chertsey, in 1996.

267 NAV1225
Boat sextant

Origin Wargrave, Berkshire, England
Maker East Berks Boat Co.
Date About 1960
Inscriptions Signed on the index bar: EBBCO
Signed on the limb: East Berks Boat Co. England
Graduation Plastic scale from -20° to 120° every 1°, measuring to 120°. Micrometer to 1′.
Dimensions Radius 8 inches (203 mm)
Description Grey plastic (Makrolon) plate; plastic handle and stiffening bar at the back. Aluminium micrometer drum and clamp on the back of the index arm. Three shades (all green); two horizon shades (both green). The index and horizon glasse are fitted with adjustment screws operated by an Allen key. Telescope (79 mm), erect image, attached to the frame by a plastic clip. Allen key. In a fitted black plastic box.
Provenance Presented by Mr F.C. Murphy, Director, P & O Strath Services Ltd, London, in 1977.
Notes Ebbco started manufacturing these plastic sextants in the 1960s; they were aimed especially at the yachting market.
Literature Watts, *The Sextant Simplified*, 97–8; Bauer, *The Sextant Handbook*, 166.

268 NAV1217
Sextant

Mate Sextant
Origin London, England
Maker Kelvin Hughes
Date About 1966
Inscriptions Signed on the limb: [KH logo] | Manufactured in Great Britain by KELVIN HUGHES DIVISION S. Smith & Sons (England) Ltd
Marked on the limb: MATE SEXTANT
Marked on the index arm: Ser. No 69842
Graduation Aluminium scale from -5° to 125° every 1°, measuring to 121°. Micrometer to 0.2″.
Dimensions Radius 6 inches (152 mm)
Description Grey-lacquered alloy three-circle-pattern frame, plastic handle containing a battery for scale illumination lamp. White plastic micrometer drum, pressure clamp, and worm screw on the back of the index arm. Four shades (three green and one grey); three horizon shades (two green and one grey). Index-glass adjustment by a screw; adjustment of horizon glass by capstan screws. Threaded telescope bracket without adjustment. Telescope (76 mm), erect image. Adjusting pin, a shaded eyepiece (green), and a dry battery. In a square fitted wooden box, containing a Kelvin Hughes certificate of examination, dated 1966.
Provenance Lent by Kelvin Hughes, London, in 1966, and presented in 1983.

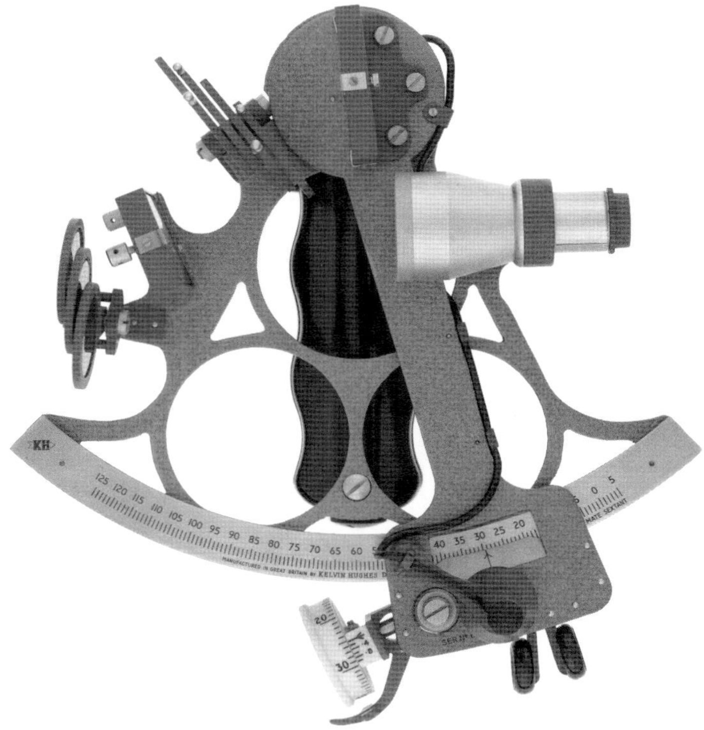

269 NAV1204
Sextant

Mate sextant
Origin London, England
Maker Kelvin Hughes
Date 1967
Inscriptions Signed on the limb: [KH logo] | Made in Great Britain by KELVIN HUGHES a Division of Smiths Industries Limited. Marked on the limb: MATE SEXTANT Marked on the index arm: Ser. N° 70660
Graduation Aluminium scale from -5° to 125° every 1°, measuring to 122°. Micrometer to 0.2′.
Dimensions Radius 6 inches (152 mm)
Description Grey crackle-lacquered aluminium three-circle-pattern frame, wooden handle. White plastic micrometer drum, pressure clamp, and worm screw on the back of the index arm. Four shades (three green and one grey); three horizon shades (two green and one grey). Index-glass adjustment by a screw; adjustment of horizon glass by capstan screws. Threaded telescope bracket without adjustment. Telescope (74 mm), erect image. Shaded eyepiece (green) and an adjusting pin. In a square fitted wooden box, containing in the lid a Kelvin Hughes certificate of examination, dated 2 August 1967.
Provenance Presented by Kelvin Hughes, Barkingside, Essex, in 1983.

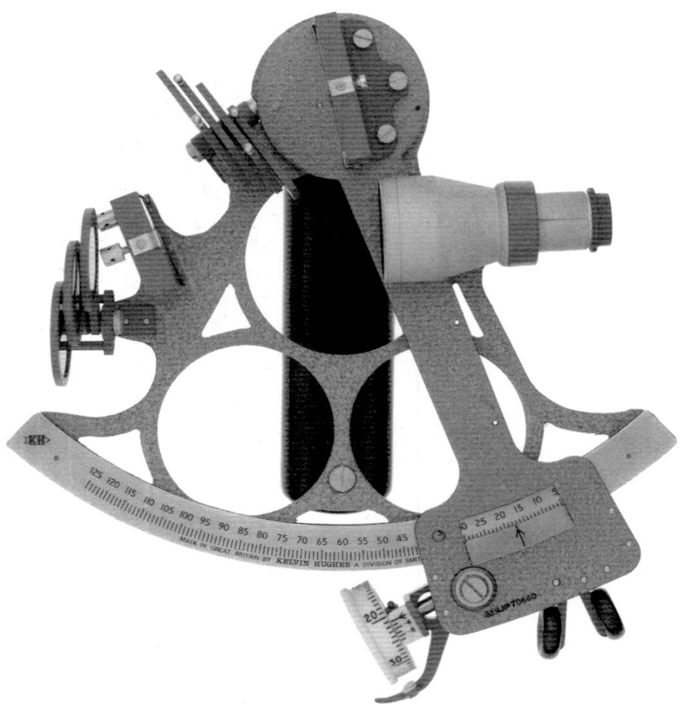

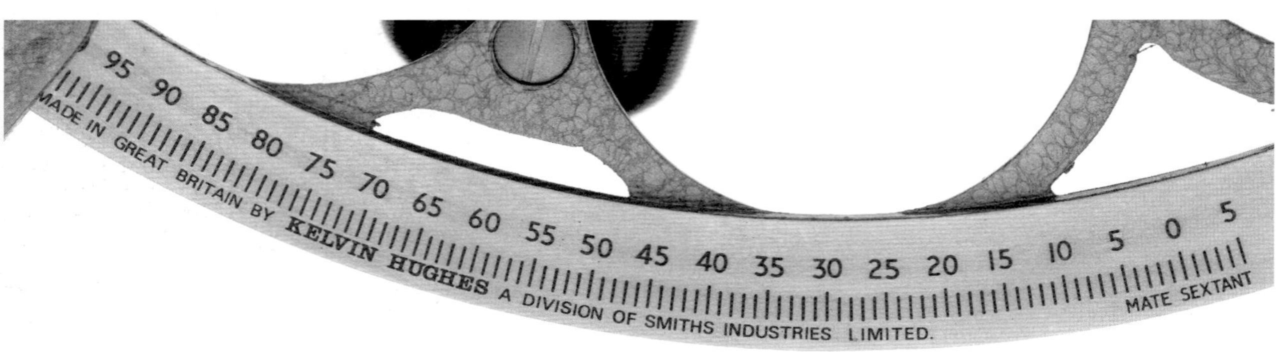

270 ACO1418
Sextant

Digital sextant
Origin London, England
Maker Henry Hughes and Son Ltd
Date About 1970
Inscriptions Signed on the limb: H.Hughes & Son Ltd. London.
Marked on the limb: 38358[broad arrow]
Marked on a plate on the index arm: Patt. 491 | Sextant | 7007 H N L 38358
Marked on the telescope: 47039[broad arrow]
Graduation Inlaid silver scale from -5° to 130° every 1°, measuring to 128°. Micrometer to 2′
Dimensions 6½ inches (165 mm)
Description Grey-lacquered brass Admiralty-pattern frame, plastic handle. White plastic micrometer drum, clamp, and worm gear on the back of the index arm. The micrometer drum is connected by a notched wheel to an electric motor made by Moore Reed and Company, Andover, England, and also connected by an electric wire to a second electric motor which is fixed on the back of the frame near the pivot. This motor is connected to the index arm by a notched wheel. The movements of the index arm and micrometer drum are relayed by wire to a computer (not present) for digital reading and presumably recording. Buttons in the handle operate the motors. Swivelling light for reading the scale and the micrometer. Four shades (all green); two horizon shades (both green). Index-glass adjustment by capped capstan screw; adjustment of horizon glass by a capstan screw. Threaded telescope bracket with perpendicular adjustment by rising-piece and a milled knob. Telescope (87 mm), erect image, large object lens (star finder); extra drawtube (69 mm). Two shaded eyepieces (both green), an oil bottle (for lubrication), and an adjusting pin. Two unidentified parts and one telescope are missing. In a rectangular fitted wooden carrying box, containing in the lid an Admiralty Compass Observatory certificate of serviceability, dated 20 April 1970.
Provenance Transferred from the Admiralty Compass Observatory in 1993.
Notes An experimental instrument that, according to documentation accompanying the donation, was never taken into production.
Literature Fanning, *Steady as She Goes*, 376–7.

271 ZBA0088
Sextant

Origin Hamburg, Germany
Maker Carl Plath
Date About 1970
Inscriptions Signed on the limb: [Plath logo] C. Plath | Hamburg
Marked on the limb: 55755 | Germany
Graduation Alloy scale from -5° to 125° every 1°, measuring to 125°. Micrometer to 0.2′.
Dimensions Radius 6¼ inches (159 mm)
Description Black-lacquered alloy ladder-pattern frame, black plastic handle containing a battery for the scale light. White plastic micrometer drum, clamping screw on a toothed rim at the bottom of the index arm. The light is incorporated into the index arm. Four shades (all orange); three horizon shades (all orange). Index glass and round horizon glass, adjustment of both by capped square-headed screws and a separate spanner. Detached prismatic telescope (120 mm, 6 × 330), erect image, mounted on a bracket without adjustment. A brush and an adjusting spanner. In a square fitted plywood box with a locking mechanism for securing the instrument when not in use, containing in the lid a certificate of examination from the Deutsches Hydrographisches Institut, Hamburg, dated 23 March 1972; also an illustrated instruction leaflet.
Provenance Transferred from the Ministry of Defence Chertsey, in 1996.
Literature Schaafhausen *et al.*, *C Plath 1862–1962,* 104–5.

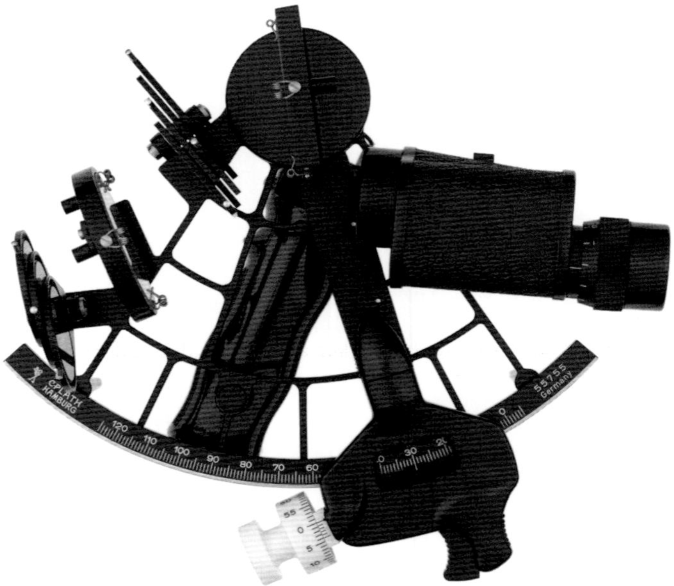

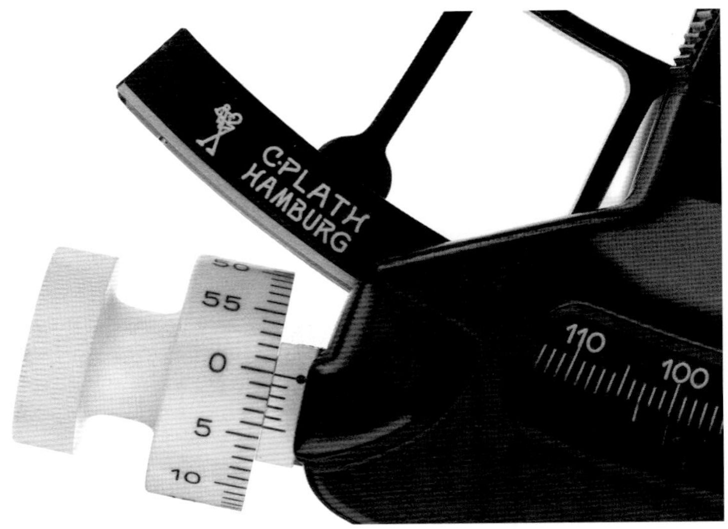

272 NAV1209
Sextant

Mate sextant
Origin London, England
Maker Kelvin Hughes
Date 1970
Inscriptions Signed on the limb: [KH logo] | Made in Great Britain by KELVIN HUGHES a Division of Smiths Industries Limited.
Marked on the limb: Mate Sextant
Marked on the index arm: Ser No 72327
Graduations Aluminium scale from -5° to 125° every 1°, measuring to 122°. Micrometer to 0.2′.
Dimensions Radius 6 inches (152 mm)
Description Grey-lacquered alloy three-circle-pattern frame, wooden handle. White plastic micrometer drum, pressure clamp, and worm screw on the back of the index arm. Four shades (three green and one grey); three horizon shades (one grey and two green). Index-glass adjustment by a screw; adjustment of horizon glass by capstan screws. Threaded telescope bracket without adjustment. Telescope (76 mm, star finder), erect image. A shaded eyepiece and an adjusting pin are missing. In a square fitted wooden box, containing in the lid a Kelvin Hughes certificate of examination, dated 29 April 1970.
Provenance Lent by Kelvin Hughes, Barkingside, Essex, in 1970, and presented in 1971.
For image see colour section plate 18.

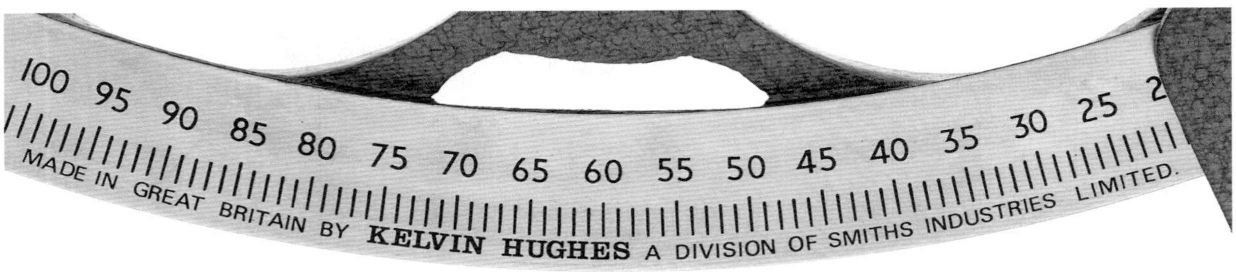

273 NAV1979
Solar sextant for submarine periscope

Type C.K.25
Origin Probably Glasgow, Scotland
Maker Probably Barr and Stroud Ltd
Date 1970–80
Inscriptions Marked on the side: SOLAR SEXTANT | for | SUBMARINE PERISCOPE | Type C.K.25 | No 438
Graduation Circular black scale with two scales, each from 0° to 100° every 1°. Black plastic micrometer to 1′.
Dimensions 215 × 100 × 101 mm (length × height × width)
Description Grey-lacquered alloy casing incorporating artificial materials, housing optical equipment. Four settings for shades (clear, dark grey, and two blue).
Provenance Presented by Barr and Stroud Ltd (Pilkington Optronics), Glasgow, in 1992 .
Notes This solar sextant was screwed onto the search periscope of a submarine, for taking sights of both the Sun and stars; this type was probably used on British conventional submarines. The provenance suggests that Barr and Stroud manufactured this instrument. For similar sextants, see Ifland, *Taking the Stars*.
Literature Ifland, *Taking the Stars*, 74–5.

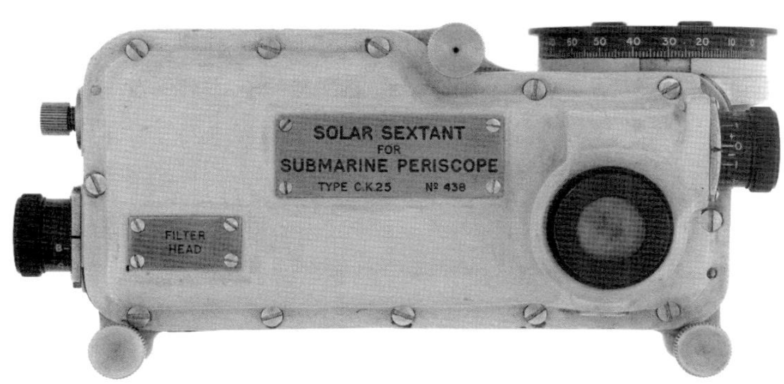

274 NAV1215
Box sextant

Small-craft precision sextant
Origin Edenbridge, Kent, England
Maker Francis Barker and Son
Date 1971
Inscriptions Signed on a label stuck on the side of the instrument: FBS | Francis Barker & Son Limited | Small Craft | Precision Sextant | Edenbridge-Kent-U.K.
Graduation Aluminium scale from -5° to 150° every 30′, measuring to 144°. Aluminium vernier to 1′, zero at the right.
Dimensions Radius 1.78 inches (45 mm); diameter 76 mm
Description Circular grey-lacquered alloy plates. The lid contains a paper table for correcting the observed altitude of the Sun's lower limb. The index arm is moved by a milled knob at the pivot point on the upper plate. Two index shades (both green) moved by levers; no horizon shades. Index- and horizon-glass adjustment by square-headed screws with a milled key fixed in the upper plate. Magnifier (single lens) on a 57 mm swivelling arm attached at the index arm's pivot. Sight vane screwed into the frame, with one pinhole and an alternative shaded sight (green) fixed in the side of the frame. A spare paper table for the Sun's semi-diameter, a paper table of natural tangents to a base of 100, a brush, and an instruction manual titled *Small Craft Precision Sextant*. In a fitted leather carrying case with a sling strap.
Provenance Lent by Francis Barker and Son Ltd, Edenbridge, Kent, in 1971, and presented in 1978.
Literature NN, *Small Craft Precision Sextant*.

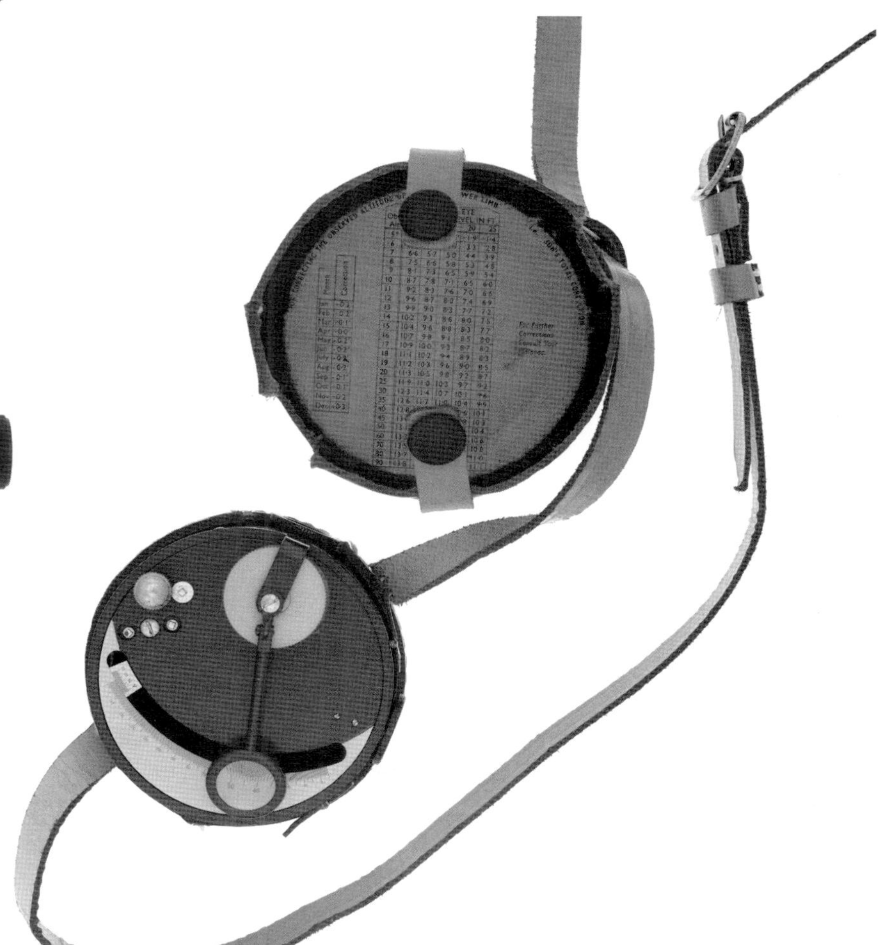

7. QUINTANTS

275 NAV 1108
Quintant

Bridge quintant
Origin London, England
Maker Matthew Berge
Date About 1810
Inscriptions Signed on the limb: Berge London | late Ramsden
Marked on the back of the frame near the pivot: 1765
Graduation Polished brass limb with inlaid gold scale from -2° to 165° every 20′, measuring to 144°. Gold vernier to 20″, zero at the right.
Dimensions Radius 8 inches (203 mm)
Description Anodized brass diamond-pattern frame with a protective frame (bridge) over the fittings and mounts, and a wooden handle with a brass-lined threaded hole for the second detached wooden handle. The tangent screw and clamping screw are on the back of the index arm. Four shades (three red and one green); three horizon shades (all green). Index-glass adjustment by a screw; adjustment of horizon glass by a screw and a capstan screw. Magnifier on an 85 mm swivelling arm. Threaded telescope bracket with perpendicular adjustment by rising-piece and a milled knob. Telescope (186 mm), inverted image, four cross-wires; telescope (72 mm), erect image; telescope (84 mm), erect image. Sight-tube (89 mm), a shaded eye-piece (red), a rotating shaded eyepiece with choice of three shades (red, orange, and green), and an adjusting pin. In a wooden keystone box, containing in the lid an MS label 'B/. Barges on Thames Southampton 1909'.
Provenance Adams collection; lent in 1936, bequeathed in 1974. The label may suggest the acquisition of this instrument by Adams in 1909.
Literature Stimson, 'The Influence of the Royal Observatory at Greenwich', 130; Clifton, *Directory*, 28.

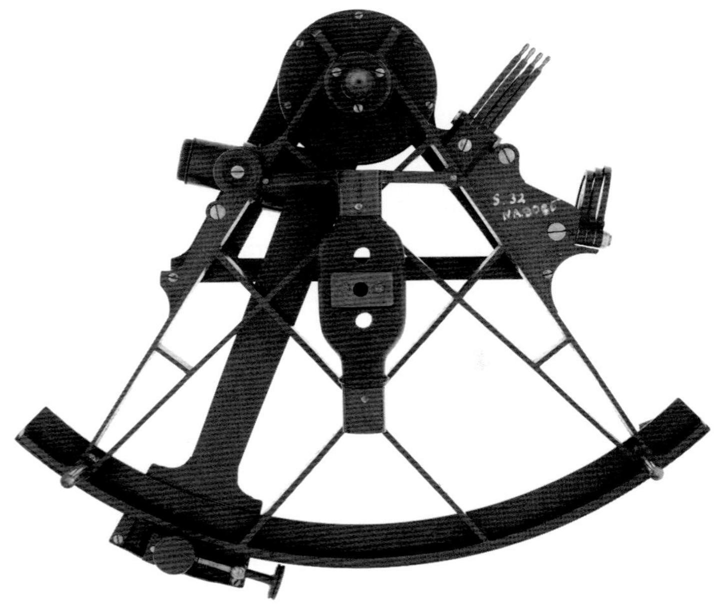

276 NAV1170
Quintant

Origin London, England
Maker Troughton and Simms
Date About 1830
Inscriptions Signed on the limb: Troughton & Simms, London
Marked on the centre strut: 843
Graduation Polished brass limb with inlaid silver scale from -5° to 160° every 10′, measuring to 137°. Silver vernier to 10″, zero at the right.
Dimensions Radius 10 inches (254 mm)
Description Anodized brass straight-bar-pattern pillar frame (24 pillars), and a wooden handle with a brass-lined hole for placing it on a stand. The tangent screw and clamping screw are on the back of the index arm. Four shades (one red and green; two are missing); three horizon shades (two red and one green). Index-glass adjustment by a screw; adjustment of horizon glass by a square-headed screw and a detached key. Magnifier on a 95 mm swivelling arm. Threaded telescope bracket in two parts, fitted for correcting collimation error; perpendicular adjustment by rising-piece and a milled knob. Shaded eyepiece (blue). Telescopes, an adjusting key, and some unidentified parts are missing. In a mahogany keystone box.
Provenance Presented by Kelvin Hughes, Barkingside, Essex, in 1961.
Notes The digits seem very modern for the date of this instrument. Edward Troughton patented the pillar and plate frame (no. 1644 of 1788).
Literature Stimson, 'The Influence of the Royal Observatory at Greenwich', 130; McConnell, *Instrument Makers to the World*; Clifton, *Directory*, 282.

277 NAV1214
Quintant

Origin London, England
Maker Nathaniel Worthington
Date About 1830
Inscriptions Signed on the left of the scale: Worthington | London
Marked (stamped) at the centre of the scale (at 45°): I[foul anchor]R
Graduation Inlaid ivory scale from -2° to 140° (lower division) and from 40° to 180° (upper division) by 30′, measuring to 120° and 160°. Vernier to 1′, zero at the right.
Dimensions Radius 10½ inches (267 mm)
Description Ebony frame and limb, with black-lacquered brass index arm and fittings, and a wooden handle; an inlaid ivory plate on the crossbar. The tangent screw and clamping screw are on the back of the index arm. Three index shades (two red and one green); no horizon shades. Index-glass adjustment by a screw; adjustment of horizon glasses by screw and capstan screws. Threaded telescope bracket without adjustment. Telescope (84 mm), erect image. Shaded eyepiece (green). In a wooden keystone box.
Provenance Purchased from Mr Frank Hawtin of the Somerset Museum, Education and Art Service, Taunton, Somerset, in 1971.
Notes Designed by Captain (later Admiral) Robert Fitzroy who had three examples made by Worthington, for hydrographic work. It allows angles up to 140° to be measured with the upper horizon glass, and up to 180°, with the lower horizon glass. The scale mark indicates that it was divided on Ramsden's engine which was later used by Worthington.
Literature Fitzroy, *Narrative*, volume 2, 274–5, 396, 637; Ifland, *Taking the Stars*, 31–2; Stimson, 'Some Board of Longitude Instruments', 112–13; Clifton, *Directory*, 305.

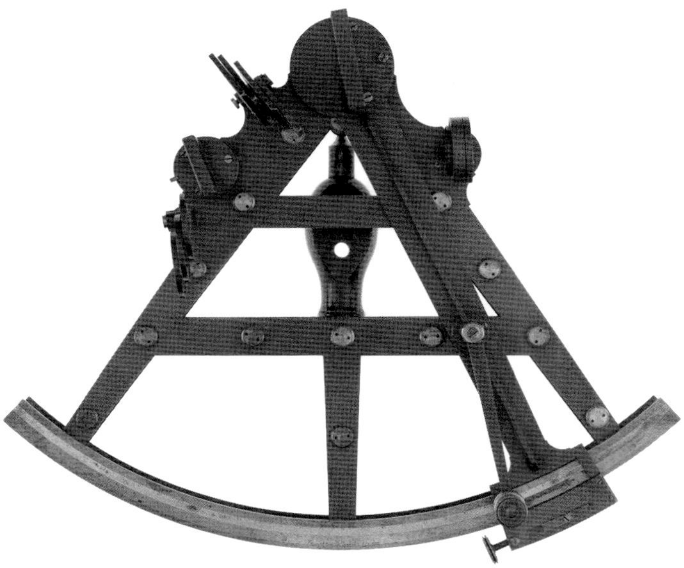

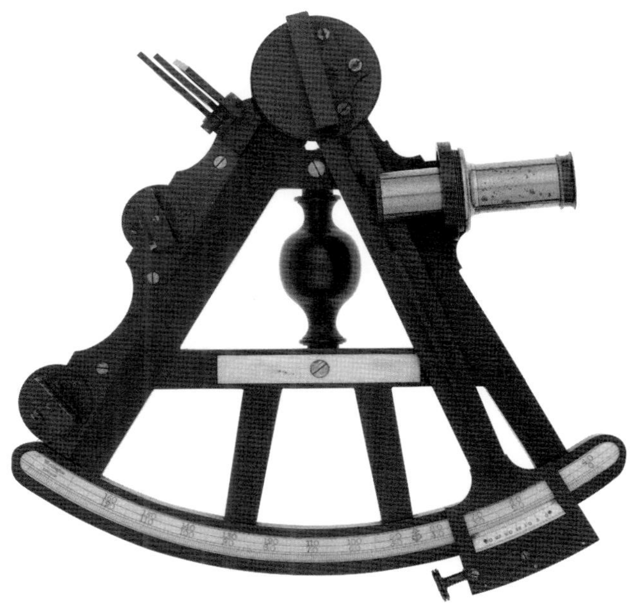

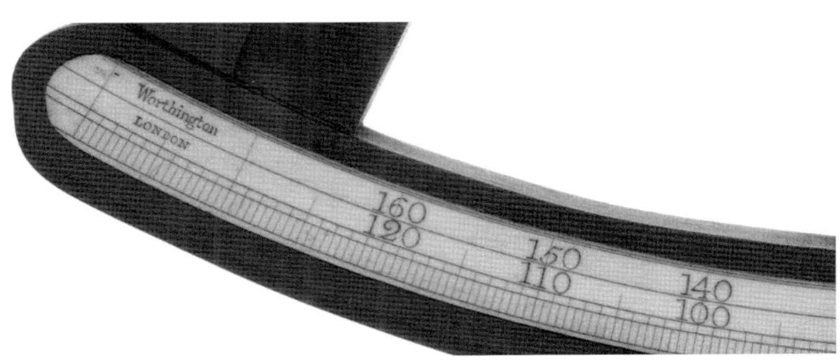

278 NAV 1227
Double sounding quintant

Origin London, England
Maker Troughton and Simms
Date About 1840
Inscriptions Signed on the frame: Troughton & Simms | London
Graduation Two polished brass limbs with inlaid silver scales, both from -5° to 185° every 20′, measuring to 150° (left) and 165° (right). Two silver verniers, both to 20′, zeros at the right and the left.
Dimensions Radius 2¾ inches (70 mm)
Description Polished brass frame, with a detached wooden handle. The tangent screws and clamping screws are on the back of the two index arms. No shades. Two index glasses, adjustment by screws; two horizon glasses, adjustment by screws and square-headed screws. Two magnifiers (single lenses), each on a 61 mm swivelling arm. L-shaped threaded telescope bracket fixed to the frame, without adjustment. Telescope (115 mm), inverted image. No box.
Provenance Purchased from David Weston Ltd, London, in 1979.
Notes Designed to measure angles on either side of an object in one observation.
Literature McConnell, *Instrument Makers to the World*; Clifton, *Directory*, 282.

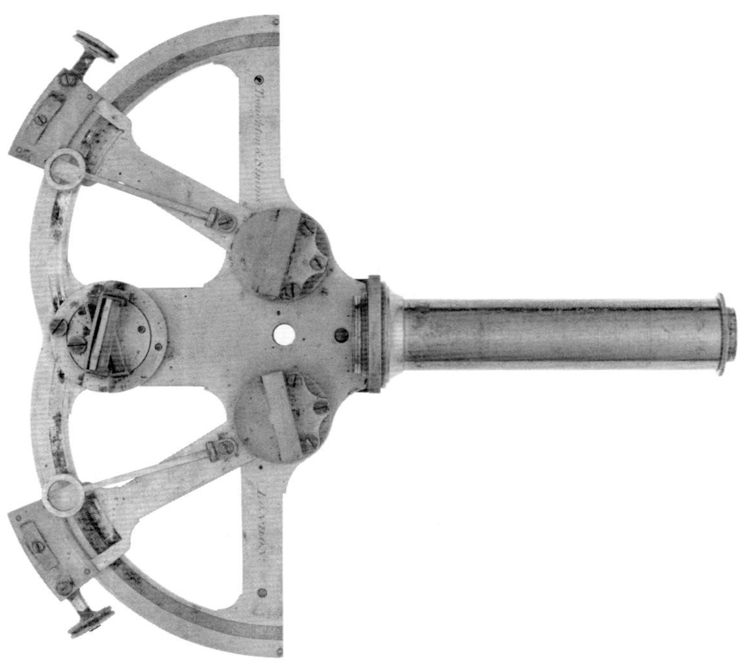

279 NAV 1122
Double sounding quintant

Origin London, England
Maker William Cary
Date About 1847
Inscriptions Signed on the limb: Cary London
Marked on the upper index arm: Da.92
Graduation Polished brass limb with two parallel inlaid silver scales; the lower, from the left, from 0° to 155° every 30′, measuring to 147°; the upper, from the right, from 0° to 150° every 30′, measuring to 144°. Two silver verniers to 1′; zero at the left (lower) and at the right (upper).
Dimensions Radius 3.15 inches (80 mm)
Description Grey – green-lacquered brass frame, with a circular brass handle. There is no tangent screw or other device for slow motion of the two index arms; the upper clamping screw is on the front of the index arm; the lower clamping screw is on the back of the arm. No shades. Two index glasses, adjustment by screws; one large horizon glass with two silvered portions and clear glass in between, adjustment by a screw and a capstan screw. Upper magnifier (single lens) on a 50 mm swivelling arm; lower magnifier (single lens) on a 60 mm swivelling arm. Threaded telescope bracket with perpendicular adjustment by rising-piece and a milled knob. Telescope (70 mm), erect image; shaded eyepiece (green). In a square fitted mahogany box, marked on a brass plate inlaid in the lid 'Da.92'.
Provenance Adams collection; lent in 1936, bequeathed in 1974.
Notes Invented by Rear Admiral Frederick William Beechey (1796–1856). The instrument can be used to measure the angles between three objects at once. When both index arms are set to 0°, the index glasses are exactly parallel. Compare with cat. no. 185. 'Da.92' possibly refers to the numbering adopted by the Hydrographic Office in February 1828, where the letter D was given to sextants.
Literature NN, 'The Double Sextant'; Stimson, 'Some Board of Longitude Instruments', 114–15; Clifton, *Directory*, 51.

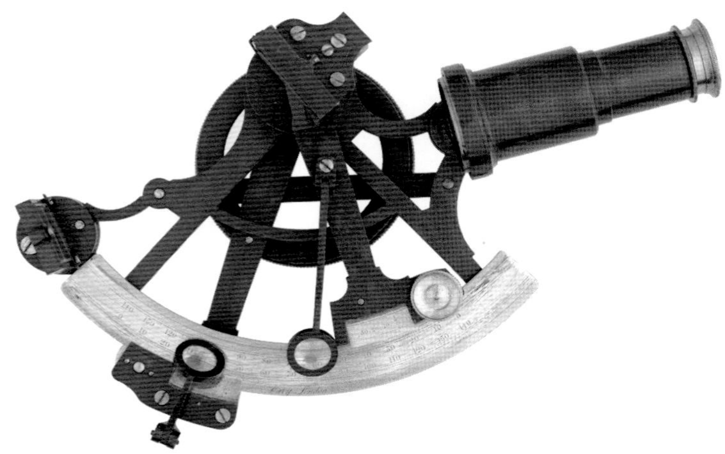

280 NAV 1119
Quintant

Origin London, England
Maker Thomas Jones
Date About 1850
Inscriptions Signed on the limb: Thomas Jones Charing Cross London
Marked on the index arm: D42
Graduation Polished brass limb with inlaid silver scale from -5° to 160° every 10′, measuring to 140°. Silver vernier to 10″, zero at the right.
Dimensions Radius 5 inches (127 mm)
Description Bronzed brass frame, with a wooden handle. The tangent screw and clamping screw are on the back of the index arm. Four shades (all grey); three horizon shades (all grey). Index-glass adjustment by a screw; adjustment of horizon glass by capstan screws. Magnifier on a 60 mm swivelling arm. Threaded telescope bracket in two parts, fitted for correcting collimation error; perpendicular adjustment by rising-piece and a milled knob. Telescope (188 mm), inverted image, two parallel cross wires; telescope (76 mm), erect image; extra drawtube (64 mm), two parallel cross-wires. Sight-tube (94 mm), a shaded eyepiece (red), and an adjusting pin. In a square mahogany box, containing in the lid a trade label for Henry Hughes and Son, 59 Fenchurch Street, London, and marked on the lid 'D42'.
Provenance Adams collection; lent in 1936, bequeathed in 1974. The presence of the Hughes label may suggest that Adams purchased this instrument from them.
Notes 'D42' was the Hydrographic Office serial numbering adopted in February 1828, where D was the letter given to sextants.
Literature Stimson, 'Some Board of Longitude Instruments', 114–15; Clifton, *Directory*, 154.

281 NAV1135
Quintant

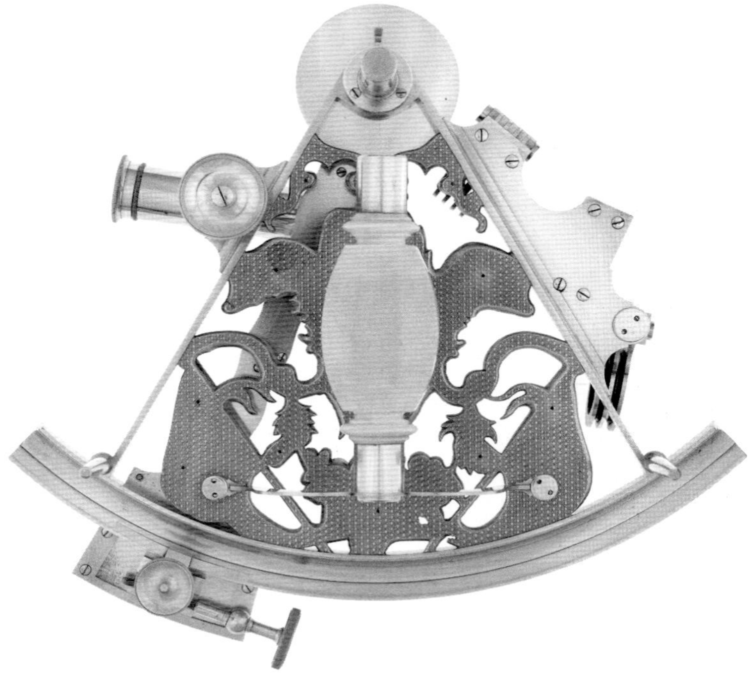

Origin London, England
Maker Janet Taylor
Date About 1850
Inscriptions Signed on the limb: Mrs. Janet Taylor, 104 . Minories London.
Graduation Silver limb with inlaid gold scale from -5° to 160° every 10′, measuring to 142°. Gold vernier to 10″, zero at the right.
Dimensions Radius 6 inches (152 mm)
Description Silver frame and fittings, with an ivory handle. Mounted at the centre of the frame in silver relief is the three-feather crest of the Prince of Wales, inscribed with his motto *Ich Dien* ('I serve') on a ribbon. The lower decoration includes a British ensign (by implication the general red, senior ensign at this date) and the Royal Standard, with thistles, roses, clover, and a daffodil symbolizing Scotland, England, Ireland, and Wales. The tangent screw and clamping screw are on the back of the index arm. Four shades (three grey and one green); three horizon shades (two grey and one green). Index-glass adjustment by a screw; adjustment of horizon glass by capstan screws. Magnifier on an 82 mm swivelling arm; a frosted glass shade. Threaded telescope bracket in two parts, fitted for correcting collimation error; perpendicular adjustment by rising-piece and a milled knob. Telescope (149 mm), inverted image, two parallel cross-wires; extra drawtube (51 mm), inverted image, two parallel cross-wires; telescope (85 mm), erect image. Sight-tube (81 mm), two shaded eyepieces (both mauve) and an adjusting pin. In a rosewood keystone box lined with blue velvet and bearing on the lid the crest of the Prince of Wales in silver.
Provenance This instrument was shown by Janet Taylor at the 1851 Great Exhibition in London and was later presented to the Prince of Wales (subsequently King Edward VII, 1841–1910). His daughter-in-law HM Queen Mary presented it in 1936.
Notes Probably a standard-pattern frame in which the inner framework was replaced by the traditional 'Prince of Wales feathers'.
Literature Croucher, 'An Exceptional Woman of Science'; Clifton, *Directory*, 275; H.C.G. Matthew, 'Edward VII (1841–1910), *king of the United Kingdom of Great Britain and Ireland, and the British dominions beyond the seas, and emperor of India*', in *Oxford Dictionary of National Biography*, Oxford, 2004, <http://www.oxforddnb.com/>
For image see colour section plate 19.

282 NAV1165
Quintant

Origin London, England
Maker John Dennett Potter
Date About 1851
Inscriptions Signed on the limb: I. D. Potter Poultry & Tower Hill, London
Graduation Polished brass limb with inlaid silver scale from -5° to 165° every 10′, measuring to 145°. Silver vernier to 10″, zero at the right.
Dimensions Radius 7.88 inches (200 mm)
Description Varnished brass diamond-pattern frame, with a wooden handle. The tangent screw and clamping screw are on the back of the index arm. Four shades (all green); three horizon shades (all green). Index-glass adjustment by a screw; adjustment of horizon glass by capstan screws and by a square-headed screw and a detached key. Threaded telescope bracket in two parts, fitted for correcting

collimation error; perpendicular adjustment by rising-piece and a milled knob. Magnifier on an 83 mm swivelling arm. Telescope (187 mm), inverted image, two parallel cross-wires; extra drawtube (79 mm), inverted image, four cross-wires; telescope (82 mm), erect image. Sight-tube (81 mm), two shaded eyepieces (both green), an adjusting pin, a milled adjusting key, and a magnifying glass marked 'E M'. In a mahogany keystone box, containing in the lid a trade label for J.B. Potter, successor to R.B. Bate, and a card from Staff Commander Edward Mourilyan RN marked in MS 'A Remembrance from the [Mourilyan] family to Edgar Tarry Adams'.

Provenance Adams collection; purchased in 1958. Apparently Mourilyan gave this instrument to Adams.

Literature Clifton, *Directory*, 221; McConnell, *R B Bate of the Poultry*, 53–4.

283 NAV1147
Pocket quintant

Origin London, England
Maker William Cary
Date About 1852
Inscriptions Signed on the crossbar: Cary, | London. Marked on the centre strut: D.86
Graduation Polished brass limb with inlaid silver scale from -5° to 165° every 20′, measuring to 144°. Silver vernier to 30″, zero at the right.
Dimensions Radius 3 inches (76 mm)
Description Anodized brass alloy frame, and a hinged wooden handle with a hole for securing a carrying string. The tangent screw and clamping screw are on the back of the index arm. Two shades (green and grey); two horizon shades (green and grey). Index-glass adjustment by a screw; adjustment of horizon glass by screws and a milled knob. Magnifier (single lens) on a 50 mm swivelling arm. Threaded telescope bracket with perpendicular adjustment by rising-piece and a milled knob. Telescope (106 mm), inverted image; telescope (70 mm), erect image. Sight-tube (70 mm), a shaded eyepiece (mauve), and a magnifying glass (missing). In a square fitted wooden box within a leather carrying case (strap missing) marked 'Admiralty HO [broad arrow] D.86'. A paper label on the lid of the box in MS 'This Sextant was used by me during the travelling Expeditions in the Arctic Regions in the years 1852, 53, 54 and was afterwards presented to me by the Hydrographer as a memento of the Service G.H.R.'
Provenance The initials 'HO' and the broad arrow indicate that this was originally an official-issue Royal Navy Hydrographic Office instrument. It was used by Captain (later Admiral Sir) George Henry Richards RN, KCB, FRS (1820–96) when commander of HMS *Assistance* in Belcher's search expedition for Sir John Franklin in 1852–54, and subsequently presented to him by the Hydrographer of the Navy. It was presented to the Museum by his descendent Captain J.B.O. Richards RN, Shepperton, Middlesex, in 1949.
Notes 'D.86' was the Hydrographic Office serial numbering adopted in February 1828, in which D was the letter given to sextants.
Literature; Clifton, *Directory*, 51; Stimson, 'Some Board of Longitude Instruments', 114–15; Hill and Paget-Tomlinson, *Instruments of Navigation*, 35; G.S. Ritchy, 'Richards, Sir George Henry (1820–96), *naval officer and hydrographer*'. in *Oxford Dictionary of National Biography*, Oxford, 2004, <http://www.oxforddnb.com/>.

284 NAV1124
Quintant

Origin England, probably London,
Maker Unknown
Date About 1860
Inscriptions Marked on the limb: H.W.W. Hope | A & N. C. S. L.
Graduation Polished brass limb with inlaid silver scale from -5° to 165° every 10′, measuring to 146°. Silver vernier to 10″, zero at the right.
Dimensions Radius 6¼ inches (159 mm)
Description Black-lacquered brass triangle-pattern frame, with a wooden handle. The tangent screw and clamping screw are on the back of the index arm. Four shades (all grey); three horizon shades (all grey). Index-glass adjustment by a screw; adjustment of horizon glass by capstan screws and by square-headed screws and a detached key. Magnifier on a 77 mm swivelling arm. Threaded telescope bracket in two parts, fitted for correcting collimation error; perpendicular adjustment by rising-piece and a milled knob. Telescope (183 mm), inverted image, four cross-wires; extra drawtube (79 mm), erect image, two parallel cross-wires; telescope (82 mm), erect image. Sight-tube (78 mm), two shaded eyepieces (red and grey), an adjusting key, an adjusting pin, and a magnifying glass in a horn cover. The upper screw holding the handle is broken. In a square fitted wooden box (repaired), with on the lid a brass plate marked 'H. W. W. Hope'.
Provenance Presented before 1937 by Admiral Sir George Price Webley Hope (1869–1959) RN, KCB, KC, MC, Chairman of the Society for Nautical Research, and a Trustee of the National Maritime Museum 1934–1952. Presumably Admiral Hope was a descendant of H.W.W. Hope.
Notes The Army and Navy Co-operative Society Ltd was a retailer in London. It is likely that this retailer obtained its quintants from a London manufacturer such as Elliott Brothers, who, like the Society, was situated in the Strand, or from Henry Hughes and Son or Heath and Co.
Literature Littlewood and Butler, *Of Ships and Stars*, Appendix 1.

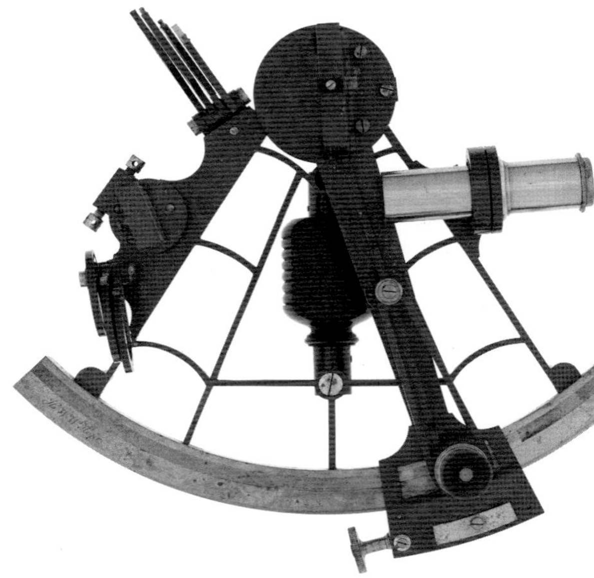

285 NAV1138
Quintant

Origin Paris, France
Maker Edmond Lorieux
Date About 1860
Inscriptions Signed on the limb: E. Lorieux à Paris (1141) | Depot chez Santi, à Marseille.
Graduation Polished brass limb with inlaid silver scale from -5° to 170° every 10′, measuring to 142°. Silver vernier to 10″, zero at the right.
Dimensions Radius 7½ inches (191 mm)
Description Polished brass Lorieux–Borda-pattern frame, with a wooden handle. The double-ended tangent screw is on the front of the index arm; the clamping screw is on the back. Four shades (one orange, two red and one green); three horizon shades (two red and one green). Index- and horizon-glass adjustment by square-headed screws and a detached key. Magnifier on a 100 mm hinged, swivelling arm; a frosted glass shade (missing). Threaded telescope bracket with perpendicular adjustment by rising-piece and a milled knob. Telescope (132 mm), erect image. The box, the adjusting key, and various unidentified parts are missing.
Provenance Presented by Sir James Caird in 1936; purchased as part of the Landau collection through Maggs Brothers.
Notes Augustin Santi was a scientific instrument maker in Marseilles from 1830 to 1884.
Literature Callender, *The Caird Collection: Supplement*, 96, no. 20; Marcelin, *Dictionnaire des fabricants français*.

286 NAV1123
Quintant

Origin France
Maker Unknown
Date About 1875
Inscriptions Marked on all fittings and the mounts: 38
Graduation Polished brass limb with inlaid silver scale from -10° to 170° every 10′, measuring to 141°. Silver vernier to 10″, zero at the right.
Dimensions Radius 7 inches (178 mm)
Description Brown-varnished brass Lorieux–Borda-pattern frame, with a wooden handle. The tangent screw working against a spring-loaded stop, is at the bottom of the index arm; the clamping screw is on the back. Three shades (red, green, and grey); three horizon shades (two red and one green). Index-glass adjustment by a screw; adjustment of horizon glass by capstan screws and by a square-headed screw and a detached key. Magnifier (single lens) on 112 mm jointed swivelling arm; a frosted glass shade. Threaded telescope bracket in two parts, fitted for correcting collimation error; perpendicular adjustment by rising-piece and a milled knob. Telescope (158 mm), inverted image, four cross-wires; extra drawtube (74 mm), erect image, two parallel cross-wires; telescope (88 mm), erect image. Sight-tube (100 mm), a milled adjusting key, and two unidentified parts (both missing). In a square fitted wooden box.
Provenance Adams collection; lent in 1936, bequeathed in 1974.

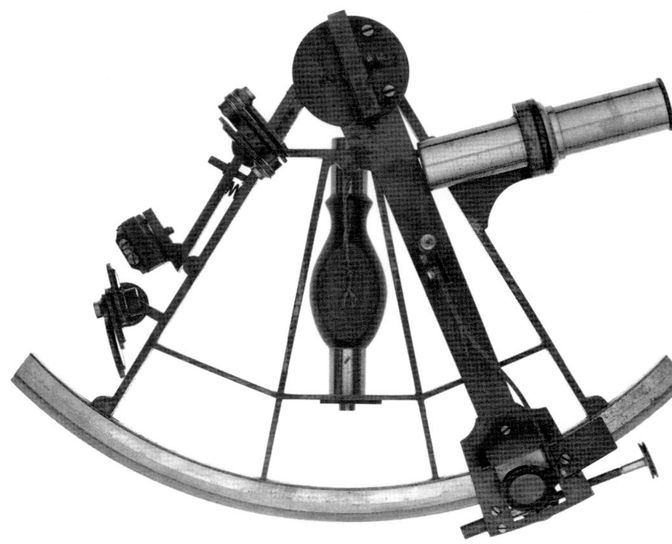

287 NAV1131
Quintant

Origin London, England
Maker Elliott Brothers
Date About 1875
Inscriptions Signed on the limb: Elliott Bros. London
Marked on the limb: Prince Edward
Graduation Polished brass limb with inlaid silver scale from -5° to 170° every 10′, measuring to 155°. Silver vernier to 10″, zero at the right.
Dimensions Radius 6¼ inches (159 mm)
Description Anodized brass triangle-pattern frame, with a wooden handle. The tangent screw and clamping screw are on the back of the index arm. Four shades (all grey); three horizon shades (all grey). Index-glass adjustment by a screw; adjustment of horizon glass by capstan screws. Magnifier on an 85 mm swivelling arm; a frosted glass shade. Threaded telescope bracket in two parts, fitted for correcting collimation error; perpendicular adjustment by rising-piece and a milled knob. Telescope (184 mm), inverted image, four cross-wires; extra drawtube (81 mm), inverted image, two parallel cross-wires; telescope (83 mm), erect image. Sight-tube (80 mm), two shaded eyepieces (red and grey), and an adjusting pin. The magnifying glass and an unidentified part are missing. In a square fitted wooden box.
Provenance Belonged to the Prince of Wales, later King Edward VII (1841–1910). Presented by his daughter-in-law HM Queen Mary in 1936.
Literature Clifton, 'An Introduction to the History of Elliott Brothers'; H.C.G. Matthew, 'Edward VII (1841–1919), *king of the United Kingdom of Great Britain and Ireland, and the British dominions beyond the seas, and emperor of India*'. in *Oxford Dictionary of National Biography*, Oxford, 2004, <http://www.oxforddnb.com/>.

288 NAV 1149
Quintant

Origin London, England
Maker John Dennett Potter
Date About 1875
Inscriptions Signed on the limb: Potter, Poultry, London. Marked on the crossbar: [broad arrow] 7 Polar Exp. 1875
Graduation Polished brass limb with inlaid silver scale from -5° to 165° every 10′, measuring to 146°. Silver vernier to 10″, zero at the right.
Dimensions Radius 6½ inches (165 mm)
Description Anodized brass straight-bar-pattern frame, with a wooden handle. The tangent screw and clamping screw are on the back of the index arm. Four shades (all green); three horizon shades (all green). Index-glass adjustment by a screw; adjustment of horizon glass by a capped milled knob. The magnifier and swivelling arm are missing. Threaded telescope bracket in two parts, fitted for correcting collimation error; perpendicular adjustment by rising-piece and a milled knob. Telescope (185 mm), inverted image, no cross-wires; shaded eyepiece (green). Other telescopes, a sight-tube, and two further eyepieces are missing. In a square fitted wooden box (with a blank brass plate on the lid), and containing in the lid a trade label for J.D. Potter, 31 Poultry, London. The right strut, part of the telescope, the shaded eyepiece, the milled knob of the rising-piece, and the knobs of the tangent screw and the clamp are covered in light chamois leather for protection of the skin in polar regions.
Provenance The broad arrow and the inscription show that this was an official-issue item. It was used by Captain (later Admiral) Sir George Strong Nares RN on the British Arctic expedition of 1875–76 in *Discovery* and *Alert*. Members of the Peary Arctic Club of New York presented it to the Museum before October 1949.
Notes The Peary Arctic Club was founded in New York, between 1899 and 1910. How the club obtained this item is unknown, and the Museum's earliest typed record of it is when it was cleaned in 1949.
Literature Clifton, *Directory*, 221; McConnell, *R B Bate of the Poultry*, 53–4; Hill and Paget-Tomlinson, *Instruments of Navigation*, 36; R.N. Rudmose Brown, 'Nares, Sir George Strong (bap. 1831, d. 1915) *Arctic explorer*',. in *Oxford Dictionary of National Biography*, Oxford, 2004, <http://oxforddnb.com/>.

289 NAV1133
Quintant

Origin London, England
Maker Elliott Brothers
Date About 1877
Inscriptions Signed on the limb: Elliott Bros, London
Marked on the limb: Prince George
Graduation Polished brass limb with inlaid silver scale from -5° to 170° every 10′, measuring to 154°. Silver vernier to 10″, zero at the right.
Dimensions Radius 6 inches (152 mm)
Description Anodized brass triangle-pattern frame, with a wooden handle. The tangent screw and clamping screw are on the back of the index arm. Four shades (all green); three horizon shades (all green). Index-glass adjustment by a screw; adjustment of horizon glass by capstan screws. Magnifier on an 89 mm swivelling arm; a frosted glass shade. Threaded telescope bracket in two parts, fitted for correcting collimation error; perpendicular adjustment by rising-piece and a milled knob. Telescope (184 mm), inverted image, four cross wires; extra drawtube (81 mm), inverted image, two parallel cross-wires; telescope (83 mm), erect image. Sight-tube (81 mm), two shaded eye-pieces (red and green), an adjusting pin, a magnifying glass (missing), and an unidentified part (missing). In a square fitted mahogany box, with a brass plate on the lid marked 'Prince George'.
Provenance Belonged to Prince George, later King George V (1865–1936), who was sent to train for the Navy in HMS *Britannia* in 1877 and used this instrument during his naval career. His widow HM Queen Mary presented it in 1936.

Literature Clifton, 'An Introduction to the History of Elliott Brothers'; H.C.G. Matthew, 'George V (1865–1936), *king of Great Britain, Ireland, and the British dominions beyond the seas, and emperor of India*',. in *Oxford Dictionary of National Biography*, Oxford, 2004, <http://www.oxforddnb.com/>.

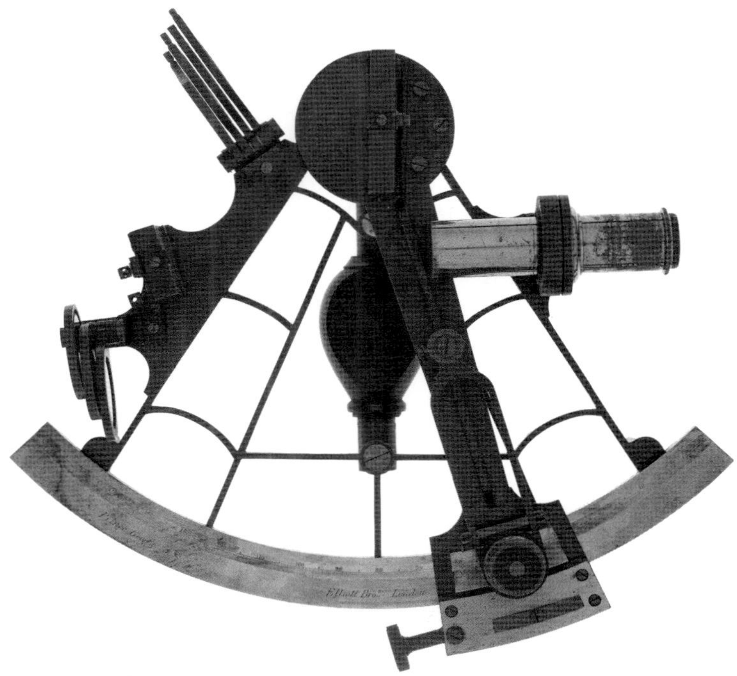

290 NAV1243
Surveyor's quintant

Origin France, possibly Paris
Maker Possibly Henri Morin
Date About 1890
Inscriptions Marked on the index arm: Coutureau Invr.
Graduation Chrome scale from -4° to 152° every 1°, measuring to 149°. Vernier 0–100 *grads*.
Dimensions Radius 2¾ inches (70 mm)
Description Polished chrome-coated bronze frame with a protective cover over the mounts. The tangent screw, working against a spring-loaded stop, and the clamping screw are on the back of the index arm. No shades. Index-glass adjustment by a screw; horizon glass without adjustment. No telescope, but a sighting arrow is inscribed on the frame. A compass (diameter 38 mm) is mounted at the centre of the frame, graduated 0–3. Fitted with an edge bar needle-with a jewelled pivot cap, and a needle locking mechanism operated by a milled knob on the back of the frame. In a red velvet-lined leather carrying case (strap missing), with a stamp in the lid for 'Instruments de Précision Nivellement Géodésie Henri [illegible] Paris'.
Provenance Purchased from Mr D. Essex, London, in 1985.
Notes The maker may have been Henri Morin, who worked in Paris from 1880; nothing has been found on Coutureau. Following the introduction of the metric system, the scale division of 400 *grads* to a circle instead of 360 degrees was introduced in France shortly after the French Revolution. Most theodolites and similar instruments from European countries are still graduated in this way.
Literature Marcelin, *Dictionnaire des fabricants français*; 'Mystery Object', in *Bulletin of the Scientific Instrument Society*, 49 (1996) 32–3.

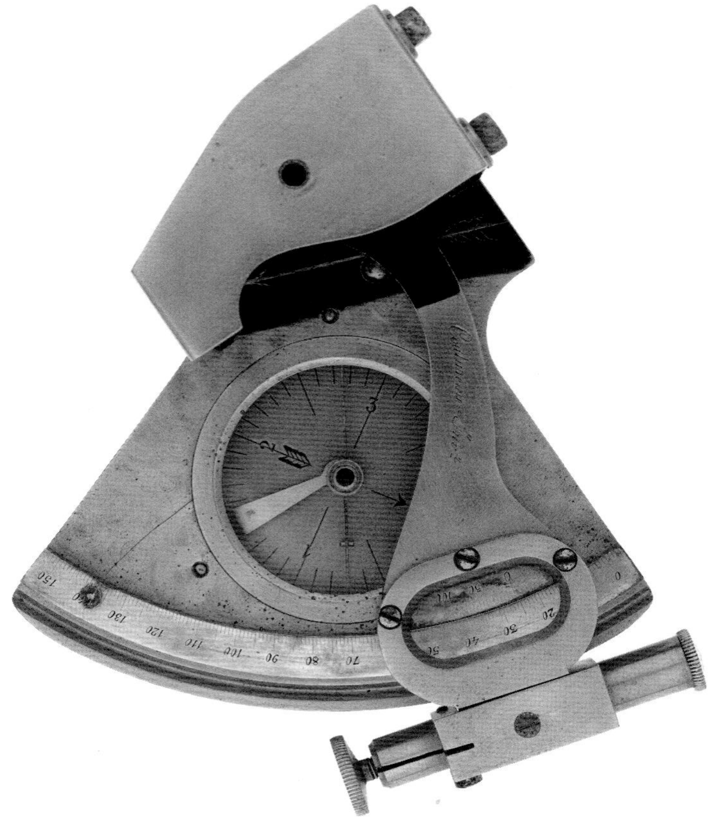

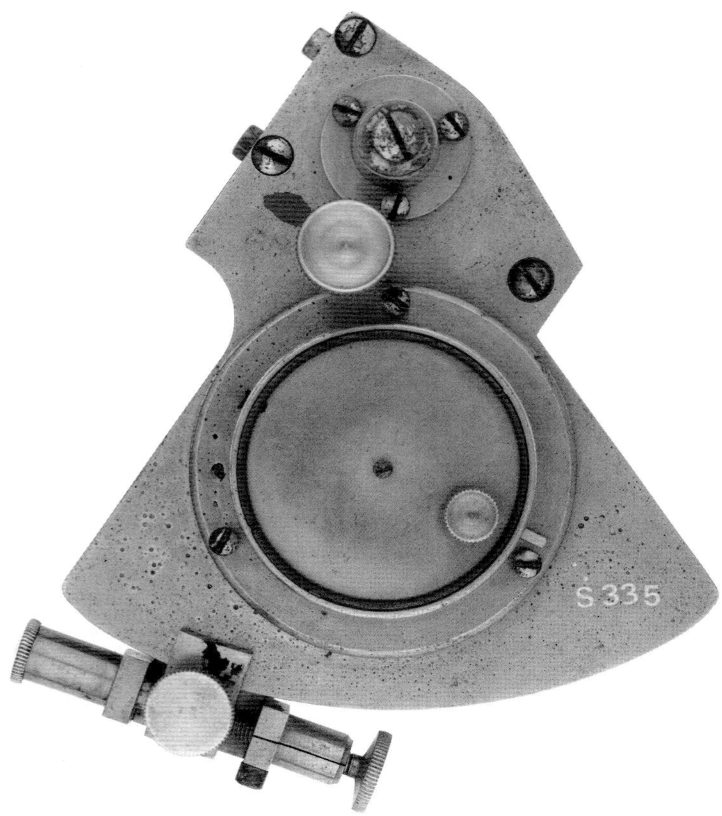

291 NAV1237
Quintant

Origin London, England
Maker Henry Hughes and Son Ltd
Date 1894
Inscriptions Signed on the limb: Henry Hughes & Son, 59, Fenchurch St., London.
Marked on the limb: 3514 | Platina & Gold
Graduation Polished aluminium limb with inlaid platinum scale from -5° to 205° every 10′, measuring to 180°. Gold vernier to 10″, zero at the right.
Dimensions Radius 6 inches (152 mm)
Description Polished aluminium diamond-pattern frame with aluminium and brass fittings, and a wooden handle with a brass-lined threaded hole for a second handle or a stand. The tangent screw is on the back of the index arm; the clamping screw is on the front. The glasses, shades, and telescope bracket are on the back of the frame. Four shades (all grey); three horizon shades (all grey). Index-glass adjustment by a screw; adjustment of horizon glass by capstan screws. Magnifier on a 165 mm swivelling arm, pivoted at the index arm pivot point. Threaded telescope bracket in two parts, fitted for correcting collimation error; perpendicular adjustment by rising-piece and a milled knob. Telescope (184 mm), inverted image, four cross-wires; extra drawtube (76 mm), two parallel cross-wires; telescope (77 mm), erect image. Sight-tube (76 mm), a rotating shaded eyepiece with a choice of three shades (all grey), an adjusting pin, a magnifying glass, and a screwdriver. In a square fitted wooden box lined with green textile, containing in the lid a Kew Observatory certificate of examination, dated July 1894.
Provenance Purchased from David Weston Ltd, London, in 1983.
Literature Chaldecott, 'Platinum and Palladium in Astronomy and Navigation'.

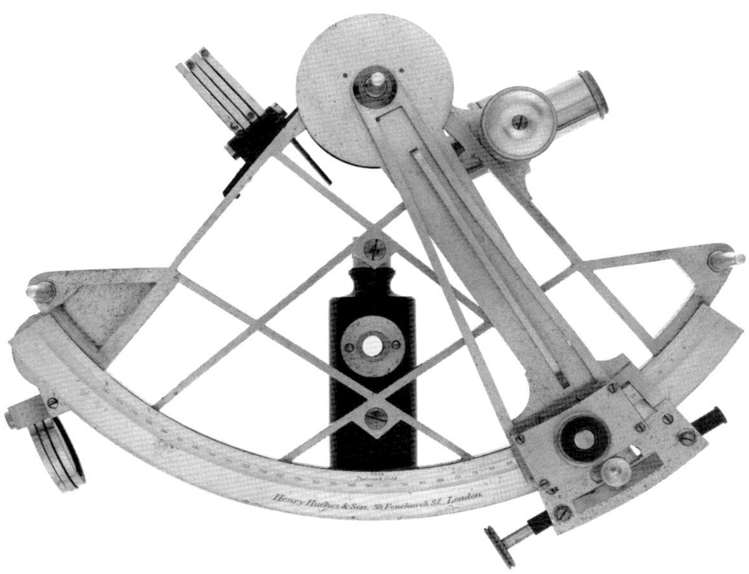

292 NAV 1130
Quintant

Origin Hamburg, Germany
Maker Carl Plath
Date About 1917
Inscriptions Signed on the index arm: C. Plath | Hamburg
Marked at the ends of the limb [Plath logo] | 7547
Marked on the index arm: [crown] | M | 1305
Graduation Polished brass limb with scale from -5° to 150° every 1°, measuring to 137°. Micrometer to 1′.
Dimensions Radius 6½ inches (165 mm)
Description Black-lacquered brass three-circle-pattern frame, with a wooden handle. Locking of index arm by sprung lever on the top of the arm; white plastic micrometer drum. Four shades (three grey and one green); two horizon shades (green and grey). Index-glass adjustment by a capstan screw; adjustment of horizon glass by screws. Threaded telescope bracket in two parts, fitted for correcting collimation error; perpendicular adjustment by rising-piece and a milled knob with a locking collar. Sight-tube (84 mm). Two L-shaped brackets and an adjusting pin. Telescopes, a shaded eyepiece, and some unidentified parts are missing. In a square fitted wooden box marked '7547 M 1305'. Contained in the lid is a metal trade label for Plath, a typed note in German on the eccentricity error, dated 26 July 1917; and a rubber stamp of the Imperial German Navy.
Provenance This instrument was taken from a German submarine during World War I, and was presented by Captain Leith RN to the Greenwich Hospital collection, as part of which it was transferred to the Museum in 1936 on indefinite loan.
Notes The crowned M (for 'Marine') refers to the Imperial German Navy.
Literature Schaafhausen *et al.*, *C Plath 1862–1962,* 104–5.

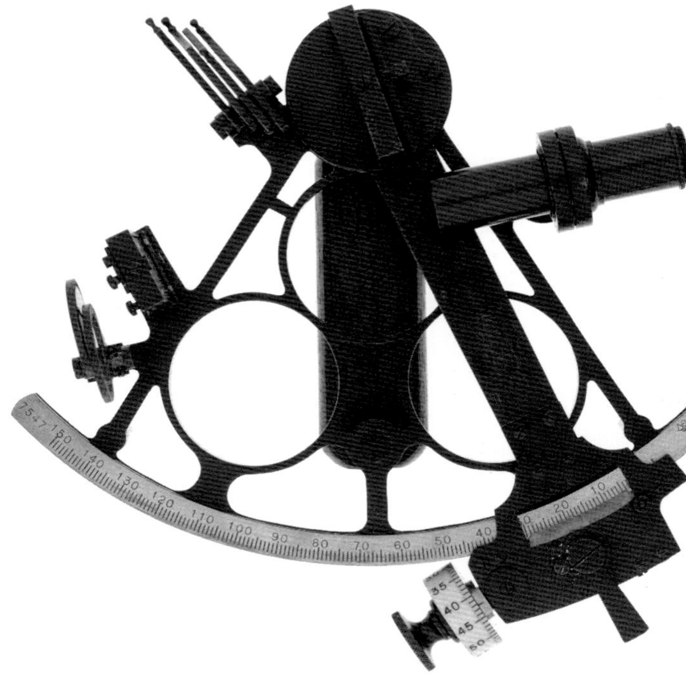

293 NAV 1197
Quintant

Origin Paris, France
Maker Albert Lepetit
Date About 1925
Inscriptions Signed on the limb: Lorieux, Lepetit. Sucr Paris. No 4885
Marked (stamped) in the telescope bracket: 13
Graduation Polished brass limb with inlaid silver scale from -5° to 155° every 10′, measuring to 144°. Silver vernier to 10″, zero at the right.
Dimensions Radius 7½ inches (191 mm)
Description Brass black-lacquered triangle-pattern frame, with a wooden handle. The double-ended tangent screw is on the front of the index arm; the clamping screw is on the back. Four shades (one purple, two green, and one orange); three horizon shades (two green and one orange). Index-glass adjustment by square-headed screw and a detached key; horizon glass (half silvered) by a milled knob and a clamp, and by a square-headed screw and a detached key. Magnifier on a 109 mm double-jointed swivelling arm; a glass shade (missing). Detached threaded telescope bracket in two parts, fitted for correcting collimation error; perpendicular adjustment by rising-piece and a milled knob. Telescope (178 mm), inverted image, four cross-wires; telescope (92 mm), erect image (star finder); telescope (225 mm), erect image (star finder). Sight-tube (73 mm), two shaded eyepieces (red and yellow), a third eyepiece (green) from another instrument, an adjusting key, and two L-shaped brackets. In a square fitted wooden box.
Provenance Purchased from Mr Peter Roberts, Epsom, Surrey, in 1966.
Notes Albert Lepetit (active from after 1921 to about 1950) succeeded Edmond Lorieux.
Literature Marcelin, *Dictionnaire des fabricants français*.

294 NAV 1182
Sounding quintant

Origin London, England
Maker Henry Hughes and Son Ltd
Date 1944
Inscriptions Signed on the limb: H. Hughes & Son Ltd. London | Made in England | 44842
Marked on the index arm: H [broad arrow] O 256 | Husun
Graduation Chromed brass scale from -5° to 150° every 1°, measuring to 150°. Micrometer to 1′.
Dimensions Radius 6 inches (152 mm)
Description Diamond-pattern alloy frame with traces of black lacquer, and a wooden handle with a brass-lined threaded hole for a second handle or a stand. Pressure clamp and worm screw at the bottom of the index arm, worm screw on the back; white plastic micrometer drum. No shades. Round index glass without adjustment; round horizon glass, adjustment by capped capstan screws. Threaded telescope bracket without adjustment. Telescope (89 mm, large objective), erect image, leather-covered. Two detached feet for screwing in to the front of the frame are missing. Several unidentified parts are missing. In a square fitted wooden box, containing in the lid a Husun certificate of examination, dated 3 November 1944; also a white plastic label for the Hydrographic Department of the Admiralty. The side of the box is marked 'H [broad arrow] O 256'.
Provenance Presented by the Hydrographer of the Navy, Cricklewood, in 1965.

295 NAV1180
Sounding quintant

Origin London, England
Maker Henry Hughes and Son Ltd
Date 1947
Inscriptions Signed on the limb: H. Hughes & Son Ltd 59 Fenchurch St. London. E.C.
Marked on the limb: 17445
Marked on the index arm: A [broad arrow] FI Pat 2274/28
Graduation Alloy scale from -5° to 160° every 1°, measuring to 148°. Micrometer to 1′.
Dimensions Radius 6 inches (152 mm)

Descriptions Black-lacquered alloy diamond-pattern frame, with a wooden handle. Pressure clamp and worm screw on the back of the index arm; white plastic micrometer drum. No shades. Round index glass, adjustment by a screw; round horizon glass, adjustment by capped capstan screws. Threaded telescope bracket without adjustment. Telescope (93 mm, large objective lens), erect image, leather-covered. Several unidentified parts are missing; one foot of the frame is broken. In a square fitted wooden box, containing in the lid a Husun certificate of examination, dated 17 November 1947.
Provenance Presented by the Hydrographer of the Navy, Cricklewood, in 1965.

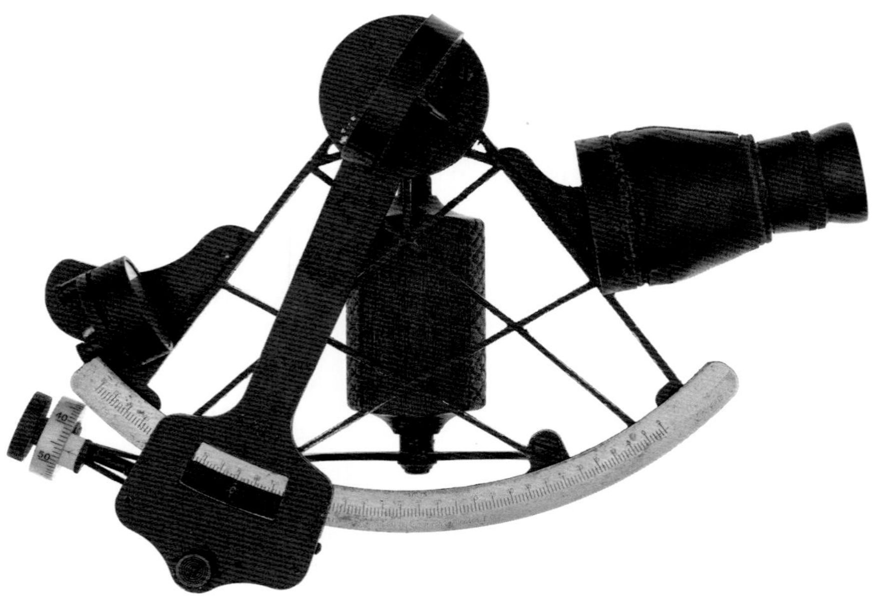

296 NAV1206
Sounding quintant

Origin New Eltham, London, England
Maker Heath and Co. Ltd
Date About 1950
Inscriptions Signed on the limb: Heath & Co New Eltham. London S.E.9. | Made in England
Marked on the index arm: H.[broad arrow]O.368 | "Hezzanith" | Endless Rapid Reader | Automatic Clamp | Patent [and] Semper | Paratus
Graduation Aluminium scale from -5° to 155° every 1°, measuring to 153°. Micrometer to 1′.
Dimensions Radius 5¾ inches (146 mm)
Description Black-lacquered aluminium diamond-pattern frame, with a wooden handle. Pressure clamp and worm screw at the bottom of the index arm; black plastic micrometer drum with white markings. No shades. Round index glass without adjustment; adjustment of the round horizon glass and the prism by capped capstan screws. Threaded telescope bracket without adjustment. Fixed telescope (118 mm), erect image. Shaded eyepiece (yellow); detached 90° prism on a bridge, to be secured to the frame by two spring clips. Adjusting pin and spanner. In a square fitted wooden box, containing in the lid a Hezzanith Instrument Works certificate of examination, dated 23 February 1954.
Provenance Presented by the Hydrographer of the Navy, Taunton, Somerset, in 1969.
Notes A regular sounding quintant to measure angles up to 153° but, when the prism is put in place, angles up to 180° can be measured, by reading the scale 'backwards'.

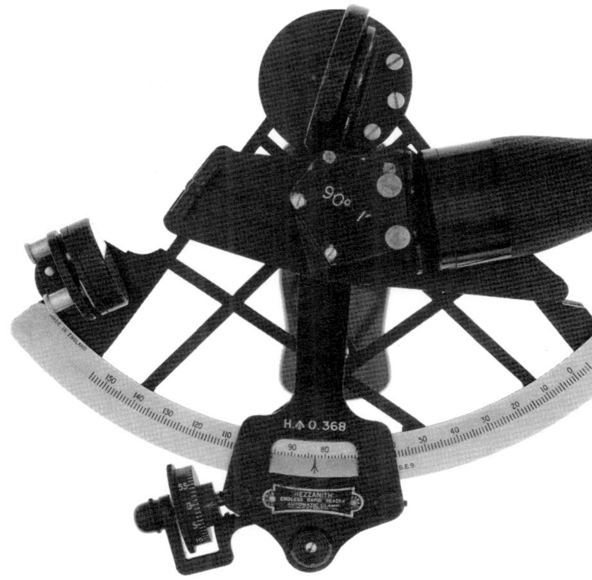

297 ACO1339
Sounding quintant

Survey Mark III
Origin London, England
Maker Henry Hughes and Son Ltd
Date 1952
Inscriptions Signed on the limb: H. Hughes & Son Ltd London | Made in Great Britain
Marked on the limb: Survey Mk III Serial No 59328
Marked on the end of the index arm: Husun | Trade Mark | 135 K.H.I./52
Graduation Black-lacquered alloy scale with white markings from -5° to 140° every 1°, measuring to 140°. Micrometer to 1′.
Dimensions Radius 6 inches (152 mm)
Description Black-lacquered alloy frame, with a black plastic handle. Pressure clamp and worm screw at the bottom of the index arm; black plastic micrometer drum with white markings. No shades. Round index glass without adjustment; round horizon glass, adjustment by capped screws. Threaded telescope bracket without adjustment. Telescope (120 mm), large objective lens. Shaded eyepiece (yellow), a bottle of oil (for lubrication), a piece of chamois leather, three detached feet for screwing into the front of the instrument, and an adjusting pin; one unidentified part is missing. In a square fitted wooden box, containing in the lid a Kelvin Hughes certificate of examination, dated 27 June 1952.
Provenance Transferred from the Admiralty Compass Observatory in 1969.
For image see colour section plate 20.

8. REFLECTING CIRCLES

298 NAV0076
Repeating circle

Borda circle
Origin London, England
Maker Probably Jesse Ramsden
Date About 1787
Inscriptions Marked on the telescope arm: I.H. de Magellan, | invenit & fieri Curavit Londini
Graduation Polished brass scale from 0° to 720° every 30′. Two brass verniers to 1′, zeros at the right.
Dimensions Radius 9½ inches (241 mm)
Description Circular blackened brass frame with bright fittings, with a detached wooden handle. The index arm carries the index glass; the telescope arm carries the horizon glass. The tangent screw and clamping screw are on the back of the index arm. A second tangent screw (without clamp) is at the ocular end of the telescope arm; the vernier is at the other end. Two detached index shades (both grey); no horizon shades. Index-glass adjustment by a thumbscrew and square-headed screws; adjustment of horizon glass by screws and by a square-headed screw with a detached key. Telescope (125 mm), inverted image, two parallel cross-wires. Two threaded telescope brackets, with perpendicular adjustment on both by rising-pieces and milled knobs. Shaded eyepiece (red). Adjusting key with a wooden handle. Magnifying glass. In a square fitted mahogany box, containing in the lid a label in MS 'Constant Angle 201° 10″. Another label in the lid reads in MS 'A/. Old Bike is ABRW.j.ea. 1911'.
Provenance Adams collection; lent in 1936, bequeathed in 1974. The second MS label may suggest that this instrument was acquired by Adams in 1911.
Notes The inscription, which translates as 'Jean Hyacinthe de Magellan invented and caused [this instrument] to be commissioned in London', is also found on several clocks with his name. As with these, it is not possible to determine what Magellan's contribution to the Borda circle was. In his article listed below, Betts recorded Ramsden as the probable manufacturer of this instrument.
Literature Magellan, *Description des octants et sextants Anglois*; Betts, 'John Hyacinth de Magellan (1722–1790) Part 1', 509, 517; Betts, 'John Hyacinth de Magellan (1722–1790) Part 2'; Hill and Paget-Tomlinson, *Instruments of Navigation*, 25.
For image see colour section plate 21.

299 NAV0077
Repeating circle

Origin London, England
Maker (Peter) Dollond
Date About 1800
Inscriptions Signed on the bottom of index arm: Dollond | London
Graduation Polished brass limb with inlaid silver scale from 0° to 720° every 20′. Two silver verniers to 20″, zeros at the centre. A polished brass scale on the inside of the limb to 180° every 5° on either side of zero.
Dimensions Diameter 6½ inches (165 mm)
Description Circular polished brass frame with a strengthening bar across, with a wood-covered handle (the cover can be unscrewed). The index arm carries the index glass; the tangent screw, working against a spring-loaded stop, and the clamping screw are on the back of the arm. Three index arms, moving with a brass scale and a vernier on the inside of the limb; they carry the telescope and the horizon glass. One arm has a tangent screw, working against a spring-loaded stop, and a clamping screw on the back of the arm. Index-glass adjustment by a screw; adjustment of horizon glass by capstan screws. Three index shades (two red and one green); three horizon shades (two red and one green). Two magnifiers on 48 mm swivelling arms; two frosted glass shades. Threaded telescope bracket with perpendicular adjustment by rising-piece and a milled knob. Telescope (158 mm), inverted image, four cross-wires; extra drawtube (52 mm), four cross-wires; extra drawtube (50 mm), four cross-wires; telescope (91 mm), erect image; extra drawtube (50 mm). Sight-tube (85 mm), an adjusting pin, a key for removing the handle cover, and a rotating shaded eyepiece with a choice of three shades (two green and one yellow). In a square fitted wooden box, containing in the lid a label in MS 'A. Ramsgate Harbour Sunset (Southampton) 1908'.
Provenance Adams collection; lent in 1936, bequeathed in 1974. The label probably refers to the acquisition of this instrument by Adams in 1908.
Notes Similar to cat. no. 310.
Literature Hill and Paget-Tomlinson, *Instruments of Navigation*, 27; Clifton, *Directory*, 87.

300 NAV0079
Reflecting circle

Origin London, England
Maker (John and Edward) Troughton
Date About 1800
Inscriptions Signed on the index arm: Troughton | 23 | London
Marked on the index arm: C.23
Graduation Polished brass limb with inlaid silver scale indicating from 160° to 0° to 160°; the entire scale is divided every 20′. Three silver verniers to 20″, zeros at the left.
Dimensions Diameter 11.3 inches (287 mm)
Description Circular polished brass frame, with a wooden handle. A detached U-shaped bracket holds a second handle with a brass-lined threaded hole for a third handle (missing), which can also be screwed on to the front of the frame. Three rigidly connected index arms on the back of the frame, and a double-ended tangent screw and a clamping screw on the back of one of the arms. Three index shades (two red and one green); three horizon shades (two red and one green). Index-glass adjustment by a screw; horizon glass by a screw and a capstan screw. A detached magnifier on a 102 mm swivelling arm, serving all three verniers. Threaded telescope bracket in two parts, fitted for correcting collimation error; perpendicular adjustment by rising-piece and a milled collar. Telescope (185 mm), inverted image, two parallel cross-wires; extra drawtube (84 mm), two parallel cross-wires; telescope (74 mm), erect image. Sight-tube (130 mm) and an adjusting pin; two shaded eyepieces (missing). In a square fitted wooden box, marked on the lid 'C.23'.
Provenance Transferred as part of the Greenwich Hospital collection in 1936, on indefinite loan.
Notes The U-shaped bracket was recommended by Mendoza Rios, and according to him first applied by Troughton. 'C.23' was the Hydrographic Office serial numbering adopted in February 1828, where C was the letter given to altitude and azimuth instruments and circles.
Literature Stimson, 'The Influence of the Royal Observatory at Greenwich', 130; Mendoza Rios, *On an Improved Reflecting Circle*, 8–9; McConnell, *Instrument Makers to the World*; Stimson, 'Some Board of Longitude Instruments', 114–15; Clifton, *Directory*, 282.

301 NAV0078
Reflecting circle

Origin London, England
Maker (John and Edward) Troughton
Date About 1800
Inscriptions Signed across the base of the index arm: Troughton | 37 | London
Graduation Polished brass limb with inlaid gold scale indicating from 160° to 0° to 160°, the entire scale is divided every 20′. Three brass verniers to 20″, zeros at the left.
Dimensions Diameter 11.6 inches (295 mm)
Description Circular polished brass frame, with a wooden handle. A detached U-shaped bracket holds a second handle with a brass-lined threaded hole for a third handle, which can also be screwed into the front of the frame. Three rigidly connected index arms on the back of the frame, and a double-ended tangent screw and a clamping screw on the back of one of the arms. Three index shades (two red and one green); three horizon shades (two red and one green). Index-glass adjustment by a screw; adjustment of horizon glass by a screw and a capstan screw. A detached magnifier on a 70 mm swivelling arm, serving all three verniers. Threaded telescope bracket in two parts, fitted for correcting collimation error; perpendicular adjustment by rising-piece and a milled collar. Telescope (185 mm), inverted image, two parallel cross-wires; extra drawtube (82 mm), two parallel cross-wires; telescope (142 mm), inverted image, four cross wires; telescope (75 mm), erect image. Sight-tube (129 mm) and two shaded eyepieces (both red). In a square fitted brassbound teak box, possibly a replacement, as teak is unusual for navigational-instrument boxes.
Provenance Adams collection; lent in 1936, bequeathed in 1974.
Notes The U-shaped bracket was recommended by Mendoza Rios, and according to him first applied by Troughton.
Literature Stimson, 'The Influence of the Royal Observatory at Greenwich', 130; Mendoza Rios, *On an Improved Reflecting Circle*, 8–9; McConnell, *Instrument Makers to the World*; Clifton, *Directory*, 282.

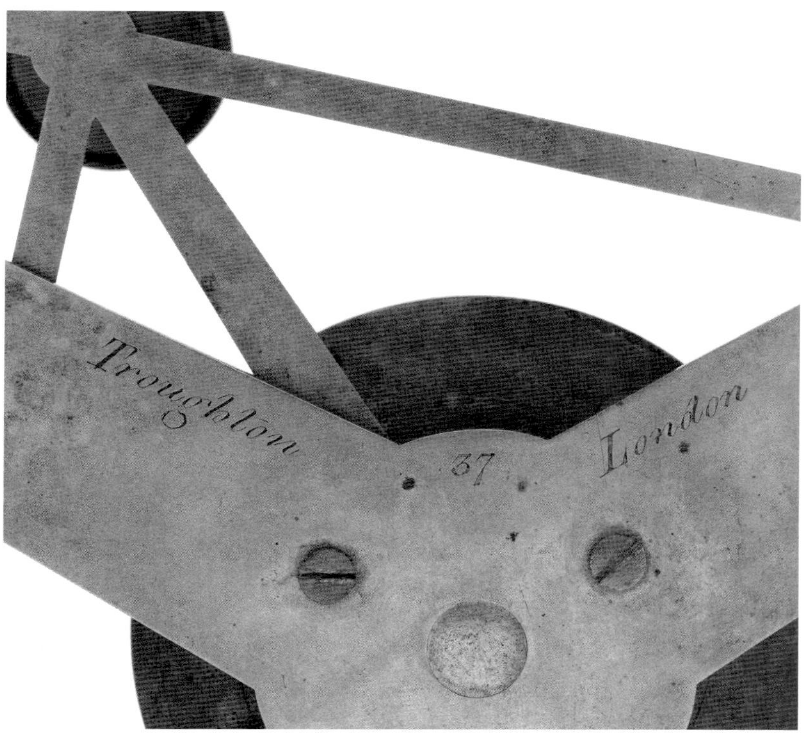

302 NAV0085
Reflecting circle

Origin London, England
Maker (John and Edward) Troughton
Date About 1803
Inscriptions Signed on the index arm: Troughton | 91 | London
Graduation Bronzed limb with inlaid silver scale indicating from 180° to 0° to 180°; the entire scale is divided by 20′. Three silver verniers to 20″, zeros at the left.
Dimensions Diameter 10¾ inches (273 mm)
Description Circular anodized brass frame, with two wooden handles. A detached U-shaped bracket holds a third handle with a brass-lined threaded hole for a fourth handle. Three rigidly connected index arms on the back of the frame, and a double-ended tangent screw and clamping screw on the back of one of the arms. Three index shades (two red and one green); three horizon shades (two red and one green). Index- and horizon-glass adjustment by screws. A detached magnifier on a swivelling arm is missing. Threaded telescope bracket in two parts, fitted for correcting collimation error; perpendicular adjustment by rising-piece and a milled collar. All telescopes and other parts are missing. No box.
Provenance Presented by Kelvin Hughes, Barkingside, in 1961.
Notes The U-shaped bracket was recommended by Mendoza Rios, and according to him first applied by Troughton.
Literature Stimson, 'The Influence of the Royal Observatory at Greenwich', 130; Mendoza Rios, *On an Improved Reflecting Circle*, 8–9; McConnell, *Instrument Makers to the World*; Clifton, *Directory*, 282.

303 NAV0080
Reflecting circle

Origin London, England
Maker (John and Edward) Troughton
Date About 1805
Inscriptions Signed on the index arm: Troughton | London
Graduation Polished limb with inlaid silver scale indicating from 150° to 0° to 150°; the entire scale is graduated every 20′. Three silver verniers to 20″, zeros at the left.
Dimensions Diameter 12.6 inches (320 mm)
Description Circular polished brass frame, with a wooden handle. A detached U-shaped bracket holds a second handle (cracked) with a brass-lined threaded hole for a third handle, which can also be screwed into the front of the frame. Three rigidly connected index arms on the back of the frame, and a double-ended tangent screw and a clamping screw on the back of one of the arms. Three index shades (red, orange, and green); three horizon shades (red, orange, and green). Index-glass adjustment by a screw; adjustment of horizon glass by a screw and a capstan screw. A magnifier on a 150-mm swivelling arm, serving all three verniers, and a detached magnifier (lenses missing) on a 65 mm swivelling arm for use on each of the verniers separately. Threaded telescope bracket with perpendicular adjustment by rising-piece and a milled collar. Telescope (180 mm), inverted image, two parallel cross-wires; extra drawtube (80 mm), two parallel cross-wires; extra drawtube (87 mm), two parallel cross-wires. Sight-tube (134 mm), a shaded eyepiece (red), an adjusting pin, two shaded eye pieces (one red, the other missing), and a magnifying glass (missing). In a square fitted box, of which the lid is missing.
Provenance Transferred as part of the Greenwich Hospital collection in 1936, on indefinite loan.
Notes This is an example of Troughton's largest reflecting circle, which he designed in 1796. The U-shaped bracket was recommended by Mendoza Rios, and according to him first applied by Troughton.
Literature Mendoza Rios, *On an Improved Reflecting Circle*, 8–9; McConnell, *Instrument Makers to the World*; Clifton, *Directory*, 282.

304 NAV0089
Repeating circle

Borda circle
Origin London, England
Maker Spencer Browning and Rust
Date About 1805
Inscriptions Signed on the limb: Spencer, Browning & Rust, London.
Graduation Polished brass scale from 0° to 720° every 30′. Two verniers to 1′, zeros at the right.
Dimensions Diameter 10¾ inches (273 mm)
Description Circular polished brass frame, with a detached wooden handle. A detached U-shaped bracket holds a second handle. The index arm carries the index glass; the telescope arm carries the horizon glass. The tangent screws and clamping screws are on the back of the index arm and on the back of the telescope arm at the objective end. Three detached shades (two red and one green); two square detached horizon shades (both red). Index-glass adjustment by a screw; adjustment of horizon glass by a screw and a capstan screw. Two threaded telescope brackets, with perpendicular adjustment by rising-pieces and milled knobs on both. Telescope (115 mm), inverted image, two parallel cross-wires. Adjusting pin; the magnifying glass is missing. In a square fitted wooden box with a blank brass plate on the lid. The box is lined with green paper and contains in the lid a trade label for George Bradford, 99 The Minories, London (1817–51), and also an unidentified London trade label, over which a piece of paper is stuck, with an MS note 'Patent Reflecting-Circle'.
Provenance Purchased from Mr T. Mackay, London, in 1981.
Notes The U-shaped bracket was recommended by Mendoza Rios, and according to him first applied by Troughton.
Literature Mendoza Rios, *On an Improved Reflecting Circle*, 8–9; Clifton, *Directory*, 36, 261.

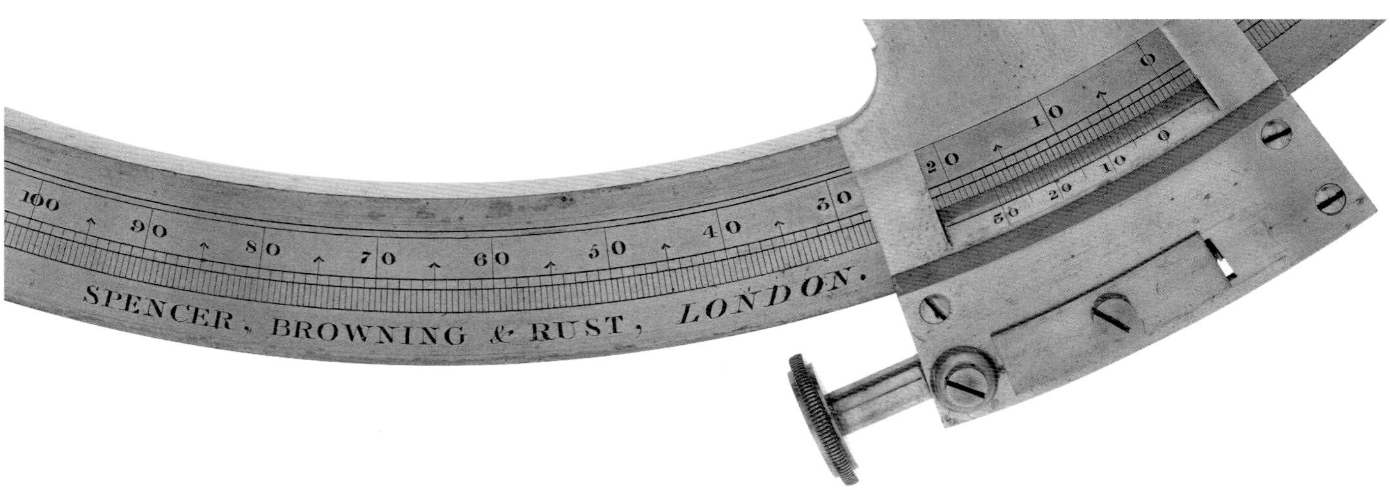

305 NAV0081

Reflecting circle

Origin London, England
Maker (Edward) Troughton
Date About 1810
Inscriptions Signed on the index arm: Troughton | 146 | London
Graduation Polished brass limb with inlaid silver scale from 160° to 0° to 160°, although the entire scale is divided every 20′. Three silver verniers to 20″, zeros at the left.
Dimensions Diameter 10.6 inches (269 mm)
Description Circular polished brass frame, with a wooden handle (cracked). A detached U-shaped bracket holds a second handle with a brass-lined threaded hole for a third handle; a fourth handle (possibly taken from another instrument) can be screwed into the front of the frame. Three rigidly connected index arms on the back of the frame, and a double-ended tangent screw and a clamping screw on the back of one of the arms. Three shades (two red and one green); three horizon shades (two red and one green). Index-glass adjustment by a screw; adjustment of horizon glass by a screw and a capstan screw. Magnifier on a 228 mm arm, swivelling at its centre and serving all three verniers, a detached second magnifier (lenses missing) on a 75 mm swivelling arm for use with each of the verniers separately. Threaded telescope bracket in two parts, fitted for correcting collimation error; perpendicular adjustment by rising-piece and a milled collar. Telescope (68 mm), inverted image, two parallel cross-wires; extra drawtube (78 mm), two parallel cross-wires; telescope (76 mm), erect image. Sight-tube (131 mm) and an adjusting pin (missing). In a square fitted wooden box.
Provenance Adams collection; lent in 1950, bequeathed in 1974.
Notes The U-shaped bracket was recommended by Mendoza Rios, and according to him first applied by Troughton.
Literature Stimson, 'The Influence of the Royal Observatory at Greenwich', 130; Mendoza Rios, *On an Improved Reflecting Circle*, 8–9; McConnell, *Instrument Makers to the World*; Clifton, *Directory*, 281.

306 NAV0075

Repeating circle

Origin London, England
Maker William Heather
Date About 1810
Inscriptions Signed on the limb: Heather London
Graduation Polished brass limb with inlaid silver scale from 0° to 720° every 20′. Silver vernier to 15″, zero at the right, and one silver sighting line.
Dimensions Diameter 9.1 inches (231 mm)
Description Circular polished brass frame, with a wooden handle on a fixed U-shaped bracket. A detached second handle can be screwed into the bracket on the front of the frame. The limb, with the scale, is on the back of the frame and can be rotated through 360° by hand; it is secured by two clamps and by the index arm clamp. The index arm carries the index glass, the telescope bracket and horizon glass are rigidly mounted on the frame. A double-ended tangent screw and a clamping screw are on the back of the index arm. Three index shades (two red, one green); three horizon shades (two red and one green). Index-glass adjustment by two screws; adjustment of horizon glass by a screw and a capstan screw. Magnifier on a 55 mm swivelling arm. Threaded telescope bracket in two parts, fitted for correcting collimation error; perpendicular adjustment by rising-piece and a milled collar. Telescope (172 mm), inverted image, four cross-wires; extra drawtube (67 mm), four cross-wires; telescope (68 mm), erect image. Sight-tube (92 mm), a shaded eyepiece (red), an adjusting pin, a screwdriver, and a magnifying glass. In a square fitted wooden box lined with green textile and containing trade labels for William Heather, 157 Leadenhall Street, London (1793–1812), London, and for H. Hughes, 59 Fenchurch Street, London (1859–75).
Provenance Adams collection; lent in 1936, bequeathed in 1974. The presence of the Hughes label may suggest that Adams purchased this instrument from them.
Notes The U-shaped bracket was recommended by Mendoza Rios, and according to him first applied by Troughton.
Literature Mendoza Rios, *On an Improved Reflecting Circle*, 8–9; Fisher, *The Makers of the Blueback Charts*, 74–83; Clifton, *Directory*, 131.

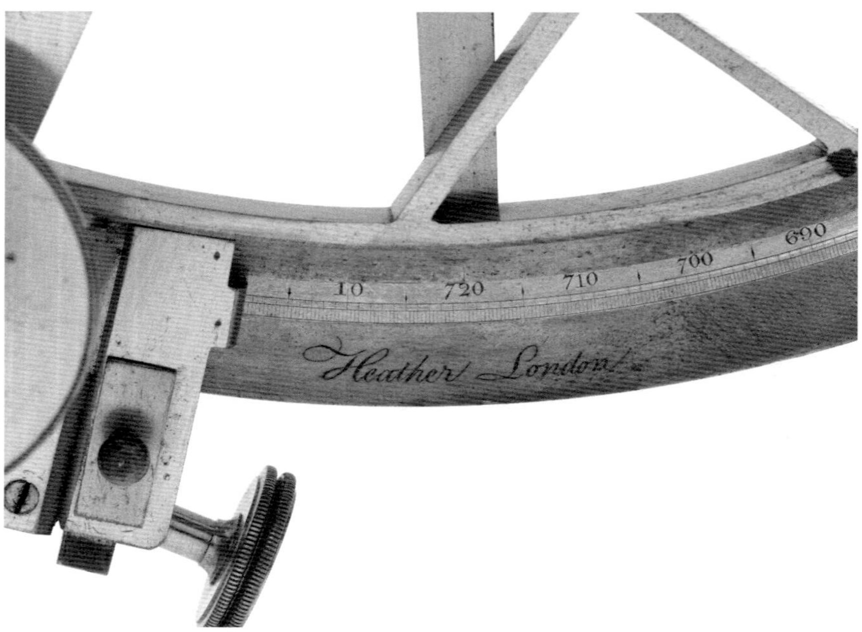

307 NAV0088

Reflecting circle

Origin London, England
Maker Matthew Berge
Date About 1810
Inscriptions Signed on the back of the plate: Berge London, late Ramsden | 1804
Marked on the front of the plate: I Brown
Graduation Polished brass limb with inlaid silver scale indicating from 140° to 0° to 140°, and from 240° to 360°, the entire scale is divided by 30′. Three silver verniers to 1″, zeros at the left.
Dimensions Diameter 2¾ inches (70 mm)
Description Circular polished brass frame (with traces of varnish), consisting of a plate and an open frame connected by steel pillars; the plate serves as a handle. Three rigidly connected index arms (engraved A, B, and C at the base of each arm) on the outside of the frame. There is no tangent screw or clamping screw. The index arm is moved by a square-headed screw and a separate key (missing) on the back of the frame. Two shades, probably for sliding on the dovetail rail inside the frame, are missing. Index-glass adjustment by a screw; adjustmnet of horizon glass by a screw and a capstan screw. Magnifier (single lens) on a 33 mm hinged swivelling arm. The detached threaded telescope bracket, without adjustment, is for mounting on the back of the frame. Telescope (82 mm), inverted image; extra drawtube (68 mm), inverted image; two oculars (one broken). Shaded eyepiece (orange), a shaded eyepiece (red) with an extra swivelling shade (red) to obtain a darker filter, and an adjusting pin. In a square fitted wooden box, of which the lid is missing.
Provenance Purchased from M. Peter Brophy, Paris, in 1971.
Notes This is a Troughton-type reflecting circle, in which the glasses are mounted within the protective frame.
Literature Stimson, 'The Influence of the Royal Observatory at Greenwich', 130; Clifton, *Directory*, 28.

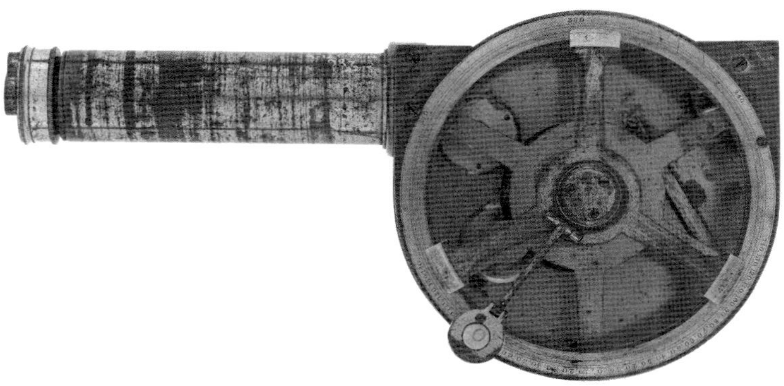

308 NAV0071
Reflecting circle

Origin London, England
Maker John and Edward Troughton
Date About 1820
Inscriptions Signed on the index arm: Troughton | 174 | London
Graduation Polished brass limb with inlaid silver scale indicating from 160° to 0° to 160°; the entire circle is divided every 20′. Three silver verniers reading to 20″, zeros at the left.
Dimensions Radius 5 inches (127 mm)
Description Circular polished brass frame, with a wooden handle. A detached U-shaped bracket holds a second handle with a brass-lined threaded hole for a third handle, which can also be screwed on to the front of the frame. Three rigidly connected index arms on the back of the frame, a double-ended tangent screw at the bottom and a clamping screw on the back of one of the arms. Three shades (two red and one green); three horizon shades (two red and one green). Index-glass adjustment by a screw; adjustment of horizon glass by a screw and a capstan screw. Two hinged magnifiers on 65 mm swivelling arms; the third magnifier (detached) is missing. Threaded telescope bracket with perpendicular adjustment by rising-piece and a milled collar. Telescope (68 mm), inverted image, four cross-wires; telescope (85 mm), erect image. Sight-tube (130 mm), two shaded eyepieces (both red), an adjusting key, and an adjusting pin. In a square fitted wooden box, containing in the lid trade labels for Troughton, 136 Fleet Street, London (1788–1804); for William Gerrard, 35 South Castle Street, Liverpool (1862–90); for Dawson and Melling, Liverpool (1837–9), and for W. Jewitt and Co. (successors to Melling and Co.), 39 Castle Street, Liverpool (1847 to after 1851). Also an oval label, in MS, 'A Bulmer Mill Turner 1907'.
Provenance Adams collection; lent in 1936, bequeathed in 1974. The Bulmer Mill Turner label presumably refers to the acquisition of the instrument by Adams in 1907.
Notes The U-shaped bracket was recommended by Mendoza Rios, and according to him first applied by Troughton.
Literature Stimson, 'The Influence of the Royal Observatory at Greenwich', 130; Mendoza Rios, *On an Improved Reflecting Circle*, 8–9; McConnell, *Instrument Makers to the World*; Clifton, *Directory*, 80, 282; Morrison-Low, *Making Scientific Instruments*, 327, 332.

309 NAV0072
Repeating circle

Borda circle
Origin Paris, France
Maker François Antoine Jecker
Date About 1820
Inscriptions Signed on a crossbar: No 499, Jecker, à Paris.
Graduation Varnished brass limb with inlaid silver scale from 0° to 720° every 20′. Brass finder arc from 155° to 0° to 145° every 1°, with two brass stops. Three silver verniers to 1′, zeros at the right.
Dimensions Diameter 11 inches (279 mm)
Description Circular varnished brass frame, with a detached wooden handle. The index arm carries the index glass; the telescope arm carries the horizon glass. A double-ended tangent screw and a clamping screw are on the back of the index arm; a second double-ended tangent screw and a clamping screw are at the ocular end of the telescope arm. Five detached shades (two red, two orange, and one green); four detached horizon shades (one green, two red, and one orange). Index- and horizon-glass adjustment by square-headed screws and a detached key. Magnifier on a 60 mm swivelling arm. Two threaded telescope brackets, with perpendicular adjustment by rising-pieces and two milled knobs on both. Telescope (142 mm), inverted image; extra drawtube (91 mm), inverted image, two parallel cross-wires; extra drawtube (84 mm), erect image. Sight-tube (82 mm), extra drawtube for the magnifier (21 mm), two shaded eye-pieces (both red), and three additional detached shades (all missing); two rectangular brackets and an adjusting key. In a square fitted mahogany box.
Provenance Adams collection; lent in 1936, bequeathed in 1974.
Literature Marcelin, *Dictionnaire des fabricants français*; Bennett, *Le Citoyen Lenoir*, no 98; Daumas, *Scientific Instruments*, 203.

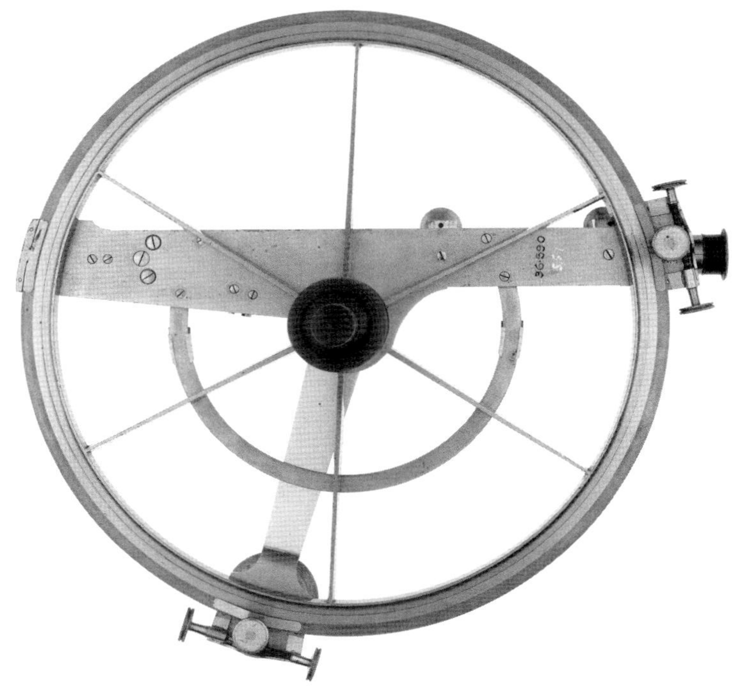

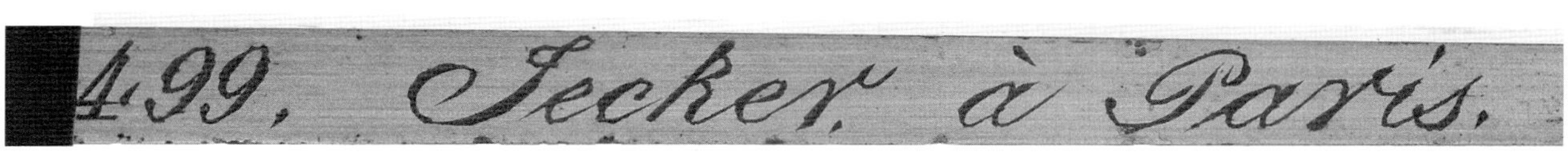

310 NAV0073
Repeating circle

Origin London, England
Maker (George) Dollond
Date About 1825
Inscriptions Signed on the index arm: Dollond, London.
Graduation Polished brass limb with inlaid silver scale from 0° to 720° every 20′. Two silver verniers to 20″, zeros at the centre. A polished brass scale on the inside of the limb to 180° every 5° on either side of zero.
Dimensions Diameter 8½ inches (216 mm)
Description Circular polished brass frame, with two wooden handles. A detached U-shaped bracket holds a third handle. The index arm carries the index glass; the tangent screw, working against a spring-loaded stop, and the clamping screw are on the back of the arm. Three index arms, moving with a brass scale and a vernier on the inside of the limb; they carry the telescope and the horizon glass. One arm has a tangent screw, working against a spring-loaded stop, and a clamping screw on the back of the arm. Index-glass adjustment by a screw; adjustment of horizon glass by capstan screws. Three hinged index shades (red, green, and orange); three horizon shades (two red and one green). Two magnifiers on 76 mm swivelling arms; two frosted glass shades. Threaded telescope bracket with perpendicular adjustment by rising-piece and a milled knob. Telescope (160 mm), inverted image, four cross-wires; telescope (91 mm), erect image; extra drawtube (46 mm), inverted image, four cross-wires. Sight-tube (84 mm), an adjusting pin, and a shaded eyepiece (green). In a square fitted wooden box, containing in the lid a trade label for Dollond and Co., 35 Ludgate Hill, 66 Old Broad Street and 5 Northumberland Avenue, London, and an oval label in MS '19B/b.07 Chapman Light Exchanged for Castolted^{m}'.
Provenance Adams collection; lent in 1936, bequeathed in 1974. The oval label presumably refers to the acquisition of the instrument by Adams, by exchange, in 1907.
Notes Similar to cat. no. 299. The U-shaped bracket was recommended by Mendoza Rios, and according to him first applied by Troughton.
Literature Mendoza Rios, *On an Improved Reflecting Circle*, 8–9; Hill and Paget-Tomlinson, *Instruments of Navigation*, 27; Barty-King, *Eyes Right*; Clifton, *Directory*, 87.

311 NAV0087
Reflecting circle

Origin London, England
Maker Edward Troughton and William Simms
Date About 1830
Inscriptions Signed on the index arm: Troughton & Simms, | 250 | London.
Graduation Polished brass limb with inlaid silver scale indicating from 140° to 0° to 140°, the entire scale is divided every 20′. Three silver verniers to 20″, zeros at the left.
Dimensions Diameter 10¾ inches (273 mm)
Description Circular polished brass frame, with a detached wooden handle. A detached U-shaped bracket holds a second handle with a brass-lined threaded hole for a third handle, which can also be screwed into the front of the frame. Three rigidly connected index arms (engraved A, B, and C at the base of each arm) on the back of the frame; a double-ended tangent screw and a clamping screw are on the back of arm A. Three index shades (two red and one green); three horizon shades (two red, one green). Index-glass adjustment by a screw; adjustment of horizon glass (glass missing) by a screw and a capstan screw. Magnifier on a 233 mm arm, swivelling on its centre at the pivot and serving all three verniers. Threaded telescope bracket with perpendicular adjustment by rising-piece and a milled collar. Sight-tube (133 mm). Telescopes and various unidentified parts are missing. In a square fitted wooden box.
Provenance Purchased from Arthur Davidson Ltd, London, in 1971.
Notes The U-shaped bracket was recommended by Mendoza Rios, and according to him first applied by Troughton.
Literature Stimson, 'The Influence of the Royal Observatory at Greenwich', 130; Mendoza Rios, *On an Improved Reflecting Circle*, 8–9; McConnell, *Instrument Makers to the World*; Clifton, *Directory*, 282.

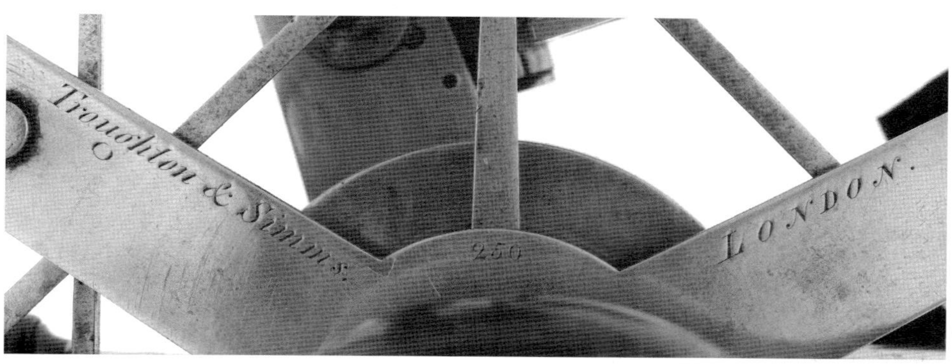

312 NAV0074
Reflecting circle

Origin Berlin, Germany
Maker Pistor and Martins
Date About 1863
Inscriptions Signed on the index arm: Pistor & Martins | Berlin | No 449
Marked on the index arm: Nederlandsche Marine
Graduation Polished brass limb with inlaid silver scale indicating from 350° to 0° to 130° and from 160° to 260°; the entire scale is divided every 10′. Two silver verniers to 10″, zeros at the centre.
Dimensions Diameter 10½ inches (267 mm)
Description Circular dark-brown-and-yellow-lacquered brass frame, with two wooden handles. A third handle can be screwed into the back of the frame. The tangent screw, working against a spring-loaded stop, and the clamping screw are at the bottom of the index arm. Four half-coloured shades (two red and one green). Index glass and horizon prism without adjustment. Magnifier on a 135 mm swivelling arm, serving both verniers; a frosted glass shade. Threaded telescope bracket in two parts, fitted for correcting collimation error; perpendicular adjustment by rising-piece and a milled knob. Telescope (157 mm), inverted image, two parallel cross-wires; extra drawtube (113 mm). Shaded eyepiece (green), a rotating eyepiece with a choice of three shades (all red), and an adjusting pin. Several unidentified parts are missing. In a square fitted mahogany box marked '449'.
Provenance Adams collection; lent in 1936, bequeathed in 1974.
Notes The inscription on the index arm indicates that the instrument originally belonged to the Royal Netherlands Navy.
Literature Brachner, 'German Nineteenth-Century Scientific Instrument Makers', 130–2, 136.

313 NAV0082
Reflecting circle

Origin Dresden, Germany
Maker Gustav Heyde
Date About 1875
Inscriptions Signed on the index arm: G. Heyde | Dresden.
Graduation Brass limb with inlaid silver scale from 0° to 180° every 10′, in four quadrants. Two silver verniers to 20″, zeros at the right.
Dimensions Diameter 7¾ inches (197 mm)
Description Circular brown-and-yellow-lacquered brass frame, with a wooden handle. The double-ended tangent screw, working against a spring-loaded stop (damaged and incomplete), and the clamping screw (missing) are on the back of the index arm. No shades. Index-glass adjustment by four screws; horizon prism without adjustment. Magnifier (lenses missing) on a 110 mm swivelling arm from the index glass mount, serving both verniers. Threaded telescope bracket in two parts, fitted for correcting collimation error by capstan screws; perpendicular adjustment by rising-piece and a milled knob. Telescopes and some unidentified parts are missing. Two shaded eyepieces (red and green). One of the feet of the frame is missing. In a square fitted wooden box partly lined with green textile, containing in the lid a piece of paper with an MS note 'Spiegelkreis No 53', followed by an instruction signed by the maker on how to store the instrument in the box. A plate is missing from the top of the lid.
Provenance Presented by Kelvin Hughes, Barkingside, Essex, in 1961.
Notes Similar in design to the reflecting circles by Pistor and Martins.
Literature Brachner, 'German Nineteenth-Century Scientific Instrument Makers', 142.

314 ACO1342
Reflecting circle

Origin Berlin, Germany
Maker Pistor and Martins
Date About 1865
Inscriptions Signed on the limb: Pistor & Martins Berlin | 824
Marked on a brass plate on the handle: [crown] | M
Graduation Brass limb with inlaid silver scale indicating from 350° to 0° to 130° and from 180° to 290°; almost the entire scale is divided every 10′. Two silver verniers to 10″, zeros at the right.
Dimensions Diameter 10.4 inches (264 mm)
Description Circular anodized brass frame and yellow-lacquered index arm, with two wooden handles. A third handle (missing) can be screwed into the back of the frame. The tangent screw (badly damaged, incomplete) is at the bottom of the index arm; the clamping screw (missing) is on the back. Six half-glass shades (two green and four red). Index glass and horizon prism without adjustment. Magnifier on an approximately 140 mm hinged swivelling arm (bent and damaged), serving both verniers. Threaded telescope bracket with perpendicular adjustment by rising-piece and a milled knob. Telescopes, unidentified parts, and the box are missing. In very poor condition.
Provenance Transferred from the Admiralty Compass Observatory in 1969.
Notes The crowned M (for Marine) indicates that the instrument originally belonged to the Imperial German Navy.
Literature Brachner, 'German Nineteenth-Century Scientific Instrument Makers', 130–2, 136.

315 NAV0083
Repeating circle

Borda circle
Origin Rostock, Germany
Maker A. Dolberg
Date About 1880
Inscriptions Signed on the index arms: A. Dolberg in Rostock No. 239
Graduation Polished brass limb with inlaid silver scale from 0° to 720° every 20′. Polished brass finder arc from 180° to 0° to 160° every 1°, with two brass stops. Two silver verniers to 20″, zeros at the right.
Dimensions Diameter 8.6 inches (218 mm)
Description Circular polished brass frame, with a wooden handle. The index arm carries the index glass; the telescope arm, with the finder arc, is semicircular and carries the horizon glass. The tangent screws and clamping screws are on the back of both index arms. Four index shades (three red and one green); three horizon shades (two red, one green). Index-glass adjustment by a screw; adjustment of horizon glasses by capstan screws. Magnifier on a hinged 115 mm swivelling arm serving both verniers; a frosted glass shade. Threaded telescope bracket with perpendicular adjustment by rising-piece and a milled knob. Telescopes, unidentified parts, and the box are missing.
Provenance Presented by Kelvin Hughes, Barkingside, in 1961.

316 NAV0070
Reflecting half-circle

Origin Berlin, Germany
Maker C. Lüttig
Date About 1898
Inscriptions Signed on the base of the index arm: No 592. | C. Lüttig in Berlin.
Graduation Polished brass limb with inlaid silver scale indicating from -10° to 140° and from 105° to 180°; the entire scale is divided every 10′, measuring to 180°. Silver vernier to 10″, zero at the right.
Dimensions Radius 6.2 inches (157 mm)
Description Brown-and-yellow-lacquered brass frame, with a wooden handle. The tangent screw, working against a spring, and the clamping screw are at the bottom of the index arm. Four half-coloured shades (two red and two green). Index glass and horizon prism without adjustment. Magnifier on a 135 mm swivelling arm; a frosted glass shade. Threaded telescope bracket in two parts, fitted for correcting collimation error; perpendicular adjustment by rising-piece and a milled knob. Telescope (162 mm), inverted image; extra drawtube (86 mm), erect image; extra drawtube (94 mm). Rotating shaded eyepiece with a choice of three shades (one green and two red), and a prismatic eyepiece for observing through the telescope at right angles to the frame. In a fitted rectangular polished mahogany box, containing in the lid a Kew Observatory certificate of examination, dated December 1898.
Provenance Adams collection; lent in 1936, bequeathed in 1974.
Notes Two prisms replace the horizon glass; one is for forward sights up to 120° and the other for back-observations, extending the instrument's range to 180°.
Literature Ifland, *Taking the Stars*, 29–32; Brachner, 'German Nineteenth-Century Scientific Instrument Makers', 131–2, 145; Hill and Paget-Tomlinson, *Instruments of Navigation*, 27–8.

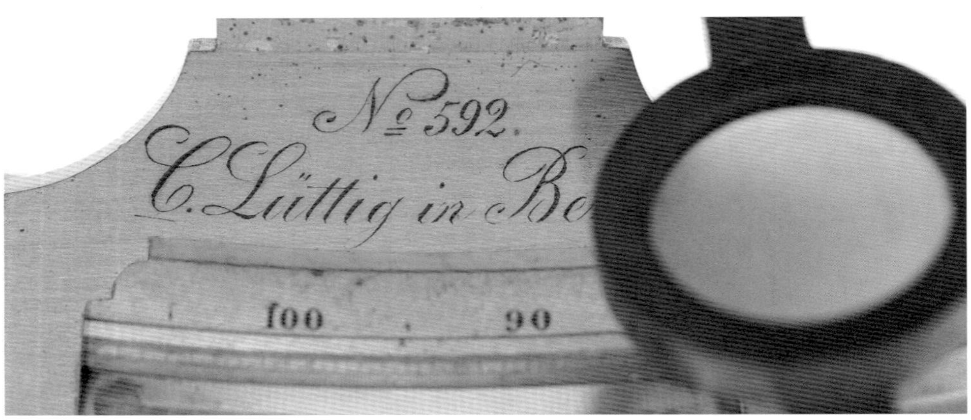

9. ARTIFICIAL HORIZONS

317 NAV0002
Artificial glass horizon

Origin England
Maker Unknown
Date About 1810
Inscriptions None
Dimensions 101 mm (diameter)
Description A circular brass case on four levelling feet covered with a reflecting but transparent glass plate, which covers spirit when the instrument is in use. The space below the glass plate is filled with spirit by use of a detached brass funnel that fits in a tube at the side of the case, covered by a threaded stop. The case should be nearly filled, leaving an air bubble that works as a level. By levelling the instrument with the feet, the user brings the air bubble under the centre of the glass plate, and the instrument is ready for use. The funnel is used for replacing the spirit and for releasing pressure when the liquid has expanded owing to warmth. In a fitted shaped mahogany box, containing in the lid printed instructions for using the instrument and, in MS, 'Presented by R. Lionel Foster, Esq.'
Provenance Presented by Mr R. Lionel Foster of Greenwich Hospital at an unknown date before December 1972.
For image see colour section plate 22.

318 NAV0019
Artificial glass horizon

Black glass artificial horizon
Origin London, England
Maker Probably Gilbert and Sons
Date About 1815
Inscriptions None
Dimensions 62 mm (diameter)
Description A circular brass case on three levelling feet covered with a blackened glass plate serving as a mirror. A spirit level is missing. In a fitted rectangular wooden box, which also contains a box sextant by Gilbert and Sons (cat. no. 164) and a pocket compass (NAV0466).
Provenance Belonged to Edward Newell Harrison, clerk of HMS *Assistance*. He went to the Arctic in 1850–51 in search of Sir John Franklin, and died in the West Indies before 1855. Presented by Miss E.V. Trenerry, Perivale, Middlesex, in 1975, a great-granddaughter of Harrison. She also presented cat. no. 164.
Literature Clifton, *Directory*, 113; Savours, *The Search for the North West Passage*, 193–4.

319 NAV0015
Artificial glass horizon

Black glass artificial horizon
Origin London, England
Maker Edward Troughton and William Simms
Date About 1830
Inscriptions Signed on the edge of the case: Troughton & Simms, London.
Dimensions 90 mm (diameter)
Description A circular brass case on three levelling feet, covered with a blackened glass plate serving as a mirror. In a fitted triangular sharkskin case containing a spirit level in a cylindrical wooden container; also a trade label for John Lilley and Son, 9 London Street and Fenchurch Street, London (1870–85), and a label in MS 'E. Comp. 5th Batt. Essex Regt. Hon. Members 2nd Prize won by E.T. Adams. 1909'.
Provenance Adams collection; lent in 1950, bequeathed in 1974.
Literature Clifton, *Directory*, 168, 282; McConnell, *Instrument Makers to the World*.

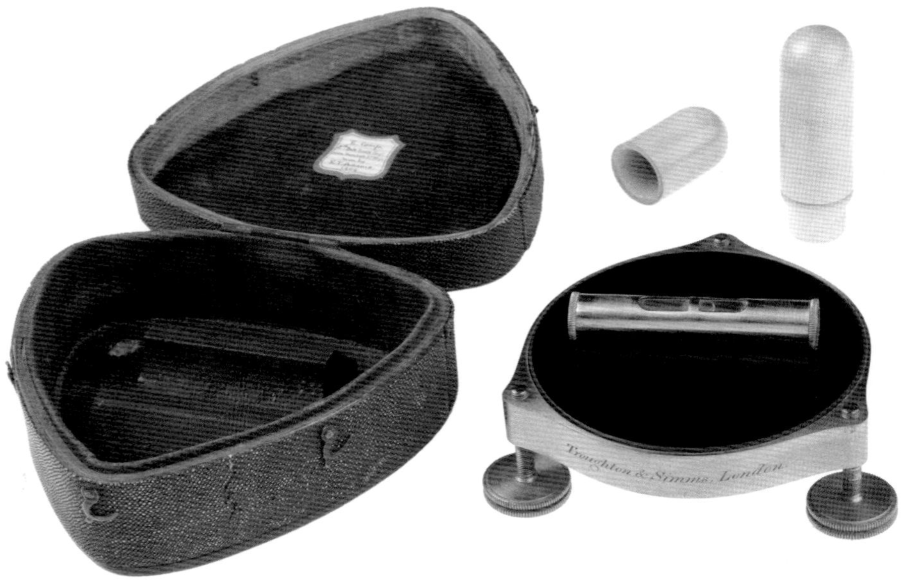

320 NAV0017
Artificial mercurial horizon

Roof artificial horizon
Origin England
Maker Unknown
Date About 1840
Inscriptions None
Dimensions 21 × 142 × 80 mm (height × length × width)
Description A rectangular mahogany trough (edge broken) for a mercurial roof artificial horizon, the rest of the instrument is missing.
Provenance John Lort Stokes (1812–85), the original owner, was an officer in the Royal Navy who spent most of his career as a hydrographic surveyor. The instrument remained in his family until purchased as part of the Lort Stokes collection through Evans, Roach and Co., Solicitors, Haverfordwest, in 1960.
Literature J.K. Laughton, 'Stokes, John Lort (1812–85) *naval officer*', in *Oxford Dictionary of National Biography*, Oxford, 2004, http://www.oxforddnb.com/.

321 NAV0018
Artificial mercurial horizon

Roof artificial horizon
Origin England, possibly London
Maker Possibly Charles Wilson
Date About 1845
Inscriptions None
Dimensions 140 × 185 × 127 mm (height × length × width)
Description A rectangular mahogany trough and a mahogany-framed, roof-shaped cover. Two sides of the cover have clear glass panels, at an angle of 90 degrees. A circular wooden jar (empty) with a threaded stopper and a bone or ivory spout, for storing the mercury when the instrument is not in use. In a fitted mahogany box, containing in the lid a trade label for Charles Wilson, late J.W. Norie and Wilson, 157 Leadenhall Street, London.
Provenance Presented by Mr J.B. Robertson, Comrie, Perthshire, in 1969.
Literature Clifton, *Directory*, 300; Fisher, *The Makers of the Blueback Charts*.

322 NAV0004
Artificial mercurial horizon

Roof artificial horizon
Origin England, possibly London,
Maker Possibly George Whitbread
Date About 1850
Inscriptions Marked (scratched) on the side of the cover: Thomas Crockford
Dimensions 115 × 158 × 86 mm (height × length × width)
Description A rectangular wooden trough and a brass-framed, roof-shaped cover. Two sides of the cover have clear glass panels, at an angle of 90 degrees. A circular wooden jar with a threaded stopper and an ivory spout, for storing the mercury when the instrument is not in use. In a fitted rectangular wooden box, with an inlaid brass plate on the lid inscribed 'Presented by Greenwich Hospital to Thomas Crockford'. The box contains a trade label for G. Whitbread, 2 Grenada Terrace, London.
Provenance Presented by Mr E.G. Dixon, London, in 1936.
Notes For the presentation of instruments to Greenwich Hospital boys, see the note with cat. no. 072.
Literature Clifton, *Directory*, 295.

323 NAV0005
Artificial mercurial horizon

Roof artificial horizon
Origin England, possibly London
Maker Possibly George Whitbread
Date About 1850
Inscriptions None
Dimensions 110 × 164 × 90 mm (height × length × width)
Description A rectangular wooden trough and a black-lacquered, brass-framed, roof-shaped cover. Two sides of the cover have clear glass panels, at an angle of 90 degrees. A circular wooden jar with a threaded stopper and an ivory spout, for storing the mercury when the instrument is not in use. In a fitted rectangular wooden box, with an inlaid brass plate on the lid inscribed 'Presented by Greenwich Hospital to Richard Godden'. Written in pencil inside the lid 'This formerly belonged to Admiral Douglas'. The lid contains a trade label for G. Whitbread, 2 Grenada Terrace, and Commercial Road, London.
Provenance Adams collection; lent in 1946, bequeathed in 1974.
Notes For the presentation of instruments to Greenwich Hospital boys, see the note with cat. no. 072. The text inside the lid possibly refers to Admiral Sir (Henry) Percy Douglas (1876–1939), a hydrographer who joined the Navy on the training vessel HMS *Britannia* in 1890.
Literature Clifton, *Directory*, 295; V. W. Baddeley, 'Douglas, Sir (Henry) Percy (1876–1939), *hydrographer*', in *Oxford Dictionary of National Biography*. Oxford, 2004, <http;//www.oxforddnb.com/>.

324 NAV0006
Artificial mercurial horizon

Roof artificial horizon
Origin London, England
Maker (George) Dollond
Date About 1850
Inscriptions Signed on one side of the cover: Dollond | London
Marked on the other side of the cover: Edmund Hope Verney, R.N.
Dimensions 101 × 127 × 43 mm (height × length × width)
Description A rectangular brass-framed, roof-shaped folding cover; the trough is missing. Two sides of the cover have clear glass panels, at an angle of 90 degrees. A cylindrical iron bottle with a threaded stopper and a pouring hole for storing the mercury when the instrument is not in use. In a fitted leather carrying case which can be attached to a leather belt.
Provenance Captain Edmund Hope Verney RN (1838–1910), the original owner, saw service in the Crimea and in the Indian Mutiny. Presented before 1946 by Admiral Sir Herbert William Richmond RN, KCB (1871–1946), a Trustee of the National Maritime Museum 1934–46.
Literature Clifton, *Directory*, 86; Barty-King, *Eyes Right*; H.E.D. Blakiston, 'Verney, Margaret Maria (1844–1930)', in: *Oxford Dictionary of National Biography*', Oxford, 2004 http://www.oxforddnb.com/; Littlewood and Butler, *Of Ships and Stars*, 70, Appendix 1.

325 NAV0007
Artificial mercurial horizon

Roof artificial horizon
Origin England
Maker Unknown
Date About 1850
Inscriptions Marked on the side of the cover:
4 E I C
Dimensions 140 × 190 × 120 mm (height × length × width)
Description A rectangular black-lacquered iron trough and an iron-framed roof-shaped cover. Two sides of the cover have clear glass panels, at an angle of 90 degrees. A circular metal jar with a threaded stopper for storing the mercury when the instrument is not in use (full). In a fitted wooden carrying box, stamped on the inside '6343'. Inside the lid in pencil, 'Ponsonby D. No 9 G. Murry E.T. Adams 1905'.
Provenance Adams collection; purchased in 1958. The pencil note presumably refers to the acquisition of the instrument by Adams in 1905.
Notes The logo 'E I C' stands for the East India Company.

326 ZBA0751
Artificial mercurial horizon

Roof artificial horizon
Origin Devonport and Plymouth, England
Maker William Charles Cox
Date About 1850
Inscriptions Signed on the side of the cover: Cox, | Devonport & Plymouth
Marked on the side of the cover: H O [broad arrow] 21
Dimensions 170 × 215 × 160 mm (height × length × width)
Description A rectangular black-lacquered steel trough with four levelling screws and a black-lacquered, brass-framed, roof-shaped cover. Two sides of the cover have clear glass panels, at an angle of 90 degrees. A circular wooden jar for storing the mercury (full) when the instrument is not in use, the threaded stopper is missing. A spirit level; a second level is missing. In a fitted carrying box, marked on the lid 'HO [broad arrow] 21' and containing a trade label for Cary and Co. (Henry Porter), 7 Pall Mall, London.
Provenance Acquired as part of the Royal Greenwich Observatory collection in 1998.
Notes The broad arrow and the initials 'HO' indicate that the instrument belonged to the Hydrographic Office of the Royal Navy.
Literature Clifton, *Directory*, 51, 69–70.

327 NAV0003
Artificial mercurial horizon

Roof artificial horizon
Origin England, possibly London
Maker possibly Elliott Brothers
Date 1861
Inscriptions None
Dimensions 148 × 188 × 131 mm (height × length × width)
Description A rectangular black-lacquered brass trough and a brass-framed, roof-shaped cover. Two sides of the cover have clear glass panels, at an angle of 90 degrees. A polished brass bottle with a threaded stopper for storing the mercury when the instrument is not in use. In a fitted mahogany box, marked on the lid '[crown] Alfred 1861', and containing a trade label for Elliott Brothers, 101 St Martin's Lane, London, successors to Watkins and Hill.
Provenance Belonged to Alfred, Duke of Edinburgh (1844–1900), a son of Queen Victoria, who had an active Naval career, became an Admiral of the Fleet, and in 1863 became Duke of Saxe-Coburg and Gotha. Elliott Brothers was at the address on the label after 1873. Presented by HM Queen Mary in 1936.
Literature Clifton, 'An Introduction to the History of Elliott Brothers'; John Van Der Kiste, 'Alfred, Prince, duke of Edinburgh (1844–1900) *duke of Saxe-Coburg and naval officer*', in *Oxford Dictionary of National Biography*, Oxford, 2004, <http://www.oxforddnb.com/>.

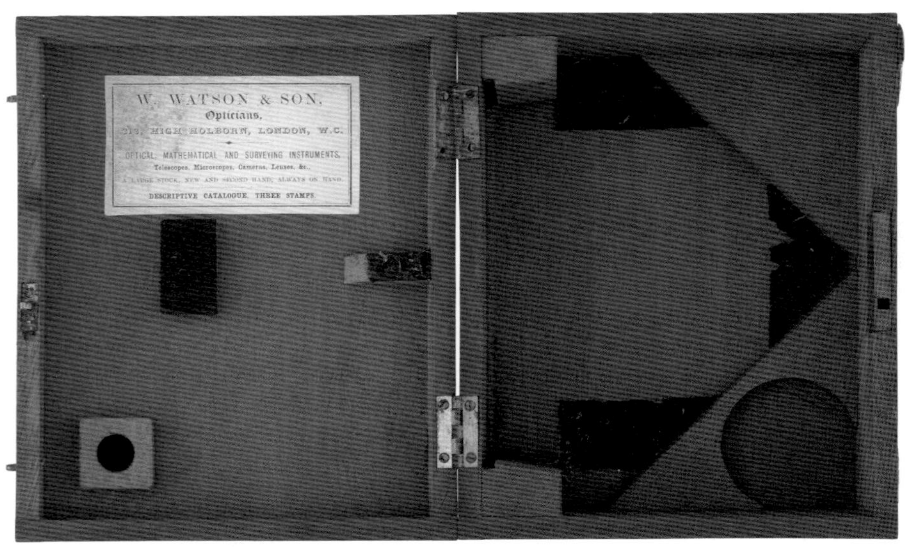

328 NAV1958
Artificial mercurial horizon

Roof artificial horizon
Origin England, possibly London
Maker Possibly William Watson and Son
Date About 1869–72
Inscriptions None
Dimensions 90 × 116 × 116 mm (height × length × width)
Description A rectangular wooden trough and a brass-framed, roof-shaped cover, both painted black. Two sides of the cover have clear glass panels, at an angle of 90 degrees. The jar for storing the mercury is missing. In a fitted wooden box, containing in the lid a trade label for W. Watson and Son, 313 High Holborn, London.
Provenance Unknown; found in the Museum's store-room during an inventory project in February 2003.
Literature Clifton, *Directory*, 292.

329 NAV1959
Artificial mercurial horizon

Roof artificial horizon
Origin England, possibly London
Maker Possibly William Watson and Son
Date About 1869–72
Inscriptions None
Dimensions 114 × 165 × 92 mm (height × length × width)
Description A rectangular wooden trough with a tin tray fitted and a brass-framed, roof-shaped cover. Two sides of the cover have clear glass panels, at an angle of 90 degrees. The jar for storing the mercury is missing. In a fitted wooden box, containing in the lid a trade label for W. Watson and Son, 313 High Holborn, London.
Provenance Unknown; found in the Museum's storeroom during an inventory project in February 2003.
Literature Clifton, *Directory*, 292.

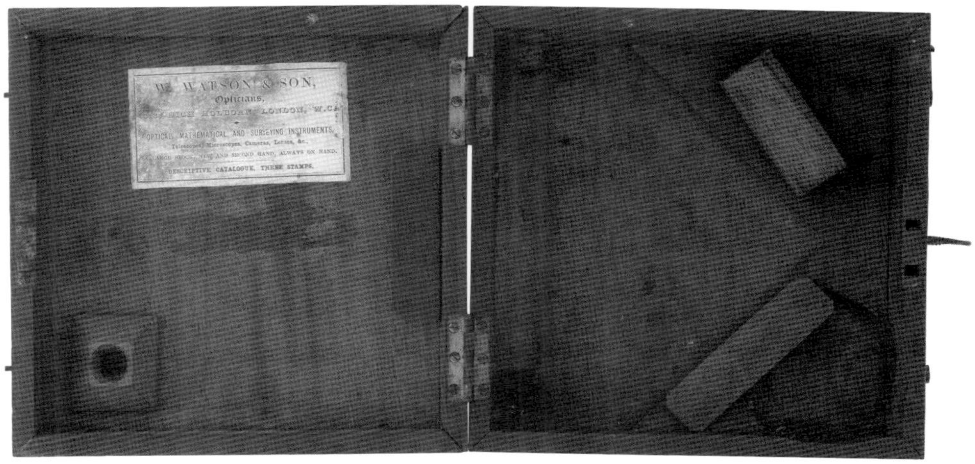

330 NAV0011
Artificial mercurial horizon

Captain George's artificial horizon
Origin London, England
Maker Cary and Porter
Date About 1870
Inscriptions Signed on the lid of the mercury reservoir: Cary London
Marked on the lid of the mercury reservoir: Capt George's Artl Horizon
Marked on the lid of the reservoir: C&P | Registered | Trade Mark | No 195
Dimensions 159 × 61 mm (length × width)
Description Two black-lacquered steel reservoirs on three levelling feet, interconnected by a tube fitted with a stopcock. One reservoir is for storing the mercury; the other has a removable clear glass cover. The mercury is let into the glass-covered reservoir for an observation. The storage reservoir is provided with a valve to admit air or allow it to escape. A spare glass cover in a chamois leather pouch, and a lever pin for operating the stopcock. In a fitted leather carrying case.
Provenance Adams collection; lent in 1950, bequeathed in 1974.
Notes Christopher George (1809–87) provisionally patented this horizon (no. 2624 of 1868). In his *Description*, he states that only H. Porter, Strand London, manufactured it. Henry Porter had been an apprentice to William Cary, 'C&P' is the trademark of Cary and Porter. George received a medal at the International Geographical Congress in Paris, 1875, for this horizon and for his double box sextant, cat. no. 182.
Literature George, *Description of a Self-Replenishing Artificial Horizon*; Ifland, *Taking the Stars*, 94–7, 194; NN, 'Captain C. George, R.N.'; Jamieson, 'Simplicity and Precision'; Clifton, *Directory*, 51.

331 ACO1514
Artificial gyroscopic horizon

Origin Paris, France
Maker A. Démichel
Date About 1890
Inscriptions Signed on top of the casing: A Demichel | Paris | No 9
Dimensions 120 × 70 mm (height × width)
Description A brass case containing a gyroscope. A piece of glass through the body of the gyroscope has on it a series of horizontal lines, which appear as a constant horizontal line of reference when the gyroscope is spun by an air pump. This artificial horizon is fitted to a sextant or quintant in the line of sight. In a fitted wooden carrying box with a brass plate on the front, inscribed 'Service Hydrographique Horizon Gyroscopique dans le vide de l'Amiral G. Fleuriais No 9, A. Demichel a Paris'.
Provenance Transferred from the Admiralty Compass Observatory in 1983.
Notes Designed by Admiral Georges Ernest Fleuriais (1840–95) in 1886 for use with a sextant or quintant (missing here). A later version of the complete instrument is in the collection, cat. no. 221.
Literature Ifland, *Taking the Stars*, 115–17; Marcelin, *Dictionnaire des fabricants français*; Mars, *Zeevaartkundige hoekmeetinstrumenten*, 12–20.

332 NAV0001
Artificial glass horizon

Black glass artificial horizon
Origin London, England
Maker Elliott Brothers
Date About 1893
Inscriptions Signed on the edge of the case: Elliott Bros London
Marked on the edge of the case: No 136
Dimensions 87 × 155 mm (length × width)
Description A rectangular black-lacquered brass case on three levelling feet, covered by a black glass mirror. In a rectangular fitted wooden box, containing a spirit level; a second spirit level, to be placed at right angles to the first, is missing. The lid of the box contains a Kew Observatory certificate of examination dated September 1893.
Provenance Adams collection; lent in 1936, bequeathed in 1974.
Literature Clifton, 'An Introduction to the History of Elliott Brothers'.

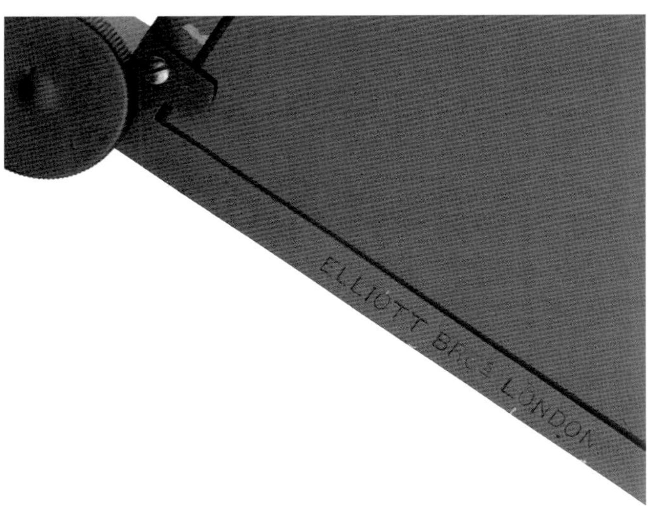

333 NAV0014
Artificial pendulum horizon

Origin Glasgow, Scotland
Maker Whyte Thomson and Co.
Date About 1893
Inscriptions Signed on the side of the weight: Whyte | Thomson & Co | Makers | Glasgow
Marked on the extension arm: Murray's Patent
Dimensions 163 mm (length of the telescope)
Description A threaded brass telescope, erect image, intended to be fixed in a sextant telescope bracket. An extension arm is attached to the telescope. The forward end of the extension arm is forked, and supports a balanced lever with a pendulum that operates in an oil bath, attached to the lower end of the extension arm. The fork at the forward end of the lever carries a cross-wire that ensures that the sextant is held horizontally, and replaces the horizon. There is a provision for clamping the pendulum to facilitate adjustment. In a rectangular fitted mahogany box with a sliding lid.
Provenance Adams collection; lent in 1950, bequeathed in 1974.
Notes James Baxter Murray patented this artificial horizon (no. 10 183 of 1893).
Literature Ifland, *Taking the Stars*, 107–9, 196; Clarke *et al.*, *Brass & Glass*, 281–7.

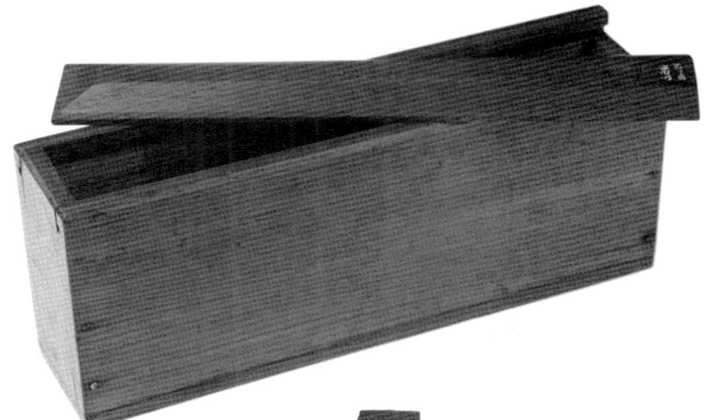

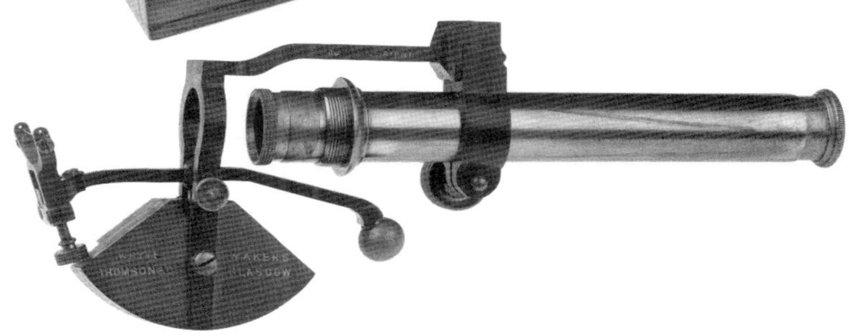

334 ACO1262
Artificial mercurial horizon

Roof artificial horizon
Origin London
Maker Dollond (Tyson Crawford)
Date About 1900
Inscriptions Signed on the side of the cover: Dollond | London Marked on the side of the cover: [broad arrow].
Dimensions 116 × 159 × 90 mm (height × length × width)
Description A rectangular cast-iron trough with a pouring hole at one corner, and a black-painted, brass-framed, roof-shaped cover. Two sides of the cover have clear glass panels at an angle of 90 degrees. A polished steel bottle with a threaded funnel-shaped cover and a threaded stopper, for storing the mercury when the instrument is not in use. In a fitted rectangular mahogany carrying box with a brass plate for securing the mercury bottle.
Provenance Transferred from the Admiralty Compass Observatory in 1969.
Literature Barty-King, *Eyes Right*.

335 NAV0010
Artificial pendulum horizon

Patent 'Hezzanith' artificial horizon
Origin London, England
Maker Heath and Co. Ltd
Date About 1900
Inscriptions Marked around the ocular: Patent "Hezzanith" Artificial Horizon
Dimensions 194 mm (length of the telescope)
Description A threaded brass telescope, erect image, with one cross-wire, intended to be fixed in a sextant telescope bracket. Halfway along the telescope tube is a balanced lever that indicates when the instrument is held in the horizontal plane. A separate oil lamp is provided for illumination at night. In a rectangular fitted wooden box with a sliding lid, containing in the lid a label for the Patent Hezzanith Artificial Horizon.
Provenance Unknown; entered in the departmental acquisition book in August 1965.
Notes George Wilson Heath patented this artificial horizon (no. 6475 of 1896).
Literature Ifland, *Taking the Stars*, 194; NN, *Heath & Co., Ltd. Patentees & Manufacturers of Marine Instruments*.

336 NAV0012
Artificial horizon

Newey's horizon finder
Origin England, probably London,
Maker Probably Henry Hughes and Son Ltd
Date About 1900
Inscriptions Marked on the chamber: Newey's Patent
Dimensions 133 mm (length of the telescope)
Description A threaded brass telescope, erect image, intended to be fixed in a sextant telescope bracket. The tube passes through a brass chamber containing two wires on gimbals, which when seen in line with a third wire, mounted in front of the telescope lens, indicates that the instrument is held horizontally. In a fitted wooden box with, on the lid, a trade label for H. Hughes and Son Ltd, 59 Fenchurch Street, London, and containing an instruction leaflet for the instrument.
Provenance Adams collection; lent in 1950, bequeathed in 1974. The presence of the Hughes label may suggest that Adams purchased this instrument from them.
Notes Master Mariner Alfred Newey of Seacombe patented this artificial horizon (no. 11 724 of 1895).
Literature Ifland, *Taking the Stars*, 109–10.

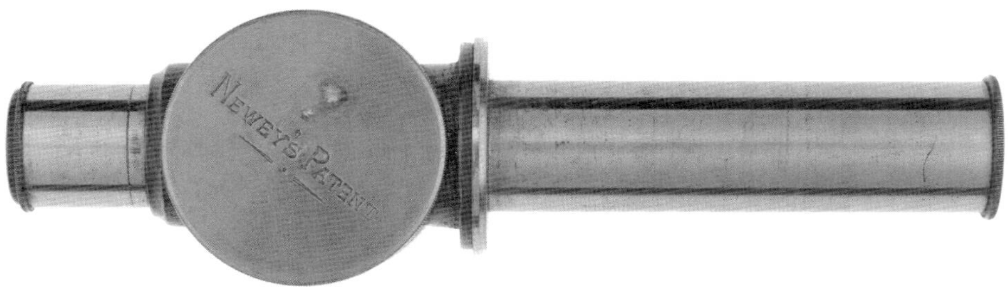

337 NAV0013
Artificial pendulum horizon

Origin England, possibly London
Maker Possibly John Lilley and Son
Date About 1900
Inscriptions Marked on the side of the arm: Beecher's | Patent
Dimensions 153 mm (length of the telescope)
Description A threaded brass sight-tube with a brass extension arm. The arm is fitted to the sight-tube, which is fixed in a sextant telescope bracket. A balance is partly submerged in a reservoir containing oil to damp its movement. The other end of the balance has a cross-wire, which ensures that the sextant is held horizontally, and replaces the horizon. In a fitted rectangular wooden box with a sliding lid, and containing a trade label for John Lilley and Son, 9 London Street and Fenchurch Street, London (1870–85).
Provenance Adams collection; lent in 1950, bequeathed in 1974.
Notes Edward Beecher patented this artificial horizon (no. 8328 of 1896).
Literature Ifland, *Taking the Stars*, 192; Clifton, *Directory*, 168.

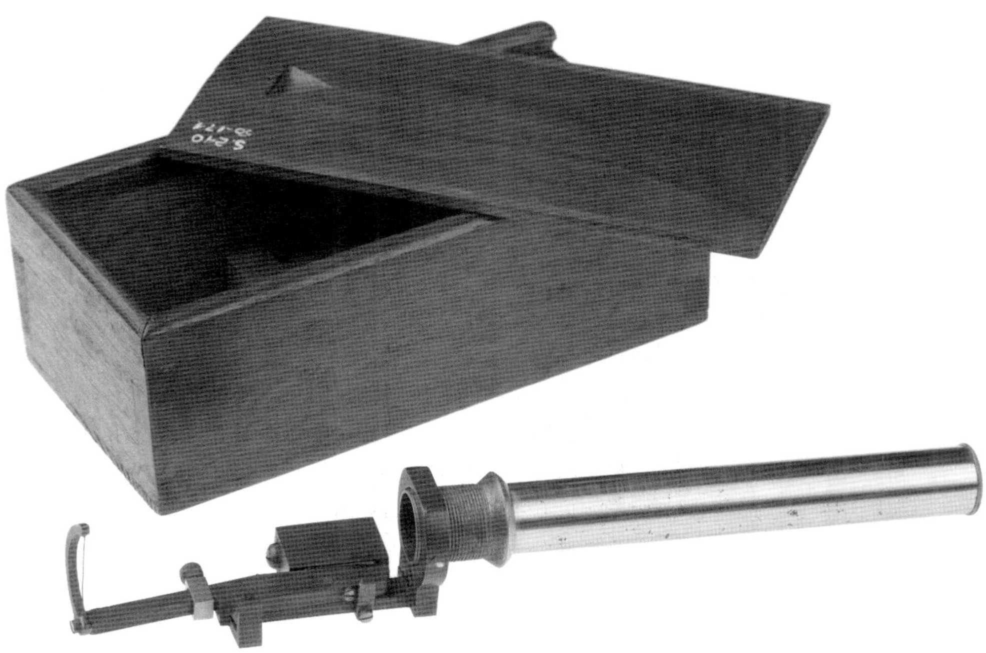

338 ACO1440
Artificial gyroscopic horizon

Origin London, England
Maker S.G. Brown Ltd
Date 1923
Inscriptions Signed on the top: No 3. | Artificial Horizon | S.G. Brown Ltd. | London W
Marked on the top: S.R.E. No CS.27.
Dimensions 355 × 315 mm (height × width)
Description Artificial horizon with a DC-driven gyroscope in a metal case. The mirror is gimballed and is contained in a gimballed cylinder within the case.
Provenance Transferred from the Admiralty Compass Observatory in 1983.

10. VARIOUS

339 NAV0008
Dip sector

Origin England, probably London
Maker Probably Henry Hughes and Son
Date Second half of the nineteenth century
Inscriptions None
Graduation Varnished brass scale every 0.5′, no digits.
Dimensions Radius 6 inches (152 mm)
Description Varnished brass frame and index arm. The index arm carries a glass (without adjustment) and is moved by a milled tangent screw; a second milled tangent screw moves a sight vane (vane missing). The screw on the back of the pivot to secure the mount of the glass is missing. The observer looks towards the horizon through the sight vane and at the same time sees the reflection of the horizon behind him, in the glass at an angle of 180 degrees. The two horizons are brought into line by moving the index arm with the tangent screw, and the dip or rise of the horizon is read from the scale by use of an indicating line.
Provenance Presented by Kelvin Hughes, Barkingside, Essex, in 1961 which suggests that Hughes manufactured this instrument.
Literature Raper, *The Practice of Navigation*, 166.

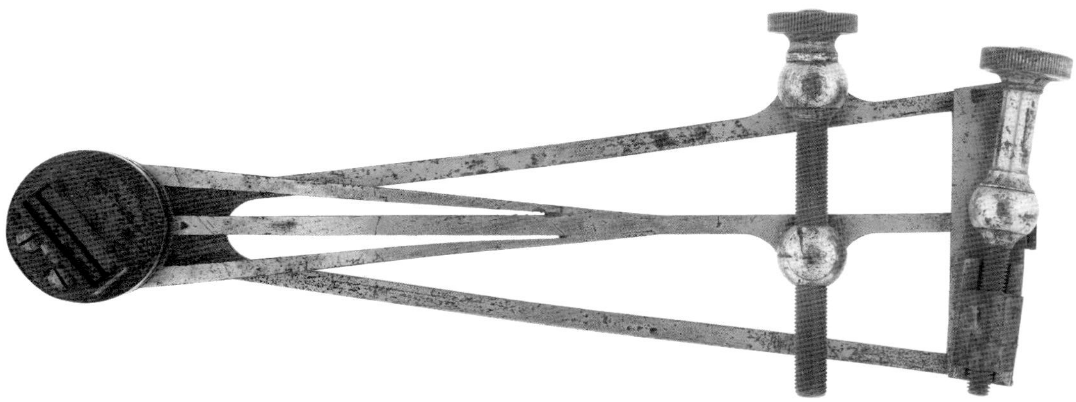

340 NAV0009
Dip sector

Origin England, probably London
Maker Probably Henry Hughes and Son
Date Second half of the nineteenth century
Inscriptions None
Graduation Varnished brass scale every 0.5′, no digits.
Dimensions Radius 6 inches (152 mm)
Description Varnished brass frame and index arm. The index arm carries a glass (without adjustment) and is moved by hand. The sight vane, with a narrow slit parallel to the horizon, is fixed to the frame, but the glass can be set by hand and secured with a milled clamping screw on the back of the frame. The observer looks towards the horizon through the sight vane and at the same time sees the reflection of the horizon behind him in the glass, at an angle of 180 degrees. The two horizons are brought into line by moving the index arm, and the dip or rise of the horizon is read from the scale by use of an indicating line.
Provenance Presented by Kelvin Hughes, Barkingside, Essex, in 1961, which suggests that Hughes manufactured this instrument.
Literature Raper, *The Practice of Navigation*, 166.

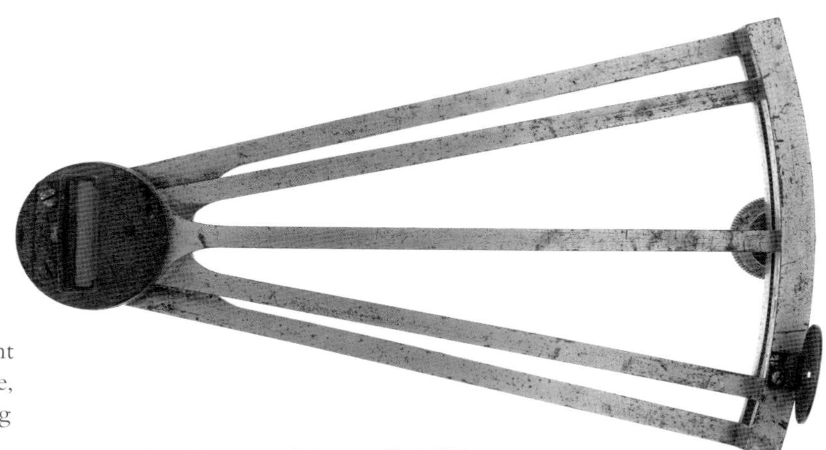

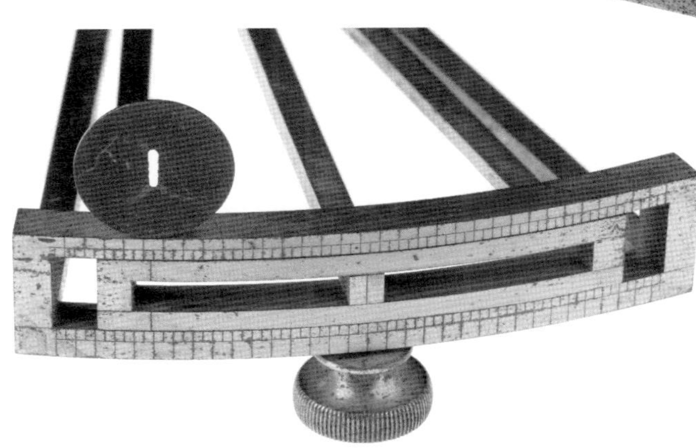

341 NAV1129
Sextant stand

Origin England
Maker Unknown
Date About 1860
Inscriptions None
Dimensions 360 mm (height)
Description Dark-brown-lacquered brass column on tripod stand, three adjustable feet for levelling by screws. Counterbalanced swinging mount for a sextant, which can be locked into any desired position. The adjusting pin is missing. In a polished wooden box. A letter in the box, written on stationery for 20, Eaton Place, SW1, reads 'I bought this sextant stand in 1894, second hand, in the city. It was used by me all the time I was carrying out navigating duties for shore sights for rating chronometers [signed] N I Oliver, 26 April 1938'. An 'O.H.M.S'. (On His Majesty's Service) Stationery Office label, dated 1924, is attached to the instrument.
Provenance Adams collection; lent in 1936, bequeathed in 1974.

342 NAV1244
Sextant tester

Origin London, England
Maker John Browning
Date About 1875
Inscriptions Signed at the centre of the vertical brass sheet: John Browning | London.
Dimensions 768 × 152 mm (height × width)
Description Black-lacquered brass and cast-iron frame, holding a vertical sheet of brass with a hole at each end. This provides an artificial planet and star, used for testing the measurement of position angles at a fixed distance. The smaller hole is for a star, and the larger for a planet. Two supports for candle lamps (both missing). The vertical sheet pivots on a vertical cast-iron column, which is mounted on a cast-iron base with three feet. In use, the image of the star is made to touch the image of the planet by adjusting the sextant from a fixed distance. The angular measurement should be the same on all sextants thus tested. A label stuck on the back of the vertical column is marked 'R.A.S., No.96'.
Provenance Purchased by the Royal Astronomical Society in 1878 and disposed of in 1984. Purchased by the Museum from Delehar Antiques, London, in 1985.
Notes John Browning worked at 111 Minories, London, from about 1864 to after 1900.
Literature Howse, 'The Royal Astronomical Society Instrument Collection'; Clifton, *Directory*, 41.

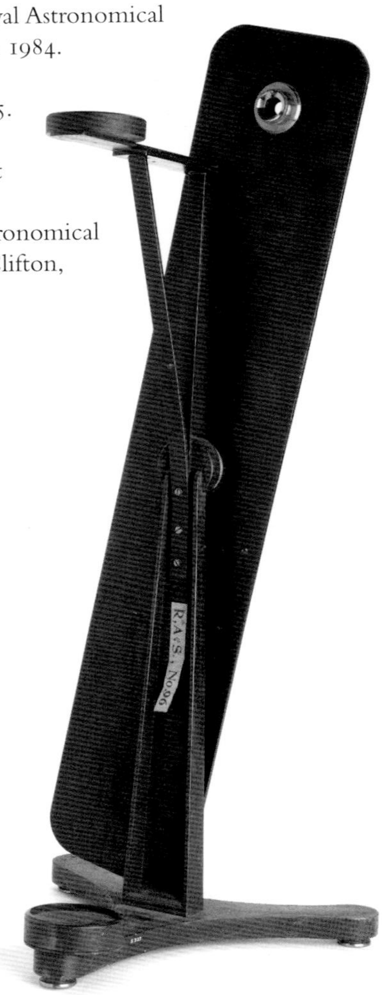

343 NAV1232
Sextant stand

Origin England, possibly London
Maker Possibly Henry Hughes and Son
Date About 1880
Inscriptions None
Dimensions 368 mm (height)
Description Varnished brass sextant stand with three folding feet and a rack and pinion adjustment for height. The upper, bowed section carries a universal counterbalanced mounting for a sextant, which can be set at any required angle. In a fitted rectangular wooden box, containing in the lid a trade label for Henry Hughes and Son, Marine Opticians, London.
Provenance Purchased from Relcy Antiques, Greenwich, London, in 1980.

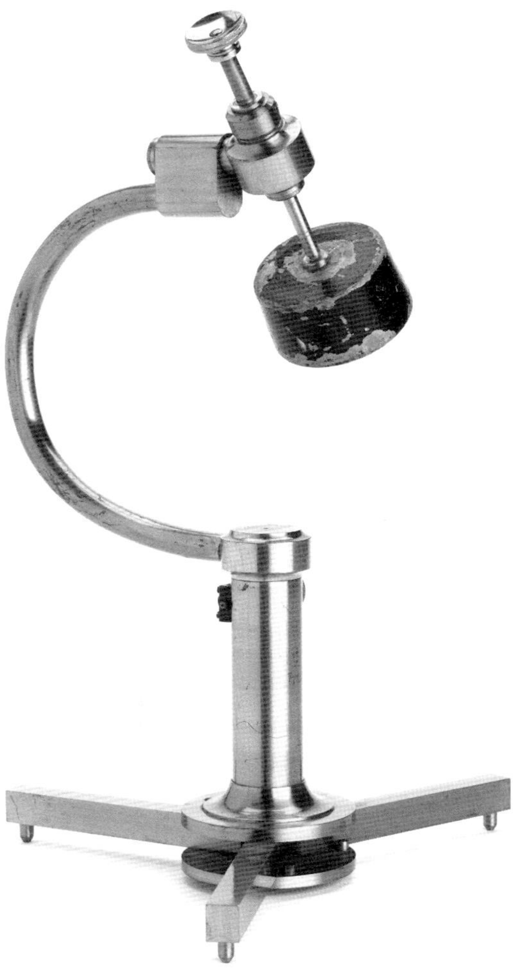

344 NAV1176
Sextant stand

Origin England, possibly London
Maker Possibly Cary Porter Ltd
Date About 1901
Inscriptions Marked on the tripod base: Ant. Ex 1901 | H.O. [broad arrow] 21
Dimensions 365 mm (height)
Description Anodized brass column on a tripod stand with three milled levelling screws, for mounting a sextant. The counterbalanced swinging mount can be locked into the desired position by milled screws. All screws are covered with leather for protection of the skin in polar regions. In a rectangular fitted wooden box with, in paint on the lid, 'Discovery 1925 Admiralty H.O. [broad arrow] 21'. The lid contains a trade label for Cary, Porter Ltd, 7 Pall Mall, London.
Provenance The instrument was used on Captain Robert Scott's National Antarctic Expedition of 1901–4 on the RSS *Discovery* and apparently remained on board when the ship was the Boy Scouts Association headquarters from 1936 to 1954. Presented by the Boy Scouts Association, ex RSS *Discovery*, London, in 1962.
Notes The broad arrow and the initials 'HO' indicate that the instrument belonged to the Hydrographic Office of the Royal Navy.
Literature Savours, *The Voyages of the Discovery*, 3–102, 300–5.

345 NAV 1094
Instrument for measuring the dip or rising of the horizon

Kimmtiefenmesser
Origin Jena, Germany
Maker Carl Zeiss
Date About 1919
Inscriptions Signed on the front of the casing: Carl Zeiss | Jena | Nr. 5421
Dimensions 140 mm (length of telescope)
Description Black-lacquered metal and chrome casing containing prisms and lenses. Through the telescope, the observer sees the reflected horizon to the left and to the right, at an angle of 180 degrees. The reflected horizons are aligned by turning a micrometer, with a white (positive) and a red (negative) scale. The dip or rise is then read from the micrometer; white indicates a rise and red indicates a dip. In a fitted leather carrying case.
Provenance This instrument came into the possession of Commodore R.P. Galer, CBE, RD, RNR while serving in HMS *Glory* in 1919, on the seizure of a Russian revolutionary battleship at Archangel. Presented by Mr J.H. Coste, Fareham, Hampshire, in 1970.
Notes Designed around 1906 by Dr Carl Pulfrich (1858–1927) who had worked for Zeiss since 1890. Although the *Kimmtiefenmesser* was made by Zeiss, it was sold by Carl Plath in 1906.
Literature Moll, *Kimmtiefenmessungen*, 13–18; NN, *Katalog über nautische Instrumente*, 27; Mars, *Zeevaartkundige hoekmeetinstrumenten*, 27–8.

346 NAV1211
Box for a bubble sextant

Origin Germany
Maker Unknown
Date About 1940
Inscriptions None
Dimensions 215 × 312 × 140 mm (length × height × width)
Description Wooden box for a German bubble sextant, containing two connections with plugs, a spring, two brushes, four light bulbs, oddments of chamois leather, and abbreviated instructions for use in German.
Provenance Presented by the Navigation and Direction Division of the Admiralty in 1961.

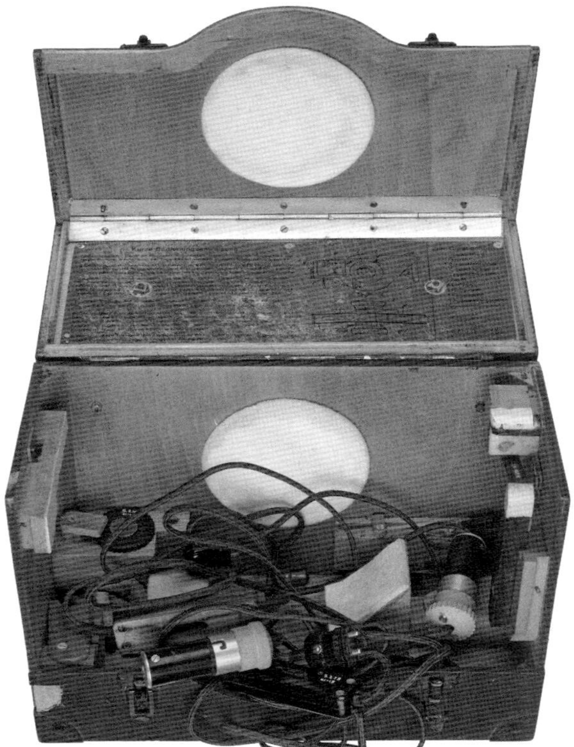

347 ACO1524
Periscope sextant stabilizing platform

Origin England
Maker Unknown
Date Possibly about 1960
Inscriptions Marked on the side of the foot: Platform Stabilising | Single Axis | 6605–99–522–9501 | Ser. N° 7049/c.o.s /3 | Weight 20 lbs.
Dimensions 190 × 260 × 210 mm (height × length × width)
Description Single-axis stabilizing platform for an artificial-horizon periscope sextant.
Provenance Transferred from the Admiralty Compass Observatory in 1983.

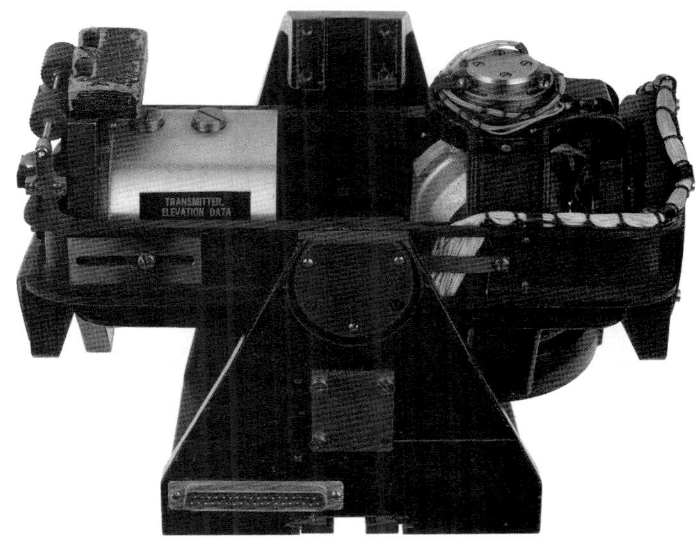

APPENDIX ONE

The Adams and other significant private collections: a checklist

Catalogue no.	object	date	maker
The Adams collection			
030	Octant	About 1750	Adams
032	Octant	About 1750	Urings
034	Octant	1753	Kley
036	Octant	About 1760	Martin
044	Octant	About 1770	Gardner
045	Octant	About 1770	Troughton
047	Octant	About 1770	Unknown
052	Octant	About 1774	Martin
058	Octant	About 1780	Chevalier
061	Octant	About 1780	Unknown
062	Octant	About 1780	Unknown
063	Octant	1783	Possibly Sandwell
064	Octant	About 1784	Unknown
067	Octant	About 1790	Spencer Browning and Rust
074	Octant	About 1800	Unknown
080	Octant	About 1810	Spencer Browning and Rust
081	Octant	About 1810	Spencer Browning and Rust
085	Octant	About 1820	Spencer Browning and Rust
089	Octant	About 1825	Cary
090	Octant	About 1825	Goy and Co (retailers)
094	Octant	About 1825	Unknown
095	Octant	About 1825	Unknown
098	Octant	About 1830	Bradford
102	Octant	About 1840	Cail
104	Octant	About 1842	Dixey
105	Octant	About 1845	Davis
109	Octant	About 1850	Unknown
118	Octant	About 1855	Cameron and Blakeney
120	Octant	About 1860	Duren
122	Octant	About 1860	Negretti and Zambra
123	Octant	About 1860	Potter

Catalogue no.	object	date	maker
124	Octant	About 1860	Unknown
125	Octant	About 1860	Norie and Wilson
140	Sextant	About 1783	Ramsden
141	Sextant	About 1785	Ramsden
142	Sextant	About 1785	Dollond
143	Sextant	About 1790	Ramsden
144	Sextant	About 1790	Ritchingham
147	Sextant	About 1791	Meredith
148	Sextant	1791	Gilbert and Wright
151	Sextant	About 1795	Ramsden
159	Sextant	About 1805	Ripley and Son
170	Sextant	About 1825	Jones
182	Double box sextant	About 1845	Cary
185	Double sounding sextant	About 1847	Worthington
186	Sextant	About 1850	Ertel and Sohn
190	Box sextant	1854–56	Elliott Brothers
193	Sextant	About 1855	Possibly Baharie
196	Sextant	About 1860	Owens
202	Sextant with artificial horizon	About 1870	Unknown
211	Sextant	About 1895	Dollond and Co
275	Quintant	About 1810	Berge
279	Double sounding quintant	About 1847	Cary
280	Quintant	About 1850	Jones
285	Quintant	About 1851	Potter
286	Quintant	About 1875	Unknown
298	Repeating circle	About 1787	Probably Ramsden
299	Repeating circle	About 1800	Dollond
301	Reflecting circle	About 1800	Troughton
305	Reflecting circle	About 1810	Troughton
306	Reflecting circle	About 1810	Heather
308	Reflecting circle	About 1820	Troughton
309	Repeating circle	About 1820	Jecker
310	Repeating circle	About 1825	Dollond
312	Reflecting circle	About 1863	Pistor and Martins
316	Reflecting half-circle	About 1898	Luttig
319	Artificial glass horizon	About 1830	Troughton and Simms
323	Artificial mercurial horizon about	1850	Possibly Whitbread
325	Artificial mercurial horizonabout	1850	Unknown
330	Artificial mercurial horizon about	1870	Cary and Potter
332	Artificial glass horizon	About 1893	Elliott Brothers
333	Artificial pendulum horizon about	1893	Whyte Thomson and Co
336	Artificial horizon	About 1900	Possibly Hughes and Son Ltd
337	Artificial pendulum horizon about	1900	Possibly Lilley and Son
341	Sextant stand	About 1860	Unknown

Adams scientific instruments not in this catalogue

NMM inventory no.	object	date	maker
NAV0020	Mariner's astrolabe	1910 (replica of cat. no. 005)	E. H. Prior and E. T. Adams
NAV0122	Gunter scale	About 1800	Unknown, English
AST0165	Equinoctial sundial	1800–50	Dring and Fage
AST0178	Universal equinoctial ring dial	1810–50	J. Newman
AST0942	Telescope	1740–60	James Short
AST0944	Telescope	About 1750	Jan van der Bildt
GLB0091	Terrestrial globe	About 1755	Nathaniel Hill
GLB0092	Celestial globe	About 1755	Nathaniel Hill
GLB0093	Terrestrial globe	About 1800	Bardin and Wright
GLB0094	Celestial globe	About 1800	Bardin and Wright
NAV0189	Slide rule	About 1825	Robert Bretsell Bate
NAV0190	Sliding Gunter rule	About 1800	W. and S. Jones
NAV0429	Azimuth compass	About 1880	Imray and Son
NAV0511	Beam compass	About 1830	Unknown
NAV0515	Camera lucida	About 1820	Unknown
NAV0559	Pantograph	About 1800	W. and S. Jones
NAV0576	Circular protractor	About 1825	William Cary
NAV0579	Circular protractor	About 1800	W. and S. Jones
NAV0602	Parallel rule	About 1854	Unknown
NAV0603	Parallel rule	About 1875	Unknown
NAV0627	Station pointer	About 1820	Troughton
NAV0856	Solar microscope	Late 18th C.	Unknown
NAV1040	Gunter quadrant	Late 17th/18th C.	Unknown
NAV1628	Pocket telescope	About 1780	Dollond
NAV1629	Portable telescope	About 1880	Edward Wood
NAV1662	Polygonal telescope	About 1770	Unknown

The Gabb collection

Catalogue no.	object	date	maker
018	Backstaff	1720	Macy
019	Backstaff	About 1720	Unknown
021	Backstaff	About 1725	Possibly Wright
022	Backstaff with artificial horizon	About 1732	Sisson
033	Octant	1753	Adams
038	Octant with artificial horizon	About 1760	Cole
042	Octant	About 1769	Goater
050	Octant	1770	Lake
055	Octant	1777	Gilbert
071	Octant	About 1800	Ripley and Son
138	Sextant with artificial horizon	About 1780	Unknown
139	Box sextant	About 1780–90	Fraser

The Landau collection

Catalogue no.	object	date	maker
056	Octant	1779	Edgworth
065	Octant	About 1785	Gregory and Wright
073	Octant	About 1800	Spencer Browning and Rust
078	Octant	About 1810	Cook
082	Octant	About 1810	Unknown
086	Octant	About 1820	Hughes
093	Octant	About 1825	Spencer and Co
106	Octant	About 1845	Whitbread
156	Surveyor's sextant	About 1800	Ekling
161	Sextant	About 1810	Unknown
195	Sextant	About 1860	Cail
285	Quintant	About 1860	Lorieux

APPENDIX TWO

Concordance between the NMM old inventory numbers, as found in *An Inventory of the Navigation and Astronomy Collections* of 1970–3, and the current numbers. The catalogue number is that which is used in this volume.

Old no.	new no.	catalogue no.
A.55	NAV0022	005
A.62 L	NAV0029	006
A.63	NAV0030	003
A.117	NAV1801	004
AS.4	NAV1929	249
AS.5	NAV1895	245
CI/N.17	NAV0140	163
Q.28	NAV1053	001
Q.37	NAV1062	002
S.1	NAV0031	020
S.2	NAV0032	025
S.3	NAV0033	027
S.4	NAV1251	039
S.5	NAV1252	030
S.6	NAV1253	035
S.7	NAV1254	036
S.8	NAV1255	044
S.9	NAV1256	066
S.10	NAV1257	047
S.11	NAV1258	052
S.12	NAV1259	104
S.13	NAV1260	080
S.14	NAV1261	067
S.15	NAV1262	081
S.16	NAV1263	074
S.17	NAV1264	090
S.18	NAV1265	107
S.19	NAV1266	125
S.20	NAV1267	089
S.21	NAV1268	105
S.22	NAV1269	120
S.23	NAV1270	123
S.24	NAV1271	114
S.25	NAV1272	076
S.26	NAV1273	113
S.27	NAV1103	140
S.28	NAV1104	141
S.29	NAV1105	143
S.30	NAV1106	151
S.31	NAV1107	155
S.32	NAV1108	275
S.33	NAV1109	147
S.34	NAV1110	136
S.35	NAV1111	144
S.36	NAV1112	148
S.37	NAV1113	142
S.38	NAV1114	193
S.39	NAV1115	211
S.40	NAV1116	170
S.41	NAV1274	118
S.42	NAV1117	196
S.43	NAV1118	182
S.44	NAV1119	280
S.45	NAV1120	185
S.46	NAV1121	186
S.47	NAV0070	316
S.48	NAV1122	279
S.49	NAV1123	286
S.51	NAV1124	284
S.52	NAV1125	206
S.53	NAV1126	199
S.54	NAV1127	209
S.56	NAV0071	308
S.57	NAV0072	309

Old no.	new no.	catalogue no.
S.58	NAV0073	310
S.59	NAV0074	312
S.60	NAV0075	306
S.61	NAV0076	298
S.62	NAV0077	299
S.63	NAV0078	301
S.64	NAV1128	202
S.65	NAV1129	341
S.66	NAV0001	332
S.67	NAV1130	292
S.68	NAV0079	300
S.69	NAV0080	303
S.70	NAV0002	317
S.73	NAV1275	053
S.74	NAV1276	068
S.77	NAV1131	287
S.78	NAV1132	203
S.79	NAV1133	289
S.80	NAV1134	197
S.81	NAV1135	281
S.82	NAV1136	213
S.83	NAV0003	327
S.84	NAV1277	198
S.91	NAV0034	014
S.92	NAV1281	056
S.93	NAV1137	195
S.94	NAV1138	285
S.95	NAV0035	028
S.96	NAV0004	322
S.97	NAV0036	018
S.98	NAV0037	019
S.99	NAV0038	021
S.100	NAV0039	022
S.101	NAV1282	033
S.102	NAV1283	055
S.103	NAV1284	050
S.104	NAV1139	153
S.105	NAV1140	149
S.106	NAV1285	071
S.107	NAV1286	082
S.108	NAV1287	073
S.109	NAV1288	078
S.110	NAV1289	086
S.111	NAV1290	093
S.112	NAV1291	038
S.113	NAV1292	106
S.114	NAV1141	161
S.115	NAV1142	156
S.116	NAV1293	116
S.117	NAV0505	007
S.118	NAV0040	011

Old no.	new no.	catalogue no.
S.119	NAV0041	023
S.120	NAV1294	043
S.121	NAV1295	054
S.124	NAV1296	138
S.125	NAV1297	098
S.126	NAV1143	159
S.127	NAV1298	064
S.128	NAV1299	102
S.129	NAV1300	061
S.130	NAV0005	323
S.131	NAV1144	162
S.132	NAV1145	189
S.133	NAV1301	079
S.134	NAV1146	171
S.135	NAV1147	283
S.136	NAV1148	139
S.137	NAV1149	288
S.138	NAV1150	187
S.139	NAV1302	099
S.140	NAV1303	029
S.141	NAV1304	097
S.142	NAV1151	175
S.143	NAV1305	096
S.144	NAV1306	042
S.145	NAV1307	121
S.147	NAV0006	324
S.148	NAV1152	208
S.149	NAV1153	192
S.150	NAV1154	176
S.151	NAV1155	165
S.152	NAV1156	166
S.153	NAV1308	072
S.154	NAV1309	059
S.156	NAV1310	087
S.157	NAV1311	091
S.158	NAV1312	100
S.159	NAV1157	177
S.161	NAV0042	024
S.162	NAV1158	200
S.163	NAV1313	122
S.164	NAV0081	305
S.165	NAV1159	207
S.166	NAV1314	085
S.168	NAV1160	188
S.169	NAV1315	128
S.170	NAV1316	062
S.171	NAV1317	063
S.172	NAV1318	045
S.173	NAV1319	034
S.174	NAV1320	058
S.175	NAV1321	095

Old no.	new no.	catalogue no.
S.176	NAV1322	094
S.177	NAV1323	124
S.178	NAV1324	032
S.179	NAV1325	065
S.181	NAV1161	134
S.182	NAV1326	112
S.183	NAV1162	201
S.184	NAV1163	191
S.186	NAV0007	325
S.187	NAV1164	190
S.188	NAV1165	282
S.189	NAV1327	109
S.190	NAV1328	110
S.191	NAV1166	222
S.192	NAV1329	092
S.193	NAV1330	111
S.194	NAV1167	254
S.195	NAV1168	212
S.196	NAV1331	069
S.197	NAV1169	226
S.198	NAV0044	026
S.198	NAV1332	077
S.200	NAV1333	083
S.201	NAV1334	119
S.202	NAV1335	108
S.203	NAV1336	127
S.204	NAV0082	313
S.205	NAV0083	315
S.206	NAV0084	117
S.207	NAV0085	302
S.208	NAV1170	276
S.209	NAV1171	261
S.211	NAV0008	339
S.212	NAV0009	340
S.213	NAV1172	262
S.216	NAV1337	088
S.217	NAV1173	214
S.218	NAV1338	084
S.219	NAV1174	174
S.220	NAV1339	101
S.221	NAV1340	103
S.222	AAA2230	178
S.223	NAV1341	046
S.224	NAV1176	344
S.225	NAV1177	133
S.226	NAV1342	041
S.227	NAV0045	012
S.229	NAV1179	137
S.230	NAV1343	070
S.231	NAV1180	295
S.231	NAV1344	048

Old no.	new no.	catalogue no.
S.232	NAV1182	294
S.233	NAV0010	335
S.234	NAV1183	256
S.235	NAV1181	255
S.236	NAV1184	257
S.237	NAV1185	258
S.238	NAV0011	330
S.239	NAV0012	336
S.240	NAV0013	337
S.241	NAV0014	333
S.242	NAV0015	319
S.243	NAV1186	131
S.244	NAV1187	132
S.245	NAV1188	205
S.254	NAV1195	181
S.256	NAV1196	218
S.257	NAV1197	293
S.259	NAV1346	040
S.260	NAV1347	037
S.261	NAV1198	167
S.262	NAV1199	157
S.263	NAV1200	204
S.264	NAV1201	172
S.265	NAV1202	241
S.267	NAV0017	320
S.268	NAV1203	168
S.269	NAV1204	269
S.270	NAV1205	215
S.271	NAV1206	296
S.272	NAV0018	321
S.274	NAV1209	272
S.277	NAV1210	242
S.278	NAV1211	346
S.279	NAV1212	243
S.280	NAV1213	220
S.284	NAV1094	345
S.285	NAV0087	311
S.286	NAV0088	307
S.287	NAV1214	277
S.288	NAV1215	274
S.290	NAV1216	154
S.292	NAV1349	115
S.293	NAV1217	268
S.294	NAV1218	146
S.295	NAV1350	031
S.296	NAV1219	231
S.297	NAV1220	164
S.298	NAV0019	318
S.301	NAV1351	060
S.302	NAV1221	223
S.303	NAV1222	216

Old no.	new no.	catalogue no.
S.304	NAV1223	184
S.305	NAV1224	210
S.307	NAV1225	267
S.308	NAV1352	075
S.309	NAV1353	057
S.310	NAV1226	145
S.311	NAV1227	278
S.312	NAV1228	227
S.313	NAV1229	235
S.314	NAV1230	129
S.316	NAV1354	126
S.317	NAV1232	343
S.319	NAV1233	173
S.321	NAV0089	304
S.323	NAV1234	158
S.325	NAV1235	160
S.326	NAV1236	135
S.327	NAV0510	008

Old no.	new no.	catalogue no.
S.328	NAV1355	169
S.329	NAV1237	291
S.330	NAV1238	248
S.331	NAV1239	221
S.333	NAV1241	180
S.334	NAV1242	150
S.335	NAV1243	290
S.336	NAV1356	051
S.337	NAV1244	342
S.338	NAV1245	194
S.339	NAV1246	179
S.340	NAV1247	263
S.341	NAV1357	049
S.342	NAV1248	228
S.343	NAV1249	152
S.344	NAV0048	013
S.345	NAV1250	229
S.346	ZBA0151	244

BIBLIOGRAPHY

ARCHIVAL MATERIAL

Cambridge University Library, Cambridge, 'Board of Longitude papers'.
Essex Record Office, 'Adams Family papers'.
National Maritime Museum, London, 'Museum's Archive'.
Royal Society, London, 'Royal Society Archive'.

PRINTED PRIMARY SOURCES

Après de Mannevillette, Jean-Baptiste-Nicolas d', *Le nouveau quartier Anglois, ou description et usage d'un nouvel instrument pour observer la latitude sur mer.* Paris, 1739.

Bird, John, *The Method of dividing Astronomical Instruments.* Published by order of the Commissioners of Longitude. London, 1767.

Blaeu, Willem Janszoon, *The Light of Navigation* [. . .]. Amsterdam, 1612.

Blaeu, Willem Janszoon, *The Sea-Mirror* [. . .]. Amsterdam, 1625.

Blundeville, Thomas, *His Exercises, containing six Treatises* [. . .]. London, 1594.

Bouguer, Jean, *Traité Complet de la Navigation* [. . .]. Paris, 1698.

Bourne, William, *A Regiment for the Sea* [. . .]. London, 1574. In: E.G.R. Taylor (ed.), *William Bourne, A Regiment for the Sea and of His Writings on Navigation.* Hakluyt Society, second series, no. CXXI. Cambridge, 1963.

Coignet, Michiel, *Nieuwe Onderwijsinghe op de Principaelste Puncten der Navigatien.* Antwerp, 1580.

Cortes, Martin, *Breve compendio de la sphera y de la arte de navegar* [. . .]. Seville, 1551.

Cuthbertson, Jonathan, *Verhandeling over het algemeen in gebruyk Octant van Hadley, met een aanhangsel, bevattende enig theoretische beschrijvingen.* Rotterdam, 1790.

Davis, John, *The Seamans Secrets.* London, 1607. In: Albert Hastings Markham (ed.), *The Voyages and Works of John Davis the Navigator.* Hakluyt Society, London, 1880.

[Dollond, John], *The Construction and Use of the Sea Quadrant, Commonly called Hadley's Quadrant.* London, 1760.

Dollond, P., and J. Dollond, *Directions for Using the New Improved Hadley's Quadrant; with a New Method of Adjusting the Glasses for the Back Observation.* London, n.d. [1772].

Gunter, Edmund, *The Description and use of The Sector. The Crosse-staffe and other Instruments* [. . .]. London, 1624.

[Hadley, George], *A Description of a New Instrument, invented by John Hadley, Esq; for taking the Latitude or other Altitudes at Sea. Printed by T. Wood, and sold by A. Nutt, at the Royal-Exchange-Gate.* N.p. [London] 1738.

Hadley, John, 'The description of a new Instrument for taking Angles'. In: *Philosophical Transactions* 37, no. 420 (1731–32) 147–57, 356.

Horsburgh, James, and Murdoch Mackenzie, *A Treatise on Marine Surveying, Corrected and Republished with a Supplement.* London, 1919.

Magellan, M.J.H. de, *Description des octants et sextants Anglois, ou quarts de cercle a reflection; avec La maniere de se servir de ces Instruments, pour prendre toutes fortes de Distances angulaires, tant fur Mer que sur Terre.* Paris/London, 1775.

Maskelyne, Nevil, 'Remarks on the Hadley's Quadrant'. In: *The Nautical Almanac and Astronomical Ephemeris, for the year 1774*. (London, 1772) 51–62.

Mayer, Tobias, (ed.), *Tabulae motuum solis et lunae novae et correctae . . . quibus accedit methodus longitudinum promota*. London, 1770.

Medina, Pedro de, *Arte de navegar* [. . .]. Valladolid, 1545.

Medina, Pedro de, *Regimiento De Navegacion* [. . .]. Seville, 1552.

Mendoza Rios, Joseph de, F.R.S., *On an Improved Reflecting Circle*. From the *Philosophical Transactions*. London, 1801.

Moore, John Hamilton, *The Seaman's Complete Daily Assistant, Being an Easy and Correct Method of Keeping a Journal at Sea*. Third edition, London, 1785.

NN, *An Explanation of E Hoppe's Improved Sextant, to which are added a few observations on the important advantages its accuracy affords to the nautical observer*. London, 1804.

Outhier, R., *Journal d'un voyage au nord en 1736 et 1737*. Amsterdam, 1747.

Pott, Constantin von, *Der doppelte Spiegel-Goniograph: Beschreibung, Gebrauch und Anwendung desselben*. Trieste, 1877.

Ramsden, Jesse, *Description of an engine for dividing strait lines on mathematical instruments*. London, 1777.

Rowe, Jacob, *Navigation Improved: In Two Books. Book I. Containing an exact Description of the Fluid Quadrant For the Latitude* [. . .]. London, 1725.

Seller, John, *Practical Navigation; or, An Introduction to the whole Art* [. . .]. London, 1680. Reproduced in Maritime History Series, published for the John Carter Brown Library, New York, 1993.

Strahan, G., W. Meadows, and J. Clarke, *The Description and Use of a New Astronomical Instrument, For taking Altitudes of the Sun and Stars at Sea, without an Horizon* [. . .]. London, 1735.

Waghenaer, Lucas Janszoon, *Spieghel der Zeevaerdt* [. . .]. Leiden, 1584.

Waghenaer, Lucas Janszoon (with additions by Anthony Ashley), *The Mariner's Mirror* [. . .]. London, 1588.

Wright, Gabriel, *The description and use of a new constructed quadrant, sextant and quintant. By his Majesty's patent, being an improvement on the Hadley's quadrant*. London, n.d. [1779].

SECONDARY SOURCES

Adams, Edgar Tarry, 'Navigating Instruments'. In: *Dictionary of British Scientific Instruments of the British Optical Manufacturers' Association*, London, 1921, 4–6.

Allen, Cecil J., *A Century of Scientific Instrument Making: 1853–1953*. London, 1953.

Andersen, Hemming, *Historic Scientific Instruments in Denmark*. Matematisk-fysiske Meddelelser 44:2. The Royal Danish Academy of Sciences and Letters. Copenhagen, 1995.

Andersen, Hemming, 'Jeppe Smith (1759–1821): A Danish Instrument-Maker'. In: R.G.W. Anderson, J.A. Bennett, and W.F. Ryan (eds.), *Making Instruments Count: Essays on Historical Scientific Instruments Presented to Gerard L'Estrange Turner*. (Aldershot, 1993) 403–17.

Banfield, Edwin, *Barometer Makers and Retailers 1660–1900*. Trowbridge, 1991.

Bartlett, Tim, 'Choosing & Using a Sextant'. In: *Sailing Today* 122 (June 2007) 68–79.

Barty-King, Hugh, *Eyes Right: The Story of Dollond & Aitchison Opticians 1750–1985*. London, 1986.

Bauer, Bruce, *The Sextant Handbook: Adjustment, Repair, Use and History*. Camden, ME, 1995.

Beaglehole, J.C., *The Life of Captain James Cook*. Stanford, CA, 1974.

Becher, A.B., 'The Pendulum Marine Artificial Horizon'. In: *The Nautical Magazine & Naval Chronicle* 5 (1844) 291–7.

Bedini, Silvio A., *Thinkers and Tinkers: Early American Men of Science*. New York, 1975.

Bedini, Silvio A., 'At the Sign of the Compass and Quadrant. The Life and Times of Anthony Lamb'. In: *Transactions of the American Philosophical Society* 74, 1 (1984) 1–84.

Bennett, J.A., *The Divided Circle: A History of Instruments for Astronomy, Navigation and Surveying*. Oxford, 1987.

Bennett, J.A., *Le Citoyen Lenoir: Scientific Instrument Making in Revolutionary France*. Cambridge, 1989.

Bennett, J.A., 'Obituary: Lt-Cdr Derek Howse'. In: *The Independent*, 4 August 1998.

Bennett, Jim, 'Catadioptrics and Commerce in Eighteenth-Century London'. In: *History of Science* XLIV (2006) 247–78.

Betts, Jonathan, 'Arnold and Earnshaw: The Practicable Solution'. In: William J.H. Andrewes (ed.), *The Quest for Longitude: The Proceedings of the Longitude Symposium, Harvard University, Cambridge, Massachusetts, November 4–6, 1993.* (Cambridge, MA, 1996) 311–28.

Betts, Jonathan, 'John Hyacinth de Magellan (1722–1790) Part 1: Horological and Scientific Agent'. In: *Antiquarian Horology* (September 2003) 509–17.

Betts, Jonathan, 'John Hyacinth de Magellan (1722–1790) Part 2: The Early Clocks'. In: *Antiquarian Horology* (June 2004) 173–83.

Blunt, E., and G.W. Blunt, *Description of E. & G. W. Blunt's Dividing Engine*. New York, 1858.

Blyzinsky, M., 'The History of the Collection'. In: E. Dekker, *Globes at Greenwich: A Catalogue of the Globes and Armillary Spheres in the National Maritime Museum, Greenwich*. (Oxford, 1999) 13–20.

Brachner, A., 'German Nineteenth-Century Scientific Instrument Makers'. In: Peter de Clercq (ed.), *Nineteenth-Century Scientific Instruments and Their Makers*. (Amsterdam/Leiden, 1985) 117–57.

Brewington, M.V., *The Peabody Museum Collection of Navigating Instruments with Notes on Their Makers*. Salem, MA, 1963.

Brooks, John, 'The Circular Dividing Engine: Development in England 1739–1843'. In: *Annals of Science* 49 (1992) 101–35.

Bruyns, see Mörzer Bruyns.

Bryant, Terence J., 'A Note on Henry Edgeworth, Mathematical, Philosophical and Optical Instrument Maker'. In: *Bulletin of the Scientific Instrument Society* 61 (1999) 18–19.

Bryden, D.J., 'Quality Control in the Making of Scientific Instruments: Kew Observatory and the Verification of Meteorological, Magnetic and Other Instruments, 1851–1899'. In: *Bulletin of the Scientific Instrument Society* 88 (2006) 48–59.

Caird, J., 'A Museum in the Making'. In: *The Syren and Shipping*, 2 January 1935.

Callender, Sir Geoffrey, *The Caird Collection of Maritime Antiquities*. 5 volumes. Greenwich, 1933–7.

Callender, Sir Geoffrey, *The Caird Collection of Maritime Antiquities: Supplement*. Greenwich, 1937.

Callender, G., 'The National Maritime Museum, Greenwich'. In: *Museums Journal* 37, 2 (May 1937) 45–57.

Cardinal, Catherine, 'Ferdinand Berthoud and Pierre Le Roy: Judgement in the Twentieth Century of a Quarrel Dating from the Eighteenth Century'. In: William J.H. Andrewes (ed.), *The Quest for Longitude: The Proceedings of the Longitude Symposium, Harvard University, Cambridge, Massachusetts, November 4–6, 1993.* (Cambridge, MA, 1996) 281–92.

Chaldecott, John A., 'Platinum and Palladium in Astronomy and Navigation: The Pioneer Work of Edward Troughton and William Hyde Wollaston'. In: *Platinum Metals Review* 31, 2 (1987) 91–100.

Chapman, Alan, *Dividing the Circle: The Development of Critical Angular Measurement in Astronomy 1500–1850*. Chichester, 1990.

Chapuis, Olivier, *A la mer comme au ciel: Beautemps-Beaupré & la Naissance de l'hydrographie moderne (1700–1850)*. Paris, 1999.

Charnock, John, *Biographica Navalis or Impartial Memoirs of the Lives and Characters of Officers of the Navy of Great Britain*. London, 1798.

Chauvenet, William, *New Method of Correcting Lunar Distances, and Improved Method of finding Error and Rate of a Chronometer by Equal Altitudes*. New York/London, 1868.

Cittert, P.H. van, *Astrolabes*. Leiden, 1954.

Clarke, T.N., A.D. Morrison-Low, and A.D.C. Simpson, *Brass & Glass: Scientific Instrument Making Workshops in Scotland as Illustrated by Instruments from the Arthur Frank Collection at the Royal Museum of Scotland*. Edinburgh, 1989.

Clercq, Peter de, *At the Sign of the Oriental Lamp: The Musschenbroek Workshop in Leiden, 1660–1750*. Rotterdam, 1997.

Clifton, Gloria, 'An Introduction to the History of Elliott Brothers up to 1900'. In: *Bulletin of the Scientific Instrument Society* 36 (1993) 2–7.

Clifton, Gloria, *Directory of British Scientific Instrument Makers 1550–1851*. London, 1995.

Clifton, Gloria, 'The London Mathematical Instrument Makers and the British Navy, 1700–1850'. In: Pieter van der Merwe (ed.), *Science and the French and British Navies, 1700–1850*. (London, 2003) 24–33.

Clifton, Gloria, 'The Adoption of the Octant in the British Isles'. In: Remmelt Daalder *et al.* (eds.), *Koersvast: Vijf eeuwen navigatie op zee*. (Zaltbommel, 2005) 85–94.

Cooper, Ernest R., FSA, 'The Davis Back-Staff or English Quadrant'. In: *The Mariner's Mirror* 30 (1944) 59–64.

Cotter, Charles H., *A History of Nautical Astronomy*. London, 1968.

Cotter, Charles H., *A History of the Navigator's Sextant*. Glasgow, 1983.

Crone, Ernst, *Cornelis Douwes 1712–1773: Zijn leven en zijn werk*. Haarlem, 1941.

Crone, Ernst, 'Pieter Holm en zijn octant'. In: *De Zee* 63 (1941) 1–7, 35–41, 76–82, 111–15, 129–36, 161–5.

Croucher, John S., 'An Exceptional Woman of Science'. In: *Bulletin of the Scientific Instrument Society* 84 (2005) 22–7.

Daumas, Maurice, *Scientific Instruments of the Seventeenth and Eighteenth Centuries and Their Makers*. London, 1972.

Davids, C.A., *Zeewezen en wetenschap: De wetenschap en de ontwikkeling van de navigatietechniek in Nederland tussen 1585 en 1815*. Amsterdam/Dieren, 1986.

Davids, Carel, 'Dutch Contributions to the Development of Navigation Technology in the 17th Century'. In: Charles Wilson and David Proctor (eds.), *The Seaborne Alliance and Diplomatic Revolution 1688*. (London, 1988) 59–74.

Dekker, E., *Globes at Greenwich: A Catalogue of the Globes and Armillary Spheres in the National Maritime Museum, Greenwich*. Oxford, 1999.

Dobson, G.M.B., 'Design of Instruments for Navigation of Aircraft'. In: *The Geographical Journal* 56, 5 (1920) 370–84.

Dunn, Richard, 'Sundials and the Arts'. In: H. Higton, *Sundials at Greenwich: A Catalogue of the Sundials, Horary Quadrants and Nocturnals in the National Maritime Museum, Greenwich*. (Oxford, 2002) 39–49.

Dunn, Richard, 'Touching and Cleaning: The Routine Work of An East London Instrument Supplier'. In: *Bulletin of the Scientific Instrument Society* 89 (2006) 21–6.

Durham, Dick, 'Starman'. In: *Yachting Monthly* (October 2006) 100–2.

Essex Record Office, *The Adams Family – Growing Up in Victorian Times: Notes*. Essex County Council, 1997.

Essex Record Office, *The Adams Family – Growing Up in Victorian Times: Further Sources*. Essex County Council, 1997.

Fanning, A.E., *Steady as She Goes: A History of the Compass Department of the Admiralty*. London, 1986.

Fauque, Danielle, 'The Introduction of The Octant in Eighteenth-Century France'. In: Remmelt Daalder *et al.* (eds.), *Koersvast: Vijf eeuwen navigatie op zee*. (Zaltbommel, 2005) 95–104.

Fisher, Susanna, *The Makers of the Blueback Charts: A History of Imray, Laurie, Norie, Wilson Ltd*. St. Ives, 2001.

FitzRoy, Robert (ed.), *Narrative of the Surveying Voyages of His Majesty's Ships Adventure and Beagle, between the Years 1862 and 1836, describing the Examination of the Southern Shores of South America, and the Beagle's Circumnavigation of the Globe*. London, 1839.

Forbes, Eric G., *The Birth of Navigational Science: The Solving in the 18th Century of the Problem of Finding Longitude at Sea*. National Maritime Museum Maritime Monographs and Reports No. 10 (1974).

George, C., Master RN, 'Description of the Improved Pocket Sextant'. In: *The Nautical Magazine & Naval Chronicle* (1844) 303–4. Also published as Paper no. 5 by The Institution of Hydronomical and Nautical Engineers, April 1867.

George, Christopher, *Description of a Self-Replenishing Artificial Horizon*. N.p., 1876.

Gould, Rupert T., *The Marine Chronometer: Its History and Development*. London, 1922.

Hewson, J.B., *A History of the Practice of Navigation*. Revised edition, Glasgow, 1963.

Hill, H.O., and E.W. Paget-Tomlinson, *Instruments of Navigation: A Catalogue of Instruments at the National Maritime Museum with Notes Upon Their Use*. London, 1958.

Hilster, N. de, 'The *Spiegelboog* (Mirror-Staff): A Reconstruction'. In: *Bulletin of the Scientific Instrument Society* 90 (2006) 6–16.

Hitchins, H.L., and W.E. May, *From Lodestone to Gyro-Compass*. Second edition, London, 1955.

Hoare, Michael Rand, *The Quest for the True Figure of the Earth: Ideas and Expeditions in Four Centuries of Geodesy*. Aldershot, 2005.

Howse, Derek, 'The Principal Scientific Instruments Taken on Captain Cook's Voyages of Exploration 1768–80'. In: *The Mariner's Mirror* 65 (1979) 119–35.

Howse, H.D., 'The Royal Astronomical Society Instrument Collection: 1827–1985'. In: *Quarterly Journal of the Royal Astronomical Society* 27 (1986) 212–36.

Howse, Derek, *Greenwich Time and the Longitude*. London, 1997.

Hughes, A.J., *Book of the Sextant with ancient and Modern Instruments of Navigation*. Glasgow, 1938.

Hughes, Arthur J., *History of Air Navigation*. London, 1946.

Hughes, E.L (ed.), *Official Catalogue of the Royal United Service Museum*. Eighth edition, London, 1932.

Hughes, H.A. '(1) The Paget Artificial Horizon for Attaching to the Sextant and Use in a Sea-way. (2) The Blish Prism'. In: *Transactions of the Optical Society* 14, 1 (1914) 59–66.

Hunt, Mary R., 'The German Gyro-Sextant'. In: *Navigation: Journal of the Institute of Navigation* 1, 6 (1947) 138–41.

Ifland, Peter, *Taking the Stars: Celestial Navigation from Argonauts to Astronauts*. Malabar, FL, 1998.

Jamieson, Annie, 'Simplicity and Precision: Distinguishing the Artificial Horizon', forthcoming.

Jardine, Lisa, 'Monuments and Microscopes: Scientific Thinking on a Grand Scale in the Early Royal Society'. In: *Notes and Records of the Royal Society of London* 55, 2 (2001) 289–308.

Jerchow, Friedrich, *From Sextant to Satellite Navigation 1837–1897: 150 Years C Plath*. Hamburg, 1987.

Kandler, Andreas, 'Nautical Instruments Made in Freiberg'. In: *JR Jena Review* 3 (1985) 141–3.

Knight, Roger, 'Afterword: Thirty Years Later'. In: Kevin Littlewood and Beverly Butler, *Of Ships and Stars: Maritime Heritage and the Founding of the National Maritime Museum Greenwich*. (London, 1998), 194–9.

Kochar, R.K., 'The Growth of Modern Astronomy in India, 1651–1960'. In: *Vistas in Astronomy* 34, 1 (1991) 69–105.

Lamb, Ursula, 'Nautical Scientists and Their Clients in Iberia (1508–1624): Science from Imperial Perspective'. In: *Centro de Estudos de História e Cartografia Antiga, Série Separatas 154*. Lisbon: Instituto de Investigação Cientifica Tropical. (Lisbon, 1985), 49–61. Reproduced in Ursula Lamb, *Cosmographers and Pilots of the Spanish Maritime Empire*. Aldershot, 1995.

Laurent, A., *Explanation, Adjustment and Use of Laurent's Patent Night Octant*. Paris, 1868.

Leback, Warren G., 'World War II: The Lifeboat Sextant'. In: *Sea Breezes* (2000) 265.

Lecky, S.T.S., *'Wrinkles' in Practical Navigation*. Revised and enlarged edition. London, 1891.

Leopold, J.H., 'Some Notes on Benjamin Ayres'. In: R.G.W. Anderson, J.A. Bennett, and W.F. Ryan (eds.), *Making Instruments Count: Essays on Historical Scientific Instruments Presented to Gerard L'Estrange Turner*. (Aldershot, 1993) 395–402.

Levathes, Louise, *When China Ruled the Seas: The Treasure Fleet of the Dragon Throne, 1405–1433*. Oxford/New York, 1994.

Lewis, David, *We, the Navigators*. Canberra, 1972.

Littlewood, Kevin, and Beverly Butler, *Of Ships and Stars: Maritime Heritage and the Founding of the National Maritime Museum Greenwich*. London, 1998.

Logsdon, Tom, *The Navstar Global Positioning System*. New York, 1992.

Losman, Arne, and Ingrid Sigurdsson, 'Äldre vetenskapliga instrument på Skokloster: Scientific Instruments of the 16th–18th Centuries at Skokloster Castle'. In: *Skokloster Studies* 10 (1975) 86–103.

Mac Lean, J., *Bijdrage tot de geschiedenis der Nederlandse instrumentmakerijen in de periode 1781–1881: De firma's Kleman en van Emden*. Leiden, 1976.

Marcelin, Franck, *Dictionnaire des fabricants français d'instruments de mesure du XVe au XIXe siècle*. Aix-en-Provence, 2004.

Markham, C.R., *On the Advantage of Forming Collections at Greenwich, with a View to Supplying the Means of Studying the Origin and Gradual Development of Various Branches of Naval Science*. N.p., 1891.

Mars, S., *Zeevaartkundige hoekmeetinstrumenten: van astrolabium en jacobsstaf tot octant met gyroscopischen horizon Fleuriais*. Batavia, 1915.

Mason, Ursula Stuart, 'Sir James Caird 1864–1954'. In: *The Mariner's Mirror* LIX (1973) 223–5.

May, W.E., 'Old Altitude Instruments'. In: *The Nautical Magazine* 151 (1944) 353–8.

May, W.E., 'Position Finding at Sea: 1, Latitude; 2, Longitude, 3, Position Lines'. In: *Sea Cadet* (March 1945) 202–4; (April 1945) 243–5; (May 1945) 278–9.

May, W.E., 'Early Reflecting Instruments'. In: *The Nautical Magazine* 153 (1945) 21–6.

May, W.E., 'The Double Altitude Problem: A Historical Note'. In: *Journal of the Institute of Navigation* 3 (1950) 416–21.

May, W.E., 'Hadley's Quadrant'. In: *The Mariner's Mirror* 41 (1955) 80–1.

May, W.E., 'Mariner's Astrolabes'. In: *Journal of the Institute of Navigation* X (1957) 95–6.

May, W.E., *A History of Marine Navigation*. Henley-on-Thames, 1973.

May, W.E., 'How the Chronometer Went to Sea'. In: *Antiquarian Horology* (1979) 638–64.

McConnell, Anita, 'Aluminium and Its Alloys for Scientific Instruments, 1855–1900'. In: *Annals of Science* 46 (1989) 611–20.

McConnell, Anita, *Instrument Makers to the World: A History of Cooke, Troughton & Simms*. York, 1992.

McConnell, Anita, *R B Bate of the Poultry 1782–1847: The Life and Times of a Scientific Instrument Maker*. London, 1993.

McConnell, Anita, 'Ramsden's Workshop: Recruiting Craftsmen from East and West'. In: Ewa Wyka, Maciej Kluza, and Anna Karolina Zawada (eds.), *East and West: The Common European Heritage: Proceedings of the XXVth Scientific Instrument Symposium*. (Krakow, 2006) 43–6.

McConnell, Anita, *Jesse Ramsden (1735–1800): London's Leading Scientific Instrument Maker*. Aldershot, 2007.

Messerschmitt, J.B., 'Ergebnisse von Sextantenprüfungen an der Deutsche Seewarte'. In: *Archive der Deutschen Seewarte* XXV (1902) 34–43.

Millburn, John R., *Retailer of the Sciences: Benjamin Martin's Scientific Instrument Catalogues, 1756–1782*. London, 1986.

Millburn, John R., *Adams of Fleet Street, Instrument Makers to George III*. Aldershot, 2000.

Miller, Robert C., 'Circular Dividing Engines in the United States before 1900'. In: *Rittenhouse: Journal of the American Scientific Instrument Enterprise* 12, 1 (1998) 12–21.

Miniati, Mara (ed.), *Museo di Storia della Scienza: Catalogo*. Florence, 1991.

Moll, E., *Kimmtiefenmessungen*. N.p. [Hamburg], 1906.

Mollan, Charles, *Irish National Inventory of Historic Scientific Instruments*. Blackrock, 1995.

Morrison-Low, Alison, *Making Scientific Instruments in the Industrial Revolution*. Aldershot, 2007.

Morton, Alan Q., and Jane A. Wess, *Public & Private Science: The King George III Collection*. Oxford, 1993.

[Mörzer Bruyns, W.F.J.], 'Instrumenten'. In: *Jaarverslag Vereeniging Nederlandsch Historisch Scheepvaart Museum 1984*. (Amsterdam, 1985) 18–19.

Mörzer Bruyns, W.F.J., 'De instrumenten van de firma Van Keulen 1680–1885'. In: E.O. van Keulen *et al.* (eds.), *'In de Gekroonde Lootsman': Het kaarten- boekuitgevers en instrumentenmakershuis Van Keulen te Amsterdam 1680–1885*. (Utrecht, 1989) 61–71.

Mörzer Bruyns, W.F.J., *The Cross-Staff: History and Development of a Navigational Instrument*. Zutphen, 1994.

Mörzer Bruyns, Willem F.J., 'The Introduction of the Elongating Glass on Sextants'. In: *Rittenhouse: Journal of the American Scientific Instrument Enterprise* 10, 3 (1996) 71–80.

Mörzer Bruyns, Willem F.J., 'The Cross-Staff Ten Years Later: An Update with Recently Found Examples'. In: *Bulletin of the Scientific Instrument Society* 80, 1 (2000) 18–23.

Mörzer Bruyns, W.F.J., *Schip Recht door Zee: De octant in de Republiek in de achttiende eeuw*. Amsterdam, 2003.

Mörzer Bruyns, Willem, 'The Importance of the Van Keulens to Navigation'. In: Dirk de Vries *et al.*, *The Van Keulen Cartography Amsterdam 1680–1885*. (Alphen aan den Rijn, 2005) 33–69.

Mörzer Bruyns, Willem F.J., 'Octants and Sextants in the Musée national de la Marine in Paris: A Report'. In: *Bulletin of the Scientific Instrument Society* 88 (2006) 41–5.

Mörzer Bruyns, W.F.J., 'Octants from the Friesian Island of Föhr for Dutch and German Whale Men, 1760–90'. In: *The Mariner's Mirror* 92, 4 (2006) 412–26.

Mörzer Bruyns, Willem F.J., 'The Navigating Instruments of the National Maritime Museum Greenwich: Research in Progress'. In: *Bulletin of the Scientific Instrument Society* 92 (2007) 23–8.

Moskowitz, Saul, 'The World's First Sextants'. In: *Navigation: Journal of The Institute of Navigation* 34, 1 (1987) 22–42.

Munday, John, 'Commander W. E. May, R.N.' In: *The Mariner's Mirror* 75 (1989) 209.

Nichols, R.L., 'Hadley's Quadrant: The Back Observation'. In: *Bulletin of the Scientific Instrument Society* 2 (1984) 8.

NN, 'Annual General Meeting of the Society for Nautical Research'. In: *The Mariner's Mirror* XIII (1927) 200–12.

NN, 'Application of a Micrometer-head to the Tangent-screw of a Sextant'. In: *The Nautical Magazine & Naval Chronicle* 16, 11 (1833) 323–4.

NN, *Beschreibung und Bedienungsvorschrift für den Sold-Libellensextanten KM* 2. Berlin, 1944.

NN, 'Captain C. George, R.N.' In: *Proceedings of the Royal Geographical Society* 9, 2 (1887) 126.

NN, *Catalogue of an Interesting Collection of Scientific Instruments: The Property of W. St. J. Pym., Esq., C.B.E.* Sotheby & Co., London, 1960.

NN, 'Commander W. E. May. Scholar of the History of Navigation'. In: *The Times*, 25 May 1989.

NN, 'The Double Sextant'. In: *The Nautical Magazine & Naval Chronicle* 7, 16 (1847) 385–6.

NN, *Exposition des Instruments et Outils d'autrefois*. Musée des Arts Décoratifs, Paris, 1936.

NN, *Heath & Co., Ltd. Patentees & Manufacturers of Marine Instruments: Sextant 20th Century Price List*. Crawford, London, *c.*1900.

NN, *An Inventory of the Navigation and Astronomy Collections in the National Maritime Museum Greenwich*. 3 volumes. London, 1970–3. Updated in 1977 and 1982.

NN, *Joseph Levi & Co Optical & Photographic Warehouse: Price List Nautical and Surveying Instruments, Barometers, Thermometers, Marine and Field Glasses, Etc.* London, 1897.

NN, *Katalog über nautische Instrumente von C. Plath*. Hamburg, 1907.

NN, 'Marine Artificial Horizon'. In: *The Nautical Magazine & Naval Chronicle* 34, III (1834) 754.

NN, 'The Micrometer Sextant'. In: *The Nautical Magazine* 121 (1929) 89–90.

NN, 'The National Maritime Museum'. In: *The Mariner's Mirror* XXIII (1937) 255–76.

NN, *National Maritime Museum Catalogue*. Greenwich, 1937.

NN, 'A New Angle Sextant'. In: *The Nautical Magazine and Journal of the Royal Navy Reserve* LXI (1892) 444, 448.

NN, 'A New Artificial Horizon for Seamen'. In: *The Nautical Magazine & Naval Chronicle* 8 (1845) 162.

NN, 'The Paget Compass'. In: *The Nautical Magazine and Journal of the Royal Navy Reserve* LXI (1892) 446–8.

NN, *Patents for Inventions. Abridgements of Specifications relating to Optical, Mathematical, and other Philosophical Instruments; including Nautical, Astronomical and Meteorological Instruments. A.D. 1636–1866*. London, 1875.

NN, *Patents for Inventions. Abridgements of Specifications. Class 97, Philosophical Instruments, [including Optical, Nautical, Surveying, Mathematical, and Meteorological Instruments]. Period 1855–1900*. London, 1905.

NN, *Pocket Bubble Sextant: Instructions for Use in Finding Latitude and Longitude*. Washington, DC, 1945.

NN, 'Rowland's Double Sextant'. In: *The Nautical Magazine & Naval Chronicle* 11, 13 (1833) 161.

NN, *Small Craft Precision Sextant*. Edenbridge, n.d. [1971].

NN, 'The Society for Nautical Research: A Brief Chronicle of the First 19 Years'. In: *The Mariner's Mirror* XVI (1930) 95–106.

NN, *Stanley: Great Turnstile, Holborn London, W.C.: J Edition* (trade catalogue). London, 1918.

NN, 'The Surveys of Graeme Spence'. In: *The Nautical Magazine & Naval Chronicle* 5 (1842) 313–19.

Parry, J.H., *The Discovery of the Sea*. London, 1975.

Phillimore, R.H., *Historical Records of the Survey of India*. London, 1945–58, volume 4.

Pipping, Gunnar, *The Chamber of Physics: Instruments in the History of Science Collections of the Royal Swedish Academy of Sciences, Stockholm*. Stockholm, 1977.

Preuss, Anne M., and Donald L. Treworgy, 'Taking a Departure. The Navigational Instruments and Artistry of Captain Charles Hervey Townshend'. In: *The Log of Mystic Seaport* 39 (1987) 21–34.

Randall, Anthony G., 'The Timekeeper that Won the Longitude Prize'. In: William J.H. Andrewes (ed.), *The Quest for Longitude: The Proceedings of the Longitude Symposium, Harvard University, Cambridge, Massachusetts, November 4–6, 1993*. (Cambridge, MA, 1996) 235–54.

Randles, W.G.L., 'The Emergence of Nautical Astronomy in Portugal in the XVth Century'. In: *Journal of Navigation* 51

(1998) 46–57. Reproduced in W.G.L. Randles, *Geography, Cartography and Nautical Science in the Renaissance: The Impact of the Great Discoveries.* Aldershot, 2000.

Randles, W.G.L., 'Portuguese and Spanish Attempts to Measure Longitude in the Sixteenth Century'. In: *The Mariner's Mirror* 81, 4 (1995) 402–8. Reproduced in W.G.L. Randles, *Geography, Cartography and Nautical Science in the Renaissance: The Impact of the Great Discoveries.* Aldershot, 2000.

Raper, Henry, *The Practice of Navigation and Nautical Astronomy*. Third edition. London, 1849.

Ritchie, G.S., *The Admiralty Chart: British Naval Hydrography in the Nineteenth Century*. London, 1967.

Savours, Ann, *The Search for the North West Passage*. London, 1999.

Savours, Ann, *The Voyages of the Discovery: The Illustrated History of Scott's Ship*. London, 1992.

Schaafhausen, F.W., Heinrich Hoffmann, and Peter Kaltenbach, *C Plath 1862–1962*. Hamburg, 1962.

Schilder, G., and W.F.J. Mörzer Bruyns, 'Navigatie'. In: *Maritieme Geschiedenis der Nederlanden*. (Bussum, 1977), volume 2, 159–99.

Schnall, Uwe, *Navigation der Wikinger: Nautische Probleme der Wikingerzeit im Spiegel der schriftlichen Quellen*. Oldenburg/Hamburg, 1975.

Sinnerstad, Ulf, 'Astronomy and the First Observatory'. In: Tore Frängsmyr (ed.), *Science in Sweden: The Royal Swedish Academy of Sciences 1739–1989*. Canton, MA, 45–71.

Stimson, Alan, *John Hadley and His Reflecting Quadrant*. National Maritime Museum Occasional Technical Paper No. 2. Greenwich, London, 1975.

Stimson, Alan, *The Sextant*. National Maritime Museum Occasional Technical Paper No. 3. Greenwich, London, 1975.

Stimson, A., 'The Influence of the Royal Observatory at Greenwich upon the Design of 17th and 18th Century Angle-Measuring Instruments at Sea'. In: *Vistas in Astronomy* 20 (1976) 123–30.

Stimson, A.N., 'Some Board of Longitude Instruments in the Nineteenth Century'. In: P.R. de Clercq (ed.), *Nineteenth-Century Scientific Instruments and Their Makers*. (Amsterdam/Leiden, 1985) 93–115.

Stimson, Alan, 'The First Marine Sextant?' In: *Bulletin of the Scientific Instrument Society* 8 (1986) 12–13.

Stimson, Alan, *The Mariner's Astrolabe: A Survey of Known, Surviving Sea Astrolabes*. Utrecht, 1988.

Stimson, Alan, 'The Old Admiralty Boardroom Carvings'. In: Remmelt Daalder *et al.* (eds.), *Koersvast: Vijf eeuwen navigatie op zee*. (Zaltbommel, 2005) 115–29.

Talbot, Stuart, 'Caleb Smith's Sea Quadrants of *c.*1735, as Constructed by Thomas Heath and George Adams Snr of London'. In: *Bulletin of the Scientific Instrument Society* 73 (2002) 2–7.

Taylor, E.G.R., *The Haven-Finding Art: A History of Navigation from Odysseus to Captain Cook*. London, 1956.

Taylor, E.G.R., *The Mathematical Practitioners of Hanoverian England 1714–1840*. Cambridge, 1966.

Taylor, E.G.R., *Navigation in the Days of Captain Cook*. National Maritime Museum Maritime Monographs and Reports No. 18 (1974).

Taylor, E.G.R., and M.W. Richey, *The Geometrical Seaman: A Book of Early Nautical Instruments*. London, 1962.

Taylor, F. Sherwood, 'Mr. G. H. Gabb'. In: *Nature* 162 (1948) 806.

Thirslund, Søren, 'Samlingen af nautiske instrumenter på Handels- og Søfartsmuseet'. In: *Handels- og Søfartsmuseet på Kronborg Arbog 1984*. (Helsingør, 1985) 133–61.

Tibbetts, G.R., *Arab Navigation in the Indian Ocean before the Coming of the Portuguese*. London, 1971.

Turner, A.J., *Early Scientific Instruments Europe 1400–1800*. London, 1987.

Turner, Anthony, 'Paris, Amsterdam, London: The Collecting, Trade and Display of Early Scientific Instruments, 1830–1930'. In: Peter de Clercq (ed.), *Scientific Instruments: Originals and Imitations: The Mensing Collection*, Museum Boerhaave. (Leiden, 2000) 23–47.

Vanvaerenbergh, Michel, and Peter Ifland, *Line of Position Navigation*. Bloomington, IN, 2003.

Warner, Deborah Jean, 'Davis' Quadrants in America'. In: *Rittenhouse: Journal of the American Scientific Instrument Enterprise* 3, 1 (1988) 23–40.

Warner, Deborah Jean, 'American Octants and Sextants: The Early Years'. In: *Rittenhouse: Journal of the American Scientific Instrument Enterprise* 3, 1 (1989) 86–112.

Warner, Deborah Jean, 'John Bird and the Origin of the Sextant'. In: *Rittenhouse: Journal of the American Scientific Instrument Enterprise* 12, 1 (1998) 1–11.

Warner, Deborah, 'Celestial Navigation Aloft: Aeronautical Sextants in the US'. In: B. Finn and B.C. Hacker (eds.), *Materializing the Military*. (London, 2005) 95–119.

Waters, D.W., *The Art of Navigation in England in Elizabethan and Early Stuart Times.* New Haven, CT, 1958.

Waters, D.W., *The Sea- or Mariner's Astrolabe.* Agrupamento de Estudos de Cartografia Antiga, Seccao de Coimbra 15. Coimbra, 1966.

Waters, D.W., *Science and the Techniques of Navigation in the Renaissance.* National Maritime Museum Maritime Monographs and Reports No. 19 (1974).

Waters, D.W., 'Derek Howse (1919–98)'. In: *British Journal for the History of Science* 33 (2000) 223–5.

Watts, Oswald, *The Sextant Simplified: A Practical Explanation of the Use of the Sextant at Sea.* London, 1985.

Weatherlake, J.G.P., *Davis Mark II Plastic Marine Sextant: Its Use, Care and Adjustment.* N.p., n.d. [*c.*1965].

Wharton, William J.L., *Hydrographical Surveying: A Description of The Means and Methods Employed in Constructing Marine Charts.* Second edition, London, 1898. Third edition, London, 1909.

Wylly, H.C. (ed.), *Official Catalogue of the Royal United Service Museum.* Seventh edition, London, 1924.

Wynter, Harriet, and Anthony Turner, *Scientific Instruments.* London, 1975.

INDEX

Ship's names are in *Italics*
Limited companies been alphabetised under their full name, such as T. Cooke and Son Ltd, Henry Hughes and Son Ltd, Harriet Wynter Ltd, etc.